Electromagnetic Noise
and Quantum Optical Measurements

Springer
Berlin
Heidelberg
New York
Barcelona
Hong Kong
London
Milan
Paris
Singapore
Tokyo

Physics and Astronomy

ONLINE LIBRARY

http://www.springer.de/phys/

Advanced Texts in Physics

This program of advanced texts covers a broad spectrum of topics which are of current and emerging interest in physics. Each book provides a comprehensive and yet accessible introduction to a field at the forefront of modern research. As such, these texts are intended for senior undergraduate and graduate students at the MS and PhD level; however, research scientists seeking an introduction to particular areas of physics will also benefit from the titles in this collection.

Hermann A. Haus

Electromagnetic Noise and Quantum Optical Measurements

With 151 Figures and 117 Problems
with 41 Selected Solutions

Solutions Manual for Instructors on Request
Directly from Springer-Verlag

 Springer

2000

Professor Hermann A. Haus
Massachusetts Institute of Technology
Department of Electrical Engineering
and Computer Sciences
Vassar Street 50, Office 36-345
Cambridge, MA 02139, USA
E-mail: haus@mit.edu

Library of Congress Cataloging-in-Publication Data

Haus, Hermann A.
    Electromagnetic noise and quantum optical measurements / Hermann Haus.
      p. cm. -- (Advanced texts in physics, ISSN 1439-2674)
    Includes bibliographical references and index.
    ISBN 3540652728 (hc. : alk. paper)
      1. Electronic circuits--Noise. 2. Electromagnetic noise--Measurement. 3. Quantum
optics--Measurement. 4. Optoelectronic devices--Noise. 5. Interference (Light) I. Title.
II. Series.

    TK7867.5 .H38 2000
    621.382'24--dc21

                                                                    99-045237

ISSN 1439-2674

ISBN 3-540-65272-8 Springer-Verlag Berlin Heidelberg New York

Springer-Verlag Berlin Heidelberg New York
a member of BertelsmannSpringer Science+Business Media GmbH

© Springer-Verlag Berlin Heidelberg 2000
Printed in Germany

Typesetting: Camera ready from the author using a Springer TeX macro package
Cover design: design & production GmbH, Heidelberg

Printed on acid-free paper   SPIN 10701179   56/3141/di   5 4 3 2 1 0

Preface

Throughout my professional career I have been fascinated by problems involving electrical noise. In this book I would like to describe aspects of electrical noise somewhat in the manner of a Russian matryoshka doll, in which each shell contains a different doll, alluding to deeper and deeper meanings hidden inside as outer appearances are peeled away.

Let us look at some dictionary definitions of noise. Surprisingly, the origin of the word in the English language is unknown. The *Oxford Universal Dictionary* (1955) has the following definition: "Noise. 1. loud outcry, clamour or shouting; din or disturbance; common talk, rumour, evil report, scandal – 1734. A loud or harsh sound of any kind; a din ... An agreeable or melodious sound. Now rare, ME. A company or band of musicians."

This is not a helpful definition of the technical meaning of noise. The Supplement to the *Oxford English Dictionary* (1989) lists the following: "Noise. 7. In scientific use, a collective term (used without the indefinite article) for: fluctuations or disturbances (usu. irregular) which are not part of a wanted signal, or which interfere with its intelligibility or usefulness."

The last definition is an appropriate one and relates to the work of Prof. Norbert Wiener who developed the mathematics of statistical functions in the 1930s and 1940s. To this day I am awed by the power of mathematical prediction of averages of outcomes of statistically fluctuating quantities. These predictions extend to the theory of and experiments on noise.

Let us look at the interpretation in other languages of the word used for the technical term "noise".

In German   *Rauschen*: rush, rustle, murmur, roar, thunder, (poet.) sough.

In Russian   *shum*: noise, hubbub, uproar; *vetra, voln*: sound of wind, waves.

In French   *bruit*: noise, din, racket, uproar, commotion, clamor; (fig.) tumult, sedition; fame, renown, reputation; *beaucoup de bruit pour rien*, much ado about nothing.

In Italian   *rumore*: noise, din, clamor, outcry, uproar; rumor.

It is interesting how different languages attach different meanings to noise. The German and Russian origins are onomatopoetic, simulating the sound of rushing water or rustling of leaves, and do not necessarily possess the connotation of unpleasantness. The French and Italian words have more abstract meanings. Surprisingly, in French, it describes characteristics of persons who stick out, are famous. In Italian it is clearly related to the word "rumor". The etymology of the word "noise" is a glimpse of the complexity and subtlety of the meanings attached to words by different cultures. In the world of physics and technology, noise is equally multifaceted.

A fascinating fact is that the ear is adjusted to have the highest allowed sensitivity without being disturbed by one of the fundamental sources of noise, thermal noise. Thermal noise is the agitation experienced by the molecules in gases, liquids, and solids at all temperatures above absolute zero (on the Kelvin scale). The molecules of air bounce around and hit the eardrums in a continuous pelting "rain" of particles. If the ear were sensitive to that bombardment, one would hear a continuous hissing noise comparable to that of the noise of a radio tuned between stations with the volume turned up. A simple computation finds that the power impinging upon the ear from this thermal noise is of the order of 0.3×10^{-12} W, a third of the threshold of hearing [1], a rather remarkable fact.

Many of us have experienced the strange sensation that is produced when a large shell is held to the ear. Popularly this is known as "hearing the ocean". In fact, this effect is due to the noise of the air particles impinging upon the ear, enhanced by the shell acting as a resonator. Thus, even a normal ear can hear the air particles impinging upon the ear when the effect is enhanced by some means. Later in this book we shall learn how resonators enhance the spectrum of noise near their resonance frequency.

My interest in noise, reflected in the content of this book, was and is mainly in electrical and optical noise. It is not hard to understand the origin of electrical noise, at least the one related to the agitation of particles. Particles with charge are surrounded by fields which, in turn, produce charge accumulation (of opposite sign) in surrounding electrodes. As the particles bounce around when driven by thermal effects or quantum effects, the charges in the electrodes are dragged along and produce spurious currents, noise currents.

Electrical communications engineers worry about noise because they have to discern signals in the presence of such background noise. In all cases in which the background noise is worrisome, the signals are weak so that amplifiers are needed to raise their power to detectable levels. Amplifiers add noise of their own to the background noise. *The ultimate source of low-frequency (including microwave) amplifier noise is the "graininess of the electrical charge"*. This fact was recognized in its full significance by Schottky in his classic paper in 1918 [2]. I quote from Schottky (my English translation):

Cascading of vacuum tube amplifiers has made possible in recent years the detection and measurement of alternating currents of exceedingly small amplitude. Many technical tasks have thereby realized a sudden benefit, but also a new field of research has been opened up. The new amplifying circuits have the same impact on electrical studies as the microscope has had for optics. Because no clear limit has appeared to date on the achievable amplification, one could hope to advance to the infinitesimally small by proper shielding, interference-free layouts, etc. of the amplifying circuits; the dream of "hearing the grass grow" has appeared achievable to mankind.

This is an allusion by Schottky to the sensory power ascribed by the brothers Grimm fairy tales to particularly endowed individuals. In the sequel he shows that the dream will not come true and I quote:

The first insurmountable obstacle is provided, remarkably, by the size of the elementary quantum of electricity (the charge of the electron).

Schottky wrote his paper a decade before the formulation of the uncertainty principle of Heisenberg. Some of the noise generated in amplifiers and recognized by Schottky can be controlled. The amplifiers can be cooled or refrigerated. The shot noise can be reduced by utilizing the mutual repulsion among the negatively charged electrons. Schottky was careful to point out in his paper that, with the current densities achievable in his day, such repulsion could be ignored. In the intervening 75 years a great deal has happened and this research led to the development of ultra-low-noise amplifiers.

The fundamental limit of the noise performance of amplifiers is ultimately determined by quantum mechanics. This was the reason why I studied optical amplification, at frequencies at which the quantum effects of the electromagnetic field are observable, and at which quantum effects are, fundamentally, responsible for the noise performance of optical amplifiers. This very property of optical amplifiers makes them ideal models of quantum measurement apparatus and permits study of the theory of quantum measurement with the aid of simple optical measurement devices. This book thus spans the range from microwave propagation and amplification to optical propagation and amplification, all the way to issues of the theory of quantum measurement.

A book based on the work of 45 years clearly rests on collaboration with many individuals. Among those I should mention with gratitude are the late Prof. Richard B. Adler, Charles Freed, Dr. James Mullen, Prof. Y. Yamamoto, Dr. J. P. Gordon, and many past and present students. Among these, credit goes to Patrick Chou, John Fini, Leaf Jiang, Thomas Murphy, Steve Patterson, Michael Watts, William Wong, and Charles Yu for the careful reading of the manuscript that led to many corrections and suggestions for improvements.

Research cannot do without financial support. Much of the early work was done with general funding by the Joint Services Electronics Program of the Research Laboratory of Electronics. More recently, as the funding became more program-specific, credit goes to the Office of Naval Research and Dr. Herschel S. Pilloff, who encouraged the research on squeezed-state generation, and Dr. Howard R. Schlossberg and the Air Force Office of Scientific Research, who funded the work on long-distance fiber communications.

I gratefully acknowledge the work by Ms. Mary Aldridge and Ms. Cindy Kopf, who typed the manuscript with exemplary patience and attention to detail. Ms. Cindy Kopf redrew and finished most of the figures in final form. I express my appreciation for the careful and thorough editing by Copy Editor Ms. Christine Tsorpatzidis.

Cambridge, Massachusetts   *Hermann A. Haus*
July 2000

Contents

Introduction

Quantitatively, the noise of a linear amplifier can be described as the noise power added by the amplifier to the signal power in the process of signal amplification. It has been found convenient to refer both the noise power and the signal power to the input of the amplifier, before amplification, because then one can make a direct comparison between the amplifier noise and the thermal noise that accompanies the signal. We have gone so far as to express the noise ascribed to the amplifier in Kelvin, namely, in terms of the thermal power that would be emitted by a thermal source if it were at this temperature.

In the 1950s, Penzias and Wilson were readying a microwave antenna for satellite communications using the latest in ultra-low-noise amplifiers. They pointed their antenna in various directions of the sky, away from the high emitters of noise such as the sun and some interstellar radio sources, and found a background noise that could not be accounted for by the noise in the amplifier. They had discovered the 3.5 K background radiation of interstellar space. (This discovery decided in favor of the big-bang theory of the origin of the universe over a rival cosmological theory.) The background noise observed by Penzias and Wilson and quoted in the book *The First Three Minutes* by Steven Weinberg [3] is roughly 1/100 of room temperature. They had to have an excellent understanding of the noise in their receiver to attribute the slight discrepancy in the observed noise power from the output of their amplifier to an unknown source of noise. Professor Bernard Burke of the MIT physics department was made aware of their discovery and brought them into contact with Prof. R. H. Dicke of Princeton, who had indicated that the background temperature of the universe should be of this magnitude if the universe indeed started from the initial big bang in a very small volume and expanded ever since. One may understand this in a somewhat simplified form as a decrease of the frequency and energy density of the original high-temperature, high-frequency electromagnetic waves as they extended over a larger and larger volume. The same would happen to the sound frequency and energy in an organ pipe in which the ends were moved continually farther and farther apart.

It is indeed remarkable that a purely technical accomplishment – the design of low-noise amplifiers, the construction of a satellite communications link, and a very good understanding of the noise in amplifiers – has provided the evidence for one of the theories of the origins of the universe. The existence of the background radiation is now well established. The number 3.5 K has been modified to 2.76 K.

At the very same time as these developments were taking place some of us practitioners were asking ourselves whether there are any fundamental lower bounds to the noise performance of an amplifier. Offhand, one might expect that the minimum amount of noise added to the signal could not be lower than the thermal background noise associated with the temperature at which the amplifier operates. But this is not the case. There is ample evidence that amplifiers can do better. Indeed, refrigerators produce locally lower temperatures than the environment in which they operate and amplifiers can perform the same feat. Further, truly super-deluxe amplifiers include refrigeration to help them reduce their noise. It looks as if there is no lower limit to the noise of an amplifier, if one is willing to pay the price of the refrigeration. Even the shot noise, which is fundamental under random emission, can be reduced by active control, at low frequencies. As the frequencies become higher and higher, such control becomes not only physically more difficult, but impossible in a more fundamental way. The intrinsic noise has a fundamental lower bound and that fundamental bound is of quantum mechanical origin. The noise of fundamental origin is proportional to the frequency of the amplifier. What makes laser noise so interesting is that it is truly fundamental; because of its enormously high level it is detectable. Before we bring up this point in more detail, let us return to noise radiation, namely the kind of radiation left over by the big bang.

Whereas it is rather clear that bouncing charged particles cause noise, why should there be an excitation of free space? The reason for its existence is the following. Free space can transmit electromagnetic radiation. Thermally agitated charged particles excite electromagnetic radiation. The radiation in turn can transfer its energy to the particles. Thus, free space containing charged particles at any temperature must contain radiation. This radiation has a very specific intensity if it is at thermal equilibrium with the thermally agitated particles, gaining as much energy per unit time from the charged particles owing to their radiation as it is losing energy per unit time to the charged particles. This radiation obeys laws very similar to the acoustic radiation caused by thermal noise.

An electromagnetic mode of frequency ν can carry energy only in units of $h\nu$, where h is Planck's constant; $h = 6.626 \times 10^{-34}$ J s. Quantum effects predominate over thermal effects when

$$hν > kT , \tag{0.1}$$

where k is Boltzmann's constant, $k = 1.38 \times 10^{-23}$ J/K. For $T = 290$ K, room temperature, the crossover occurs in the far-infrared regime at a frequency $\nu = 6 \times 10^{12}$ Hz, that is, much higher than conventional microwave frequencies. At frequencies below the limit imposed by (0.1), shot noise, thermal noise and related sources of noise predominate, at higher frequencies quantum noise is predominant. Quantum noise has its origin in the graininess of electromagnetic radiation, somewhat as shot noise has its origin in the graininess of electric charge. According to quantum theory, electromagnetic energy is a phenomenon that can be both particle-like and wave-like, the principle of duality. Each particle, i.e. each photon, carries an energy $h\nu$, this energy being higher the higher the frequency ν. For a given amount of power received, the number of particles received decreases with increasing frequency, making their graininess more noticeable. For this very reason, amplifiers of optical radiation are much noisier than amplifiers of microwave or lower-frequency radiation.

In 1973 A. Hasegawa and G. Tappert at Bell Telephone Laboratories suggested [4] that optical fibers could propagate solitons. An optical fiber made of silicon dioxide glass is dispersive in that the velocities of travel of sinusoidal optical waves of different wavelengths are different. It is nonlinear owing to the so-called Kerr effect: the index of refraction of the optical material depends upon the intensity of the optical wave. This effect is named after John Kerr, like Maxwell a Scot. (It turns out that W. C. Roentgen of X-ray fame also discovered the effect, but Kerr published first.)

Optical pulses that maintain their shape as they propagate (solitons) can form in glass fibers if the dispersion and Kerr effect balance. The Kerr effect is called positive if the index increases with increasing intensity, negative if it decreases with increasing intensity. The dispersion is called positive if the velocity increases with wavelength λ, negative if it changes in the opposite direction. The Kerr effect in glass is positive. Negative dispersion and a positive Kerr effect can balance each other to allow for soliton propagation. Hence, to see solitons in fibers one must excite them at wavelengths at which silicon dioxide has negative dispersion. This is the case for wavelengths longer than 1.3 μm (although fiber dispersion can be affected by core-cladding design). Optical fibers have one other remarkable property: at a wavelength of 1.5 μm they have extremely low loss; they are extraordinarily transparent. Light at this wavelength loses only a few percent of its power when propagating over a 1 km fiber. For this reason, optical fibers are a particularly felicitous medium for signal propagation.

It was the stability of the soliton pulses that motivated Hasegawa in 1984 to propose long-distance optical communications using soliton pulses [5]. The signal would be digital, made up of pulses (solitons) and empty time intervals, symbolizing a string of ones and zeros. Over a trans-Atlantic distance of 4800 km, the optical signal would have to be amplified to compensate for the loss.

At the present time, most practical amplifiers for fiber transmission are made of rare-earth-doped fibers (the rare earth being erbium) "pumped" by a source at a wavelength in an absorption band of the dopant. The optical pumping is done by light from an optical source, a laser with photons of energy $h\nu_p$. The dopant atoms (erbium in the case of the fiber) absorb the pump photons and are excited to higher-lying energy levels which decay rapidly and nonradiatively to the upper laser level. When an atom in the upper laser level is stimulated by signal photons of energy $h\nu$, the atom makes a transition from the upper laser level to the lower laser level, emitting a photon. This so-called stimulated emission increases the signal, i.e. amplifies it.

Stimulated emission is not the only radiation emitted by the excited atoms. As already pointed out by Einstein, an excited atom eventually decays radiatively to a lower-lying level by spontaneous emission even in the absence of stimulating radiation. This emission is independent of the stimulated emission. It masks the signal and is experienced as "noise" after detection.

At the time of Hasegawa's proposal, long-distance optical signal transmission was more complicated: the signal (pulse or no pulse) was detected, regenerated and reemitted in so-called "repeaters" spaced every 100 km or so. In this way the intervening loss was compensated but, equally importantly, the noise added to the signal by random disturbances was removed. Digital signals transmitted via repeaters were thus particularly immune to noise. One disadvantage of this robust scheme of communications in transoceanic cable transmission is that, once the cable has been laid, the format of transmission cannot be changed, because the repeaters are designed to handle only one particular format. Hasegawa's bold move would do away with repeaters and replace them with simple optical amplifiers. Once a cable of this type is installed, it is not tied to a particular signaling format. The pulse rate could be changed at the transmission end and the receiver at the reception end, but no changes would have to be made in the cable and amplifier "pods" at the bottom of the ocean.

The implementation of Hasegawa's idea took some time. The first question was whether the solitons propagating along a fiber would be sufficiently immune to the spontaneous-emission noise "added" in the optical amplifiers. In 1984, while on sabbatical at AT&T Bell Laboratories, the author, with J. P. Gordon, showed [6] that the noise in the amplifiers would change the carrier wavelength of the solitons in a random way. Since the speed of the solitons is a function of the carrier wavelength, the arrival time of the pulses would acquire a random component; the solitons may end up in the wrong time slots, causing errors [6]. This effect is now known as the Gordon–Haus effect. With the parameters of the fiber proposed by Hasegawa, his "repeaterless" scheme could not have spanned the Atlantic. The analysis clearly demonstrated the dependence of the effect on the parameters of the fiber. But with a redesign of the fiber, the Atlantic could be spanned!

L. F. Mollenauer and his group at AT&T Bell Laboratories [7] made pioneering experiments in which they verified many of the predicted properties of soliton propagation. Since a fiber 4800 km long would cost of the order of $100 million, they used a loop of the order of 100 km in length, with three amplifiers, in which they launched a pseudorandom sequence of solitons (ones) and empty intervals (zeros) and recirculated them as many times as they wished, thus simulating long distance propagation. They confirmed the Gordon–Haus effect.

Noise is a familiar phenomenon accompanying any measurement. The numerical values of the quantity measured differ from measurement to measurement. In undertaking a measurement, the experimentalist starts from the assumption that a sequence of measurements on identically prepared systems will arrive at a set of outcomes that will have an average, the value of which will be identified with the average value of the quantity measured. (This assumes of course that the measurement is not distorting the average value as often happens when the measurement apparatus is nonlinear.) Measurements in quantum theory fit into this general view of measurement. The ideal apparatus of quantum measurements does not have nonlinear distortions; the average value of the measurements on an observable is indeed its expectation value. The individual outcomes of the measurements, in general, exhibit scatter, just as they do for a classical signal in the presence of noise

Bell of "Bell's inequality" fame was disturbed by the interpretation of a quantum measurement, in particular by the von Neumann postulate by which every measurement projects the wave function of the observable into an eigenstate of the measurement apparatus [8]. He saw the postulate as a graft onto the standard quantum description. He considered quantum theory incomplete, like Einstein before him, but in a different sense. As an example of a complete theory, he cited Maxwell's theory of electromagnetism. The equations that describe the electromagnetic field also contain in them the rules for the measurement of the field. In contrast, the von Neumann postulate has to be invoked in interpreting the outcome of a quantum measurement.

In the last chapter in this book, we attack the problem of quantum measurements in the optical domain, since quantum formalisms for optical apparatus will be well developed at that point. We shall discuss "quantum nondemolition" (QND) measurements that leave the measured observable unchanged. A QND measurement can be used to "derive" the von Neumann postulate through the study of two QND measurements in cascade. One can show that the conditional probability of measuring the same value of an observable in the second setup as in the first can be made unity through proper design of the apparatus. We consider this a direct derivation from quantum mechanics of the von Neumann postulate, in response to Bell's criticism.

Bell was questioning the placement of the boundary between the quantum and classical domains [9]: "Now nobody knows where the boundary between the classical and quantum domain is situated." We shall argue that the boundary can be placed in most situations by virtue of the nature of all measurement apparatus. A measurement apparatus has to deliver a result that can be interpreted classically [10], such as the position of the needle of a meter or a trace on a scope. For this to be possible, the measurement apparatus, even though described quantum mechanically, must have lost, at its output, quantum coherences that have no interpretation in terms of positive probabilities. This is the point of Zurek [11] and others [12–14], who have shown that macroscopic systems lose coherence extremely rapidly.

It is appropriate that the subject of noise should lead us to ask some fundamental questions in quantum theory. Quantum theory predicts the behavior of an ensemble of identically prepared systems. The statistical theory of noise does likewise. The fluctuations in the observations made on a quantum system can be, and should be, interpreted as noise. It is, in this writer's opinion, futile to search for a means to predict the outcome of one single measurement. Statistical mechanics makes only probabilistic predictions about a system, because of a lack of complete knowledge of the system's initial conditions. Quantum mechanics raises the lack of knowledge of the initial conditions to the level of a principle. Hence the statistical character of the description of nature by quantum mechanics is unavoidable.

At the outset, a disclaimer is in order. This book is not a synopsis of the excellent work on electrical noise, optical communications, squeezed states, and quantum measurement that has appeared in the literature. Instead, it is a personal account of the author's and his coworkers' work over a career spanning 45 years. Such an account has a certain logical consistency that has didactic merit, a feature that would be sacrificed if an attempt had been made to include the excellent work of other authors in such a way as to do it justice. For the same reason, the literature citations will be found to be deficient. Yet the author hopes that despite these deficiencies, and maybe even on account of them, the reader will find this to be a coherent presentation from a personal point of view of a very fascinating field.

The first three chapters provide the background necessary to understand the basic concepts used in the remainder of the book: power flow, electromagnetic energy, group velocity, and group velocity dispersion; modes in waveguides and resonators; resonators as multiports and their impedance matrix and scattering-matrix description; and single-mode fibers, the optical Kerr effect, and polarization coupling in fibers. Most concepts and laws will be familiar to the reader. The first three chapters thus serve mainly as a convenient reference for the later developments.

Chapter 4 derives the probability distribution for the carriers of a current exhibiting shot noise and arrives at the spectrum of the current. Next, the thermal noise on a transmission line is derived from the equipartition theorem. From this analysis of a reversible (lossless) system it is possible, surprisingly, to derive Nyquist's theorem that describes the emission of noise from a resistor, an irreversible process. The noise associated with linear loss at thermal equilibrium calls for the introduction of Langevin noise sources. Finally, we derive the probability distribution of photons on a waveguide (one-dimensional system) at thermal equilibrium, the so called Bose–Einstein distribution.

With the background developed in Chap. 4 we enter the discussion of classical noise in passive and active multiports. If the multiports are linear, their noise can be described fully by associated Langevin sources. At thermal equilibrium, these possess some very simple properties. In particular, the spectral density matrix, appropriately weighted, forms the so-called characteristic noise matrix. For a passive network at thermal equilibrium, this matrix is proportional to the identity matrix. In the more general case of a linear passive network not at equilibrium, or a linear active network, such as a linear amplifier, the characteristic noise matrix contains all the information necessary to evaluate the optimum noise performance of the network, the noise performance that leads to the maximum signal-to-noise ratio at large gain. This optimum noise performance is described, alternatively, as the minimum excess noise figure at large gain, or the minimum noise measure. The optimization is studied with the simple example of a microwave field effect transistor (FET).

Chapter 6 develops the background for the treatment of quantum noise. The electromagnetic field is expressed in terms of a superposition of modes whose amplitudes obey simple-harmonic-oscillator equations. The field is quantized by quantization of the harmonic-oscillator amplitudes. The quantum noise of a laser oscillator below threshold is derived. The Heisenberg description of operator evolution is adhered to, in which the operators evolve in time. Langevin operator noise sources are introduced in the equations for passive and active waveguides (an example of the latter is erbium-doped-fiber amplifiers). The role of the noise sources is to ensure conservation of commutators, which are a fundamental attribute of the modes in the waveguide. The noise of a typical fiber amplifier is derived. Through much of the text, the quantum noise will appear additive to the "classical" c-number signal. Laser amplifiers are well described in this way. However, in general, the quantum noise is not represented so simply. The Wigner function is the quantum equivalent of a probability distribution. In contrast to a classical probability distribution, the Wigner function is not positive definite. In order to gain a better understanding of peculiar forms of quantum noise, we study the Wigner distribution as applied to a so-called Schrödinger cat state, a quantum state of macroscopic character. This analysis is followed up in Chap. 7 by the quantum description of linear multiports. The formalism is presented in the Heisenberg representation, which displays the correspondence with the classical network description. The Schrödinger representation, in which the wave functions, rather than the operators, evolve in time, is introduced and a comparison between the two descriptions is made. The concept of entangled states is introduced. A strong analogy is found to exist between the classical characteristic noise matrix and its quantum counterpart. It is found that the commutator relations determine the characteristic noise matrix of a quantum network. This is the manifestation of a fundamental law, first explicitly stated by Arthurs and Kelly [15], that requires all linear phase-insensitive amplifiers to add noise to the amplified signal, if the amplification is phase-insensitve.

Chapter 8 analyzes detection of microwave signals and optical signals. The former can be treated classically; the latter require a quantum description. Direct, homodyne, and heterodyne detection are described. The latter two provide gain. Heterodyne detection provides phase-insensitive gain and thus behaves like any other linear amplifier that must add noise to the signal. Homodyne detection is phase-sensitive and it is found that, in principle, it need not add noise to the signal.

Chapter 9 looks in detail at high-bit-rate optical-communication detection via optical preamplification followed by direct detection. In the process, we find the full photon probability distributions for ideal amplifiers as well as for the practical case of an erbium-doped-fiber amplifier. The analysis is based on a quantum description of amplifiers developed by J. A. Mullen and the author in 1962 [16]. The statistics of the photodetector current are determined by the photon statistics, from which the bit-error rate is derived. The minimum number of photons per pulse required for a bit-error rate of 10^{-9} is determined. The analysis is backed up by recently obtained experimental data from Lucent Technologies, Bell Laboratories. Engineering practice has introduced a definition of a so-called noise figure for the characterization of the noise performance of optical amplifiers. This definition is in conflict with the definition of the noise figure used for the description of low-frequency and microwave amplifiers as standardized by the Institute of Electrical and Electronic Engineers. In concluding the chapter we construct a definition that is consistent with the IEEE definition [17].

Chapter 10 studies soliton propagation along optical fibers. Solitons possess particle-like properties as well as wave-like properties: one may assign to them position and momentum, and amplitude and phase. In the quantum theory of solitons, these four excitations are quantized in the same way as they are quantized for particles on one hand and waves on the other hand. The perturbation theory of solitons is established and from it we derive the timing jitter of solitons in long-distance propagation, which is the main source of error in a long-haul soliton communication system. Means of controlling this effect are described. We show that periodically amplified solitons shed so-called continuum that limits the allowed spacing between amplifiers. In long-distance communications, the noise added by the amplifiers is always so large that the system operates at a power level much larger than that of the minimum photon number derived in Chap. 9.

Chapter 11 treats phase-sensitive amplification. One important example is the laser above threshold, in which a fluctuation component in phase with the signal sees a different amplification from the one seen by a fluctuation in quadrature with the signal. The Schawlow–Townes linewidth [18] is derived. Next, we turn to parametric amplification. This amplification is produced via a pump excitation of a medium with a so-called second-order nonlinearity, a nonlinearity with a response that is quadratic in the exciting fields. The amplification can be nondegenerate or degenerate. In the former case, the amplification is closely analogous to linear phase-insensitive amplification. Degenerate parametric amplification is phase-sensitive and thus need not add noise to the signal. In the quantum description of such an amplifier we find that it produces so-called squeezed states: the quantum noise in one phase with respect to the "pump" is amplified, and the quantum noise in quadrature is attenuated. Degenerate parametric amplifiers can produce "squeezed vacuum". We show how squeezed vacuum can be used in an interferometer to improve the signal-to-noise ratio of a phase measurement.

Squeezed vacuum can also be produced by a third-order nonlinearity, such as the optical Kerr effect. Fibers are particularly convenient for the use of the Kerr nonlinearity because of their small mode volume and small loss. The theory of the generation of squeezed vacuum in a fiber loop is presented in Chap. 12. Experiments are described that have generated squeezed vacuum, leading to a reduction of noise by 5.1 dB below shot noise. Further, a phase measurement is described that used the squeezed vacuum so generated for an improved signal-to-noise ratio. Chapter 13 discusses the squeezing of solitons. Solitons behave as particles and waves as outlined in Chap. 10. The squeezing that can be achieved can address both the particle and the wave nature of the soliton.

The last chapter takes up the issue of the theory of quantum measurement using optical measurements as an example. At this point, we can use the formalism developed in the book to present a full quantum analysis of the measurement process. We take the point of view that physical reality can be assigned to an observable only with a full description of the measurement apparatus, which in turn is a quantum system obeying quantum laws. Further, we go through the analysis of a quantum measurement and the evolution of the density matrix of the observable as it proceeds through the measurement apparatus. We show that the reduced density matrix obtained by tracing the density matrix over the measurement apparatus "collapses" into diagonal form, an observation consistent with, yet different from, the von Neumann postulate of the collapse of the wave function of the observable into an eigenstate of the measurement apparatus. Pursuing this point further, we analyze the effect of a cascade of two measurements of the photon number of a signal. We show that with proper design of the measurement apparatus, the conditional probability of observing m photons in the second measurement if n photons have been measured in the first approaches a Kronecker delta, δ_{nm}. This is again consistent with, yet somewhat different from, the von Neumann postulate that the measurement apparatus projects the state of the observable into an eigenstate of the measurement apparatus. Finally we address the Schrödinger cat paradox, using an optical realization of the measurement apparatus, and show that the cat does not end up in a superposition state of "dead" and "alive."

1. Maxwell's Equations, Power, and Energy

This book is about fluctuations of the electromagnetic field at microwave and optical frequencies. The fluctuations take place in microwave and optical structures. Hence a study of electromagnetic-field fluctuations requires the terminology and analytic description of structures excited by microwave or optical sources. The equipartition theorem of statistical mechanics used in Chap. 4 in the derivation of Nyquist's theorem is formulated in terms of energy. Hence, in the application of the equipartition theorem, an understanding of the concept of energy is necessary. When media are present, the medium stores energy as well. The excitation of a mode of the electromagnetic field, as discussed in Chap. 2, involves both the energy of the electromagnetic field and the energy in the excited medium.

We start with Maxwell's equations, which characterize electromagnetic fields at all frequencies. Media are described by constitutive laws which must obey certain constraints if the medium is to be conservative (lossless). Such media store energy when excited by an electromagnetic field. Poynting's theorem relates the temporal rate of change of stored-energy density to the divergence of the power flow. The characterization of dispersive media is straightforward in the complex formulation, with frequency-dependent susceptibilities. The energy density in the medium involves the susceptibility tensor and its derivative with respect to frequency. Finally, we look at the reciprocity theorem, which provides relations among the scattering coefficients of a multiport network. The chapter contains topics from [19–24].

1.1 Maxwell's Field Equations

The first two of Maxwell's equations, in their familiar differential form, relate the curl of the electric field E to the time rate of change of the magnetic flux density B, and the curl of the magnetic field H to the sum of the electric current density J and the time rate of change of the displacement flux density D.

Faraday's law is

$$\nabla \times E = -\frac{\partial B}{\partial t} \ . \tag{1.1}$$

Ampère's law is

$$\nabla \times \boldsymbol{H} = \boldsymbol{J} + \frac{\partial \boldsymbol{D}}{\partial t} \ . \tag{1.2}$$

One may take the fields $\boldsymbol{E}$ and $\boldsymbol{H}$ as the fundamental fields, and the vectors $\boldsymbol{B}$ and $\boldsymbol{D}$ as the hybrid fields that contain both the fundamental fields and properties of the medium. Alternately, one may define $\boldsymbol{E}$ and $\boldsymbol{B}$ as fundamental and consider $\boldsymbol{D}$ and $\boldsymbol{H}$ as hybrid. The former point of view is that of the so called Chu formulation; the latter is more widely accepted by the physics community. It has been shown [19] that the two points of view give the same physical answers and thus one is free to choose either. The difference between the two formulations is hardly noticeable in a discussion of stationary media. However, when moving media and forces are taken into account, the difference is both profound and subtle. While the issue involved does not affect the discussion in the remainder of this book, the author nevertheless takes the opportunity to discuss some of its aspects, since it played an important role in his research in the 1960s, and the way the issue was eventually resolved is typical of any fundamental research. Professor L. J. Chu modeled magnetization by representing magnetic dipoles by two magnetic charges of equal magnitude and opposite sign. In this way, a perfect analogy was established between polarizable and magnetizable media. The formulation of moving dielectric media, as developed by Panofsky and Phillips [20], could be applied to moving magnetic media in a way that was consistent with relativity. Further, this point of view established an analogy between the electric field $\boldsymbol{E}$ and the polarization density $\boldsymbol{P}$ on one hand, and the magnetic-field intensity $\boldsymbol{H}$ and the magnetization density $\boldsymbol{M}$ on the other hand. Soon after the publication of this approach in a textbook on electromagnetism [21], the approach was criticized by Tellegen [22]. He pointed out that magnetic dipoles ought to be represented by circulating currents, because such currents are the sources of magnetism at the fundamental level. More seriously, the force on a circulating current was shown to be different from that on a magnetic dipole in the presence of time-varying electric fields. It turned out that the difference between the force on a magnetic dipole and the force on a current loop with the same dipole moment as found by Tellegen was small, involving relativistic terms. However, if there were such a difference, the replacement of magnetic dipoles by magnetic charge pairs would be flawed. The argument seemed valid at the time. It led Prof. P. Penfield and the author to study the problem more carefully. We assumed that Chu's approach was valid, and that there must exist a subtle error in Tellegen's derivation of the force on a magnetic dipole formed from a current loop. This "hunch" proved correct. It turned out that a magnetic dipole made up of a current loop in a self-consistent way, such as a current flowing in a superconducting wire loop, undergoes changes in a time-varying electric field, changes that were omitted by Tellegen. The charges induced by the electric field create currents when the field is time-varying. These currents, when exposed to the magnetic field, are acted upon by a force that cancels the critical term found by Tellegen [22]. The force on a magnetic dipole made up of two magnetic charges or of a circulating current was indeed the same, except that in the case of the current model relativistic effects had to be included in the rest frame of the loop, because there is motion in the rest frame of the loop. Thus, Chu's model was not only correct, but much simpler, since it did not need to consider relativistic issues in the rest frame of the magnetic dipole. A full account of this investigation is presented in [19]. As happens so often, related work went on at the same time, resulting in publications by Shockley and James [24] and Coleman and van Vleck [25].

Returning to the discussion at hand, we shall opt for Chu's approach, in which E and H are considered fundamental field quantities, whereas D and B are hybrid quantities containing the polarization and magnetization of the medium. In addition to Faraday's law (1.1) and Ampère's law (1.2), which relate the curl of the electric and magnetic fields to their vector sources, we have the two Maxwell's equations which relate E and H to their scalar sources by two divergence relationships.

Gauss's law for the electric field is

$$\nabla \cdot D = \rho \, , \tag{1.3}$$

where ρ is the charge density other than the polarization charge density. Gauss's law for the magnetic field is

$$\nabla \cdot B = 0 \, . \tag{1.4}$$

The equation of continuity

$$\nabla \cdot J = - \frac{\partial \rho}{\partial t} \tag{1.5}$$

is a consequence of (1.2) and (1.3). The vectors and scalars appearing in (1.1)–(1.5) are, in general, all functions of time and space. We use rationalized mks units. The electric field E is given in V/m; H is given in A/m. A convenient unit for the magnetic flux density B is V s/m^2, the current density J is given in A/m^2.

The medium acts as a source of electromagnetic fields via its polarization density P and magnetization density M:

$$D = \epsilon_o E + P \, , \tag{1.6}$$

$$B = \mu_o (H + M) \, . \tag{1.7}$$

Equations (1.1)–(1.7) by themselves do not yet determine the fields. In addition one has to know the relations between M and H, and between P and E, and the relation between the fields and the current density J. These are the so-called constitutive relations. Once the constitutive relations are available the set of equations is complete and the equations can be solved subject to appropriate boundary conditions.

In the case of a linear anisotropic dielectric medium, the polarization $\boldsymbol{P}$ is related to the electric field by linear equations:

$$P_x = \epsilon_o(\chi_{xx}E_x + \chi_{xy}E_y + \chi_{xz}E_z) \,, \tag{1.8a}$$
$$P_y = \epsilon_o(\chi_{yx}E_x + \chi_{yy}E_y + \chi_{yz}E_z) \,, \tag{1.8b}$$
$$P_z = \epsilon_o(\chi_{zx}E_x + \chi_{zy}E_y + \chi_{zz}E_z) \,. \tag{1.8c}$$

These three equations are written succinctly in tensor notation:

$$\boldsymbol{P} = \epsilon_o\overline{\overline{\chi}}_e \cdot \boldsymbol{E} \,. \tag{1.9}$$

It is convenient to combine the constitutive law (1.9) with the definition of the displacement flux density (1.4) and write it in the form

$$\boldsymbol{D} = \overline{\overline{\epsilon}} \cdot \boldsymbol{E} \tag{1.10}$$

with $\overline{\overline{\epsilon}}$ defined as the dielectric tensor

$$\overline{\overline{\epsilon}} \equiv \epsilon_o \left(\overline{\overline{1}} + \overline{\overline{\chi}}_e\right) \,, \tag{1.11}$$

where $\overline{\overline{1}}$ is the identity tensor. The dielectric permeability tensor $\overline{\overline{\epsilon}}$ is symmetric, as will be proved later.

Analogous relations may be written between the magnetization $\boldsymbol{M}$ and the magnetic field intensity $\boldsymbol{H}$. Since there is symmetry between polarization effects and magnetization effects in the Chu formulation, it is easy to treat magnetization effects by analogy. One writes for the magnetic field

$$\boldsymbol{B} = \overline{\overline{\mu}} \cdot \boldsymbol{H} \,, \tag{1.12}$$

where $\overline{\overline{\mu}}$ is the permeability tensor. At optical frequencies, magnetic effects are generally negligible, except in the case of the Faraday effect.

In the special case of an isotropic medium, the tensors $\overline{\overline{\mu}}$ and $\overline{\overline{\epsilon}}$ reduce to scalars μ and ϵ times the identity tensor. Finally, in the absence of any matter the constants ϵ and μ assume particular values, which are worth remembering

$$\epsilon_o \simeq \frac{1}{36\pi} \times 10^{-9} \frac{\text{A s}}{\text{V m}} = \text{mho s/m} \,, \tag{1.13}$$

$$\mu_o = 4\pi \times 10^{-7} \frac{\text{V s}}{\text{A m}} = \text{ohm s/m} \,. \tag{1.14}$$

The product of ϵ_o and μ_o has a fundamental significance:

$$\mu_o\epsilon_o = \frac{1}{c^2} \text{ s}^2/\text{m}^2 \,, \tag{1.15}$$

where c is the light velocity in free space. The value of ϵ_o is adjusted to provide the correct value of the speed of light; it changes as the speed of light is determined more and more accurately.

If the only currents in the medium considered are due to conduction and if the medium is linear, we have the simple relation for the current density J

$$J = \sigma E \; , \tag{1.16}$$

where σ is the conductivity of the medium in mho/m. This is the field-theoretical form of Ohm's law. A form of Ohm's law more general than (1.16) applies to anisotropic linear conducting media. In such media the current density J and field E are related by a tensor relation analogous to (1.9):

$$J = \bar{\bar{\sigma}} \cdot E \; , \tag{1.17}$$

where $\bar{\bar{\sigma}}$ is a tensor. In general, $\bar{\bar{\sigma}}$ is not symmetric. However, in Sect. 1.3 we shall show that $\bar{\bar{\sigma}}$ must be a symmetric tensor if the material is resistive in the true sense of the word.

Equations (1.1)–(1.7) in conjunction with (1.10), (1.12), and (1.17) are sufficient to find the electromagnetic field in a linear medium, provided proper boundary conditions are stated.

Before concluding this section, we note that Maxwell's equations are time-reversible if they do not contain a conduction current J and there is no free charge ρ. Indeed, suppose we have found a solution $E(r,t)$ and $H(r,t)$ to Maxwell's equations (1.1) and (1.2), with the constitutive laws (1.10) and (1.12) determining $D(r,t)$ and $B(r,t)$. Then, if we switch from t to $-t$, from $E(r,t)$ to $E(r,-t)$, $H(r,t)$ to $-H(r,-t)$, $D(r,t)$ to $D(r,-t)$, and $-B(r,t)$ to $-B(r,-t)$, it is easy to verify that (1.1) and (1.2) are obeyed automatically, along with the constitutive laws (1.10) and (1.12). The new solution is called the time-reversed solution. It is obtained from the evolution of the forward-running solution as if the movie reel on which the evolution is recorded were run backwards. The B and H fields are, of course, reversed.

1.2 Poynting's Theorem

In radiation problems or in problems of electromagnetic propagation, we are often interested in the transmission of power from one region of space to another. It is, therefore, important to clarify all concepts relating to power and energy. Poynting's theorem accomplishes this. Poynting's theorem is a mathematical identity which can be endowed with profound physical significance. We start with Maxwell's equation (1.1) and dot-multiply by H. We take (1.2) and dot-multiply by E. Subtracting the two relations and making use of a well-known vector identity, we obtain

$$\nabla \cdot (E \times H) + E \cdot J + H \cdot \frac{\partial B}{\partial t} + E \cdot \frac{\partial D}{\partial t} = 0 \; . \tag{1.18}$$

Equation (1.18) is the differential form of Poynting's theorem. Integrating over a volume $\mathcal{V}$, bounded by a surface S, we obtain

$$\oint_S (E \times H) \cdot dS + \int_{\mathcal{V}} E \cdot J d\mathcal{V} + \int_{\mathcal{V}} \left(H \cdot \frac{\partial B}{\partial t} + E \cdot \frac{\partial D}{\partial t} \right) d\mathcal{V} = 0 .$$

$$(1.19)$$

In (1.19) we have made use of Gauss's theorem. Equation (1.19) is the integral form of Poynting's theorem. Let us turn to an interpretation of (1.19). The integral $\int_{\mathcal{V}} E \cdot J d\mathcal{V}$ is the power imparted to the current flow J inside the volume $\mathcal{V}$. This power may be consumed in the ohmic loss of the material within which the current flows; or, for example, if the current is due to a flow of electrons in free space, the power goes into the time rate of increase of the kinetic energy of the electrons. The second volume integral in (1.19) is interpreted as the power that is needed to change the electric and magnetic fields. Part of it may be used up in the magnetization or polarization processes, the rest goes into storage. With the integral $\int E \cdot J d\mathcal{V}$ interpreted as the power imparted to the current flow and the last integral in (1.19) as the power needed to change the fields in the medium, there is only one interpretation for the first term in (1.19) on the basis of the principle of energy conservation. The integral $\oint E \times H \cdot dS$ over the surface enclosing the volume must be the electromagnetic power flow out of the volume. Indeed, from the principle of energy conservation we have to postulate that

(a) the power flowing out of the volume, through the surface enclosing the volume,
(b) the power imparted to the current flow, and
(c) the power that goes into the changes of the fields in the medium (and vacuum where there is no medium)

should all add up to zero. One may attach the meaning of density of electromagnetic power flow to the vector $E \times H$, often denoted by S, the so-called Poynting vector. The second volume integral in (1.19) can be separated into a field part and a material part, using (1.6) and (1.7):

$$\int_{\mathcal{V}} \left(E \cdot \frac{\partial D}{\partial t} + H \cdot \frac{\partial B}{\partial t} \right) d\mathcal{V}$$

$$= \frac{d}{dt} \int_{\mathcal{V}} \left(\frac{1}{2} \epsilon_o E^2 + \frac{1}{2} \mu_o H^2 \right) d\mathcal{V}$$

$$(1.20)$$

$$+ \int_{\mathcal{V}} \left(E \cdot \frac{\partial P}{\partial t} + H \cdot \frac{\partial}{\partial t} \mu_o M \right) d\mathcal{V} ,$$

where we have replaced the partial time derivative $\partial/\partial t$ by d/dt, since the volume integral is independent of r. The first part of the right-hand side, involving the time derivative of $\frac{1}{2}\epsilon_o E^2 + \frac{1}{2}\mu_o H^2$, can be considered to be the rate of change of the energy stored in the electric and magnetic fields, and the second part the rates at which energy is imparted to the polarization and magnetization. Whether the energy imparted to the polarization is stored or not depends upon whether $\boldsymbol{E} \cdot d\boldsymbol{P}$ integrated from a value $\boldsymbol{P} = 0$ to a value $\boldsymbol{P} = \boldsymbol{P}$ is independent of the path of integration in $\boldsymbol{P}$ space. Indeed, consider the energy imparted to $\boldsymbol{P}$ per unit volume. If $\boldsymbol{P} = 0$ at $t = -\infty$ and $\boldsymbol{P} = \boldsymbol{P}$ at t, we have

$$\int_{-\infty}^{t} dt\, \boldsymbol{E} \cdot \frac{\partial \boldsymbol{P}}{\partial t} = \int_{o}^{\boldsymbol{P}} \boldsymbol{E} \cdot d\boldsymbol{P} . \tag{1.21}$$

If $\boldsymbol{P}$ returns to zero at $t = t'$, then

$$\int_{-\infty}^{t} dt\, \boldsymbol{E} \cdot \frac{\partial \boldsymbol{P}}{\partial t} = \oint \boldsymbol{E}(\boldsymbol{P}) \cdot d\boldsymbol{P} , \tag{1.22}$$

where the last expression is an integral over a closed contour in $\boldsymbol{P}$ space, with $\boldsymbol{E}$ treated as a function of $\boldsymbol{P}$. If the integral $\int_{o}^{\boldsymbol{P}} \boldsymbol{E} \cdot d\boldsymbol{P}$ is independent of the path of integration in $\boldsymbol{P}$ space, then $\oint \boldsymbol{E} \cdot d\boldsymbol{P} = 0$ and no energy has been consumed in raising $\boldsymbol{P}$ from zero to some value $\boldsymbol{P}$ and returning it back to zero. In this case, the integral $\int \boldsymbol{E} \cdot d\boldsymbol{P}$ can be interpreted as energy stored in the polarization. Analogous statements can be made about the magnetic contribution $\boldsymbol{H} \cdot d(\mu_o \boldsymbol{M})$.

In a linear medium, it is more convenient to add the field part of the imparted-energy differential, $d(\frac{1}{2}\epsilon_o E^2)$, to the polarization part, $\boldsymbol{E} \cdot d\boldsymbol{P}$, identifying the total-energy differential, dW_e, with

$$dW_e = \boldsymbol{E} \cdot d\epsilon_o \boldsymbol{E} + \boldsymbol{E} \cdot d\boldsymbol{P} = \boldsymbol{E} \cdot d\boldsymbol{D} . \tag{1.23}$$

In the next section we shall take advantage of this identification.

The physical conclusions drawn from Poynting's theorem will enable us to evaluate the electromagnetic power that passes through a given cross section in space, say the cross section of a waveguide. However, Poynting's theorem, as a mathematical identity, can be used for purposes other than the evaluation of power flow. An illustration of one of these applications is the so-called uniqueness theorem of Sect. 1.4.

1.3 Energy and Power Relations and Symmetry of the Tensor $\bar{\bar{\epsilon}}$

In Sect. 1.1 we introduced the dielectric tensor and the magnetic permeability tensor as descriptive of the response of a linear medium. These tensors must obey symmetry and positive-definiteness conditions imposed by energy considerations that follow from Poynting's theorem, derived in the preceding section. From Poynting's theorem we know that the energy per unit volume supplied to the field and polarizable medium is

$$W_e = \int_0^{D} \boldsymbol{E} \cdot d\boldsymbol{D} \ . \tag{1.24}$$

In the above integral, the electric field is considered a function of $\boldsymbol{D}$. The energy is obtained as a line integral of a field $\boldsymbol{E}$ in the space of D_x, D_y, and D_z. Hence, the energy is naturally a function of the displacement density $\boldsymbol{D}$. In the case of a linear medium, however, it is more convenient to use $\boldsymbol{E}$ as the independent variable. When the constitutive relation (1.10) between $\boldsymbol{D}$ and $\boldsymbol{E}$ is introduced, we obtain

$$d\boldsymbol{D} = \bar{\bar{\epsilon}} \cdot d\boldsymbol{E} \ . \tag{1.25}$$

We thus have for the electric energy density, (1.25),

$$W_e = \int_0^{E} \boldsymbol{E} \cdot \bar{\bar{\epsilon}} \cdot d\boldsymbol{E} \ . \tag{1.26}$$

The integral (1.26) is best visualized by considering it as a line integral in a space within which the three components of the electric field are used as the coordinates (see Fig. 1.1). Now suppose that we apply an electric field to the dielectric material and then remove it. In doing so we obtain for the integral (1.26)

$$\oint \boldsymbol{E} \cdot \bar{\bar{\epsilon}} \cdot d\boldsymbol{E} = 0 \ , \tag{1.27}$$

where the contour integral is carried out over a closed path in the space of $\boldsymbol{E}$. The contour integral (1.27) must be zero. The problem is identical to the problem of defining a conservative force field $\boldsymbol{F}(r)$ in the three-dimensional space $r(x, y, z)$. If the contour integral $\oint_C \boldsymbol{F} \cdot d\boldsymbol{r}$ over any closed contour C vanishes, then the force field is conservative. By Stokes' theorem, the contour integral can be converted into a surface integral over a surface S spanning the contour C

$$\oint_C \boldsymbol{F} \cdot d\boldsymbol{r} = \int_S \text{curl} \, \boldsymbol{F} \cdot d\boldsymbol{a} \ ,$$

where the curl is given in Cartesian coordinates by

$$\text{curl} \, \boldsymbol{F} = \nabla \times \boldsymbol{F} = \begin{bmatrix} \dfrac{\partial F_z}{\partial y} - \dfrac{\partial F_y}{\partial z} \\[2mm] \dfrac{\partial F_x}{\partial z} - \dfrac{\partial F_z}{\partial x} \\[2mm] \dfrac{\partial F_y}{\partial x} - \dfrac{\partial F_x}{\partial y} \end{bmatrix} \ .$$

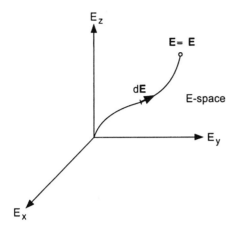

Fig. 1.1. Integration path in E space

Since the integral vanishes over any arbitrary contour, a conservative field has to be curl-free. This analogy can be used to obtain constraints on the tensor ϵ. The argument is cast into the space of coordinates E_x, E_y, and E_z. The "force field" is

$$(E \cdot \bar{\bar{\epsilon}})_x = \epsilon_{xx} E_x + \epsilon_{yx} E_y + \epsilon_{zx} E_z \ ,$$

$$(E \cdot \bar{\bar{\epsilon}})_y = \epsilon_{xy} E_x + \epsilon_{yy} E_y + \epsilon_{zy} E_z \ , \tag{1.28}$$

$$(E \cdot \bar{\bar{\epsilon}})_z = \epsilon_{xz} E_x + \epsilon_{yz} E_y + \epsilon_{zz} E_z \ .$$

This "force field" has to be curl-free in the Cartesian "space" of E, where the partial derivatives are with respect to E_x, E_y, and E_z:

$$\nabla_E \times (E \cdot \bar{\bar{\epsilon}}) = 0 \ . \tag{1.29}$$

It follows from (1.29) that

$$\epsilon_{yz} = \epsilon_{zy} \ , \tag{1.30a}$$
$$\epsilon_{xz} = \epsilon_{zx} \ , \tag{1.30b}$$
$$\epsilon_{yx} = \epsilon_{xy} \ . \tag{1.30c}$$

The $\bar{\bar{\epsilon}}$ tensor must be symmetric.

Next, we turn to the evaluation of the energy. We note that for a symmetric $\bar{\bar{\epsilon}}$ tensor the order in the multiplication

$$E \cdot \bar{\bar{\epsilon}} \cdot dE$$

is immaterial. But, since

$$d(E \cdot \bar{\bar{\epsilon}} \cdot E) = dE \cdot \bar{\bar{\epsilon}} \cdot E + E \cdot \bar{\bar{\epsilon}} \cdot dE \ ,$$

we have

$$d(\boldsymbol{E} \cdot \bar{\bar{\epsilon}} \cdot \boldsymbol{E}) = 2\boldsymbol{E} \cdot \bar{\bar{\epsilon}} \cdot d\boldsymbol{E} \ . \tag{1.31}$$

Using the above expression, we can find immediately for the stored-energy density

$$W_e = \int_0^E \boldsymbol{E} \cdot \bar{\bar{\epsilon}} \cdot d\boldsymbol{E} = \frac{1}{2} \int_0^E d(\boldsymbol{E} \cdot \bar{\bar{\epsilon}} \cdot \boldsymbol{E}) = \frac{1}{2} \boldsymbol{E} \cdot \bar{\bar{\epsilon}} \cdot \boldsymbol{E} \ . \tag{1.32}$$

Since the stored-energy density must be a positive quantity for any field $\boldsymbol{E}$, the elements of $\bar{\bar{\epsilon}}$ have to form a positive-definite matrix. A matrix is positive-definite if all determinants of the principal minors of the matrix are positive. In particular,

$$\epsilon_{xx} > 0, \quad \epsilon_{yy} > 0, \quad \text{and} \quad \epsilon_{zz} > 0$$

is necessary but not sufficient.

The preceding proof started from the postulate that the integral (1.26) carried out over a closed contour must yield zero so that the medium returns all the energy supplied to it in a process which starts with zero field and ends up with zero field. In fact, an integral over a closed contour must always yield zero if we do not permit the medium to generate power. Indeed, if the integral happened to come out positive when the contour was followed in one sense, indicating power consumption, then reversal of the sense would result in a negative value, i.e. energy generation. Hence, the contour integral must yield zero for all passive media. But, then, the medium is dissipation-free. Therefore, one may state unequivocally that a linear dielectric which responds instantaneously to the field, as in (1.10), is dissipation-free.

In the special case of an isotropic medium, where the tensor $\bar{\bar{\epsilon}}$ can be replaced by a scalar ϵ (or rather by the identity tensor multiplied by the scalar ϵ), (1.32) reduces to

$$W_e = \frac{1}{2} \epsilon \boldsymbol{E}^2 \ . \tag{1.33}$$

In a very similar manner one can arrive at the conclusion that the permeability tensor $\bar{\bar{\mu}}$ is symmetric and that linear materials fulfilling (1.12) are lossless, and one can obtain the expression for the magnetic energy stored per unit volume:

$$W_m = \int_0^B \boldsymbol{H} \cdot d\boldsymbol{B} \tag{1.34}$$

is the energy supplied by the magnetic field in order to produce the magnetic flux density $\boldsymbol{B}$. The similarity of (1.34) and (1.24) shows that all mathematical steps performed in connection with the treatment of a linear dielectric medium are applicable to linear magnetic media. For the density of magnetic energy storage in a linear medium, we have

$$W_m = \frac{1}{2} \boldsymbol{H} \cdot \overline{\overline{\mu}} \cdot \boldsymbol{H} \ . \tag{1.35}$$

As was found in the case of a dielectric medium, the elements of $\overline{\overline{\mu}}$ have to form a positive definite matrix. Again, for an isotropic medium (1.35) reduces to

$$W_m = \frac{1}{2} \mu H^2 \ . \tag{1.36}$$

Finally, consider briefly the power dissipated in a conducting medium characterized by (1.17). The power per unit volume P is

$$P = \boldsymbol{E} \cdot \boldsymbol{J} = \boldsymbol{E} \cdot \overline{\overline{\sigma}} \cdot \boldsymbol{E} \ .$$

Only the symmetric part of the conductivity tensor contributes to the power dissipation. Indeed, it is easy to show that for an antisymmetric tensor, $\overline{\overline{\sigma}}^{(a)}$,

$$\boldsymbol{E} \cdot \overline{\overline{\sigma}}^{(a)} \cdot \boldsymbol{E} = 0 \ .$$

If the medium is passive, the power must always be dissipated (and not generated), and P must always be positive, regardless of the applied field $\boldsymbol{E}$. Accordingly, the elements of the symmetric $\overline{\overline{\sigma}}$ tensor must form a positive definite matrix.

The Poynting theorem (1.19) was stated generally, and no assumption about the linearity of the medium had been made. If we introduce (1.32) and (1.34), we have

$$\boldsymbol{E} \cdot \frac{\partial \boldsymbol{D}}{\partial t} = \boldsymbol{E} \cdot \overline{\overline{\epsilon}} \cdot \frac{\partial \boldsymbol{E}}{\partial t} = \frac{1}{2} \frac{\partial}{\partial t} (\boldsymbol{E} \cdot \overline{\overline{\epsilon}} \cdot \boldsymbol{E})$$

and

$$\boldsymbol{H} \cdot \frac{\partial \boldsymbol{B}}{\partial t} = \boldsymbol{H} \cdot \overline{\overline{\mu}} \cdot \frac{\partial \boldsymbol{H}}{\partial t} = \frac{1}{2} \frac{\partial}{\partial t} (\boldsymbol{H} \cdot \overline{\overline{\mu}} \cdot \boldsymbol{H}) \ .$$

Introducing these two expressions into (1.19), we have for a linear dielectric medium

$$\oint (\boldsymbol{E} \times \boldsymbol{H}) \cdot d\boldsymbol{S} + \int \boldsymbol{E} \cdot \boldsymbol{J} d\mathcal{V} + \frac{d}{dt} \int \frac{1}{2} (\boldsymbol{E} \cdot \overline{\overline{\epsilon}} \cdot \boldsymbol{E} + \boldsymbol{H} \cdot \overline{\overline{\mu}} \cdot \boldsymbol{H}) d\mathcal{V} = 0 \ . \tag{1.37}$$

In an isotropic medium within which $\overline{\overline{\epsilon}}$ reduces to scalars, (1.37) assumes the form

$$\oint (\boldsymbol{E} \times \boldsymbol{H}) \cdot d\boldsymbol{S} + \int \boldsymbol{E} \cdot \boldsymbol{J} d\mathcal{V} + \frac{d}{dt} \int \frac{1}{2} (\epsilon E^2 + \mu H^2) d\mathcal{V} = 0 \ . \tag{1.38}$$

In free space, in the absence of currents, $\boldsymbol{J} = 0$, $\epsilon = \epsilon_o$, $\mu = \mu_o$, and (1.37) reduces to

$$\oint \boldsymbol{E} \times \boldsymbol{H} \cdot d\boldsymbol{S} + \frac{d}{dt} \int \frac{1}{2} (\epsilon_o E^2 + \mu_o H^2) \, d\mathcal{V} = 0 \ . \tag{1.39}$$

1.4 Uniqueness Theorem

In the analysis of electromagnetic fields it is necessary to know what intitial conditions and what boundary conditions are necessary to determine the fields. It is also of interest to know whether a set of initial and boundary conditions determines the fields uniquely. Energy conservation theorems or their generalizations often serve to provide the proof of uniqueness. In this section we use Poynting's theorem to determine the necessary and sufficient boundary conditions and initial conditions to describe the evolution of a field uniquely.

Consider a volume $\mathcal{V}$ enclosed by the surface S. The volume is assumed to be filled with a linear medium characterized by (1.10), (1.12), and (1.17). The quantities $\bar{\bar{\epsilon}}, \bar{\bar{\mu}}$, and $\bar{\bar{\sigma}}$ may be functions of position. Suppose that at the time $t = 0$ the magnetic field and the electric field are completely specified throughout the volume $\mathcal{V}$. Assume further that for all time the tangential E field is specified over the part S' of the surface S, and the tangential H field is specified over the remaining part S''. The uniqueness theorem then states that the E and H fields through the entire volume are specified uniquely through all time by these initial and boundary conditions.

The best way of proving the theorem is to suppose that it is not fulfilled. When this supposition leads to a contradiction, the proof is accomplished. Thus, suppose that, for given initial E and H fields throughout the volume, and for tangential E and H fields over the surface given for all time, two different solutions exist inside the volume. We denote the two different solutions by the subscripts 1 and 2. Since Maxwell's equations in the presence of linear materials are linear, the difference of the two solutions is also a solution. Thus, consider the difference solution

$$H_d = H_1 - H_2 , \tag{1.40}$$

$$E_d = E_1 - E_2 , \tag{1.41}$$

with

$$H_d(t = 0) = E_d(t = 0) = 0 \tag{1.42}$$

and

$$n \times E_d = 0 \quad \text{on } S', \quad n \times H_d = 0 \quad \text{on } S'' \quad \text{for all } t . \tag{1.43}$$

The difference field must fulfill Poynting's theorem, (1.37), applied to the volume enclosed by the surface S:

$$\oint_S E_d \times H_d \cdot dS + \int_{\mathcal{V}} E_d \cdot \bar{\bar{\sigma}} \cdot E_d \, d\mathcal{V} \\ + \frac{d}{dt} \int_{\mathcal{V}} \frac{1}{2} (E_d \cdot \bar{\bar{\epsilon}} \cdot E_d + H_d \cdot \bar{\bar{\mu}} \cdot H_d) d\mathcal{V} = 0 . \tag{1.44}$$

The surface integral in (1.44) vanishes for all time by virtue of (1.43), and the volume integrals vanish at $t = 0$ by virtue of (1.42). The volume integral has the form of an energy storage of the difference solution, a positive definite quantity since the matrices of $\bar{\bar{\epsilon}}$ and $\bar{\bar{\mu}}$ are positive definite (Sect. 1.3). Since the initial energy storage of the difference solution is equal to zero at $t = 0$, the time derivative of the second volume integral in (1.44) can only be positive (or zero). The first volume integral in (1.44) can only be positive (or zero). It follows that the E field and H field of the difference solution must remain zero through all time. Therefore, the original solutions 1 and 2, by assumption different, must actually be identical. The uniqueness theorem is proved. Once a solution of Maxwell's equations is obtained for a linear medium which fulfills the initial conditions and the boundary conditions over all time, one can conclude from the uniqueness theorem that the solution obtained is the only possible solution.

1.5 The Complex Maxwell's Equations

In the study of electromagnetic processes in linear media, processes with sinusoidal time variation at one single (angular) frequency ω are of particular importance. The reason for this is the following. Microwave and optical frequencies are extremely high. Any modulation of a carrier is usually at a frequency low compared with the carrier frequency. Thus, in most cases, a modulated microwave or optical process can be treated as a slow succession of steady states, each at one single frequency. More generally, even if the process cannot be treated as a slow succession of steady states, any arbitrary time-dependent process can be treated as a superposition of sinusoidal processes by Fourier analysis.

In a linear medium a steady-state excitation at a single frequency ω produces responses that are all at the same frequency. A field vector depends sinusoidally upon time if all three of its orthogonal coordinates are sinusoidally time dependent. The three components of a vector are scalars. The use of complex scalars for sinusoidally time-varying scalars is well known. The following treatment of complex vectors is based on this knowledge.

Thus, suppose that we write the electric and magnetic fields in complex form:

$$E(r,t) = \mathrm{Re}(\bar{E}\,e^{-i\omega t}) = \frac{1}{2}(\bar{E}\,e^{-i\omega t} + \bar{E}^*e^{+i\omega t})\,, \tag{1.45}$$

$$B(r,t) = \mathrm{Re}(\bar{B}\,e^{-i\omega t}) = \frac{1}{2}(\bar{B}\,e^{-i\omega t} + \bar{B}^*e^{+i\omega t})\,, \tag{1.46}$$

where the asterisk indicates the complex conjugate. Let us introduce the expressions for E and B into Maxwell's equation (1.1). We obtain

$$\nabla \times (\bar{\boldsymbol{E}} e^{-i\omega t} + \bar{\boldsymbol{E}}^* e^{+i\omega t}) = i\omega (\bar{\boldsymbol{B}} e^{-i\omega t} - \bar{\boldsymbol{B}}^* e^{+i\omega t}) \ . \tag{1.47}$$

Equation (1.47) must apply at an arbitrary time. Setting the time to $t = 0$, we obtain

$$\nabla \times (\bar{\boldsymbol{E}} + \bar{\boldsymbol{E}}^*) = i\omega (\bar{\boldsymbol{B}} - \bar{\boldsymbol{B}}^*) \ . \tag{1.48}$$

Setting $\omega t = -\pi/2$, we obtain

$$\nabla \times (i\bar{\boldsymbol{E}} - i\bar{\boldsymbol{E}}^*) = i\omega (i\bar{\boldsymbol{B}} + i\bar{\boldsymbol{B}}^*) \ . \tag{1.49}$$

Dividing (1.49) by i and adding the result to (1.48), we finally have

$$\nabla \times \bar{\boldsymbol{E}} = i\omega \bar{\boldsymbol{B}} \ . \tag{1.50}$$

In (1.50), the time does not enter. This equation is an equation for functions of space only. The introduction of complex notation has thus enabled us to separate out the time dependence and obtain equations involving spatial dependence only. Thus far we have indicated the complex fields $\bar{\boldsymbol{E}}$ and $\bar{\boldsymbol{B}}$, which are functions of $\boldsymbol{r}$, by an overbar. Henceforth we shall dispense with this special notation. It will be obvious from the context whether the fields are real and time-dependent or complex and time-independent.

In a similar manner we obtain for all Maxwell's equations

$$\nabla \times \boldsymbol{E} = i\omega \boldsymbol{B} \ , \tag{1.51}$$

$$\nabla \times \boldsymbol{H} = \boldsymbol{J} - i\omega \boldsymbol{D} \ , \tag{1.52}$$

$$\nabla \cdot \boldsymbol{B} = 0 \ , \tag{1.53}$$

$$\nabla \cdot \boldsymbol{D} = \rho \ , \tag{1.54}$$

$$\boldsymbol{B} = \bar{\bar{\mu}} \cdot \boldsymbol{H} \ , \tag{1.55}$$

$$\boldsymbol{D} = \bar{\bar{\epsilon}} \cdot \boldsymbol{E} \ , \tag{1.56}$$

$$\nabla \cdot \boldsymbol{J} = i\omega \rho \ . \tag{1.57}$$

The quantities in (1.51)–(1.57) are complex vector or scalar quantities and are functions of space only.

The complex form of Maxwell's equations can treat dispersive media in a simple way that is not possible with the real, time-dependent form of

Maxwell's equations. The polarization of dispersive polarizable media is related to the electric field by a differential equation in time. Complex notation in the Fourier transform domain replaces differential equations in time with algebraic equations with frequency-dependent coefficients. For an instantaneous response, the polarization is related to the electric field by a susceptibility tensor $\overline{\overline{\chi}}$ as shown in (1.9). In a dispersive dielectric medium, the dielectric susceptibility simply becomes a function of frequency, $\overline{\overline{\chi}}_e = \overline{\overline{\chi}}_e(\omega)$:

$$P = \epsilon_o \overline{\overline{\chi}}_e(\omega) \cdot E . \tag{1.58}$$

The dielectric tensor $\overline{\overline{\epsilon}}$ becomes frequency-dependent through the definition (1.11), $\overline{\overline{\epsilon}} = \overline{\overline{\epsilon}}(\omega)$. The same holds for a dispersive magnetic medium; the magnetic suceptibility tensor becomes frequency-dependent, $\overline{\overline{\chi}}_m = \overline{\overline{\chi}}_m(\omega)$. The magnetization density is given by

$$M = \overline{\overline{\chi}}_m(\omega) \cdot H . \tag{1.59}$$

The magnetic permeability tensor $\overline{\overline{\mu}}$ also becomes frequency dependent, $\overline{\overline{\mu}} = \overline{\overline{\mu}}(\omega)$.

In Sect. 1.1 we mentioned the time reversibility of Maxwell's equations in their real, time-dependent form, in the absence of free charges and conduction current. Time reversibility can also be extracted from the complex form of Maxwell's equations. Replacing ω by $-\omega$ effectively turns the time evolution around. This reversal of the sign of frequency leaves (1.51), (1.52), (1.55), and (1.56) unchanged if E^*, D^*, $-H^*$, and $-B^*$ are accepted as the new field solutions, and the susceptibility and permeability tensors obey the relation

$$\overline{\overline{\chi}}_e(\omega) = \overline{\overline{\chi}}_e^*(-\omega) , \tag{1.60a}$$

$$\overline{\overline{\chi}}_m(\omega) = \overline{\overline{\chi}}_m^*(-\omega) . \tag{1.60b}$$

The relations (1.60a) and (1.60b) are the consequence of the fact that P, M, E, and H are real, time-dependent vectors. For this condition to hold

$$P^*(-\omega) = \overline{\overline{\chi}}_e^*(-\omega) \cdot E^*(-\omega) = \overline{\overline{\chi}}_e^*(-\omega) \cdot E(\omega) = \overline{\overline{\chi}}_e(\omega) \cdot E(\omega) .$$

Since $E(\omega)$ can be adjusted arbitrarily, it follows that $\overline{\overline{\chi}}_e^*(-\omega) = \overline{\overline{\chi}}_e(\omega)$.

Another aspect of time reversibility is of importance. Note that $-B^*$ replacing B implies also the reversal of any d.c. magnetic field present. If this is not done, the field solutions are not time-reversible. This is the case in the Faraday effect.

1.6 Operations with Complex Vectors

In order to get a better understanding of what is involved in complex-vector operations, we shall study a few special cases. As an example, consider the dot product of a complex vector E with itself. Splitting the complex vector into its real and imaginary parts, we can write

$$\boldsymbol{E} \cdot \boldsymbol{E} = [\mathrm{Re}(\boldsymbol{E}) + \mathrm{i}\,\mathrm{Im}(\boldsymbol{E})] \cdot [\mathrm{Re}(\boldsymbol{E}) + \mathrm{i}\,\mathrm{Im}(\boldsymbol{E})]$$
$$= \mathrm{Re}(\boldsymbol{E}) \cdot \mathrm{Re}(\boldsymbol{E}) - \mathrm{Im}(\boldsymbol{E}) \cdot \mathrm{Im}(\boldsymbol{E}) + 2\mathrm{i}\,\mathrm{Re}(\boldsymbol{E}) \cdot \mathrm{Im}(\boldsymbol{E}) \,. \tag{1.61}$$

Equation (1.61) indicates an interesting feature of complex vectors. It is quite possible for the dot product of a complex vector with itself to be equal to zero without the vector itself being zero. (This feature should be contrasted with a dot product of a real vector with itself. If this dot product turns out to be zero, one must conclude that the vector itself is a zero vector.) Indeed, looking at (1.61) we find that its right-hand side can be equal to zero if the following two conditions are fulfilled:

$$\mathrm{Re}(\boldsymbol{E}) \cdot \mathrm{Re}(\boldsymbol{E}) = \mathrm{Im}(\boldsymbol{E}) \cdot \mathrm{Im}(\boldsymbol{E}) \,, \tag{1.62}$$

$$\mathrm{Re}(\boldsymbol{E}) \cdot \mathrm{Im}(\boldsymbol{E}) = 0 \,. \tag{1.63}$$

The first of the two above equations requires that the real part of the vector be equal in magnitude to its imaginary part. The second of the two equations requires that the real part of the complex vector be perpendicular to its imaginary part. A complex vector whose dot product with itself is equal to zero corresponds to a time-dependent vector with circular polarization.

Next let us study another interesting dot product of a complex vector with itself, that is, with its own complex conjugate. In detail, we have

$$\boldsymbol{E} \cdot \boldsymbol{E}^* = [\mathrm{Re}(\boldsymbol{E}) + \mathrm{i}\,\mathrm{Im}(\boldsymbol{E})] \cdot [\mathrm{Re}(\boldsymbol{E}) - \mathrm{i}\,\mathrm{Im}(\boldsymbol{E})]$$
$$= [\mathrm{Re}(\boldsymbol{E})]^2 + [\mathrm{Im}(\boldsymbol{E})]^2 \,. \tag{1.64}$$

We find that the product $\boldsymbol{E} \cdot \boldsymbol{E}^*$ is equal to the sum of the squares of the real and imaginary parts of the vector. This important product is referred to as the square of the magnitude of the complex vector. If $\boldsymbol{E} \cdot \boldsymbol{E}^*$ vanishes, $\boldsymbol{E}$ is a zero vector.

Equations (1.61) and (1.64) show how the rules of vector multiplication and multiplication of complex numbers are combined in operations involving complex vectors. Applying these same rules, one obtains easily

$$\boldsymbol{E} \times \boldsymbol{E} = 0 \,. \tag{1.65}$$

The cross product of a complex vector with itself is zero. This result is identical with the result obtained from cross multiplication of real vectors. Next, considering the cross product of a complex vector $\boldsymbol{E}$ with its own conjugate, we obtain

$$\boldsymbol{E} \times \boldsymbol{E}^* = [\mathrm{Re}(\boldsymbol{E}) + \mathrm{i}\,\mathrm{Im}(\boldsymbol{E})] \times [\mathrm{Re}(\boldsymbol{E}) - \mathrm{i}\,\mathrm{Im}(\boldsymbol{E})]$$
$$= 2\mathrm{i}\,\mathrm{Im}(\boldsymbol{E}) \times \mathrm{Re}(\boldsymbol{E}) \,. \tag{1.66}$$

This product is not automatically equal to zero. It is zero if, and only if, the real and imaginary parts of the vector E are parallel to each other. We conclude that the product $E \times E^*$ is equal to zero if, and only if, the time-dependent vector $E(r, t)$ is linearly polarized.

In (1.61)–(1.66) we have studied various products of a complex vector with itself or its own complex conjugate. Next, we look at products of two different complex vectors. We start with $E(r, t) \times H(r, t)$. Introducing complex notation, we obtain

$$E(r, t) \times H(r, t) = \frac{1}{2}[E(r)e^{-i\omega t} + E^*(r)e^{+i\omega t}]$$

$$\times \frac{1}{2}[H(r)e^{-i\omega t} + H^*(r)e^{+i\omega t}]$$

$$= \frac{1}{4}[E(r) \times H^*(r) + E^*(r) \times H(r)]$$

$$+ \frac{1}{4}[E(r) \times H(r)e^{-2i\omega t} + E^*(r) \times H^*(r)e^{+2i\omega t}] .$$

(1.67)

Two terms have resulted on the right-hand side of (1.67). The first term does not involve time. The second term is a sinusoidally time-dependent vector varying at double the frequency. If we take a time average of (1.67), the second term drops out and there remains

$$\frac{1}{T} \int_0^T E(r, t) \times H(r, t) dt = \frac{1}{2} \text{Re}(E \times H^*) , \qquad (1.68)$$

where

$$T = \frac{2\pi}{\omega} .$$

If $E(r, t)$ is identified with the electric field and $H(r, t)$ with the magnetic field of an electromagnetic process sinusoidally varying with time at the frequency ω, we have found that the time average of the power flow density is equal to $\frac{1}{2} \text{Re}(E \times H^*)$.

In a similar manner, one can show, for two sinusoidally time-dependent vectors $A(r, t)$ and $B(r, t)$,

$$\frac{1}{T} \int_0^T A(r, t) \cdot B(r, t) dt = \frac{1}{2} \text{Re}[A(r) \cdot B^*(r)] . \qquad (1.69)$$

Equation (1.69) has an important physical significance. Set $A(r, t) = E(r, t)$, the sinusoidal time-varying electric field in an anisotropic nondispersive dielectric. Replace the vector $B(r, t)$ in (1.69) with $\bar{\bar{\epsilon}} \cdot E(r, t) = D(r, t)$, the displacement flux density set up by the sinusoidal time-varying electric field. We then have

$$\frac{1}{T} \int_0^T \boldsymbol{E}(\boldsymbol{r},t) \cdot \bar{\bar{\epsilon}} \cdot \boldsymbol{E}(\boldsymbol{r},t) dt = \frac{1}{2} \mathrm{Re}[\boldsymbol{E}(\boldsymbol{r}) \cdot \bar{\bar{\epsilon}} \cdot \boldsymbol{E}^*(\boldsymbol{r})] \ . \tag{1.70}$$

The complex-conjugate sign has been omitted on the tensor $\bar{\bar{\epsilon}}$, since $\bar{\bar{\epsilon}}$ is real if the medium is nondispersive. Since $\bar{\bar{\epsilon}}$ is a symmetric tensor, we have

$$\boldsymbol{E}^*(\boldsymbol{r}) \cdot \bar{\bar{\epsilon}} \cdot \boldsymbol{E}(\boldsymbol{r}) = \boldsymbol{E}(\boldsymbol{r}) \cdot \bar{\bar{\epsilon}} \cdot \boldsymbol{E}^*(\boldsymbol{r}) \ . \tag{1.71}$$

The product of $\boldsymbol{E} \cdot \bar{\bar{\epsilon}} \cdot \boldsymbol{E}^*$ is equal to its own complex conjugate according to (1.71) and is, therefore, real. Instead of (1.70) we may then write

$$\frac{1}{2} \frac{1}{T} \int_0^T \boldsymbol{E}(\boldsymbol{r},t) \cdot \bar{\bar{\epsilon}} \cdot \boldsymbol{E}(\boldsymbol{r},t) dt = \frac{1}{4} \boldsymbol{E}(\boldsymbol{r}) \cdot \bar{\bar{\epsilon}} \cdot \boldsymbol{E}^*(\boldsymbol{r}) \ . \tag{1.72}$$

Equation (1.72) expresses the time average of the electric energy storage in terms of the complex electric-field vector. We obtain in a similar manner, for the time average of the magnetic energy storage,

$$\frac{1}{2} \frac{1}{T} \int_0^T \boldsymbol{H}(\boldsymbol{r},t) \cdot \bar{\bar{\mu}} \cdot \boldsymbol{H}(\boldsymbol{r},t) dt = \frac{1}{4} \boldsymbol{H}(\boldsymbol{r}) \cdot \bar{\bar{\mu}} \cdot \boldsymbol{H}^*(\boldsymbol{r}) \ . \tag{1.73}$$

Having gained some experience with operations on complex vectors, we are now able to derive various theorems involving products among complex vectors. One such theorem is Poynting's theorem, which is important for the identification of power flow and energy density in dispersive media.

1.7 The Complex Poynting Theorem

We have mentioned before that the amplitude and phase information of a real, time-dependent vector is contained in its complex counterpart. We have also mentioned that it is often useful to gain an understanding of relations existing among the complex vectors themselves. In this way we can often obtain interpretations of physical processes without having to go back into the real, time domain. The complex Poynting theorem is one of the theorems that can be proved using the complex, time-independent vectors.

The conventional form of the theorem is obtained by assuming the frequency ω to be real. A more general theorem is obtained if one assumes ω to be complex, as we shall do here [26]. In particular, we shall replace $-i\omega$ in (1.51) and (1.52) by s and set

$$\mathrm{Re}(s) = \alpha \ , \qquad \mathrm{Im}(s) = -\omega \ . \tag{1.74}$$

Thus

$$\nabla \times \boldsymbol{E} = -s\boldsymbol{B} , \tag{1.75}$$

$$\nabla \times \boldsymbol{H} = \boldsymbol{J} + s\boldsymbol{D} . \tag{1.76}$$

The use of a complex value for the frequency s means that one is considering sinusoidal processes that grow or decay exponentially with time. In order to interpret physically the expressions in the Poynting theorem that are obtained in this way, it is necessary to restrict α to small values

$$|\alpha| \ll |\omega| . \tag{1.77}$$

Indeed, the term $(1/2)\text{Re}(\boldsymbol{E} \times \boldsymbol{H}^*)$ can be interpreted as the time-averaged electromagnetic power density only if the amplitudes of $\boldsymbol{E}(\boldsymbol{r}, t)$ and $\boldsymbol{H}(\boldsymbol{r}, t)$ vary sufficiently slowly in time that an average over one period can still yield unequivocal results.

Starting with (1.75), we dot-multiply it by $\boldsymbol{H}^*$. Further, we dot-multiply the complex conjugate of (1.76) by $\boldsymbol{E}$. By subtracting the two resulting equations from each other and using a well-known vector identity, we have

$$\nabla \cdot (\boldsymbol{E} \times \boldsymbol{H}^*) + \boldsymbol{E} \cdot \boldsymbol{J}^* + s\boldsymbol{B} \cdot \boldsymbol{H}^* + s^*\boldsymbol{D}^* \cdot \boldsymbol{E} = 0 . \tag{1.78}$$

The integral form of the Poynting theorem is obtained by integrating (1.78) over a chosen volume $\mathcal{V}$ enclosed by a surface S, and making use of Gauss's theorem to transform the divergence term into a surface integral. Since the divergence of $\boldsymbol{E} \times \boldsymbol{H}^*$ is essentially the surface integral of $\boldsymbol{E} \times \boldsymbol{H}^*$ over a small volume divided by the volume, we may conduct all power and energy arguments on the basis of the differential form of the Poynting theorem. In order to obtain a physical meaning for (1.78), it is convenient to separate out explicitly the terms corresponding to the polarization of matter. Introducing the polarization $\boldsymbol{P}$, we may write for $\boldsymbol{D}$

$$\boldsymbol{D} = \epsilon_o \boldsymbol{E} + \boldsymbol{P} . \tag{1.79}$$

A time rate of change of the polarization leads to a motion of charge that is equivalent to an electric current density, so far as its effects upon the field are concerned:

$$\boldsymbol{J}_p = s\boldsymbol{P} . \tag{1.80}$$

In the same way a time rate of change of the magnetization produces an effect analogous to a current density of magnetic charge:

$$\boldsymbol{J}_m = s\mu_o \boldsymbol{M} . \tag{1.81}$$

The polarization current is completely equivalent to an electric current. It is convenient to add the polarization current density to the free current density so as to obtain a total electric current density $\boldsymbol{J}_e$:

$$\boldsymbol{J}_p + \boldsymbol{J} = \boldsymbol{J}_e \ . \tag{1.82}$$

Introducing (1.79)–(1.82) into (1.78), we may write for the complex Poynting theorem

$$\nabla \cdot (\boldsymbol{E} \times \boldsymbol{H}^*) + \boldsymbol{E} \cdot \boldsymbol{J}_e^* + \boldsymbol{H}^* \cdot \boldsymbol{J}_m + 4\langle (sW_m^o + s^*W_e^o) \rangle = 0 \ . \tag{1.83}$$

Here we have introduced the symbols W_m^o and W_e^o for the magnetic and electric energy densities in free space:

$$\langle W_e^o \rangle = \frac{1}{4}\epsilon_o \boldsymbol{E} \cdot \boldsymbol{E}^* \ , \tag{1.84}$$

$$\langle W_m^o \rangle = \frac{1}{4}\mu_o \boldsymbol{H} \cdot \boldsymbol{H}^* \ . \tag{1.85}$$

The angle brackets indicate a time average. In the real, time-dependent form, the scalar product of $\boldsymbol{E}$ and $\boldsymbol{J}_e$ is the power per unit volume imparted to the electric current density. Analogously, the scalar product of $\boldsymbol{H}$ and $\boldsymbol{J}_m$ gives the power per unit volume supplied by the magnetic field to the magnetic current density. It is reasonable, therefore, to introduce the following definition for the complex power density:

$$P + iQ = \frac{1}{2}(\boldsymbol{E} \cdot \boldsymbol{J}_e^* + \boldsymbol{H}^* \cdot \boldsymbol{J}_m) \ , \tag{1.86}$$

where P is the time-averaged power density and Q is the so-called reactive power density. When we introduce the definition (1.86) into (1.83) and split the latter into its real and imaginary parts, we obtain

$$\nabla \cdot \frac{1}{2}\mathrm{Re}(\boldsymbol{E} \times \boldsymbol{H}^*) + P + 2\alpha \langle W_m^o + W_e^o \rangle = 0 \tag{1.87}$$

and

$$\nabla \cdot \frac{1}{2}\mathrm{Im}(\boldsymbol{E} \times \boldsymbol{H}^*) + Q - 2\omega \langle W_m^o - W_e^o \rangle = 0 \ . \tag{1.88}$$

Equation (1.87) contains the divergence of $(1/2)\,\mathrm{Re}(\boldsymbol{E} \times \boldsymbol{H}^*)$. This is the divergence of the time-dependent Poynting vector averaged over one period of the (slowly growing, $\alpha > 0$) sinusoidal processes. It shows that the electromagnetic power delivered per unit volume is equal to the time-averaged power density P supplied to the medium and the time rate of growth of the free-space energy density. Equation (1.88) contains phase information on the divergence of the complex Poynting vector that cannot be obtained simply from the time-dependent form of the Poynting theorem.

We shall now consider a medium that does not support a free current density, so that $\boldsymbol{J} = 0$. Thus, the current density $\boldsymbol{J}_e$ is made up fully by the polarization current density $\boldsymbol{J}_p$. Introducing the constitutive laws (1.58) and

(1.59) into the expression for the real and reactive power densities supplied to the medium (1.86), we obtain the following:

$$P + iQ = \frac{1}{2}(\boldsymbol{E} \cdot s^* \boldsymbol{P}^* + \boldsymbol{H}^* \cdot s\mu_o \boldsymbol{M})$$

$$= \frac{1}{2}[(\alpha + iw)\boldsymbol{E} \cdot \epsilon_o \overline{\overline{\chi}}_e^*(w + i\alpha) \cdot \boldsymbol{E} \qquad (1.89)$$

$$+ (\alpha - iw)\boldsymbol{H}^* \cdot \mu_o \overline{\overline{\chi}}_m(w + i\alpha) \cdot \boldsymbol{H}] \,.$$

If the medium is lossless, then $P = 0$ in the steady state when $s = iw$. From this requirement we find from (1.89) for a lossless medium that

$$\mathrm{Re}\left[iw\boldsymbol{E} \cdot \epsilon_o \overline{\overline{\chi}}_e^*(w) \cdot \boldsymbol{E}^*\right] = 0 \,. \qquad (1.90)$$

This condition is met when the χ tensor is Hermitian, that is, when

$$\overline{\overline{\chi}}_e^\dagger(w) = \overline{\overline{\chi}}_e(w) \,, \qquad (1.91)$$

where the dagger † indicates complex conjugate transposition of the tensor. Thus, we conclude that a lossless dielectric medium possesses a Hermitian χ tensor. This is a generalization of the condition of symmetry found for the ϵ tensor earlier, when we required that the energy be a single-valued function of the integration path in the space of E_x, E_y, and E_z. In that case we dealt with a real ϵ, i.e. a real $\overline{\overline{\chi}}_e$. The polarization responded instantaneously to the applied field. The same symmetry holds for the magnetic susceptibility tensor $\overline{\overline{\chi}}_m$.

In general, the susceptibility tensors in (1.89) have to be evaluated for the complex frequency $s = \alpha - iw$. The inequality (1.77) permits a Taylor expansion of the susceptibility tensors up to first order in α, so that we obtain

$$P + iQ = \frac{1}{2}\left[(\alpha + iw)\boldsymbol{E} \cdot \epsilon_o\left(\overline{\overline{\chi}}_e^* - i\alpha\frac{\partial \overline{\overline{\chi}}_e^*}{\partial w}\right) \cdot \boldsymbol{E}^*\right]$$

$$+ \frac{1}{2}\left[(\alpha - iw)\bar{\boldsymbol{H}}^* \cdot \mu_o\left(\overline{\overline{\chi}}_m + i\alpha\frac{\partial \overline{\overline{\chi}}_m}{\partial w}\right) \cdot \boldsymbol{H}\right]$$

$$= -\frac{1}{2}iw[\boldsymbol{H}^* \cdot \mu_o\overline{\overline{\chi}}_m \cdot \boldsymbol{H} - \boldsymbol{E} \cdot \epsilon_o\overline{\overline{\chi}}_e^* \cdot \boldsymbol{E}^*] \qquad (1.92)$$

$$+ \frac{1}{2}\alpha\left[\boldsymbol{H}^* \cdot \mu_o\left(\overline{\overline{\chi}}_m + w\frac{\partial \overline{\overline{\chi}}_m}{\partial w}\right)\boldsymbol{H}\right.$$

$$\left. + \boldsymbol{E} \cdot \epsilon_o\left(\overline{\overline{\chi}}_e^* + w\frac{\partial \overline{\overline{\chi}}_e^*}{\partial w}\right) \cdot \boldsymbol{E}^*\right] \,.$$

In evaluating $\overline{\overline{\chi}}_e^*(\omega + i\alpha)$ one must note that an expansion of $\overline{\overline{\chi}}_e$ to first order in $\Delta\omega = i\alpha$ is made first, and then the complex conjugate is taken. When the medium is lossless, the first of the terms in the last expression of (1.92) is pure imaginary. Further, when the frequency is complex, $s = -i\omega + \alpha$, then the field amplitudes grow with a time dependence $\exp(\alpha t)$. The energy is proportional to products of fields and thus has the time dependence $\exp(2\alpha t)$ and the rate of growth of the energy is 2α. When this fact is taken into account, and it is noted that P is the power density needed to supply the rate of growth of energy, we find from (1.92) for the energy density in the medium, W_M,

$$W_M = \frac{1}{4}\epsilon_o \boldsymbol{E} \cdot \left(\overline{\overline{\chi}}_e^* + \omega\frac{\partial\overline{\overline{\chi}}_e^*}{\partial\omega}\right) \cdot \boldsymbol{E}^* + \frac{1}{4}\mu_o \boldsymbol{H}^* \cdot \left(\overline{\overline{\chi}}_m + \omega\frac{\partial\overline{\overline{\chi}}_m}{\partial\omega}\right) \cdot \boldsymbol{H} .$$

$$(1.93)$$

In a dispersive medium, the energy density involves the derivative of the susceptibility tensor.

A simple example may illustrate the identification of energy density. A neutral plasma made up of light electrons and heavy ions, excited by a sinusoidal electric field, experiences displacement of the electrons, whereas the ions may be considered stationary. The system is isotropic and hence the susceptibility tensor is a scalar. Denote the density of the electrons by N, their charge by q, their mass by m, and their displacement by δ. The equation of motion for the displacement is

$$m\frac{d^2\delta}{dt^2} = qE .$$

The displacement is $\delta = -qE/m\omega^2$. The effective polarization density produced is

$$P = Nq\delta = -\frac{q^2}{m}\frac{E}{\omega^2} .$$

Hence, the susceptibility is

$$\chi_e = -\frac{q^2 N}{m\epsilon_o\omega^2} ,$$

and is negative. If one had naively identified the energy density as $(\epsilon_o/4)\chi_e E^2$, one would have obtained a negative answer. Using the correct expression, one finds

$$W_M = \frac{\epsilon_o}{4}\left(\chi_e + \omega\frac{\partial\chi_e}{\partial\omega}\right)E^2 = \frac{1}{4}\frac{q^2 N}{m\omega^2}E^2 .$$

It is easy to identify this energy density as the time-averaged kinetic-energy density of the electrons:

$$W_M = \frac{1}{2}\frac{mv^2}{2}N = \frac{1}{4}m\omega^2\delta^2 N ,$$

where the additional factor of $1/2$ comes from the time averaging.

1.8 The Reciprocity Theorem

In Sects. 1.1 and 1.5 we showed that solutions of Maxwell's equations are time-reversible if the system contains only loss-free media and all d.c. magnetic fields, if present, are reversed. As we shall see later, time reversibility also implies reciprocity. Reciprocity imposes constraints on the form that scattering matrices and impedance matrices of a linear system can assume. However, a system can be reciprocal even when it contains loss. In this section we prove the reciprocity theorem for electromagnetic fields. In Chap. 2 we shall use it to arrive at symmetry conditions for impedance and scattering matrices.

Consider a general volume $\mathcal{V}$ enclosed by a surface S, and filled with a linear medium characterized by a conductivity σ and susceptibility tensor $\bar{\bar{\epsilon}}$, which are, in general, functions of position. If we specify the tangential $\boldsymbol{E}$ field over the part S' of the surface S and the tangential $\boldsymbol{H}$ field over the remaining part S'' of the surface, we can solve Maxwell's equations (1.51)–(1.57) and obtain a unique solution inside the volume. Suppose that one such boundary condition has been specified. We shall denote the solution corresponding to it by the superscript 1. Next, suppose that another boundary condition over the surface S enclosing the volume V is given. The tangential $\boldsymbol{E}$ and $\boldsymbol{H}$ fields over the surface corresponding to the second boundary condition should be different from those of the first one. Denote the solution of Maxwell's equations corresponding to this boundary condition by the superscript 2. Let us write down Maxwell's equations for these two solutions:

$$\nabla \times \boldsymbol{E}^{(1)} = \mathrm{i}\omega \boldsymbol{B}^{(1)} \,, \tag{1.94}$$

$$\nabla \times \boldsymbol{H}^{(1)} = \sigma \boldsymbol{E}^{(1)} - \mathrm{i}\omega \boldsymbol{D}^{(1)} \,, \tag{1.95}$$

$$\nabla \times \boldsymbol{E}^{(2)} = \mathrm{i}\omega \boldsymbol{B}^{(2)} \,, \tag{1.96}$$

$$\nabla \times \boldsymbol{H}^{(2)} = \sigma \boldsymbol{E}^{(2)} - \mathrm{i}\omega \boldsymbol{D}^{(2)} \,. \tag{1.97}$$

Now, dot-multiplying (1.94) by $\boldsymbol{H}^{(2)}$, we obtain

$$\nabla \times \boldsymbol{E}^{(1)} \cdot \boldsymbol{H}^{(2)} = \mathrm{i}\omega \boldsymbol{B}^{(1)} \cdot \boldsymbol{H}^{(2)} \,. \tag{1.98}$$

Dot-multiplying (1.95) by $\boldsymbol{E}^{(2)}$, (1.96) by $\boldsymbol{H}^{(1)}$, and (1.97) by $\boldsymbol{E}^{(1)}$, we obtain three further equations:

$$\nabla \times \boldsymbol{H}^{(1)} \cdot \boldsymbol{E}^{(2)} = \sigma \boldsymbol{E}^{(2)} \cdot \boldsymbol{E}^{(1)} - \mathrm{i}\omega \boldsymbol{D}^{(1)} \cdot \boldsymbol{E}^{(2)} \,, \tag{1.99}$$

$$\nabla \times \boldsymbol{E}^{(2)} \cdot \boldsymbol{H}^{(1)} = \mathrm{i}\omega \boldsymbol{B}^{(2)} \cdot \boldsymbol{H}^{(1)} \,, \tag{1.100}$$

$$\nabla \times \boldsymbol{H}^{(2)} \cdot \boldsymbol{E}^{(1)} = \sigma \boldsymbol{E}^{(2)} \cdot \boldsymbol{E}^{(1)} - \mathrm{i}\omega \boldsymbol{D}^{(2)} \cdot \boldsymbol{E}^{(1)} \ . \tag{1.101}$$

Now let us recall that we are dealing with linear media characterized by permeability and susceptibility tensors. Using this fact, adding (1.98) to (1.99) and subtracting the result from the sum of equations (1.100) and (1.101), and using a well-known vector identity, we finally have

$$\nabla \cdot (\boldsymbol{E}^{(1)} \times \boldsymbol{H}^{(2)}) - \nabla \cdot (\boldsymbol{E}^{(2)} \times \boldsymbol{H}^{(1)})$$
$$= \mathrm{i}\omega[\boldsymbol{H}^{(1)} \cdot (\overline{\overline{\mu}} - \overline{\overline{\mu}}_t) \cdot \boldsymbol{H}^{(2)} + \boldsymbol{E}^{(2)} \cdot (\overline{\overline{\epsilon}} - \overline{\overline{\epsilon}}_t) \cdot \boldsymbol{E}^{(1)}] \ , \tag{1.102}$$

where the subscript "t" indicates transposition. If the medium is characterized by symmetric $\overline{\overline{\epsilon}}$ and $\overline{\overline{\mu}}$ tensors, the right hand side of (1.102) vanishes and we have

$$\nabla \cdot (\boldsymbol{E}^{(1)} \times \boldsymbol{H}^{(2)}) = \nabla \cdot (\boldsymbol{E}^{(2)} \times \boldsymbol{H}^{(1)}) \ . \tag{1.103}$$

Integrating (1.103) over the volume V enclosed by the surface S, we obtain the theorem

$$\oint_S \boldsymbol{E}^{(1)} \times \boldsymbol{H}^{(2)} \cdot d\boldsymbol{S} = \oint_S \boldsymbol{E}^{(2)} \times \boldsymbol{H}^{(1)} \cdot d\boldsymbol{S} \tag{1.104}$$

for

$$\overline{\overline{\mu}} = \overline{\overline{\mu}}_t \ ,$$

$$\overline{\overline{\epsilon}} = \overline{\overline{\epsilon}}_t \ .$$

The theorem (1.104) is the so-called reciprocity theorem. We shall have occasion to use it when discussing properties of microwave junctions and optical couplers. If the system contains lossless media with Hermitian dielectric and permeability tensors that are not symmetric, the reciprocity theorem does not apply. Such media are important for the construction of nonreciprocal structures such as circulators and Faraday isolators.

1.9 Summary

We have presented Maxwell's equations, both in their time-dependent form and in the complex form as applicable to excitations at one single frequency. The time-dependent form of the constitutive laws must be written in terms of differential equations in time if the response of the medium is noninstantaneous. In the complex form, the constitutive laws become simple linear relations between the polarization and the electric field, and between the magnetization and the magnetic field.

An understanding of power flow and energy density is a prerequisite for the analysis of thermal noise in electromagnetic structures as carried out in

Chap. 4. We learned that the energy density is determined by the energy storage in the field and in the polarizable and magnetizable medium. We were able to derive a simple expression for the energy storage in terms of the susceptibilities and their derivatives with respect to frequency. In fibers, the energy storage in the material (silica) is an important part of the net energy storage and determines the dispersion of the fiber.

A medium is dispersive if its polarization and/or magnetization does not follow the electric and/or magnetic field instantaneously. A consequence of dispersion is that the group velocity, which is also the velocity of energy propagation, becomes frequency-dependent, as we shall see in the next chapter. This kind of dispersion is an important characteristic of optical fibers.

Problems[1]

1.1* All vector identities used in this book are derivable from the following relations of vector algebra.

(a) In a triple scalar product $A \times B \cdot C$ one may interchange the $\cdot$ and $\times$ without changing the product. A cyclic interchange of the order of the vector factors leaves the product unchanged.

(b) $A \times (B \times C) = (A \cdot C)B - (A \cdot B)C$ $\hspace{2cm}$ (1)

(c) The chain rule holds

$$\nabla(fg) = f\nabla g + g\nabla f \; . \hspace{2cm} (2)$$

Here, g and f can be replaced by vectors and the multiplication by a vector multiplication.

Using these facts, prove

$$\nabla \times (\nabla \times A) = \nabla(\nabla \cdot A) - \nabla^2 A \; ,$$

$$\nabla \cdot (E \times H) = (\nabla \times E) \cdot H - (\nabla \times H) \cdot E \; .$$

Do not use decomposition into components in a coordinate system.

---

[1] Solutions are given for problems with an asterisk.

1.2* A symmetric susceptibility tensor expressed in one particular coordinate system can be put into diagonal form by expressing it in a new coordinate system that is rotated with respect to the original one. To gain some understanding of these transformations and to keep the analysis simple, we shall confine ourselves to a two-dimensional example. Show that the transformation of the components of a vector in coordinate system (1) into the coordinate system (2) rotated by an angle θ obeys the law

$$E' = ME \ ,$$

where E' is the column matrix $\begin{bmatrix} E_{x'} \\ E_{y'} \end{bmatrix}$, E is the column matrix $\begin{bmatrix} E_x \\ E_y \end{bmatrix}$, and the matrix M is

$$M = \begin{bmatrix} \cos\theta & \sin\theta \\ -\sin\theta & \cos\theta \end{bmatrix}$$

(see Fig. P1.2.1).

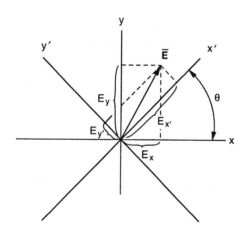

Fig. P1.2.1. E field in two coordinate systems

(a) Show that the tensor transformation obeys the law $\bar{\bar{\epsilon}}' = M\bar{\bar{\epsilon}}M^{-1}$ with the components of the dielectric tensor treated as components of a square matrix.

(b) Show that a symmetric tensor can be put into diagonal form by proper choice of θ. Find θ in terms of the tensor components.

1.3 Find the major and minor axes of the polarization ellipse represented by the complex vector

$$E = i_x + i(i_x - ai_y) \ .$$

1.4 Determine the energy densities W_e and W_m, and the Poynting vector $\boldsymbol{E} \times \boldsymbol{H}$ for a plane wave $\boldsymbol{i}_x E_o \cos(\omega t - kz)$ propagating in free space. Check that Poynting's theorem (1.38) is satisfied.

1.5 Construct the complex vector expression for the electric field of a right-handed circularly polarized plane wave at frequency ω propagating in free space in the $+z$ direction with its peak amplitude E_o occurring at $z = 0, t = 0$. Determine the complex magnetic field and the complex Poynting vector.

1.6 In Sect. 1.7, the example is given of a plasma of charged particles moving within a neutralizing background. It is shown that the energy density formula for a dispersive medium includes the kinetic energy of the plasma. Generalize the example to a charge distribution that is bound to its unperturbed position by a spring constant k. The equation of motion of each of the charges is

$$m \frac{d^2}{dt^2} \delta + k\delta = qE .$$

Determine all the energies and show that the energy density formula contains all pertinent energies.

Solutions

1.1 The del operator can be treated as a vector, as long as it is noted that differentiation is implied. Further, note that the del operator commutes with itself. Using the first equation and identifying $\boldsymbol{A}$ and $\boldsymbol{B}$ with ∇, we obtain

$$\nabla \times (\nabla \times \boldsymbol{A}) = (\nabla \cdot \boldsymbol{A})\nabla - \nabla^2 \boldsymbol{A} .$$

As written, this equation does not make sense, since the del operator must operate on a function. However, a scalar and a vector commute and thus the above equation can also be written

$$\nabla \times (\nabla \times \boldsymbol{A}) = \nabla(\nabla \cdot \boldsymbol{A}) - \nabla^2 \boldsymbol{A} ,$$

which is the desired result. Consider next the second expression. We use differentiation by parts, and then use the fact that the cross and dot can be interchanged in a triple scalar product. In this way we obtain a recognizable vector operation:

$$\nabla \cdot (\boldsymbol{E} \times \boldsymbol{H}) = (\nabla \times \boldsymbol{E}) \cdot \boldsymbol{H} - (\nabla \times \boldsymbol{H}) \cdot \boldsymbol{E} .$$

1.2

(a) Multiplication of a vector by a tensor produces a new vector. Thus, for example, the displacement flux density $\boldsymbol{D}$ results from the multiplication of the E field by the dielectric tensor $\bar{\bar{\epsilon}}$: $\boldsymbol{D} = \bar{\bar{\epsilon}} \cdot \boldsymbol{E}$. When expressed in Cartesian coordinates, the product can be written in terms of matrix multiplications. Without changing notation, we write for the $\boldsymbol{D}$ vector in the new coordinate system $\boldsymbol{D}' = M\boldsymbol{D} = M\bar{\bar{\epsilon}}M^{-1}M\boldsymbol{E} = M\bar{\bar{\epsilon}}M^{-1}\boldsymbol{E}' = \bar{\bar{\epsilon}}'\boldsymbol{E}'$.

(b) Consider the tensor transformation as matrix multiplication. Note: $\epsilon_{xy} = \epsilon_{yx}$.

$$M\bar{\bar{\epsilon}}M^{-1} = \begin{bmatrix} \cos\theta & \sin\theta \\ -\sin\theta & \cos\theta \end{bmatrix} \begin{bmatrix} \epsilon_{xx} & \epsilon_{xy} \\ \epsilon_{yx} & \epsilon_{yy} \end{bmatrix} \begin{bmatrix} \cos\theta & -\sin\theta \\ \sin\theta & \cos\theta \end{bmatrix} = \bar{\bar{\epsilon}}\,' ,$$

$$\epsilon'_{xx} = \epsilon_{xx}\cos^2\theta + \epsilon_{yy}\sin^2\theta + \epsilon_{xy}\sin 2\theta ,$$

$$\epsilon'_{xy} = \epsilon'_{xy} = \epsilon_{xy}\cos 2\theta - \frac{1}{2}(\epsilon_{xx} - \epsilon_{yy})\sin 2\theta ,$$

$$\epsilon'_{yy} = \epsilon_{yy}\cos^2\theta + \epsilon_{xx}\sin^2\theta + \epsilon_{xy}\sin 2\theta .$$

The tensor is put into diagonal form by a rotation by the angle θ, where

$$\theta = \frac{1}{2}\tan^{-1}\frac{2\epsilon_{xy}}{\epsilon_{xx} - \epsilon_{yy}} .$$

2. Waveguides and Resonators

The preceding chapter introduced general properties of Maxwell's equations. It identified power flow and energy density and derived the uniqueness theorem and the reciprocity theorem. This background is necessary for the analysis of metallic waveguides and resonators as used in microwave structures. In this chapter, we analyze the modes of waveguides with perfectly conducting cylindrical enclosures. We determine the mode patterns and the dispersion relations, i.e. the phase velocity as a function of frequency. We derive the velocity of energy propagation and show that it is equal to the group velocity, i.e. the velocity of propagation of a wavepacket formed from a superposition of sinusoidal excitations within a narrow band of frequencies. Then we study the modes in an enclosure, a so-called cavity resonator. We determine the orthogonality properties of the modes. Next, resonators coupled to the exterior via "ports of access" are analyzed. Their impedance matrix description is obtained and the reciprocity theorem is applied to the impedance matrix. This analysis is in preparation for the study of noise in multiports, which begins in Chap. 5. Finally, we look at resonators in a general context. The analysis is based solely on the concept of energy conservation and time reversal. The derivation is applicable to any type of resonator, be it microwave, optical, acoustic, or other. Most of the results obtained here are contained in the literature [21, 27–30]. The concepts of the waveguide mode and of resonant modes are necessary for the quantization of electromagnetic systems. Even though the analysis in this chapter concentrates on waveguides and resonators in perfectly conducting enclosures, the generic approach to resonance is independent of the details of the electromagnetic mode and is based solely on the concept of losslessness and time reversibility. This is the approach used in the analysis and quantization of the modes of optical resonators.

2.1 The Fundamental Equations
of Homogeneous Isotropic Waveguides

A uniform waveguide consists of a conducting envelope surrounding a uniform, in general lossy, medium. The cross section of the waveguide does not change along its longitudinal axis. For the purpose of analysis we shall assume that the conducting envelope forming the waveguide is lossless, that is, perfectly conducting. The assumption of a lossy conductor would lead, on one hand, to prohibitive mathematical difficulties; on the other hand, the disregard of loss in the walls is always a good approximation. The loss per wavelength in waveguides at microwave frequencies is small and can be disregarded to first order. It can be taken into account a posteriori by simple methods of perturbation theory.

We shall assume that the medium filling the waveguide is uniform and isotropic and characterized by a (scalar) conductivity σ, permeability μ, and dielectric constant ϵ. The region inside the waveguide is not necessarily singly connected, i.e. we can allow for longitudinal conductors inside the conducting envelope. In this way we can treat coaxial cables, multiconductor systems, and hollow-pipe waveguides by one and the same theory (Fig. 2.1).

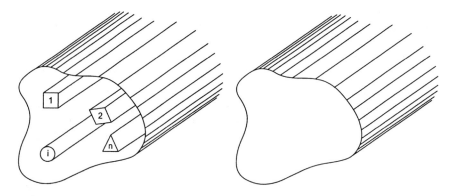

Fig. 2.1. Examples of waveguide geometries

We shall be concerned with the steady-state, sinusoidally time-varying solutions inside the waveguide. Thus, we can make use of the complex Maxwell equations. Under the assumption made about the medium filling the waveguide, we have

$$\nabla \times \boldsymbol{E} = \mathrm{i}\omega\mu\boldsymbol{H} \ , \tag{2.1}$$

$$\nabla \times \boldsymbol{H} = (\sigma - \mathrm{i}\omega\epsilon)\boldsymbol{E} \ . \tag{2.2}$$

In addition to (2.1) and (2.2), we need the divergence relations, which, under the assumption of a charge-free, uniform medium, reduce to

$$\nabla \cdot \boldsymbol{E} = 0 \ , \tag{2.3}$$

$$\nabla \cdot \boldsymbol{H} = 0 \ . \tag{2.4}$$

Equations (2.1)–(2.4), in conjunction with the boundary condition that the E field tangential to the envelope is zero, determine the E and H fields completely. We now turn to a formal solution of these equations. It is expedient to introduce an auxiliary parameter into (2.1) and (2.2) so as to enhance their symmetry. We define the propagation constant in the medium characterized by σ, ϵ, μ by

$$k = \sqrt{\omega\mu(\omega\epsilon + i\sigma)} \, . \tag{2.5}$$

The quantity k is the propagation constant of an infinite, parallel, plane wave, at the frequency ω, within an infinite medium characterized by the conductivity σ, dielectric constant ϵ, and permeability μ. We further define the impedance parameter ξ by

$$\xi = \sqrt{\frac{i\omega\mu}{i\omega\epsilon - \sigma}} \, . \tag{2.6}$$

ξ is the ratio between the E and H fields of an infinite, parallel, plane wave in the medium under consideration. For its inverse we use the symbol η:

$$\eta = \frac{1}{\xi} = \sqrt{\frac{i\omega\epsilon - \sigma}{i\omega\mu}} \, . \tag{2.7}$$

The square roots in (2.5)–(2.7) are defined so as to give positive real parts of the corresponding expressions. With the aid of these auxiliary parameters, we can write (2.1) and (2.2) in the form

$$\nabla \times \boldsymbol{E} = ik\xi\boldsymbol{H} \, , \tag{2.8}$$

$$\nabla \times \boldsymbol{H} = -ik\eta\boldsymbol{E} \, . \tag{2.9}$$

Taking the curl of (2.8) and using (2.9) gives the Helmholtz equation for the electric field,

$$\nabla^2\boldsymbol{E} + k^2\boldsymbol{E} = 0 \, . \tag{2.10}$$

In a similar way one obtains the Helmholtz equation for the magnetic field,

$$\nabla^2\boldsymbol{H} + k^2\boldsymbol{H} = 0 \, . \tag{2.11}$$

At this point, we can proceed with the solution of the Helmholtz equation for the electric or magnetic field. Since the structure is uniform along one axis, say the z axis, one has to expect that the z components of the fields and the z dependence of the field will play an important role in the final solution. In order to single out the z components of the E and H fields, it is expedient to break up the fields into transverse and longitudinal components. This is done by multiplying the equations both scalarly and vectorially by the unit vector along the z direction, i_z.

Considering first the dot multiplication of (2.8) by i_z, we have

$$i_z \cdot \nabla \times E = \mathrm{i}k\xi H \cdot i_z . \tag{2.12}$$

We separate the transverse and longitudinal components of the electric field E and the magnetic field H in the manner shown below.

$$E = E_T + i_z E_z ,$$

$$H = H_T + i_z H_z \tag{2.13}$$

The subscript T indicates a vector that lies entirely in the plane transverse to the z axis. The subscript z indicates the z component of the vector (a scalar). In a similar manner we can split the ∇ operator into a transverse and longitudinal part:

$$\nabla = \nabla_T + i_z \frac{\partial}{\partial z} , \tag{2.14}$$

where, in Cartesian coordinates,

$$\nabla_T = i_x \frac{\partial}{\partial x} + i_y \frac{\partial}{\partial y} . \tag{2.15}$$

Introducing the definitions (2.13) and (2.14) into (2.12), and noting that

$$i_z \cdot \left(\nabla_T + i_z \frac{\partial}{\partial z} \right) \times (E_T + i_z E_z) = i_z \cdot \nabla_T \times E_T ,$$

we have the simple result

$$i_z \cdot \nabla_T \times E_T = \mathrm{i}k\xi H_z . \tag{2.16}$$

The dot multiplication of (2.8) by i_z reduced its left-hand side to a transverse derivative of the transverse E field alone. On the right-hand side only the z component of the H field remains. In a similar manner we obtain, by dot multiplication of (2.9) by i_z,

$$i_z \cdot \nabla_T \times H_T = -\mathrm{i}k\eta E_z . \tag{2.17}$$

Next, let us cross multiply (2.8) by i_z. For its left-hand side we obtain, using the definitions (2.13) and (2.14)

$$i_z \times (\nabla \times E) = i_z \times \left[\left(\nabla_T + i_z \frac{\partial}{\partial z} \right) \times (E_T + i_z E_z) \right]$$

$$= i_z \times (\nabla_T \times E_T) + i_z \times \left(i_z \times \frac{\partial E_T}{\partial z} \right) \tag{2.18}$$

$$- i_z \times (i_z \times \nabla_T E_z) .$$

Noting that $\nabla_T \times \boldsymbol{E}_T$ is z-directed and making use of the expression for triple vector multiplication, we have

$$\boldsymbol{i}_z \times (\nabla \times \boldsymbol{E}) = \nabla_T E_z - \frac{\partial}{\partial z} \boldsymbol{E}_T \ . \tag{2.19}$$

Introducing (2.19) into (2.8) cross multiplied by $\boldsymbol{i}_z$, we finally have

$$\nabla_T E_z - \frac{\partial}{\partial z} \boldsymbol{E}_T = \mathrm{i}k\xi(\boldsymbol{i}_z \times \boldsymbol{H}_T) \ . \tag{2.20}$$

In a similar manner we obtain from (2.9)

$$\nabla_T H_z - \frac{\partial}{\partial z} \boldsymbol{H}_T = -\mathrm{i}k\eta(\boldsymbol{i}_z \times \boldsymbol{E}_T) \ . \tag{2.21}$$

Equations (2.16), (2.17), (2.20), and (2.21) contain the same information as the original equations (2.8) and (2.9). Whereas the two operations performed on (2.8) and (2.9) can be performed on any system, the result is useful only when looking for solutions whose boundary conditions are independent of z, i.e. solutions in a uniform waveguide. In the treatment of uniform waveguides, these operations lead to a systematic analysis that underscores properties which are independent of the waveguide cross section.

The z components of (2.10) and (2.11) are

$$\nabla_T^2 E_z + \frac{\partial^2}{\partial z^2} E_z = -k^2 E_z \ , \tag{2.22}$$

$$\nabla_T^2 H_z + \frac{\partial^2}{\partial z^2} H_z = -k^2 H_z \ . \tag{2.23}$$

Since the Laplace operator can be written using definition (2.14) as

$$\nabla^2 = \nabla_T^2 + \frac{\partial^2}{\partial z^2} \ , \tag{2.24}$$

independent equations hold for the longitudinal component of the electric field and the longitudinal component of the magnetic field. If the waveguide had instead been filled by a medium that was nonuniform throughout the cross section, i.e. a function of x and y, or was anisotropic and, therefore, characterized by a tensor dielectric susceptibility and a tensor magnetic permeability, a mutual coupling would have existed between the two equations for the longitudinal fields.

A simple solution of (2.23) is $H_z = 0$. Accordingly, there are solutions for the electromagnetic field inside the waveguide which have no longitudinal H field, provided we are able to match all boundary conditions. Similarly $E_z = 0$ is a solution of (2.22). Accordingly, there are solutions of Maxwell's equations inside a uniform waveguide which do not possess a longitudinal electric field, provided that all boundary conditions can be matched with the fields thus found.

2.2 Transverse Electromagnetic Waves

In the preceding section we separated Maxwell's equations into longitudinal and transverse components directed along and across a guiding structure enclosed by perfectly conducting walls. In this section we look at solutions for electric and magnetic fields that are purely transverse, for which both E_z and H_z are zero. Not all kinds of structure can support waves of this transverse character, as we shall find in the course of the analysis.

When we set $H_z = 0$, we find from (2.16) that

$$\nabla_T \times \boldsymbol{E}_T = 0 \,. \tag{2.25}$$

Hence, the transverse electric field must be derivable from a potential. We attempt separation of variables, expressing the solution as a product of a function of z and a function of the transverse coordinates:

$$\boldsymbol{E}_T = -V(z)\nabla_T \Phi(x,y) \,, \tag{2.26}$$

where $\Phi(x, y)$ is a scalar. Since the electric field is divergence-free, we must have

$$\nabla \cdot \boldsymbol{E}_T = 0 = -V(z)\nabla_T^2 \Phi \,, \tag{2.27}$$

and thus the potential function $\Phi(x, y)$ must be a solution of Laplace's equation. The potential has to be constant on a perfect conductor so as not to allow fields that are tangential to the conductor. A solution of Laplace's equation cannot possess extrema in the region of its validity. Thus if the guide consists of a hollow, perfectly conducting pipe, the only possible solution is $\Phi = \text{const}$, which does not give rise to an electric field. Hence we conclude that hollow, conducting pipes cannot support TEM waves. On the other hand, a coaxial cable consisting of concentric cylindrical conductors of radii r_a and r_b, as shown in Fig. 2.2, supports the simple solution of Laplace's equation

$$\Phi = \frac{1}{\ln(r_a/r_b)}\ln\frac{r}{r_b} \,. \tag{2.28}$$

If we introduce the ansatz (2.26) into the Helmholtz equation for the electric field (2.10), we find that the function $V(z)$ has to obey

$$\frac{d^2}{dz^2}V = -k^2 V \,, \tag{2.29}$$

which is the one-dimensional wave equation. If the potential is normalized as in (2.28), the value of $V(z)$ gives the line integral of the electric field from the inner conductor to the outer conductor; it is the voltage as measured in a transverse plane. Note, however, that the electric field is not curl-free globally, and hence a voltage can be defined unequivocally in terms of only a

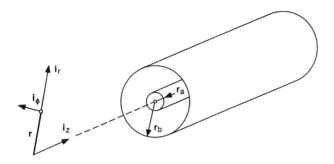

Fig. 2.2. Example of a waveguide geometry

line integral in the transverse cross section. We denote the normalized electric field in the transverse plane by

$$e_T(x, y) = -\nabla_T \Phi(x, y) , \tag{2.30}$$

and write, for the electric-field solution in general,

$$\mathbf{E} = V(z) e_T(x, y) . \tag{2.31}$$

The solution for the magnetic field can also be written as a product of a function of z alone and a function of the transverse coordinates alone:

$$\mathbf{H} = I(z) h_T(x, y) . \tag{2.32}$$

We have, from (2.21),

$$\frac{d}{dz} I = ik\eta \frac{1}{K} V \tag{2.33}$$

with

$$h_T = K(i_z \times e_T) , \tag{2.34}$$

where K is a normalization constant. Similarly, from (2.20),

$$\frac{d}{dz} V = ik\xi K I \tag{2.35}$$

with

$$e_T = -\frac{1}{K}(i_z \times h_T) , \tag{2.36}$$

which is consistent with (2.34). We have found two coupled first-order differential equations for $V(z)$ and $I(z)$. Elimination of either $V(z)$ or $I(z)$ from the two coupled equations leads to the wave equation for either $V(z)$, as in

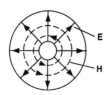

Fig. 2.3. Field patterns in coaxial cable

(2.29), or $I(z)$. Figure 2.3 shows the electric- and magnetic-field patterns in a cross section of constant z for the coaxial cable.

The normalization constant K can be chosen by insisting that the complex power flow be equal to the product of V and I^*. The Poynting flux integrated over one cross section reduces to

$$\int_{\substack{\text{cross}\\\text{section}}} dS\, \boldsymbol{i_z} \cdot \boldsymbol{E} \times \boldsymbol{H}^* = V I^* \int_{\substack{\text{cross}\\\text{section}}} dS\, \boldsymbol{i_z} \cdot \boldsymbol{E_T} \times \boldsymbol{H_T^*}$$

$$= V I^* K \int_{\substack{\text{cross}\\\text{section}}} dS\, |\nabla_T \Phi|^2 \ . \tag{2.37}$$

Using integration by parts in two dimensions, we find

$$\int_{\substack{\text{cross}\\\text{section}}} dS\, |\nabla_T \Phi|^2 = \oint ds\, \boldsymbol{n} \cdot \Phi^* \nabla_T \Phi - \int_{\substack{\text{cross}\\\text{section}}} dS\, \Phi^* \nabla_T^2 \Phi$$

$$= \oint dS\, \boldsymbol{n} \cdot \Phi^* \nabla_T \Phi \ , \tag{2.38}$$

where $\boldsymbol{n}$ is the unit vector normal to the contours of the coaxial-cable cross section in the x–y plane. If the potential on the outer conductor is set equal to zero, and that on the inner conductor is set equal to one, then the integral is found to be the flux of the electric field per unit length and unit voltage. If one introduces the capacitance per unit length C with

$$C = \epsilon \oint ds\, \boldsymbol{n} \cdot \Phi^* \nabla_T \Phi \tag{2.39}$$

and one requires the power to be equal to the voltage–current product, one finds from (2.38) and (2.37), that $V I^* K C / \epsilon = V I^*$. Thus, $K = \epsilon / C$, and one may write for (2.33) and (2.35)

$$\frac{d}{dz} I = (\mathrm{i}\omega C - G) V \ , \tag{2.40}$$

$$\frac{d}{dz} V = \mathrm{i}\omega L I \ . \tag{2.41}$$

The conductance G is given by $(\sigma/\epsilon)C$, and $L = \mu K$ is the inductance per unit length, obeying the constraint

$$LC = \epsilon\mu \; . \tag{2.42}$$

The fact that the inductance per unit length L is indeed equal to μK follows directly from an evaluation of the flux per unit length associated with the current I. This flux Ψ is given by an integral between the inner and outer conductor, from point (1) on the inner conductor to point (2) on the outer conductor

$$\Psi = \int_{(1)}^{(2)} i_z \times ds \cdot \mu \boldsymbol{H} = - \int_{(1)}^{(2)} i_z \times ds \cdot K\mu(i_z \times \nabla_T\Phi)I \tag{2.43}$$

$$= -\mu K I \int_{(1)}^{(2)} ds \cdot \nabla_T\Phi = \mu K I(\Phi_1 - \Phi_2) = \mu K I \; .$$

Equations (2.40) and (2.41) are the well known transmission line equations in complex form. In the absence of conduction, $\sigma = 0$, their solutions can be written

$$V = \sqrt{2Y_o}\left(ae^{i\beta z} + be^{-i\beta z}\right) \; , \tag{2.44}$$

$$I = \sqrt{2Z_o}\left(ae^{i\beta z} - be^{-i\beta z}\right) \; , \tag{2.45}$$

with $\beta = \omega\sqrt{\mu\epsilon}$ and $Z_o = \sqrt{L/C} = 1/Y_o$ and where a and b are the forward and backward wave amplitudes so normalized that the time-averaged power carried by the waves is given by

$$\frac{1}{2}\mathrm{Re}\int_{\substack{\text{cross} \\ \text{section}}} dS\, i_z \cdot \boldsymbol{E} \times \boldsymbol{H}^* = |a|^2 - |b|^2 \; . \tag{2.46}$$

2.3 Transverse Magnetic Waves

Transverse electromagnetic (TEM) waves propagate only in structures that have two conductors. In a hollow pipe, the modes must possess either a longitudinal E field or a longitudinal H field. In this section we derive the equations for modes with longitudinal E fields. For E_z we assume a product solution of the form

$$E_z = \exp(i\beta z)\Phi(x, y) \; , \tag{2.47}$$

which is in the form of a wave with the propagation constant β. When this ansatz is entered into (2.22) we obtain

$$\nabla_T^2 \Phi = (\beta^2 - k^2)\Phi \ . \tag{2.48}$$

This is the scalar Helmholtz equation of the form

$$\nabla_T^2 \Phi + p^2 \Phi = 0 \ , \tag{2.49}$$

with p^2 defined as

$$p^2 = k^2 - \beta^2 \tag{2.50}$$

and subject to the boundary condition that $E_z = \Phi$ vanish on the wall, the so-called Dirichlet boundary condition. It may be worth mentioning that the same two-dimensional Helmholtz equation governs the displacement of a membrane of uniform tension tied to a drumhead with the same cross section as the waveguide. The frequencies of vibration of the membrane are found by the solution of this eigenvalue problem.

The Helmholtz equation has solutions only for discrete values of p^2, with p^2 real and positive, as we now proceed to show. By integration by parts one may derive the following Green's theorem for two scalar functions Φ and Ψ:

$$\int_{\substack{\text{cross} \\ \text{section}}} dS \, \Psi \nabla_T^2 \Phi = \oint ds \, \boldsymbol{n} \cdot \Psi \nabla_T \Phi - \int_{\substack{\text{cross} \\ \text{section}}} dS \, \nabla_T \Psi \cdot \nabla_T \Phi \ . \tag{2.51}$$

Now set $\Phi = \Phi$ and $\Psi = \Phi^*$. Using (2.51) and the boundary condition obeyed by Φ, one finds

$$\int_{\substack{\text{cross} \\ \text{section}}} dS \, \Phi^* \nabla_T^2 \Phi = -p^2 \int_{\substack{\text{cross} \\ \text{section}}} dS \, |\Phi|^2 = - \int_{\substack{\text{cross} \\ \text{section}}} dS \, |\nabla_T \Phi|^2 \tag{2.52}$$

or

$$p^2 = \frac{\int_{\text{cross section}} dS \, |\nabla_T \Phi|^2}{\int_{\text{cross section}} dS \, |\Phi|^2} \ . \tag{2.53}$$

Thus, the eigenvalue p^2 is indeed real and positive. This fixes immediately the dispersion relation for the propagation constant β. If the medium is lossless, (2.50) gives for the propagation constant

$$\beta = \pm\sqrt{\omega^2 \mu \epsilon - p^2} \ . \tag{2.54}$$

The dispersion diagram is shown in Fig. 2.4 for the case of a lossless medium that is nondispersive (ϵ and μ independent of frequency). For frequencies below the so called cutoff frequency, the propagation constant is imaginary; the modes are decaying or growing. Above the cutoff frequency, the modes are traveling waves. Since the square root has two values, two waves are associated with each eigensolution, i.e. with each mode. If one takes

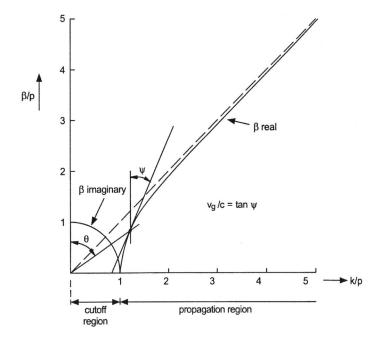

Fig. 2.4. Dispersion diagram for lossless waveguide

a rectangular waveguide as an example, one finds the following solutions for Φ:

$$\Phi = \sin\left(\frac{m\pi}{a}x\right)\sin\left(\frac{n\pi}{b}y\right) \; , \tag{2.55}$$

where m and n are integers. Figure 2.5a shows the potential surface $\Phi(x,y)$ of the lowest-order TM mode. The lines of steepest descent are the lines of the transverse electric field; the lines of equal height are the lines of the transverse magnetic field. The latter are divergence-free (see Figs. 2.5b,c).

The electric field acquires longitudinal components that peak in intensity at the center of the guide. The total electric field, transverse and longitudinal, is of course divergence-free. An infinite number of solutions exists, each with its own dispersion relation. The eigenvalues $p_{mn}^2 = (m\pi/a)^2 + (n\pi/b)^2$ increase with increasing order, i.e. increasing m and n. It is easy to prove the orthogonality of the solutions in the case of a rectangular waveguide. It is of greater interest to show that two solutions with different values of p^2, say p_μ^2 and p_ν^2, are orthogonal, where the Greek subscripts stand for the double subscript mn. We use Green's theorem (2.51) for each of the two solutions and subtract the results. The contour integrals vanish when the boundary conditions are taken into account, and thus one obtains

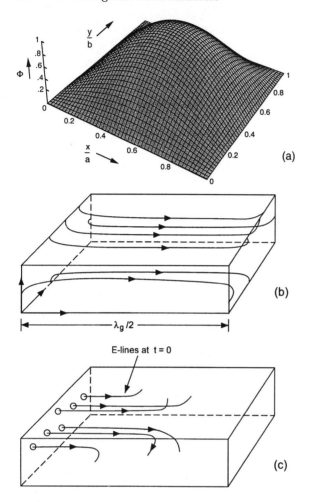

Fig. 2.5. Field patterns of some lowest-order TM modes of rectangular waveguide. (a) Plot of potential Φ for mode $m = n = 1$. (b) The E field of the propagating wave for $m = n = 1$. (c) The E field of the cutoff wave for $m = n = 1$

$$\int_{\substack{\text{cross} \\ \text{section}}} dS\, \Phi_\mu^* \nabla_T^2 \Phi_\nu - \int_{\substack{\text{cross} \\ \text{section}}} dS\, \Phi_\nu^* \nabla_T^2 \Phi_\mu$$

$$= \left(p_\mu^2 - p_\nu^2\right) \int_{\substack{\text{cross} \\ \text{section}}} dS\, \Phi_\mu^* \Phi_\nu = 0 \, . \tag{2.56}$$

The integral of the product of the field profiles vanishes for solutions with different eigenvalues. Now, let us proceed to find the transverse fields. From (2.21), we find for the transverse H field

$$-\mathrm{i}\beta \boldsymbol{H}_T = -\mathrm{i}k\eta(\boldsymbol{i}_z \times \boldsymbol{E}_T) \, . \tag{2.57}$$

From (2.20), we obtain

$$e^{i\beta z}\nabla_T\Phi - i\beta E_T = ik\xi(i_z \times H_T) = -i\frac{k^2}{\beta}E_T \qquad (2.58)$$

or

$$E_T = i\frac{\beta}{p^2}e^{i\beta z}\nabla_T\Phi \ . \qquad (2.59)$$

The transverse E field is proportional to the gradient of Φ, or E_z, with the sign determined by the sign of the propagation constant. The magnetic-field is obtained from the electric field using (2.57); the magnetic-field lines are perpendicular to the electric-field lines.

It is convenient to define a transverse field pattern $e_T(x,y)$ as

$$e_T(x,y) \propto i\nabla_T\Phi \qquad (2.60)$$

and normalize it so that

$$\int_{\substack{\text{cross} \\ \text{section}}} dS\,|e_T(x,y)|^2 = 1 \ . \qquad (2.61)$$

Correspondingly, one defines the normalized magnetic-field pattern as

$$h_T(x,y) = i_z \times e_T(x,y) \ . \qquad (2.62)$$

We shall specialize the discussion in the remainder of this section to lossless media with $\sigma = 0$. In this case the characteristic admittance η and characteristic impedance ξ are real and the propagation constant β is either real (above cutoff) or imaginary (below cutoff). For each mode above cutoff, we may write the general field solution as the superposition of a forward wave and a backward wave. If one defines a characteristic admittance of the mode, Y_o,

$$Y_o \equiv \frac{k\eta}{\sqrt{k^2 - p^2}} = \frac{\omega\epsilon}{\sqrt{\omega^2\mu\epsilon - p^2}} \ , \qquad (2.63)$$

the electric field and magnetic field can be written in terms of forward- and backward-wave amplitudes a and b:

$$E_T = \sqrt{2/Y_o}\left(ae^{i\beta z} + be^{-i\beta z}\right)e_T(x,y) \ , \qquad (2.64)$$

$$H_T = \sqrt{2Y_o}\left(ae^{i\beta z} - be^{-i\beta z}\right)h_T(x,y) \ . \qquad (2.65)$$

The amplitudes are so normalized that the difference of their squares is equal to the power flow

$$\frac{1}{2} \int_{\substack{\text{cross} \\ \text{section}}} da \, \text{Re}(\boldsymbol{E}_T \times \boldsymbol{H}_T^*) = |a|^2 - |b|^2 \ . \tag{2.66}$$

We shall use this normalization of the forward and backward waves throughout the book. The modes possess power orthogonality, i.e. the powers of the different modes add; there are no cross terms. This is shown easily via the orthogonality relation (2.56), but can also be seen on purely physical grounds. Suppose one considers a solution made up of two waves of different modes with different propagation constants β_μ and β_ν. If cross terms existed, the power would vary with distance as $\exp \pm i(\beta_\mu - \beta_\nu)z$. Since power is conserved, this is impossible, and the cross terms must be zero.

Figure 2.5b shows the field pattern of a propagating wave for the $m = 1, n = 1$ lowest-order TM mode in a rectangular waveguide. We shall find it useful to write the transverse electric and magnetic fields in terms of what we shall call a voltage $V(z)$ and a current $I(z)$. Thus, (2.64) and (2.65) could be written alternately as

$$\boldsymbol{E}_T = V(z)\boldsymbol{e}_T(x, y) \ , \tag{2.67}$$

$$\boldsymbol{H}_T = I(z)\boldsymbol{h}_T(x, y) \ . \tag{2.68}$$

The ratio $V(z)/I(z)$ defines an impedance $Z(z) = V(z)/I(z)$, which can be related to the reflection coefficient $\Gamma(z)$, defined by

$$\Gamma(z) \equiv \frac{b}{a} e^{-i2\beta z} \ . \tag{2.69}$$

By comparing (2.67) and (2.68) on one hand with (2.64) and (2.65) on the other hand, one finds the relation

$$\Gamma(z) = \frac{Z(z) - Z_o}{Z(z) + Z_o} \ , \tag{2.70}$$

with $Z_o = 1/Y_o$. We shall find these relations useful further on.

Equation (2.66) is only valid for modes above cutoff. Mode solutions below cutoff possess transverse electric and magnetic fields that are 90 degrees out of phase and hence do not propagate power by themselves. Power is transmitted only when the growing and decaying wave solutions are excited simultaneously. The power is due to the cross terms between the fields of these two waves, which are z-independent as required by power conservation.

Before we conclude this section, it is important to note that the orthogonality condition (2.56) implies orthogonality of the transverse electric fields. Indeed

$$\boldsymbol{e}_{T\nu} \cdot \boldsymbol{e}_{T\mu}^* \propto \nabla_T \Phi_\nu \cdot \nabla_T \Phi_\mu^* = \nabla_T \cdot (\Phi_\nu \nabla_T \Phi_\mu^*) - \Phi_\nu \nabla_T^2 \Phi_\mu^* \ .$$

Thus, if we evaluate the integral

$$\int dS \, e_{T\nu} \cdot e_{T\mu}^* \propto \int dS \, \nabla_T \Phi_\nu \cdot \nabla_T \Phi_\mu^*$$

$$= \int dS \, \nabla_T \cdot (\Phi_\nu \nabla_T \Phi_\mu^*) - \int dS \, \Phi_\nu \nabla_T^2 \Phi_\mu^*$$

over the cross section of the waveguide, use Gauss's theorem, and take into account the boundary conditions, we find

$$\int dS \, e_{T\nu} \cdot e_{T\mu}^* \propto - \int dS \, \Phi_\nu \nabla_T^2 \Phi_\mu^* = p_\nu^2 \int dS \, \Phi_\nu \Phi_\mu^* \, .$$

The right-hand side vanishes for $p_\nu \neq p_\mu$. In this way we have proven the orthogonality of the transverse electric-field patterns of modes with different eigenvalues as well. Orthogonality of the transverse electric-field patterns implies orthogonality of the transverse magnetic-field patterns since $e_{T\nu} \cdot e_{T\mu}^* = h_{T\nu} \cdot h_{T\mu}^*$.

2.4 Transverse Electric Waves

The analysis of transverse electric (TE) waves proceeds completely analogously to that of transverse magnetic waves. Now, a longitudinal magnetic field is assumed, and E_z is set equal to zero. One assumes solutions of the form

$$H_z = \exp(\mathrm{i}\beta z)\Psi(x,y) \, . \tag{2.71}$$

The scalar function Ψ must obey boundary conditions different from those obeyed by the function Φ of TM waves. On the perfect conductor, the tangential electric field must be zero. Since the curl of the magnetic field is proportional to the electric field, we must set

$$\frac{\partial \Psi}{\partial n} = 0 \, . \tag{2.72}$$

Indeed, if this derivative did not vanish, if H_z changed within a distance Δn from the surface of the conductor as shown in Fig. 2.6, there would be a nonzero line integral around the closed contour, the component of the curl normal to the plane of the contour would not vanish, and there would be a tangential electric field at the surface of the conductor.

Except for the change from Dirichlet boundary conditions to Neumann boundary conditions (2.72), the analysis proceeds as in the case of the TM waves. In a rectangular waveguide one finds the solutions

$$\Psi = \cos\left(\frac{m\pi}{a}x\right) \cos\left(\frac{n\pi}{b}y\right) \, . \tag{2.73}$$

In the present case, meaningful solutions are obtained with either $m = 0$ or $n = 0$. Thus the lowest eigenvalues pertaining to TE waves are smaller than

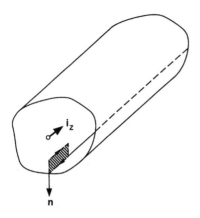

Fig. 2.6. Derivation of boundary condition for Ψ

those of TM waves. The mode with the lowest cutoff frequency is a TE wave. This is true for any waveguide cross section since the Neumann boundary condition (2.72) is less confining than the Dirichlet boundary condition. The drumhead analogy may be helpful. The Neumann boundary condition allows the membrane to move vertically along the rim of the drumhead without friction, only its slope is confined to be zero.

The normalized dispersion diagram of TE waves is the same as that for TM waves. Green's theorem (2.51) can be used to prove the orthogonality relation

$$\left(p_\nu^2 - p_\mu^2\right) \int_{\substack{\text{cross}\\\text{section}}} da\, \Psi_\nu^* \Psi_\mu = 0 \,. \tag{2.74}$$

The general solution is the superposition of forward and backward traveling (or decaying, if cutoff) waves. For one single propagating mode in a lossless waveguide, one has

$$\boldsymbol{E}_T = \sqrt{2/Y_o} \left(a e^{\mathrm{i}\beta z} + b e^{-\mathrm{i}\beta z}\right) \boldsymbol{e}_T(x, y) \,, \tag{2.75}$$

$$\boldsymbol{H}_T = \sqrt{2Y_o} \left(a e^{\mathrm{i}\beta z} - b e^{-\mathrm{i}\beta z}\right) \boldsymbol{h}_T(x, y) \,, \tag{2.76}$$

where $\boldsymbol{h}_T(x, y) \propto \mathrm{i}\nabla_T \psi$ (compare (2.60)) and $\boldsymbol{h}_T(x, y) = \boldsymbol{i}_z \times \boldsymbol{e}_T(x, y)$. Now, the characteristic admittance is defined by

$$Y_o \equiv \frac{\sqrt{k^2 - p^2}}{k\xi} = \frac{\sqrt{\omega^2 \mu \epsilon - p^2}}{\omega \mu} \,. \tag{2.77}$$

Some field patterns of TE modes are shown in Fig. 2.7. The power flow for TM waves is as in (2.66). The power of modes with different eigenvalues is

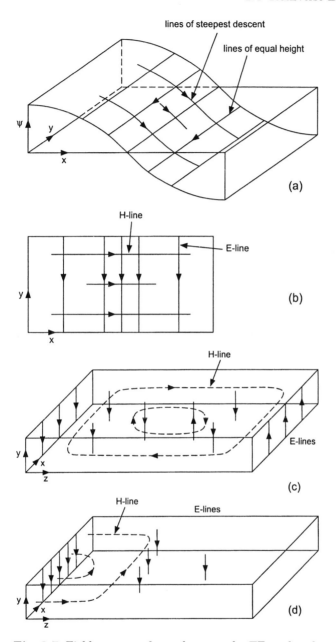

Fig. 2.7. Field patterns of some lowest-order TE modes of a rectangular waveguide. (a) Plot of potential Ψ for $m = 1, n = 0$ mode. (b) The transverse field patterns of the $m = 1, n = 0$ mode. (c) The E and H lines of the propagating $m = 1, n = 0$ wave. (d) The E and H lines of the cutoff wave $m = 1, n = 0$: the fields are 90° out of phase

additive; cross terms vanish. To prove this one may use the orthogonality condition (2.74), or again resort to Poynting's theorem and the z-independence of the time-averaged power flow.

The TE modes and TM modes are power-orthogonal, even if their eigenvalues coincide. This follows from the fact that the transverse electric field of a TM mode is the gradient of a potential, and the transverse magnetic field of a TE mode is a gradient as well. Thus, consider the contribution of the transverse electric field of the TM mode and the transverse magnetic field of a TE mode:

$$\int_{\substack{\text{cross}\\\text{section}}} dS\, \mathbf{i}_z \cdot \mathbf{e}_{T,TM} \times \mathbf{h}^*_{T,TE} = \mathbf{i}_z \cdot \int_{\substack{\text{cross}\\\text{section}}} dS\, \nabla_T \Phi \times \nabla_T \Psi^*$$

$$= \mathbf{i}_z \cdot \int_{\substack{\text{cross}\\\text{section}}} dS\, \nabla_T \times (\Phi \nabla_T \Psi^*) - \mathbf{i}_z \cdot \int_{\substack{\text{cross}\\\text{section}}} dS\, \Phi(\nabla_T \times \nabla_T \Psi^*) \, .$$

(2.78)

The first integral is a curl that can be evaluated as a contour integral around the boundary of the waveguide, on which Φ vanishes. Thus this contribution is zero. The second integral contains the curl of a gradient and thus its kernel is zero. Hence we can conclude that TE and TM modes are power-orthogonal even if they possess the same eigenvalues.

The orthogonality relation (2.74) refers to the scalar functions Ψ_ν and Ψ_μ. Just as at the end of Sect. 2.3, a simple manipulation shows that orthogonality of the transverse field patterns is implied as well. With proper normalization one has

$$\int dS\, \mathbf{h}_{T\nu} \cdot \mathbf{h}^*_{T\mu} = \delta_{\mu\nu} \, .$$

As mentioned earlier, the orthogonality of the magnetic-field patterns implies the orthogonality of the electric field patterns and vice versa.

2.4.1 Mode Expansions

We have found that a conducting enclosure supports an infinite number of modes. We also found that not all the modes are propagating modes at a given frequency of excitation $\omega = k/\sqrt{\mu\epsilon}$ (for a lossless medium). When $k < p_\nu$, the mode is cut off. The existence of mode cutoff is important from a practical point of view. If an excitation consisting of several propagating modes travels down the guide, the different modes interfere differently at different waveguide cross sections. If the excitation is composed of a band of frequencies, the interference of the different modes is a function of distance along the guide. Such a behavior is unacceptable if the signal propagation is to be distortion-free. Hence, in most practical applications frequencies of excitation are used that are in the band in which only the dominant TE

mode is above cutoff, and all other modes are below cutoff. In the case of a
two-conductor transmission line this mode is, of course, the TEM mode.

We now look at the excitation of the modes at one cross section. In order
to determine the excitation of the modes, mode orthogonality conditions have
to be invoked. This exercise is then a good example of the use of orthogonality
conditions, which will be applied again in Sect. 2.9. The intent is to find the
power radiated from a wire across the waveguide carrying a current I_o at
frequency ω. The current can be thought to be produced by excitation of the
waveguide by a coaxial cable as shown in Fig. 2.8.

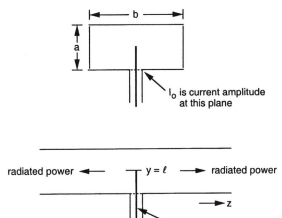

Fig. 2.8. A waveguide fed by a coaxial cable

The problem can be formulated in terms of an excitation by a current
sheet $K(r_T)$ across the waveguide at one cross section, say at $z = 0$. The
field to either side of the sheet is a superposition of an infinite set of E and
H modes, propagating or decaying away from the sheet in the $+z$ and $-z$
directions. The transverse fields are given by

$$E_T(0_+) = \sum_\nu \sqrt{\frac{2}{Y_{0\nu}}} a_\nu e_T(x, y) \,,$$

$$H_T(0_+) = \sum_\nu \sqrt{2Y_{0\nu}}\, a_\nu h_T(x, y) \quad \text{for } z > 0 \tag{2.79}$$

and

$$E_T(0_-) = \sum_\nu \sqrt{\frac{2}{Y_{0\nu}}} b_\nu e_T(x, y) \,,$$

$$\boldsymbol{H}_T(0_-) = -\sum_\nu \sqrt{2Y_{0\nu}} \, b_\nu \boldsymbol{h}_T(x,y) \quad \text{for } z < 0 . \tag{2.80}$$

Across the plane of the current sheet, the magnetic field experiences a discontinuity (see Fig. 2.8)

$$\boldsymbol{i}_z \times [\boldsymbol{H}(0_+) - \boldsymbol{H}(0_-)] = \boldsymbol{K} , \tag{2.81}$$

whereas the electric field is continuous:

$$\boldsymbol{i}_z \times [\boldsymbol{E}(0_+) - \boldsymbol{E}(0_-)] = 0 . \tag{2.82}$$

After multiplication by the mode patterns $\boldsymbol{e}_{T\mu}$ and $\boldsymbol{h}_{T\mu}$ and integrating over the cross section, one finds, using the orthogonality conditions,

$$a_\mu + b_\mu = -\frac{1}{\sqrt{2Y_{o\mu}}} \int dS \, \boldsymbol{i}_z \times \boldsymbol{K} \cdot \boldsymbol{h}_{T\mu} \tag{2.83}$$

and $a_\mu - b_\mu = 0$. From these two equations one may compute the amplitudes of the modes. If all modes but the dominant mode (denoted by $\nu = 0$) are below cutoff, the power radiated in both directions by the current sheet is simply

$$\text{power} \; = 2|a_o|^2 = \frac{1}{Y_{00}} \left| \int dS \, \boldsymbol{i}_z \times \boldsymbol{K} \cdot \boldsymbol{h}_{T0} \right|^2 . \tag{2.84}$$

Let us now specialize to the problem at hand. The dominant mode of a rectangular waveguide has the normalized transverse magnetic-field pattern

$$\boldsymbol{h}_{T0} = \sqrt{\frac{2}{ab}} \boldsymbol{i}_x \cos\left(\frac{\pi x}{a}\right) . \tag{2.85}$$

A current in a thin wire is composed of waves with a propagation constant equal to the free-space propagation constant of a plane wave in a medium characterized by μ, ϵ. (To understand why this is the case, think first of a coaxial cable with a thin center conductor. The propagation constant is $k = \omega\sqrt{\mu\epsilon}$. Now remove the outer conductor. If the center wire is thin, the energy storage near the wire outweighs the energy storage farther away, and hence the current distribution is not affected by the removal of the outer conductor.) If the wire terminates, as in an antenna, the distribution is $[-I_o/\sin(k\ell)]\sin[k(y - \ell)]$ if I_o is the current at $y = 0$ and ℓ is the length of the wire. Hence

$$\boldsymbol{K} = -\boldsymbol{i}_y \frac{I_o}{\sin(k\ell)} \sin[k(y - \ell)]\delta(x - a/2) . \tag{2.86}$$

Combining (2.84), (2.85), and (2.86), we find

$$\text{power} \; = \frac{2}{k^2 ab Y_{00}} = \frac{[1 - \cos(k\ell)]^2}{\sin^2 k\ell} . \tag{2.87}$$

2.5 Energy, Power, and Energy Velocity

The complex Poynting theorem of Sect. 1.7 developed expressions for the energy stored in matter when the medium is dispersive. The approach was in the form of a thought experiment in which excitations were applied that grew or decayed at a rate α. The constitutive relations were evaluated to first order in the growth or decay rate. In this way derivatives with respect to frequency appeared in the energy density expressions. In this section we follow a somewhat different approach. Derivatives are taken with respect to frequency of the complex form of Maxwell's equations and identities are derived therefrom. One of the findings is that the group velocity of a waveguide mode is the velocity of energy propagation.

2.5.1 The Energy Theorem

In the analysis of waveguide modes, we assumed that the waveguide was filled uniformly with an isotropic medium with a scalar dielectric constant and magnetic permeability. Had we not made this assumption, we would have found that TE waves and TM waves do not exist independently, but are coupled by the medium and/or the boundary conditions. The derivation in this section is not more difficult if tensor media are included. Hence, we shall develop the formalism in this, more general, context. Inside the volume $\mathcal{V}$ of a waveguide, formed by the waveguide walls and two cross sections at $z = z_1$ and $z = z_2$, Maxwell's equations hold:

$$\nabla \times \boldsymbol{E} = i\omega \overline{\overline{\mu}} \cdot \boldsymbol{H} , \tag{2.88}$$

$$\nabla \times \boldsymbol{H} = -i\omega \overline{\overline{\epsilon}} \cdot \boldsymbol{E} . \tag{2.89}$$

The fields are functions of position $\boldsymbol{r}$ and of frequency ω. First we take the derivatives with respect to ω of (2.88) and (2.89):

$$\nabla \times \frac{\partial \boldsymbol{E}}{\partial \omega} = i \left(\overline{\overline{\mu}} + \omega \frac{\partial \overline{\overline{\mu}}}{\partial \omega} \right) \cdot \boldsymbol{H} + i\omega \overline{\overline{\mu}} \cdot \frac{\partial \boldsymbol{H}}{\partial \omega} , \tag{2.90}$$

$$\nabla \times \frac{\partial \boldsymbol{H}}{\partial \omega} = -i \left(\overline{\overline{\epsilon}} + \omega \frac{\partial \overline{\overline{\epsilon}}}{\partial \omega} \right) \cdot \boldsymbol{E} - i\omega \overline{\overline{\epsilon}} \cdot \frac{\partial \boldsymbol{E}}{\partial \omega} . \tag{2.91}$$

Then we dot-multiply (2.90) by $\boldsymbol{H}^*$, (2.91) by $-\boldsymbol{E}^*$, the complex conjugate of (2.88) by $\partial \boldsymbol{H}/\partial \omega$, and the complex conjugate of (2.89) by $-\partial \boldsymbol{E}/\partial \omega$. We then add the resulting equations and cancel terms (noting that $\overline{\overline{\mu}}$ and $\overline{\overline{\epsilon}}$ are Hermitian tensors). We obtain

$$\nabla \cdot \left(\frac{\partial \boldsymbol{E}}{\partial \omega} \times \boldsymbol{H}^* + \boldsymbol{E}^* \times \frac{\partial \boldsymbol{H}}{\partial \omega} \right)$$

$$= \mathrm{i} \left[\boldsymbol{H}^* \cdot \left(\overline{\overline{\mu}} + \omega \frac{\partial \overline{\overline{\mu}}}{\partial \omega} \right) \cdot \boldsymbol{H} + \boldsymbol{E}^* \cdot \left(\overline{\overline{\epsilon}} + \omega \frac{\partial \overline{\overline{\epsilon}}}{\partial \omega} \right) \cdot \boldsymbol{E} \right] .$$

Integrating over the volume $\mathcal{V}$ and using Gauss's theorem, we have

$$\oint \left(\frac{\partial \boldsymbol{E}}{\partial \omega} \times \boldsymbol{H}^* + \boldsymbol{E}^* \times \frac{\partial \boldsymbol{H}}{\partial \omega} \right) \cdot d\boldsymbol{S}$$

$$= \mathrm{i} \int \left[\boldsymbol{H}^* \cdot \left(\overline{\overline{\mu}} + \omega \frac{\partial \overline{\overline{\mu}}}{\partial \omega} \right) \cdot \boldsymbol{H} + \boldsymbol{E}^* \cdot \left(\overline{\overline{\epsilon}} + \omega \frac{\partial \overline{\overline{\epsilon}}}{\partial \omega} \right) \cdot \boldsymbol{E} \right] dV . \tag{2.92}$$

We may identify the integral on the right-hand side of (2.92) as four times the time-averaged stored energy w in the volume $\mathcal{V}$:

$$w = \frac{1}{4} \int \left[\boldsymbol{H}^* \cdot \left(\overline{\overline{\mu}} + \omega \frac{\partial \overline{\overline{\mu}}}{\partial \omega} \right) \cdot \boldsymbol{H} + \boldsymbol{E}^* \cdot \left(\overline{\overline{\epsilon}} + \omega \frac{\partial \overline{\overline{\epsilon}}}{\partial \omega} \right) \cdot \boldsymbol{E} \right] dV . \tag{2.93}$$

Using this fact, we may write for (2.94)

$$\oint \left(\frac{\partial \boldsymbol{E}}{\partial \omega} \times \boldsymbol{H}^* + \boldsymbol{E}^* \times \frac{\partial \boldsymbol{H}}{\partial \omega} \right) \cdot d\boldsymbol{S} = 4\mathrm{i}w . \tag{2.94}$$

Equation (2.94) is the energy theorem. It relates frequency derivatives of the electric and magnetic fields on the surface S to the energy stored in the volume enclosed by S.

2.5.2 Energy Velocity and Group Velocity

The theorem (2.94) can now be used to find a relation for the energy velocity. We shall identify the fields $\boldsymbol{E}$ and $\boldsymbol{H}$ in (2.94) with a single wave solution, with the propagation constant β, in a uniform waveguide filled with an isotropic, uniform medium characterized by μ, ϵ. The surface S is formed by the waveguide walls and two reference planes at z_1 and z_2, a distance $z_2 - z_1 = L$ apart. We have for the left-hand side of (2.93)

$$\oint \left(\frac{\partial \boldsymbol{E}}{\partial \omega} \times \boldsymbol{H}^* + \boldsymbol{E}^* \times \frac{\partial \boldsymbol{H}}{\partial \omega} \right) \cdot d\boldsymbol{S}$$

$$= \int_{\substack{\text{cross} \\ \text{section}}} \left(\frac{\partial \boldsymbol{E}_T}{\partial \omega} \times \boldsymbol{H}_T^* + \boldsymbol{E}_T^* \times \frac{\partial \boldsymbol{H}_T}{\partial \omega} \right) \cdot \boldsymbol{i}_z dS \Bigg|_{z_1}^{z_2} . \tag{2.95}$$

The transverse electric and magnetic fields of a single wave of amplitude a of either a TE, a TM, or a TEM mode can be written

$$E_T = \sqrt{2/Y_o}\,ae^{i\beta z}e_T(x,y) \,, \tag{2.96}$$

$$H_T = \sqrt{2Y_o}\,ae^{i\beta z}h_T(x,y) \,. \tag{2.97}$$

The integral on the right-hand side of (2.95) involves only quantities that change between the two cross sections. The only z-dependent quantity introduced by the frequency derivative derives from the factor $\exp(i\beta z)$:

$$\frac{\partial}{\partial \omega}\left(e^{i\beta z}\right) = iz\frac{d\beta}{d\omega}e^{i\beta z} \,. \tag{2.98}$$

Therefore, using (2.94) and (2.98) in (2.95), we find

$$4i\left(z_2 - z_1\right)\frac{d\beta}{d\omega}|a|^2\int_{\substack{\text{cross}\\\text{section}}}da\,i_z\cdot e_T\times h_T^* = 4iw \,. \tag{2.99}$$

Since the field patterns are normalized, and the power p in the wave is equal to $|a|^2$, we conclude that

$$\frac{d\omega}{d\beta} \equiv v_g = \frac{p}{w/L} \,. \tag{2.100}$$

The derivative $d\omega/d\beta$ is an energy velocity, the ratio of power flow to energy per unit length. The same quantity is also known as the group velocity, the velocity of propagation of a wavepacket with a spectrum consisting of a narrow band of frequencies. Indeed, if one adds two waves with the dependences $\exp[i\beta(\omega)z]\exp(-i\omega t)$ and $\exp[i\beta(\omega + \Delta\omega)z]\exp[-i(\omega + \Delta\omega)t]$ one obtains

$$\exp\left\{i\left[\beta(\omega)z - \omega t\right]\right\} + \exp\left\{i\left[\beta(\omega + \Delta\omega)z - (\omega + \Delta\omega)t\right]\right\}$$
$$= \exp\left\{i\left[\beta(\omega)z - \omega t\right]\right\}\left\{1 + \exp\left[i\Delta\omega\left(\frac{d\beta}{d\omega}z - t\right)\right]\right\} \,. \tag{2.101}$$

The wavepacket has an envelope that goes periodically to zero at distances $\Lambda = 2\pi(d\omega/d\beta)/\Delta\omega$ and travels at the so called group velocity $d\omega/d\beta$. The energy theorem has shown that the group velocity is also the velocity of energy propagation. This is not surprising. If one constructs a wavepacket in the manner of (2.101), the fields vanish periodically at the nodes of the wavepacket. No power can cross the cross sections of zero field. Hence the energy of the excitation is trapped between the nulls and travels at the speed of the nulls, at the group velocity.

2.5.3 Energy Relations for Waveguide Modes

Next, we derive a property of waves in uniform waveguides that follows directly from the complex Poynting theorem, (1.88) of Chap. 1, which is repeated below:

$$\nabla \cdot \frac{1}{2}\mathrm{Im}(\boldsymbol{E} \times \boldsymbol{H}^*) + Q - 2\omega\left(W_m^o - W_e^o\right) = 0 . \tag{2.102}$$

Q is the so-called reactive power generated per unit volume and is given in (1.92). (Note that $\overline{\overline{\chi}}_e$ and $\overline{\overline{\chi}}_m$ are Hermitian tensors.)

$$Q = \frac{1}{2}i\omega(\boldsymbol{E}^* \cdot \epsilon_o\overline{\overline{\chi}}_e \cdot \boldsymbol{E} - \boldsymbol{H}^* \cdot \mu_o\overline{\overline{\chi}}_m \cdot \boldsymbol{H}) . \tag{2.103}$$

Combining (2.102) and (2.103) and using the definitions of the dielectric and permeability tensors, we obtain

$$\nabla \cdot \frac{1}{2}\mathrm{Im}(\boldsymbol{E} \times \boldsymbol{H}^*) - 2\omega(\boldsymbol{H}^* \cdot \overline{\overline{\mu}} \cdot \boldsymbol{H} - \boldsymbol{E} \cdot \overline{\overline{\epsilon}} \cdot \boldsymbol{E}^*) = 0 \tag{2.104}$$

or

$$\nabla \cdot \frac{1}{2}\mathrm{Im}(\boldsymbol{E} \times \boldsymbol{H}^*) - 2\omega(\mu|\boldsymbol{H}|^2 - \epsilon|\boldsymbol{E}|^2) = 0 \tag{2.104a}$$

for the case of an isotropic medium as analyzed in the present chapter. We integrate this equation over a volume of length L, between the cross sections z_1 and z_2 in the waveguide. We further assume that the fields are those of a single traveling wave and the guide is lossless so that all quantities in (2.104a) are z-independent. The integral of the divergence is zero, because the Poynting fluxes through the two end faces of the volume cancel, and there is no Poynting flux at the waveguide walls. In fact, the theorem that follows is not restricted to waveguides with metallic walls. The only requirement is that no Poynting flux exit radially through a cylinder enveloping the waveguide. In this more general sense, the theorem applies to dielectric waveguides and optical fibers, as discussed in the next chapter. We obtain

$$\int_{\substack{\text{waveguide} \\ \text{cross section}}} \mu|\boldsymbol{H}|^2 dS = \int_{\substack{\text{waveguide} \\ \text{cross section}}} \epsilon|\boldsymbol{E}|^2 dS . \tag{2.105}$$

This equation can be interpreted as stating that the time-averaged electric and magnetic energies per unit length in a traveling wave are equal to each other. This interpretation holds only for a nondispersive medium. The energy storages in a dispersive medium are more complicated and, as pointed out in the example of a plasma, consist of both the field energies and the energies stored in the excitation of the medium. In this more general case (2.105) still holds, but cannot be interpreted so simply.

2.5.4 A Perturbation Example

Before we conclude this section we introduce some concepts of perturbation theory, which we shall employ throughout the text. We can test the results against the equations obtained for modes in metallic waveguides. As in the case of (2.105), the perturbation theory developed here is applicable to any waveguide that is uniform along the z direction, such as a dielectric waveguide or an optical fiber.

Modes of uniform, lossless waveguides always appear in pairs: a forward-propagating wave a of spatial dependence $\exp(i\beta z)$ is paired with a backward wave b of dependence $\exp(-i\beta z)$, where β, the propagation constant, is real. This is the direct consequence of time reversibility of Maxwell's equations in their complex form, as discussed in Sect. 1.1. (Note, however, the "caveat" concerning the reversal of a d.c. magnetic field in the case of the Faraday effect.) Let us concentrate on the solution for the forward wave a of a particular mode ν. We shall omit the subscript denoting this particular mode in the subsequent analysis for simplicity. Clearly, the amplitude a obeys the differential equation

$$\frac{da}{dz} = i\beta a . \tag{2.106}$$

Suppose next that the waveguide has some small loss. The loss will introduce attenuation, and (2.106) is modified to

$$\frac{da}{dz} = [-\operatorname{Im}(\beta) + i\operatorname{Re}(\beta)]a , \tag{2.107}$$

where $\operatorname{Im}(\beta)$ is the attenuation constant. The attenuation constant can be computed from an energy conservation argument: the spatial rate of change of power p along the waveguide must be equal to the power dissipated per unit length:

$$\frac{dp}{dz} = -2\operatorname{Im}(\beta)p = -\text{power dissipated per unit length} . \tag{2.108}$$

Since the power is quadratic in the fields, its spatial rate of decay is twice that of the fields. Now, the power p is equal to the product of the group velocity and the energy w per unit length. Hence, we find from (2.108) and (2.100)

$$2\operatorname{Im}(\beta) = \frac{d\left[\operatorname{Re}(\beta)\right]}{d\omega} \frac{\text{power dissipated per unit length}}{w/L} , \tag{2.109}$$

where we note that the group velocity is to be evaluated for a lossless guide for which $\operatorname{Re}(\beta) = \beta$. This equation determines $\operatorname{Im}(\beta)$. Let us determine how it works in the case of a lossy, dissipative waveguide. In this case we can find the complex propagation constant directly from (2.50), which is applicable to both transverse electric and transverse magnetic waves:

$$\beta = \sqrt{k^2 - p^2} = \sqrt{\omega^2\mu\epsilon + i\omega\mu\sigma - p^2}$$

$$\approx \sqrt{\omega^2\mu\epsilon - p^2} + \frac{i\omega\mu\sigma}{2\sqrt{\omega^2\mu\epsilon - p^2}} . \tag{2.110}$$

Thus, we have found that

$$2\,\mathrm{Im}(\beta) = \frac{\omega\mu\sigma}{\mathrm{Re}(\beta)} \; . \tag{2.111}$$

Next, we show that (2.111) is consistent with (2.109). The inverse group velocity is

$$\frac{d\beta}{d\omega} = \frac{2\omega\mu\epsilon + \omega^2(\mu\partial\epsilon/\partial\omega + \epsilon\partial\mu/\partial\omega)}{\beta} \; . \tag{2.112}$$

The energy per unit length is

$$\begin{aligned}
\frac{w}{L} &= \frac{1}{4}\Bigg[\left(\epsilon + \omega\frac{\partial\epsilon}{\partial\omega}\right)\int_{\substack{\text{waveguide}\\\text{cross section}}} |\boldsymbol{E}|^2 dS \\
&\quad + \left(\mu + \omega\frac{\partial\mu}{\partial\omega}\right)\int_{\substack{\text{waveguide}\\\text{cross section}}} |\boldsymbol{H}|^2 dS\Bigg] \\
&= \frac{1}{4}\left[\left(\epsilon + \omega\frac{\partial\epsilon}{\partial\omega}\right) + \frac{\epsilon}{\mu}\left(\mu + \omega\frac{\partial\mu}{\partial\omega}\right)\right]\int_{\substack{\text{waveguide}\\\text{cross section}}} |\boldsymbol{E}|^2 dS \; ,
\end{aligned} \tag{2.113}$$

where we have made use of the theorem (2.105). The power dissipated per unit length is equal to

$$\text{power dissipated per unit length} \; = \frac{1}{2}\sigma\int_{\substack{\text{waveguide}\\\text{cross section}}} |\boldsymbol{E}|^2 dS \; . \tag{2.114}$$

Combining (2.112), (2.113), and (2.114) with (2.109), we see that the perturbational formula gives a result consistent with the direct derivation of the attenuation constant. The important fact to remember is that the incorporation of loss by means of a perturbation formula is a powerful method applicable in all practical cases, since waveguides that have a large loss, a large change of amplitude per wavelength, are of little practical use.

2.6 The Modes of a Closed Cavity

The problem of a microwave cavity fed by a number of incoming waveguides can be conveniently formulated and solved by considering first the case of a perfectly closed cavity, i.e. a region of space completely enclosed by perfectly conducting walls. The present section is devoted to a study of the resonant modes in such a closed cavity. Consider a region of space filled with a uniform medium that is isotropic and characterized by a scalar dielectric permittivity ϵ, magnetic permeability μ, and conductivity σ. This region has a volume $\mathcal{V}$ and the surface S bounding the volume is formed from lossless walls. For the sake of generality, we shall assume that part of the surface, S', is formed from a perfect electric conductor and the part S'' is formed from a perfect magnetic conductor. The inclusion of magnetic walls in the analysis gives it greater flexibility, which will be of use later.

Inside the volume $\mathcal{V}$ enclosed by the surface S the electromagnetic fields have to satisfy the source-free Maxwell equations

$$\nabla \times \boldsymbol{E} = -\mu \frac{\partial}{\partial t} \boldsymbol{H} , \tag{2.115}$$

$$\nabla \times \boldsymbol{H} = \sigma \boldsymbol{E} + \epsilon \frac{\partial}{\partial t} \boldsymbol{E} , \tag{2.116}$$

$$\nabla \cdot \epsilon \boldsymbol{E} = 0 . \tag{2.117}$$

$$\nabla \cdot \mu \boldsymbol{H} = 0 \tag{2.118}$$

In the above equations the vectors $\boldsymbol{E}$ and $\boldsymbol{H}$ are space- and time-dependent. This means that the dielectric constant ϵ, the conductivity σ, and the permeability μ must all be constants, independent of time. Hence the analysis in this section assumes that the medium in the resonator is nondispersive. This is only a temporary restriction. We shall find out that the modes derived in this section have purely geometric properties and hence are not medium-specific. We shall be able to use them in an analysis of resonators containing dispersive media. The fields in (2.115)–(2.118) have to satisfy the boundary conditions

$$\boldsymbol{n} \times \boldsymbol{E} = 0, \quad \boldsymbol{n} \cdot \boldsymbol{H} = 0 \quad \text{on } S' ; \tag{2.119a}$$
$$\boldsymbol{n} \cdot \boldsymbol{E} = 0, \quad \boldsymbol{n} \times \boldsymbol{H} = 0 \quad \text{on } S'' . \tag{2.119b}$$

Combining (2.115) and (2.116) one finds

$$\nabla \times (\nabla \times \boldsymbol{E}) + \mu\epsilon \frac{\partial^2}{\partial t^2} \boldsymbol{E} + \mu\sigma \frac{\partial \boldsymbol{E}}{\partial t} = 0 . \tag{2.120}$$

We attempt a solution by separation of variables. The electric field is written as a product of a function of time and a function of space:

$$\boldsymbol{E} = V(t)\boldsymbol{e}(\boldsymbol{r}) . \tag{2.121}$$

The function of space has to satisfy the appropriate boundary conditions of the enclosing surface and is assumed to obey the eigenvalue equation

$$\nabla \times [\nabla \times \boldsymbol{e}(\boldsymbol{r})] = p^2 \boldsymbol{e}(\boldsymbol{r}) . \tag{2.122}$$

The mode pattern $\boldsymbol{e}(\boldsymbol{r})$ has nonzero curl. It is convenient to assign zero divergence to this pattern, leaving the representation of fields with divergence to a different set of modes. The modes with nonzero curl are called "solenoidal". Since the identity

$$\nabla \times [\nabla \times e(r)] = \nabla[\nabla \cdot e(r)] - \nabla^2 e(r)$$

holds, we may rewrite the eigenvalue equation as

$$\nabla^2 e(r) + p^2 e(r) = 0 . \tag{2.123}$$

Equation (2.123), subject to the boundary conditions (2.119a) and (2.119b), has solutions that are functions solely of the geometry of the resonator walls and are independent of the medium filling the resonator. One very simple example is a cavity made of a rectangular waveguide of side lengths a and b and shorted with two conducting planes at $z = 0$ and $z = c$. One may pick standing wave solutions for the waveguide modes and choose the propagation constant so that the tangential electric field at the two shorting planes is zero. In this case one finds a triply infinite set of eigenvalues given by

$$p_{mnq}^2 = \left(\frac{m\pi}{a}\right)^2 + \left(\frac{n\pi}{b}\right)^2 + \left(\frac{q\pi}{c}\right)^2 ,$$

where m, n, and q are integers. In order to satisfy (2.120), the function of time must satisfy the equation

$$\epsilon\mu \frac{d^2}{dt^2} V + \sigma\mu \frac{d}{dt} V + p^2 V = 0 . \tag{2.124}$$

The divergence-free modes found thus far are called solenoidal. If the cavity contains free charges, the solenoidal modes are not sufficient to characterize the field. There must exist modes with divergence and no curl, the so-called divergence modes. They are derivable from a potential:

$$e(r) = -\nabla\Phi(r) . \tag{2.125}$$

The potential can be chosen to obey the Helmholtz equation:

$$\nabla^2\Phi + p^2\Phi = 0 . \tag{2.126}$$

This eigenvalue problem, subject to the boundary conditions on Φ and the normal derivative of Φ on the surfaces S' and S'', respectively, has an infinite number of solutions.

Before we study the eigenvalue equations (2.122) and (2.126), it is of interest to show that one could have proceeded by solving Maxwell's equations in terms of the magnetic field. One could have set

$$H(r, t) = I(t)h(r) . \tag{2.127}$$

Elimination of the electric field from (2.115) and (2.116) leads to an equation for the magnetic field of the same form as (2.120):

$$\nabla \times (\nabla \times H) + \epsilon\mu \frac{\partial^2}{\partial t^2} H + \sigma\mu \frac{\partial}{\partial t} H = 0 . \tag{2.128}$$

The magnetic field pattern $h(r)$ is chosen to obey the eigenvalue equation

$$\nabla \times [\nabla \times h(r)] = p^2 h(r) ,\qquad (2.129)$$

and the differential equation for the time-dependent amplitude $I(t)$ of the magnetic field is

$$\epsilon\mu\frac{d^2}{dt^2}I + \sigma\mu\frac{d}{dt}I + p^2 I = 0 .\qquad (2.130)$$

Since the time dependence of $I(t)$ as predicted from (2.130) must be the same as that of $V(t)$ in (2.124), it follows that the eigenvalue p^2 in (2.122) and (2.129) must be the same.

There are also divergence solutions for the magnetic field. They are the gradient of a potential

$$h(r) = -\nabla\Psi .\qquad (2.131)$$

Ψ can be chosen to obey the scalar Helmholtz equation

$$\nabla^2\Psi + p^2\Psi = 0 ,\qquad (2.132)$$

with the boundary condition that $d\Psi/dn = 0$ on S' and $\Psi = 0$ on S''.

In concluding this section, we reemphasize that the eigenvalue equations (2.122), (2.126), (2.129), and (2.132), with the associated boundary conditions, involve only the geometry of the resonator and are independent of the uniform material filling it. Thus, the modes obtained by solving the eigenvalue equations can be utilized for an expansion of the fields in a resonator of the same geometry, but filled with an arbitrary medium; and, more generally, in a resonator driven by sources.

2.7 Real Character of Eigenvalues and Orthogonality of Modes

The divergence-free electric- and magnetic-field patterns, $e(r)$ and $h(r)$ derived in the preceding section can be shown to satisfy certain orthogonality relations. The proof of the orthogonality relations bears a close resemblance to the previously presented proofs of the orthogonality properties of the waveguide modes. One makes use of a three-dimensional vector Green's theorem, which we now proceed to derive. One starts with Gauss's theorem:

$$\int \nabla \cdot D \, dV = \oint D \cdot dS .\qquad (2.133)$$

Here D is an arbitrary three-dimensional complex vector function of space, restricted only by the stipulation that it be once differentiable. We substitute for D the expression

$$D = A \times (\nabla \times B) , \tag{2.134}$$

where, again, A and B are arbitrary complex three-dimensional vector functions of the spatial variable, restricted only by the stipulation that they are once and twice differentiable, respectively. Making use of the following vector identity (differentiation by parts),

$$\nabla \cdot [A \times (\nabla \times B)] = (\nabla \times A) \cdot (\nabla \times B) - A \cdot \nabla \times (\nabla \times B) , \tag{2.135}$$

we obtain, combining (2.133)–(2.135),

$$\int [(\nabla \times A) \cdot (\nabla \times B) - A \cdot \nabla \times (\nabla \times B)] \cdot d\mathcal{V}$$
$$= \oint A \times (\nabla \times B) \cdot dS . \tag{2.136}$$

Equation (2.136) is the first vector Green's theorem. The second vector Green's theorem can be obtained from this by interchanging the functions A and B and subtracting the resulting relation from (2.136). Thus we obtain the second vector Green's theorem.

$$\int [B \cdot \nabla \times (\nabla \times A) - A \cdot \nabla \times (\nabla \times B)] \, d\mathcal{V}$$
$$= \oint [A \times (\nabla \times B) - B \times (\nabla \times A)] \cdot dS \tag{2.137}$$

Now let us substitute for A in (2.136) an electric-field pattern E_ν which is a solution to (2.122) pertaining to a particular eigenvalue p_ν. For B, we substitute its complex conjugate:

$$A = e_\nu , \quad B = e_\nu^* . \tag{2.138}$$

The equation satisfied by e_ν is

$$\nabla \times (\nabla \times e_\nu) = p_\nu^2 e_\nu . \tag{2.139}$$

We interpret the integration in (2.136) as being carried out over the entire volume of the closed cavity enclosed by the lossless wall. By virtue of the boundary conditions satisfied by e_ν on S' and by $\nabla \times e_\nu$, which is proportional to h_ν, on S'', (2.119a) and (2.119b), the surface integral on the right-hand side of (2.136) vanishes:

$$\oint_{S',S''} e_\nu \times (\nabla \times e_\nu^*) \cdot dS = 0 . \tag{2.140}$$

Solving the remaining expression for p_ν^2, we obtain

$$p_\nu^2 = \frac{\int (\nabla \times e_\nu) \cdot (\nabla \times e_\nu)^* dV}{\int e_\nu \cdot e_\nu^* dV} \ .$$ (2.141)

Equation (2.141) shows that the eigenvalue p_ν^2 of (2.139) must be real and positive for fields satisfying the boundary conditions (2.119a) and (2.119b).

Next, we turn to the proof of the orthogonality relationships. In (2.137) we make the substitutions

$$A = e_\nu , \quad B = e_\mu^* .$$ (2.142)

Again, extending the integral over the entire volume of the closed cavity, the surface integral on the right-hand side of (2.137) vanishes and there results

$$(p_\nu^2 - p_\mu^2) \int e_\nu \cdot e_\mu^* dV = 0 ,$$ (2.143)

where we have taken into account that the fields e_ν and e_μ have to satisfy equations of the form of (2.139) with real eigenvalues p^2. From (2.143) we conclude that

$$\int e_\nu \cdot e_\mu^* dV = 0 , \quad p_\nu \neq p_\mu .$$ (2.144)

Field patterns pertaining to different eigenvalues are orthogonal in the sense of (2.144). The case in which two distinct field patterns have the same eigenvalue p is called degeneracy. In such a case, orthogonality is not automatically assured. It is possible, however, to construct an orthogonal set of field patterns even in a degenerate case by using linear combinations of the degenerate modes. Assuming that such an orthogonalization has been carried out on the entire set of modes, one may express the orthogonality condition in the form

$$\int e_\nu \cdot e_\mu^* dV = V\delta_{\nu\mu} ,$$ (2.145)

where in addition it has been assumed that the field patterns have been normalized so that the volume integral of the square of the field pattern is equal to the volume of the cavity. Analogous orthogonality conditions can be proved for the magnetic-field patterns.

It is clear that the magnetic-field patterns are proportional to the curl of the electric-field patterns. Setting

$$\nabla \times e_\nu(r) = p_\nu h_\nu(r) ,$$ (2.146)

one finds that the magnetic-field patterns are automatically normalized. Indeed, introducing (2.146) into (2.136), we find

$$p_\nu p_\mu \int h_\nu \cdot h_\mu^* dV - p_\mu^2 \int e_\nu \cdot e_\mu^* dV = 0 .$$ (2.147)

This leads to the orthogonality condition

$$\int h_\nu \cdot h_\mu^* \, d\mathcal{V} = \mathcal{V} \delta_{\nu\mu} \,. \tag{2.148}$$

Since the eigenvalue p^2 of (2.139) is real, it is always possible to choose the field patterns e_μ to be real, i.e. linearly polarized at every point in the cavity. The direction of the polarization may vary from point to point. Indeed, suppose that we have found a complex solution of (2.139). Then both the real and the imaginary parts of the complex solution must be solutions of (2.139).

An analogous analysis can be carried out for the divergence modes of the electric and magnetic fields. One finds that they have real, positive eigenvalues p^2 and also obey orthogonality conditions.

The solenoidal modes are orthogonal to the divergence modes as well. We shall prove this in the case of the modes of the E field. We denote the solenoidal mode by the subscript ν and the divergence mode by the subscript α, and evaluate the volume integral $\int e_\nu \cdot e_\alpha \, d\mathcal{V}$ over the volume of the resonator. The divergence mode can be expressed as the gradient of a scalar potential. We thus have

$$\int e_\nu \cdot e_\alpha \, d\mathcal{V} = -\int e_\nu \cdot \nabla \Phi_\alpha \, d\mathcal{V}$$

$$= -\oint_{S'+S''} \Phi_\alpha e_\nu \cdot dS + \int \Phi_\alpha \nabla \cdot e_\nu \, d\mathcal{V} \,. \tag{2.149}$$

Both integrals on the right-hand side of (2.149) are zero. The second integral contains the divergence of the solenoidal mode, which is zero by definition. The surface integral contains no contribution from the surface S'', over which the electric field is tangential to the surface. The contribution from the surface S' looks, at first, as though it is not equal to zero. However, since the potential must be constant on S' to satisfy the boundary condition, this integral is proportional to the net flux of the mode ν passing through the surface S'. This flux must be zero for a solenoidal mode, for which field lines do not appear or disappear. Since no flux can escape through S'', no net flux can pass through S'. Hence, we have proven the orthogonality of solenoidal and divergence modes. An analogous analysis can be applied to the modes of the magnetic field to prove that the solenoidal and divergence modes are orthogonal.

The proof that the eigenvalues of (2.123) and (2.126) are real and the proof of the orthogonality of the eigensolutions is the framework for the mode expansion of any electromagnetic field in a closed cavity. It is also the starting point for the quantization of electromagnetic fields, as is done in Chap. 6. In anticipation of the quantization, we shall limit the subsequent analysis to lossless closed resonators, $\sigma = 0$. In a closed cavity containing no sources, the divergence modes remain unexcited, since they require electric and magnetic charge distributions for their existence. The electric field in the resonator can be written

$$E(r, t) = \sum_\mu V_\mu(t) e_\mu(r) , \qquad (2.150)$$

and the magnetic field can be expressed by

$$H(r, t) = \sum_\mu I_\mu(t) h_\mu(r) . \qquad (2.151)$$

The E field patterns $e_\mu(r)$ and the magnetic-field patterns $h_\mu(r)$ are, of course, related to each other. If one inserts (2.150) and (2.151) into (2.115) and (2.116) one obtains

$$\sum_\mu V_\mu(t) \nabla \times e_\mu(r) = -\mu \sum_\mu \frac{d}{dt} I_\mu(t) h_\mu(r) \qquad (2.152)$$

and

$$\sum_\mu I_\mu(t) \nabla \times h_\mu(r) = -\epsilon \sum_\mu \frac{d}{dt} V_\mu(t) e_\mu(r) . \qquad (2.153)$$

From (2.139) and (2.146), one sees that $\nabla \times h_\mu(r) = p_\mu e_\mu(r)$. One then finds

$$\epsilon \frac{d}{dt} V_\mu = p_\mu I_\mu . \qquad (2.154)$$

Further, from (2.146) and (2.152) it follows that

$$\mu \frac{d}{dt} I_\mu = -p_\mu V_\mu , \qquad (2.155)$$

with

$$\nabla \times h(r) = p_\mu e(r) . \qquad (2.156)$$

Equations (2.154) and (2.155) are the equations of a harmonic oscillator, with V_μ identifiable with the position and I_μ with the momentum. This is the starting point for the quantization of the electromagnetic field.

If the fields are specified at $t = 0$, then the use of the orthogonality conditions provides the initial values for the coefficients $V_\mu(0)$ and $I_\mu(0)$:

$$V_\mu(0) = \int dV\, E(0, r) \cdot e_\mu(r)/\mathcal{V}; \quad I_\mu(0) = \int dV\, H(0, r) \cdot h_\mu(r)/\mathcal{V} . \qquad (2.157)$$

In a closed resonator, modes can always be defined so that their field patterns are real. Then the electric field and the magnetic field of a mode are 90° out of phase. Ring resonators formed from waveguides closing on themselves propagate traveling waves. These can be constructed from two standing waves that are spatially displaced by a quarter wavelength. The preceding equations are equally applicable, but note has to be taken that the field patterns $e(r)$ and $h(r)$ are complex functions of space.

2.8 Electromagnetic Field
Inside a Closed Cavity with Sources

Next, we determine the fields inside a closed cavity within which there are electric and magnetic current distributions varying sinusoidally with time. The current distributions are assumed to be specified. They play the role of driving currents, capable of supplying power so that a sinusoidal steady state is established within the cavity. The power dissipated by the losses in the cavity is supplied by the current distribution. This idealized problem will find an application in the next section when solving for the fields in an open cavity driven through a waveguide. Denote the electric current density distribution by J_e and the magnetic current density distribution by J_m. These current distributions are complex vector functions of the spatial variable r. The electric and magnetic fields inside the cavity satisfy the equations

$$\nabla \times E = i\omega\mu H - J_m , \tag{2.158}$$

$$\nabla \times H = \sigma E - i\omega\epsilon E + J_e . \tag{2.159}$$

The cavity is assumed to be filled with an isotropic medium characterized by a scalar dielectric permittivity ϵ, magnetic permeability μ, and conductivity σ. Since the equations are cast in the frequency domain, the material parameters may be functions of frequency in the analysis to follow. We shall take advantage of this fact at an appropriate stage.

The electric field has, further, to satisfy the divergence relation

$$\nabla \cdot \epsilon E = \rho . \tag{2.160}$$

Since we have assumed a magnetic current distribution, we must include the possibility of the existence of a magnetic charge density. The magnetic field has to satisfy the equation

$$\nabla \cdot \mu H = \rho_m . \tag{2.161}$$

In order to solve the present problem we take advantage of the complete set of solenoidal and divergence modes found for the empty, undriven cavity. We expand the electric- and magnetic-field patterns within the cavity in terms of these:

$$E = \sum_{\kappa=\alpha,\nu} V_\kappa e_\kappa , \tag{2.162}$$

$$H = \sum_{\kappa=\alpha,\nu} I_\kappa h_\kappa . \tag{2.163}$$

We distinguish solenoidal modes from divergence modes by the subscripts ν for the former and α for the latter. In these equations V_ν, I_ν, V_α, and I_α are complex expansion coefficients, as yet undetermined. In contrast to the expansion carried out in Sect. 2.7, these coefficients do not have the time dependence e^{st} which is natural to any one of these individual modes when undriven. Indeed, now the time dependence is sinusoidal by assumption, at the frequency ω of the driving current distribution.

Introducing (2.162) and (2.163) into (2.158) and (2.159), one obtains expressions for the expansion coefficients V_κ and I_κ. It is convenient here to separate the analysis of the solenoidal-mode expansion from that of the divergence-mode expansion. Dot-multiplying (2.158) by h_ν^*, using (2.146), integrating over the volume of the cavity, and using the mode orthogonality property, one obtains

$$p_\nu V_\nu = i\omega\mu I_\nu - \frac{1}{V} \int J_m \cdot h_\nu^* \, dV \,. \tag{2.164}$$

Similarly, using the expansions (2.162) and (2.163) in (2.159), dot-multiplying by e_ν^*, and integrating over the volume of the cavity, one obtains

$$p_\nu I_\nu = (\sigma - i\omega\epsilon)V_\nu + \frac{1}{V} \int J_e \cdot e_\nu^* \, dV \,. \tag{2.165}$$

Similar expressions can be obtained for the expansion coefficients of the divergence modes:

$$0 = i\omega\mu I_\alpha - \frac{1}{V} \int J_m \cdot h_\alpha^* \, dV \,, \tag{2.166}$$

$$0 = (\sigma - i\omega\epsilon)V_\alpha + \frac{1}{V} \int J_e \cdot e_\alpha^* \, dV \,. \tag{2.167}$$

Equations (2.164)–(2.167) suggest the equivalent circuits that are shown in Fig. 2.9. It should be noted that the expansion coefficients V_ν and I_ν that play the role of voltage and current in the equivalent circuits are interconnected by (2.164) and (2.165). In contrast, the coefficients I_α and V_α are independent and, correspondingly, the equivalent circuits of (2.166) and (2.167) are independent. Solving for V_ν and I_ν separately, from (2.164) and (2.165) one obtains

$$V_\nu = \frac{(i\omega\mu/V) \int J_e \cdot e_\nu^* \, dV - (p_\nu/V) \int J_m \cdot h_\nu^* \, dV}{p_\nu^2 - k^2} \,, \tag{2.168}$$

$$I_\nu = \frac{(p_\nu/V) \int J_e \cdot e_\nu^* dV - [(\sigma - i\omega\epsilon)/V] \int J_m \cdot h_\nu^* dV}{p_\nu^2 - k^2} \,. \tag{2.169}$$

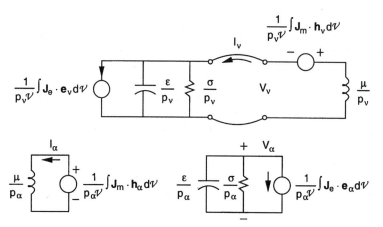

Fig. 2.9. Equivalent circuits for modes of driven cavity

Correspondingly, we have for I_α and V_α

$$I_\alpha = \frac{1}{i\omega\mu\mathcal{V}} \int \boldsymbol{J}_m \cdot \boldsymbol{h}_\alpha^* d\mathcal{V} \,, \tag{2.170}$$

$$V_\alpha = -\frac{1}{(\sigma - i\omega\epsilon)\mathcal{V}} \int \boldsymbol{J}_e \cdot \boldsymbol{e}_\alpha^* d\mathcal{V} \,. \tag{2.171}$$

The equations developed in this section will find direct application in the analysis of an open cavity driven through a number of input waveguides.

2.9 Analysis of Open Cavity

The analysis in the preceding sections was devoted to the study of completely closed cavities. The case of an undriven cavity was taken up first. Then cavities containing driving current density distributions were studied. The case of a driven cavity was an application of the mode analysis of the undriven cavity. The study of the open cavity to be undertaken in this section can be reduced to the previously analyzed problem of a closed, driven cavity. This we now proceed to show.

An open microwave cavity is a metallic enclosure with one or more holes, through which electromagnetic energy may be supplied to, or extracted from, the cavity via feeding waveguides. Consider the cavity of Fig. 2.10. It is fed by N waveguides, in which we choose convenient reference planes. Now, form a closed cavity from the open cavity of Fig. 2.10 by placing at all reference planes in the incoming waveguides perfect magnetic shorts. The closed cavity possesses a complete set of solenoidal and divergence modes. Any field inside

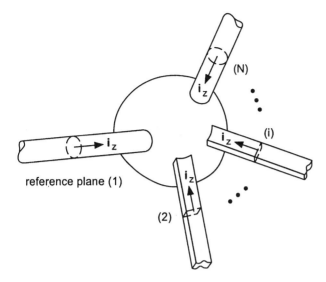

Fig. 2.10. Cavity fed from N waveguides

the region of the closed cavity that satisfies all the boundary conditions can be expanded in terms of these modes. The fields in the open cavity have a nonzero tangential magnetic field on the reference cross sections and, therefore, violate the boundary conditions imposed on the fields (and modes) in the closed cavity. These fields cannot be expanded directly in terms of the modes of the closed cavity. However, we may adapt the fields of the open cavity so that they can be expanded in terms of the complete set of modes of the closed cavity by constructing an artificial field which is identical, in every respect, to the actual physical field in the open cavity throughout the volume of the cavity, but has a tangential magnetic field that vanishes on the reference planes. Accordingly, the tangential magnetic field of the artificial field experiences a discontinuity at the reference planes. Denote the field of the open cavity at the ith reference plane by $\boldsymbol{H}_{Ti}$. At the ith reference plane, the artificial field constructed from the field of the open cavity changes from $\boldsymbol{H}_{Ti}$ to zero within a very small (theoretically infinitesimally small) region in front of the reference plane. Such a discontinuity is created by an electric surface current of magnitude

$$\boldsymbol{K}_i = -\boldsymbol{n} \times \boldsymbol{H}_{Ti} \,, \tag{2.172}$$

where $\boldsymbol{n}$ is the normal to the reference plane pointing outwards from the cavity. The artificial field is expandable in terms of the closed-cavity modes. It is a field in the closed cavity driven by the current distributions on the various reference planes. We have reduced the problem of the analysis of an open cavity fed by incoming waveguides to the problem of a closed cavity driven by surface current distributions in front of the N reference planes of the original open cavity, with the reference planes themselves replaced by perfect magnetic conductors.

Equations (2.168)–(2.171), developed for the case of a closed, current-driven cavity are thus directly applicable to our present problem. The electric current distribution density J_e consists of surface current density distributions (2.172) at the N waveguide cross sections. The surface current density distributions (2.172) can in turn be related to the fields existing in the feeding waveguides. Indeed, the transverse magnetic field appearing on the right-hand side of (2.172) must be expressible in terms of the waveguide modes of the ith waveguide. One has

$$H_{Ti} = \sum_n I_{n,i} h_{Tn,i} , \tag{2.173}$$

where the subscript n denotes the nth mode in the ith waveguide and the origin of the z coordinate is chosen conveniently at the reference cross section in the ith waveguide with the z axis directed into the cavity. Using (2.172) and (2.173) in (2.168)–(2.171), one has

$$V_\nu = -\frac{i\omega\mu}{(p_\nu^2 - k_\nu^2)\mathcal{V}} \sum_{n,j} I_{n,j} \int e_{Tn,j} \cdot e_\nu^* \, dS , \tag{2.174}$$

$$V_\alpha = \frac{1}{(\sigma - i\omega\epsilon)\mathcal{V}} \sum_{n,j} I_{n,j} \int e_{Tn,j} \cdot e_\alpha^* \, dS . \tag{2.175}$$

In (2.174) and (2.175) we have made use of the fact that

$$n = -i_z \quad \text{and} \quad i_z \times h_{Tn,j} = -e_{Tn,j} . \tag{2.176}$$

In order to find a relation between the amplitudes $V_{n,j}$ and $I_{n,j}$ of the modes in the feeding waveguides, we have to express the electric field at the reference cross section in terms of the waveguide modes on one hand, and in terms of the cavity modes on the other hand. One has

$$\sum_n V_{n,j} e_{T,n,j} = \sum_{\kappa=\alpha,\nu} V_\kappa e_\kappa . \tag{2.177}$$

Using the orthogonality condition on the transverse field patterns of the magnetic field in the waveguide, $e_{Tn,j}$, one obtains from (2.177)

$$V_{m,j} = \sum_{\kappa=\alpha,\nu} V_\kappa \sum_j \int e_\kappa \cdot e_{Tm,j}^* \, dS , \tag{2.178}$$

where the integration is carried out over the jth reference cross section. Introducing (2.178) into (2.174) and (2.175), one has

$$V_{m,j} = \sum_{n,i} Z_{mn;ji} I_{n,i} \tag{2.179}$$

with

$$Z_{mn;ji} = \sum_{\nu} \frac{-i\omega\mu}{(p_\nu^2 - k^2)\mathcal{V}} \int_j e_{Tm,j}^* \cdot e_\nu \, dS \int_i e_{Tn,i} \cdot e_\nu^* \, dS$$

$$+ \sum_{\alpha} \frac{1}{(\sigma - i\omega\epsilon)\mathcal{V}} \int_j e_{Tm,j}^* \cdot e_\alpha \, dS \int_i e_{Tn,i} \cdot e_\alpha^* \, dS \, . \tag{2.180}$$

Equations (2.179) and (2.180) will be exploited in Chap. 5 in connection with the analysis of multiports. Equation (2.179) is the impedance matrix description of a multiterminal network. The terminal "voltages" are proportional to the terminal "currents"; the proportionality constants are the elements of an impedance matrix. The matrix is of order $M \times N$, where M is the number of cavity modes and N is the total number of waveguide modes in all waveguides coupled to the cavity. In principle, there is an infinite number of resonator modes; in practice it suffices to include only a few in the analysis.

2.10 Open Cavity with Single Input

We illustrate the general formalism that led to (2.179) and (2.180) with the example of a resonator connected to a single waveguide within the frequency regime in which only one dominant mode propagates in the waveguide. It is a rich example which connects with the energy theorem, serves as another illustration of perturbation theory, and leads to the definition of the dimensionless quality factors, in terms of which resonances can be defined irrespective of whether they are electromagnetic, acoustic, or descriptive of any other resonant phenomenon. Equation (2.179) reduces to a simple impedance relation of a two-terminal-pair element. One has

$$V = ZI \, , \tag{2.181}$$

where the impedance Z is given by

$$Z = \sum_{\nu} \frac{-i\omega\mu}{(p_\nu^2 - k^2)\mathcal{V}} \left| \int e_T \cdot e_\nu^* \, dS \right|^2$$

$$+ \sum_{\alpha} \frac{1}{(\sigma - i\omega\epsilon)\mathcal{V}} \left| \int e_T \cdot e_\alpha^* \, dS \right|^2 \, . \tag{2.182}$$

An equivalent circuit representing the impedance (2.182) is shown in Fig. 2.11.

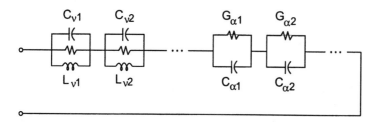

Fig. 2.11. Equivalent circuit of cavity with single port of access

Equation (2.182) has an interesting structure. It has poles in the lower half of the complex ω plane, located symmetrically with respect to the imaginary axis. In the absence of loss, i.e. $\sigma = 0$, $k^2 = \omega^2 \mu\epsilon$, the poles move onto the real axis. If the impedance $\mathrm{Im}(Z)/\sqrt{\mu/\epsilon} = X/\sqrt{\mu/\epsilon}$ is plotted against frequency (Fig. 2.12), it goes from negative infinity to positive infinity, crossing the abscissa in between. The slope of the function is negative throughout.

2.10.1 The Resonator and the Energy Theorem

The dependence upon frequency of the impedance illustrated in Fig. 2.12 is a direct consequence of the energy theorem (2.94). Let us write the electric and magnetic fields at the waveguide reference plane in the form

$$E = V e(r_T) \,, \tag{2.183}$$

$$H = I h(r_T) \,. \tag{2.184}$$

Next, we note that the field patterns $e(r_T)$ and $h(r_T)$ in a metallic waveguide are frequency-independent. Finally, we assume an excitation with a magnetic-field amplitude I that is also frequency independent. Then, in (2.94),

$$\oint \left(\frac{\partial E}{\partial \omega} \times H^* + E^* \times \frac{\partial H}{\partial \omega} \right) \cdot dS = |I|^2 \frac{\partial Z}{\partial \omega} \int e_T \times h_T^* \cdot dS = 4iw \,, \tag{2.185}$$

and, using the fact that $h_T = i_z \times e_T$, we find

$$\frac{\partial X}{\partial \omega} = -\frac{4w}{|I|^2 \int e_T \cdot e_T^* \, dS} \,. \tag{2.186}$$

The derivative of the impedance is proportional to the stored energy, and is negative. It is this negative definiteness of the derivative that gives rise to the form of the graph in Fig. 2.12.

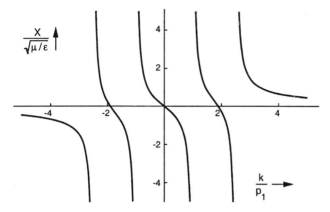

Fig. 2.12. Plot of normalized X, i.e. $X_n \equiv \mathcal{V}p_1/(\sqrt{\mu/\epsilon}|\int e_T \cdot e_{\nu=1}^* \, dS|^2)$ versus k/p_1; $p_2/p_1 = 2.5$; $p_1|\int e_T \cdot e_{\nu=2} \, dS|^2 = 2p_2|\int e_T \cdot e_{\nu=1} \, dS|^2$

2.10.2 Perturbation Theory and the Generic Form of the Impedance Expression

Equation (2.182) was derived for a particular choice of reference plane in the incoming waveguide. A different choice of reference plane leads, in general, to a different set of values of ω_ν and of $|\int e_T \cdot e_\nu^* \, da|^2$. Indeed, these quantities are characteristic of the modes of the closed cavity formed from the open cavity by placing a magnetic short at the reference plane. Clearly, any choice of reference plane has to lead to an impedance Z with the correct dependence upon frequency as viewed from the chosen reference plane. Among the various choices of reference planes, *one* is particularly convenient and, therefore, is usually the one taken: the choice which makes all terms in the two summations in (2.182) negligible in the neighborhood of one particular resonance frequency except for a single one, say, the one pertaining to the resonance at the frequency ω_μ. For such a choice of reference plane, (2.182) assumes a simple form in the neighborhood of the frequency ω_μ:

$$Z \simeq -\frac{i\omega_\mu \mu}{(p_\mu^2 - k^2)\mathcal{V}}|\int e_T \cdot e_\mu^* \, dS|^2 . \qquad (2.187)$$

This particularly simple form of the impedance of the resonator in the vicinity of the resonance frequency ω_μ can be put into "generic" form by making use of its physical implications. Let us first assume that the waveguide presents an open circuit (infinite impedance) to the resonator. This could be accomplished by placing a perfectly conducting shorting plane a quarter wavelength away from the reference plane. In this case, the denominator has to vanish. This leads to an equation for the frequency of resonance of the closed resonator. We obtain

$$k^2 = p_\mu^2 = \omega^2 \mu\epsilon + i\omega\mu\sigma \ . \tag{2.188}$$

In general, the loss represented by σ will be small if the structure is to act as a resonator. Thus, the frequency of resonance ω_μ will be given approximately by (2.188) with $\sigma = 0$:

$$\omega_\mu^2 \mu(\omega_\mu)\epsilon(\omega_\mu) = p_\mu^2 \ . \tag{2.189}$$

Here we have been careful to indicate that the material may be dispersive and thus the dielectric constant and magnetic permeability have to be evaluated at the pertinent frequency. Equation (2.189) defines the frequency of resonance. Now, we introduce the loss and assume that the frequency shifts to $\omega_\mu + \Delta\omega$ owing to the loss. Introducing this ansatz and the definition of ω_μ into (2.188), we find, to first order in $\Delta\omega$ and σ,

$$2\omega_\mu \Delta\omega \, \mu\epsilon + \omega_\mu^2 \Delta\omega \frac{\partial\epsilon}{\partial\omega}\mu + \omega_\mu^2 \Delta\omega \, \epsilon\frac{\partial\mu}{\partial\omega} + i\omega_\mu\mu\sigma = 0 \ . \tag{2.190}$$

Solving for $\Delta\omega$, we find

$$2\frac{\Delta\omega}{\omega_\mu} = -i\frac{\sigma}{(1/2)\omega_\mu[(\epsilon + \omega_\mu\partial\epsilon/\partial\omega) + (\epsilon/\mu)(\mu + \omega_\mu\partial\mu/\partial\omega)]} \ . \tag{2.191}$$

The frequency is negative imaginary. The field decays owing to the loss. In fact, (2.191) could have been derived by standard perturbation theory. The time-averaged power dissipated in the resonator is

$$\text{time-averaged dissipated power} = p_d = \frac{|V_\mu|^2}{2}\sigma \int e_\mu \cdot e_\mu^* \, dV \ . \tag{2.192}$$

The energy storage in the resonator is

$$\text{stored energy} = w$$

$$= \frac{|V_\mu|^2}{4}\left[\left(\epsilon + \omega_\mu\frac{\partial\epsilon}{\partial\omega}\right) + \frac{\epsilon}{\mu}\left(\mu + \omega_\mu\frac{\partial\mu}{\partial\omega}\right)\right] \tag{2.193}$$

$$\times \int e_\mu \cdot e_\mu^* \, dV \ ,$$

where we have used (2.154) to express the magnitude of $|I_\mu|$ in terms of $|V_\mu|$, with $p_\mu = \omega_\mu\sqrt{\mu\epsilon}$. If the field decays at the rate $-\text{Im}(\Delta\omega)$, the energy decays at twice this rate. The decay of the energy accounts for the dissipated power:

$$2\,\text{Im}(\Delta\omega) = -\frac{p_d}{w} \ . \tag{2.194}$$

Combining (2.192), (2.193), and (2.194), we arrive at (2.191). But now the relation is the consequence of a standard perturbation theory that can be applied to resonators of all kinds, acoustic, mechanical, etc. It is in this general context that one associates a so-called "unloaded" quality factor Q with the resonance, which is defined by

$$Q_{o\mu} = \frac{\omega_\mu w}{p_d} = \frac{(1/2)\omega_\mu[(\epsilon\omega_\mu\partial\epsilon/\partial\omega) + \epsilon/\mu(\mu + \omega_\mu\partial\mu/\partial\omega)]}{\sigma} . \tag{2.195}$$

The adjective "unloaded" refers to the fact that the output waveguide has been closed off and thus does not load the resonator. In terms of this Q factor, the rate of decay is given directly by

$$-2\frac{\mathrm{Im}(\Delta\omega)}{\omega_\mu} = \frac{1}{Q_{o\mu}} . \tag{2.196}$$

Next, consider the resonator when it is connected to a matched guide so that $Z = -Z_o$ in (2.187). An initial excitation in the resonator will decay more rapidly, since energy is lost not only to the conduction process but also to the power escaping through the output port. Instead of (2.190), we find from (2.187)

$$2\omega_\mu\Delta\omega\,\mu\epsilon + \omega_\mu^2\Delta\omega\frac{\partial\epsilon}{\partial\omega}\mu\epsilon + \omega_\mu^2\Delta\omega\frac{\partial\mu}{\partial\omega}\epsilon + i\omega_\mu\mu\sigma$$
$$+ i\omega_\mu\mu Y_o| \int e_T \cdot e_\mu \, dS|^2/\mathcal{V} = 0 ,$$

which, solved for $\Delta\omega$, gives

$$2\frac{\Delta\omega}{\omega_\mu} = -i\frac{\sigma + Y_o| \int e_T \cdot e_\mu^* \, dS|^2/\mathcal{V}}{(1/2)\omega_\mu[(\epsilon + \omega_\mu\partial\epsilon/\partial\omega) + (\epsilon/\mu)(\mu + \omega_\mu\partial\mu/\partial\omega)]} . \tag{2.197}$$

It is clear that the rate of decay has increased owing to the coupling to the resonator mode. A power p_e escapes from the resonator, contributing to the rate of decay. Now that the decay is caused by both the dissipated power p_d and the power escaping from the waveguide, we must have

$$2\,\mathrm{Im}(\Delta\omega) = -\frac{p_d}{w} - \frac{p_e}{w} . \tag{2.198}$$

Comparing this expression with (2.197), we find that the power escaping from the resonator is given by

$$p_e = \frac{|V_\mu|^2}{2} \left(Y_o \int e_\mu \cdot e_\mu^* \, d\mathcal{V} \left| \int e_T \cdot e_\mu^* \, dS \right|^2 /\mathcal{V} \right) . \tag{2.199}$$

We may define a Q factor analogous to (2.195) which expresses the rate of decay due to the escaping power:

$$Q_{e\mu} = \frac{\omega_\mu w}{p_e} = \frac{(1/2)\omega_\mu[\epsilon + \omega_\mu\partial\epsilon/\partial\omega + (\epsilon/\mu)(\mu + \omega_\mu\partial\mu/\partial\omega)]}{Y_o| \int e_T \cdot e_\mu^* \, dS|^2/\mathcal{V}} . \tag{2.200}$$

The impedance relation (2.187) can be written in generic form, using the definitions of Q and the expansion of $p_\mu^2 - k^2$ to first order in $\Delta\omega$ as in (2.190). One finds the very simple relation

$$\frac{Z}{Z_o} = \frac{Q_{o\mu}/Q_{e\mu}}{-i(2\Delta\omega/\omega_\mu)Q_{o\mu} + 1} . \tag{2.201}$$

This relation involves $\Delta\omega/\omega_\mu$ and the Qs. The relation can be made even more generic by removing the reference to normalized impedance and replacing it with the reflection coefficient, which has a more general meaning. We have

$$\Gamma = \frac{Z - Z_o}{Z + Z_o} = \frac{1/Q_{e\mu} - 1/Q_{o\mu} + i(2\Delta\omega/\omega_\mu)}{1/Q_{e\mu} + 1/Q_{o\mu} - i(2\Delta\omega/\omega_\mu)} . \tag{2.202}$$

Suppose that the cavity is excited by an oscillator of adjustable frequency ω, well "padded" by an isolator which ensures that the oscillator emits a forward-traveling wave a unaffected by the impedance presented to the oscillator by the cavity. The power absorbed by the cavity is

$$|a|^2 - |b|^2 = |a|^2(1 - |\Gamma|^2) = \frac{4/Q_{e\mu}Q_{o\mu}}{(1/Q_{e,\mu} + 1/Q_{o,\mu})^2 + (2\Delta\omega/\omega_\mu)^2}|a|^2 . \tag{2.203}$$

This expression contains the sum of the inverse external Q and the inverse unloaded Q:

$$\frac{1}{Q_{o\mu}} + \frac{1}{Q_{e\mu}} = \frac{1}{Q_{L\mu}} , \tag{2.204}$$

which defines the inverse "loaded" Q. The name stems from the fact that the loaded Q determines the rate of decay of a resonance at the frequency ω_μ set up in the cavity when the source is removed from the cavity and replaced by a matched load. Equation (2.203) is in a form entirely independent of an equivalent circuit or of the specific electromagnetic example. A reflection coefficient, the ratio of the reflected and incident waves, is a general concept applicable to any system propagating waves. The Qs were defined in terms of decay rates for different terminations of the resonator. These rates, again, need not be specifically associated with electromagnetic fields but could be acoustic, such as those associated with surface acoustic waves (SAWs), or purely mechanical.

A measurement of the frequency separation of the half-power points determines the loaded Q (2.204). A measurement of the reflection coefficient at resonance gives the ratio of the unloaded Q to the external Q. Thus, from these two measurements the external Q and the unloaded Q of the μth resonance can be determined.

2.11 Reciprocal Multiports

In the analysis of resonators, we arrived at an impedance matrix description of a resonator connected to a number of waveguides. The fields inside a perfect enclosure were expanded in terms of the modes of a lossless resonator. The result was an impedance matrix for a resonator containing a conducting medium described by a conductivity σ. Active systems with gain are obtained when the conductivity is made negative.

The expansion in terms of resonator modes gave the full frequency dependence of the impedance matrix. One need not go to this degree of detail to obtain some important relations among the elements of the impedance matrix describing a multiport. If an electromagnetic system has N ports of access via N single-mode waveguides, or via waveguides with a total number of N propagating modes, then one may describe the excitation in each of the waveguides or modes by the amplitudes of the electric and magnetic fields at reference planes sufficiently far from the structure that the cutoff modes excited at and near the connection to the system are of negligible amplitude at the reference planes. The electric fields can be expressed in terms of their mode amplitudes V_j, the magnetic fields in terms of their amplitudes $I_j, j = 1, 2, \ldots N$. By the uniqueness theorem, the excitation is described fully by the tangential electric field over the part S' of the surface enclosing the volume of interest, and the tangential magnetic field over the remaining part S''. If the system is in a perfectly conducting enclosure, $S = S'$, then the tangential electric field vanishes over the perfectly conducting enclosure S', and the tangential magnetic field is fully described by the amplitudes I_j of the waveguide modes over the surface S' containing the reference cross sections. From the knowledge of the magnetic fields across the reference cross sections, the electric fields can be determined uniquely. This means one must have a linear relation between the I_j and the V_j:

$$V_j = Z_{jk} I_k \,, \tag{2.205}$$

where the Z_{jk}s are complex coefficients representing the network in terms of an impedance matrix description. If the network is lossless, then one must have

$$\sum_j (V_j I_j^* + V_j^* I_j) = \boldsymbol{I}^\dagger \boldsymbol{Z} \boldsymbol{I} + \boldsymbol{I}^\dagger \boldsymbol{Z}^\dagger \boldsymbol{I} = 0 \,. \tag{2.206}$$

Since the currents can be chosen arbitrarily, one must have

$$\boldsymbol{Z} + \boldsymbol{Z}^\dagger = 0 \,. \tag{2.207}$$

One may determine other constraints on the impedance matrix imposed by the reciprocity theorem of Sect. 1.8. The reciprocity theorem for a structure containing media with symmetric dielectric and magnetic permeability tensors is

$$\oint \boldsymbol{E}^{(1)} \times \boldsymbol{H}^{(2)} \cdot d\boldsymbol{S} = \oint \boldsymbol{E}^{(2)} \times \boldsymbol{H}^{(1)} \cdot d\boldsymbol{S} \; . \tag{2.208}$$

If we now consider two different excitations of the structure, indicated by superscripts (1) and (2), (2.208) can be written

$$\sum_i V_j^{(1)} I_j^{(2)} = \sum_j V_j^{(2)} I_j^{(1)} \tag{2.209a}$$

or

$$\boldsymbol{I}_t^{(1)} \boldsymbol{Z}_t \boldsymbol{I}^{(2)} = \boldsymbol{I}_t^{(2)} \boldsymbol{Z}_t \boldsymbol{I}^{(1)} = \boldsymbol{I}_t^{(1)} \boldsymbol{Z} \boldsymbol{I}^{(2)} \; , \tag{2.209b}$$

where the subscript "t" indicates transposition of a matrix. Since the excitations are arbitrary, one must have

$$\boldsymbol{Z}_t = \boldsymbol{Z} \; . \tag{2.210}$$

The impedance matrix of a structure obeying the reciprocity theorem must be symmetric. The impedance matrix (2.180) of a resonator with multiple ports of access obeys the reciprocity theorem if the mode patterns of the cavity and waveguide are taken to be real. Then the proper phase relation is established between the E fields and the voltages and between the H fields and the currents.

2.12 Simple Model of Resonator

The preceding analysis was a formal derivation from Maxwell's equations of the terminal characteristics of a resonator. At optical frequencies, the physical conductors (metals) that model adequately the behavior of a perfect conductor at microwave frequencies are too lossy to provide loss-free enclosures. Instead, open dielectric structures are used for resonators at optical frequencies. An optical Fabry–Pérot resonator may be formed from dielectric mirrors that capture free-space Hermite Gaussian modes as described in the next chapter. These share many properties of enclosed structures.

Further, resonators occur in other realizations than perfectly conducting enclosures. They may be acoustic resonators. There is a generic commonality to all these that can be brought out using only three principles: (a) energy conservation, (b) time reversibility, and (c) perturbation theory. In this section we use these principles to arrive at the equation of a resonator coupled to incoming and outgoing waves [31].

Denote the amplitude of a mode in a closed resonator by $U(t)$. It obeys the following differential equation in time:

$$\frac{dU}{dt} = -\mathrm{i}\omega_o U \; , \tag{2.211}$$

where ω_o is the resonance frequency. We normalize the amplitude so that $|U(t)|^2$ is the energy in the mode. When the resonator is opened by connecting to it a waveguide, or by making the mirrors partially transmissive in the case of a Fabry–Pérot resonator, the amplitude of the mode must decay at the rate $1/\tau_e$ because of the escaping radiation. Equation (2.211) changes into

$$\frac{dU}{dt} = -(i\omega_o + 1/\tau_e)U \ . \tag{2.212}$$

The time rate of change of the energy is

$$\frac{d|U|^2}{dt} = -\frac{2}{\tau_e}|U|^2 \ . \tag{2.213}$$

In the spirit of Sect. 2.10, we may define an external Q which relates the rate of decay of the mode due to coupling to a waveguide to the resonance frequency:

$$\frac{1}{Q_e} = \frac{2}{\omega_o \tau_e} \ . \tag{2.214}$$

Thus far we have studied a resonance and its decay due to escaping radiation when there is no excitation of the resonator from the waveguide. Next we study the case of excitation of the resonator by an incident wave. Denote the amplitude of the incident wave by a. As usual, we normalize a so that its square is equal to the power. The system is linear, and thus the excitation through a can be expressed by modifying (2.212):

$$\frac{dU}{dt} = -(i\omega_o + 1/\tau_e)U + \kappa a \ , \tag{2.215}$$

where κ is a coupling coefficient. One may ask why we have chosen to express the coupling in terms of a, rather than its time derivative or integral. This choice is justified for all systems that have high Q. Indeed, if the Q is high, only excitations at and near the resonance frequency can produce a response. If the coupling is due to da/dt, one may replace it by $-i\omega_o a$ to lowest order, and incorporate the factor $-i\omega_o$ into the coupling coefficient. A similar argument applies to coupling to the integral of a.

We may solve (2.215) for an excitation a proportional to $\exp(-i\omega t)$:

$$U = \frac{\kappa a}{i(\omega_o - \omega) + 1/\tau_e} \ . \tag{2.216}$$

Now, let us revisit the case of the unexcited resonance as it decays by coupling into the external waveguide. We assume that the incoming waveguide propagates only one mode. It is clear that the escaping energy excites an outgoing wave of complex amplitude b whose power is equal to the rate of decay of the energy:

$$\frac{d|U|^2}{dt} = -\frac{2}{\tau_e}|U|^2 = -|b|^2 \ . \tag{2.217}$$

Next, consider the time-reversed solution. Decay becomes growth, and an outgoing wave b becomes an incoming wave a (whose amplitude is made to grow exponentially). The frequency of the excitation is $\omega_o + i/\tau_e$. We introduce this frequency into (2.216) and find

$$|U|^2 = \frac{|\kappa|^2 |a|^2}{(2/\tau_e)^2} . \tag{2.218}$$

Since the outgoing wave became an incoming wave, we have, from (2.217),

$$|a|^2 = \frac{2}{\tau_e} |U|^2 . \tag{2.219}$$

Comparing (2.218) and (2.219), we find for the coupling coefficient

$$\kappa = \sqrt{\frac{2}{\tau_e}} . \tag{2.220}$$

We can set this real by proper choice of the reference plane in the waveguide. Thus, combining (2.215) and (2.220), the equation of the open resonator coupled to an input waveguide becomes

$$\frac{dU}{dt} = -(i\omega_o + 1/\tau_e)U + \sqrt{\frac{2}{\tau_e}} a . \tag{2.221}$$

Finally, consider the relation for the reflected wave b. The system is linear, and thus we must have

$$b = c_a a + c_u U . \tag{2.222}$$

Again, we skirt the possibility that the relationship is in terms of derivatives or integrals by noting that in the narrow frequency interval of interest these operators can be replaced by multipliers. We already have the results of the thought experiment for $a = 0$, the decay of the mode into the waveguide. Thus we may set $a = 0$ in (2.222) and use (2.217), with the result

$$b = c_u U = \sqrt{\frac{2}{\tau_e}} U , \tag{2.223}$$

and thus

$$c_u = \sqrt{\frac{2}{\tau_e}} . \tag{2.224}$$

We dispose of a phase factor by noting that the phase of the mode U is arbitrary and can be chosen so as to make the coefficient c_u real. The coefficient c_a is determined by power conservation. We have from (2.215)

$$|a|^2 - |b|^2 = \frac{d|U|^2}{dt} = -\frac{2}{\tau_e}|U|^2 + \sqrt{\frac{2}{\tau_e}}(aU^* + a^*U)$$

or

$$|a|^2 - |c_a|^2|a|^2 - |c_u|^2|U|^2 - (c_a c_u^* aU^* + c_a^* c_u a^*U)$$
$$= -\frac{2}{\tau_e}|U|^2 + \sqrt{\frac{2}{\tau_e}}(aU^* + a^*U) , \qquad (2.225)$$

for an arbitrary a. Using the value (2.224) for c_u, we find $c_a = -1$, and thus the relation between the incident and reflected waves becomes

$$b = -a + \sqrt{\frac{2}{\tau_e}}U . \qquad (2.226)$$

The equations can be modified to include internal loss by supplementing the decay rate $1/\tau_e$ due to the escape of radiation into the coupling waveguide by a decay rate $1/\tau_o$ due to the internal loss:

$$\frac{dU}{dt} = -(i\omega_o + 1/\tau_e + 1/\tau_o)U + \sqrt{\frac{2}{\tau_e}}a . \qquad (2.227)$$

Equation (2.226) remains unaffected. Equations (2.226) and (2.227) fully define the behavior of the resonator in the neighborhood of its resonance frequency. It is this pair of resonator equations that connects classical electromagnetic fields to quantum fields. Not surprisingly, it is also the appropriate quantum description of phononic excitations. If we ask for the reflection coefficient Γ as a function of frequency of excitation, we find

$$\Gamma = \frac{(1/\tau_e) - (1/\tau_o) + i(\omega - \omega_o)}{(1/\tau_e) + (1/\tau_o) - i(\omega - \omega_o)} . \qquad (2.228)$$

This is the same result as obtained from the formal analysis (2.202), with the identification of the unloaded Q as

$$\frac{1}{Q_o} = \frac{2}{\omega_o \tau_o} \qquad (2.229)$$

and the external Q as that given by (2.214). The analysis can be generalized to multiple resonances in one cavity with one input. An equation of the form of (2.227) is written for each resonance:

$$\frac{dU_j}{dt} = -i(\omega_{o,j} - i/\tau_{e,j} - i/\tau_{o,j})U_j + \sqrt{2/\tau_{e,j}}a . \qquad (2.230)$$

The coupling between the forward and backward waves is generalized to

$$b = c_a a + \sum_j \sqrt{\frac{2}{\tau_{e,j}}}U_j . \qquad (2.231)$$

The reflection coefficient can be written as

$$\Gamma = \frac{b}{a} = c_a + \sum_j \frac{2/\tau_{e,j}}{1/\tau_{e,j} + 1/\tau_{o,j} - i(\omega - \omega_{o,j})} \ . \tag{2.232}$$

The reflection coefficient in the absence of loss, i.e. for $1/\tau_{o,j} = 0$, must be of unity magnitude. This gives a relation for the coefficient c_a.

2.13 Coupling Between Two Resonators

The preceding section developed the equations for the excitation of a resonator from an input waveguide using the constraints of time reversal and energy conservation. It also established the formalism necessary to develop the equations for a transmission resonator, a task left for one of the problems.

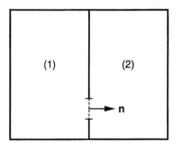

Fig. 2.13. Two resonators coupled by a hole

When two resonators are coupled by a hole, such as the two resonators shown in Fig. 2.13, the formalism is slightly different, and in some ways simpler than in the preceding section. For the purpose of the analysis we assume that the two resonators are lossless, their resonance frequencies are real. Loss can be taken into account by choosing complex frequencies, as has already been done in Sect. 2.12. The derivation of the equations for the modes of the two resonators requires only energy conservation considerations. It is clear that the evolution of the mode in resonator (1) is affected by resonator (2). If the coupling is weak, one may supplement the equation for the uncoupled resonator (1) by a coupling term proportional to the excitation in resonator (2):

$$\frac{dU_1}{dt} = -i\omega_1 U_1 + \kappa_{12} U_2 \ . \tag{2.233}$$

In a similar way one may describe the excitation of resonator (2):

$$\frac{dU_2}{dt} = -\mathrm{i}\omega_2 U_2 + \kappa_{21} U_1 .$$
(2.234)

These are the coupled-mode equations of the two resonator modes. The coupling coefficients depend on the geometry of the coupling hole. Energy conservation imposes a constraint. Indeed, from energy conservation we have

$$\frac{d|U_1|^2}{dt} + \frac{d|U_2|^2}{dt} = 0$$
(2.235)

$$= \kappa_{12} U_2 U_1^* + \kappa_{21} U_1 U_2^* + \kappa_{12}^* U_2^* U_1 + \kappa_{21}^* U_1^* U_2 .$$

Since the amplitudes U_1 and U_2 can be chosen arbitrarily, one must require

$$\kappa_{12} = -\kappa_{21}^* \equiv \kappa .$$
(2.236)

If we assume a time dependence $\exp(-\mathrm{i}\omega t)$ for the amplitudes U_1 and U_2 and use (2.233), (2.234), and (2.236), we obtain the determinantal equation for the frequency

$$(\omega - \omega_1)(\omega - \omega_2) - |\kappa|^2 = 0 ,$$
(2.237)

with the solution

$$\omega = \frac{\omega_1 + \omega_2}{2} \pm \sqrt{\left(\frac{\omega_1 - \omega_2}{2}\right)^2 + \kappa^2} .$$
(2.238)

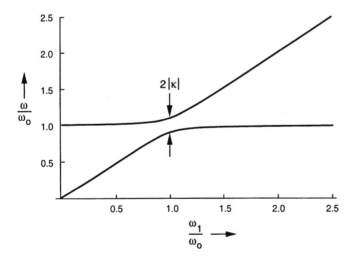

Fig. 2.14. The solutions to the determinantal equation (2.237); $|\kappa| = 0.1$

Note that the frequencies of the coupled system depend only upon the magnitude of the coupling coefficient, not its phase. One can imagine a situation in which one of the cavities, say cavity (1), is tuned by a plunger and its frequency is varied, while the second cavity remains unchanged. Then one may construct a diagram, as shown in Fig. 2.14, for the frequency w of the coupled system as a function of ω_1, with ω_2 fixed. The greatest deviation from the natural frequencies of the two resonators occurs in the case of degeneracy, $\omega_1 = \omega_2$, where we find that the two frequencies of the coupled system are separated by $2|\kappa|$. The solutions for the amplitudes U_1 and U_2 in the degenerate case are

$$U_1 = \left[A_+ e^{-i|\kappa|t} + A_- e^{+i|\kappa|t} \right] e^{-i\omega_o t} , \qquad (2.239)$$

$$U_2 = i\frac{\kappa}{|\kappa|}(A_+ e^{-i|\kappa|t} - A_- e^{i|\kappa|t})e^{-i\omega_o t} , \qquad (2.240)$$

where $\omega_o = \omega_1 = \omega_2$.

From the nature of the solutions one may draw conclusions as to the phase of the coupling coefficient in some specific cases. Take for example the case of two identical resonators coupled by a hole between them in a structure with a symmetry plane containing the hole. The mode solutions must be either symmetric or antisymmetric. From (2.240) we conclude that the coupling coefficient must be pure imaginary.

For a better understanding of the coupled-mode formalism it is helpful to look at the analysis of the electromagnetic fields, as was done in Sect. 2.9 for the impedance matrix of the open resonator. We start with the example in Fig. 2.13 and define the modes in the uncoupled resonators by placing a perfect magnetic short across the hole. Now that the coupling is removed, the tangential magnetic fields vanish across the hole. We denote the electric- and magnetic-field patterns of the uncoupled modes of resonance frequencies ω_1 and ω_2 by $e_1(r), e_2(r), h_1(r)$, and $h_2(r)$. The magnetic fields of the uncoupled modes have zero tangential components at the hole. When the hole is opened, mode (2) causes a nonzero tangential magnetic field to appear in resonator (1). Denote this field by $I_2 h_2^{(p)}(r)$. It is clearly proportional to the amplitude of the magnetic-field pattern in resonator (2). As in the treatment of the open cavity, the appearance of this field in resonator (1) is represented by a surface current

$$K_e = -n \times H = -I_2 n \times h_2^{(p)}(r) \qquad (2.241)$$

inside the closed resonator (1), over the surface of the hole. This case has been treated in Sect. 2.8, and the resulting equations for the amplitudes of the electric and magnetic fields are

$$\epsilon\frac{dV_1}{dt} = p_1 I_1 - \frac{I_2}{\mathcal{V}} \int_{\text{hole}} n \cdot e_1^* \times h_2^{(p)} dS , \qquad (2.242)$$

$$\mu\frac{dI}{dt} = -p_1V_1 \quad \text{with} \quad p_1 = \omega_1\sqrt{\mu_o\epsilon_o} \,. \tag{2.243}$$

In a similar way we may write down equations for the perturbed mode of resonator (2). Next we introduce the canonical amplitudes U_1 and U_2, which, in the absence of coupling, reduce the second-order differential equation of each resonator to uncoupled first-order differential equations. Note that U_1 and U_2 have the unperturbed time dependences $\exp(-i\omega_1 t)$ and $\exp(-i\omega_2 t)$. The transformations are

$$U_j \propto \sqrt{\epsilon}V_j + i\sqrt{\mu}I_j; \quad j = 1,2 \,. \tag{2.244}$$

When (2.242) and (2.243) are put into canonical form and only the term with positive frequency is retained in the coupling term, since the excitation by the coupling term with negative frequency is off-resonance and can be neglected, we find

$$\frac{dU_1}{dt} = -i\omega_1 U_1 + i\frac{U_2}{\mathcal{V}\sqrt{\mu\epsilon}}\int_{\text{hole}} \boldsymbol{n}\cdot\boldsymbol{e}_1^* \times \boldsymbol{h}_2^{(p)}dS \,. \tag{2.245}$$

Comparison with (2.233) shows that the coupling coefficient is

$$\kappa_{12} = \frac{i}{\mathcal{V}\sqrt{\mu\epsilon}}\int_{\text{hole}} \boldsymbol{n}\cdot\boldsymbol{e}_1^* \times \boldsymbol{h}_2^{(p)}dS \,. \tag{2.246}$$

Similarly, we find for the coupling coefficient κ_{21}

$$\kappa_{21} = -\frac{i}{\mathcal{V}\sqrt{\mu\epsilon}}\int_{\text{hole}} \boldsymbol{n}\cdot\boldsymbol{e}_2^* \times \boldsymbol{h}_1^{(p)}dS \,. \tag{2.247}$$

Note that the coupling coefficients are imaginary when $\boldsymbol{e}_2, \boldsymbol{e}_1, \boldsymbol{h}_2^{(p)}$, and $\boldsymbol{h}_1^{(p)}$ are real, as pointed out earlier on the basis of symmetry of the mode solutions. Energy conservation requires, according to (2.236),

$$\int_{\text{hole}} \boldsymbol{n}\cdot\boldsymbol{e}_1^* \times \boldsymbol{h}_2^{(p)}dS = -\left(\int_{\text{hole}} \boldsymbol{n}\cdot\boldsymbol{e}_2^* \times \boldsymbol{h}_1^{(p)}dS\right)^* \,. \tag{2.248}$$

This is a constraint on the perturbation fields. For symmetric resonators, this constraint is automatically satisfied. However, the interesting fact is that it holds for asymmetric resonators as well.

2.14 Summary

This chapter was a brief introduction to the theory of modes in microwave waveguides and resonators. The emphasis was on modes and mode expansions. In microwave design it is common to use coaxial cables or waveguides within the frequency regime within which only the TEM mode or the dominant waveguide mode, respectively, propagates, while all other modes are below their cutoff frequency. The response of the system simplifies to that of one represented by simple equivalent circuits.

We derived the dispersion relations and considered energy and power. These concepts are fundamental to the analysis of thermal noise and quantum noise, since energy considerations are the basis of statistical physics and thermodynamics.

The analysis of waveguides and resonators included the presence of media in the enclosure. The media could be dispersive and lossy. They could also be made active if the conductivity σ was made negative. Thus, the analysis includes the description of active devices such as amplifiers and lasers, as discussed in connection with noise performance in Chap. 5 and subsequent chapters. The structures could be equipped with many input waveguides and thus are electromagnetic models of multiports.

The exact analysis of waveguides and resonators filled with a uniform medium was helpful in gaining an understanding of perturbation methods, which, on one hand, gave the attenuation constant of a waveguide mode due to loss and, on the other hand, derived the equations of a resonator at and near one of its resonance frequencies using power conservation and time reversibility. These perturbation approaches are particularly useful and accurate in optical structures, because in such structures the losses per wavelength, or per cycle, have to be small if the structures are to be of any practical use.

Problems

2.1 Monolithic microwave integrated circuits (MMICs) contain transmission line structures with piecewise uniform dielectric media as shown in Fig. P2.1.1. The purpose of this problem is to show that such structures cannot support TEM waves.

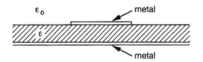

Fig. P2.1.1. A transmission line in an MMIC

(a) Prove that the electric field obeys the following differential equation in a (piecewise) uniform dielectric medium:

$$\nabla^2 \boldsymbol{E} + \omega^2 \mu_o \epsilon \boldsymbol{E} = 0 . \tag{1}$$

(b) In an axially uniform structure solutions exist that have the z dependence $e^{i\beta z}$. Then (1) reduces to

$$\nabla_T^2 E + (\omega^2 \mu_o \epsilon - \beta^2) E = 0 . \tag{2}$$

A TEM wave has no longitudinal component of E and H. Thus E and H are purely transverse. $E = E_T$, $H = H_T$.

(c) Prove that $\nabla_T \times E_T = 0$. Therefore $E_T = -\nabla_T \Phi(x, y) e^{i\beta z}$.

(d) Show that the divergence relation in a piecewise uniform dielectric reduces to

$$\nabla_T^2 \Phi = 0 . \tag{3}$$

(e) Prove that a conductor pair in a piecewise uniform dielectric system cannot support a TEM wave, unless ϵ is constant throughout all of space.

2.2* In a square waveguide, the modes E_{mn} are degenerate with the modes E_{nm}, and the modes H_{nm} with the modes H_{mn}.

(a) Show that the H_{mn} mode with

$$\Psi_{mn} = -\cos\frac{m\pi}{a}x \cos\frac{n\pi}{b}y \tag{1}$$

is orthogonal to the mode with

$$\Psi_{nm} = \cos\frac{n\pi}{a}x \cos\frac{m\pi}{b}y , \tag{2}$$

for $m \neq n$, even when $b = a$.

(b) Consider the mode Ψ_{10}. Construct the new function $\Psi_{10} + \Psi_{01}$. Sketch the H_T field and E_T field of the mode.

(c) Find another linear combination giving a mode that is orthogonal to that of part (b). Sketch the H_T field and E_T field.

2.3 Find the power radiated in one direction by a short wire at the center of a rectangular waveguide of dimensions a, b, i.e. the extension of the center conductor of a "feeder" coaxial cable (see Fig. 2.8). The waveguide is shorted at a distance $\lambda_g/4$, where $\lambda_g = 2\pi/\beta$, and

$$K = i_y I_o \frac{\sin k(\ell - y)}{\sin k\ell} \delta\left(x - \frac{a}{2}\right) \text{ for } y < \ell ,$$

$$K = 0 \qquad\qquad\qquad \text{for } y > \ell .$$

2.4* A resistive sheet of 1000 Ω square (i.e. σ times the thickness θ is 10^{-3} S; S stands for siemens or mho) is to be used in an attenuator. For an attenuation of 10 dB, evaluate the length of the sheet required at 10 GHz (see Fig. P2.4.1). Use a perturbation approach. Compute the loss from $\frac{1}{2}\int \sigma|E|^2 dV$ over the volume of the sheet using the unperturbed field.

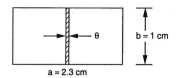

Fig. P2.4.1. An attenuator

2.5

(a) Write down the potential functions for all modes E_{mnp} and H_{mnp} of a rectangular cavity resonator (see Fig. P2.5.1).

(b) If $b < a < \ell$, which mode has the lowest resonance frequency?

(c) Describe the $\boldsymbol{E}$ and $\boldsymbol{H}$ field patterns of this mode in the x, y plane at $z = 0$ and $z = \ell/2$.

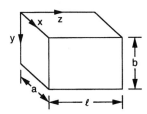

Fig. P2.5.1. A rectangular cavity

2.6* A waveguide partially filled with an anisotropic medium does not support TE or TM waves. However, if the medium does not change along the waveguide axis, the z axis, the waveguide possesses translational symmetry and propagates waves with the dependence $\exp(i\beta z)$.

Show that energy velocity for such modes is still equal to $d\omega/d\beta$.

2.7 A cavity at resonance presents a reflection coefficient $\Gamma_{\text{res}} = +0.33$. The frequencies at which the power absorbed by the cavity is half of that at resonance lie 10 MHz apart. The resonant frequency of the cavity is 5000 MHz. Find the unloaded Q and the external Q. Neglect the losses far off resonance.

2.8* Generalize equations (2.227) and (2.226) to a resonator with two inputs. You can shut off one port at a time, reducing the resonator to a one-port, and obtain the parameters of the two-port in this way. You should permit two, in general different, decay rates τ_{e1} and τ_{e2} for the two ports.

Derive the power transmitted through the resonator for an incident wave $a_1 = A \exp -i\omega t$ as a function of frequency. When is the power transmission through the resonator 100%?

2.9 A lossless "Y", as shown in Fig. P2.9.1, is a three-port. The three-port can be matched from port (1) by slow tapering. Show that if it is matched as seen from port (1), it cannot appear matched as seen from ports (2) and (3). Find the scattering matrix.

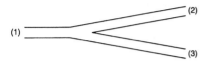

Fig. P2.9.1. A tapered "Y"

2.10 Consider a lossless propagation system formed from a multimode waveguide that transforms incident waves a into transmitted waves b via the transfer matrix T.

(a) Prove that $T^\dagger T = T T^\dagger = 1$.
(b) Consider excitations at ω and $\omega + \Delta\omega$, with $a(\omega + \Delta\omega) = a(\omega)$. Using the energy theorem, show that

$$T^\dagger \frac{dT}{d\omega} = iW \,,$$

where W is a positive definite Hermitian matrix.

Solutions

2.2

(a) The product of the potential functions can be written as

$$\Psi_{mn}\Psi_{nm} = \cos\left(\frac{m\pi}{a}x\right)\cos\left(\frac{n\pi}{a}x\right)\cos\left(\frac{m\pi}{a}y\right)\cos\left(\frac{n\pi}{a}y\right)$$

$$= \frac{1}{4}\left[\cos\left(\frac{(m-n)\pi}{a}x\right) + \cos\left(\frac{m+n)}{a}x\right)\right] \tag{1}$$

$$\times\left[\cos\left(\frac{(m-n)\pi}{a}y\right) + \cos\left(\frac{m+n)\pi}{a}y\right)\right].$$

The integrals with respect to x and y extend over an interval a. They vanish because of the periodicity of the functions.

(b) Figure S2.2.1a shows the potential surface. The lines of equal height are the E lines, the lines of steepest descent are the H lines.

(c) Figure S2.2.1b shows the potential surface for the orthogonal mode, $\Psi_{10} - \Psi_{01}$.

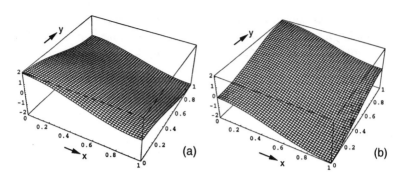

Fig. S2.2.1. (a) Plot of $\Psi_{10} + \Psi_{01}$; (b) plot of $\Psi_{10} - \Psi_{01}$

2.4 For an E field of complex amplitude E_o in the center of the guide, the power dissipated per unit length is

$$P_d = \frac{1}{2}\sigma\theta|E_o|^2 b . \tag{1}$$

The power flow in the waveguide is

$$P_o = \frac{1}{4}\sqrt{\frac{\epsilon_o}{\mu_o}}\sqrt{1 - \frac{p^2}{\omega^2\mu_o\epsilon_o}}|E_o|^2 ab . \tag{2}$$

The power decays with an attenuation constant γ, which is given by

$$\gamma = \frac{P_d}{P_o} = \frac{2\sigma\theta}{a}\sqrt{\frac{\mu_o}{\epsilon_o}}\frac{1}{\sqrt{1 - p^2/\omega^2\mu_o\epsilon_o}} . \tag{3}$$

The net attenuation over a length L is $\exp(\gamma L)$. We find $\gamma = 0.432$ cm^{-1} and $L = 5.33$ cm.

2.6 By superimposing two modes of differentially different frequencies ω and $\omega + \Delta\omega$ with equal amplitudes, one may construct a wavepacket whose fields go to zero at distances spaced by $(2\pi/\Delta\omega)d\omega/d\beta$. The energy stored in this packet cannot escape and the packet travels at the group velocity. Thus the argument that the energy travels at the group velocity is a very general argument and only breaks down when the propagation constant cannot be differentiated with respect to frequency.

2.8 If there are two ports of access, each port causes a decay of the mode, and each port feeds the mode. The generalization of (2.227) is

$$\frac{dU}{dt} = -(i\omega_o + 1/\tau_o + 1/\tau_{e1} + 1/\tau_{e2})U + \sqrt{\frac{2}{\tau_{e1}}}a_1 + \sqrt{\frac{2}{\tau_{e2}}}a_2 . \tag{1}$$

There are two reflected waves, each of which can be evaluated from time reversal and energy conservation

$$b_i = -a_i + \sqrt{\frac{2}{\tau_{ei}}} U; \quad i = 1, 2 . \tag{2}$$

When $a_1 = A \exp(-i\omega t)$ and $a_2 = 0$ we find from the above, for the power escaping from port (2)

$$|b_2|^2 = \frac{2|U|^2}{\tau_{e2}} = \frac{4|A|^2 \tau^2 / \tau_{e1}\tau_{e2}}{(\omega - \omega_o)^2 \tau^2 + 1} , \tag{3}$$

where

$$\frac{1}{\tau} = \frac{1}{\tau_{e1}} + \frac{1}{\tau_{e2}} + \frac{1}{\tau_o} . \tag{4}$$

All of the power is transferred if, and only if $1/\tau_o = 0$ and $1/\tau_{e1} = 1/\tau_{e2}$. The resonator must be loss-free and the two external Qs must be the same.

3. Diffraction, Dielectric Waveguides, Optical Fibers, and the Kerr Effect

Physical conductors (metals) that model adequately the behavior of a perfect conductor at microwave frequencies are too lossy at optical frequencies to provide low-loss enclosures. The same holds for reflectors. Whereas a metallic reflector is perfectly adequate at microwave frequencies, at optical frequencies reflectors have to be constructed using layered dielectrics of the proper thickness and dielectric constant. Total internal reflection is utilized in the construction of dielectric waveguides at microwave frequencies as well as at optical frequencies. At optical frequencies these dielectric waveguides are realized as fibers. Optical beams can also be contained in free space, if periodically refocused by lenses or mirrors. Optical resonators can be built with two or more curved mirrors that balance the diffraction of the beam bouncing back and forth and maintain a resonance mode in the space between the mirrors. The modes in dielectric waveguides and the modes of optical resonators share many of the properties of microwave waveguides and resonators discussed in the preceding chapter.

We start with a discussion of optical beams propagating in free space, the so-called Gaussian and Hermite Gaussian beams. We discuss the modes in optical fibers and derive their dispersion relations, i.e. the propagation constants as functions of frequency. We present both the standard derivation in terms of coupled TE and TM waves and the simplified linearly polarized (LP) approach. This is followed by the derivation of the perturbation formula for the change of the propagation constant due to an index change of the fiber. We study the propagation of waves in the presence of group velocity dispersion. We look at the coupling of two waves of orthogonal polarization in an optical fiber.

The detailed study of wave propagation in fibers is preliminary to the study of optical-fiber communications in Chaps. 9 and 10. High-bit-rate optical communications have made enormous progress in recent years. The low loss and low dispersion of optical fibers make the fiber an ideal transmission medium, permitting much higher bit rates than is possible with microwave transmission. Recently, designs for repeaterless transoceanic fiber cables have been implemented with a bit rate of 5 Gbit/s. The loss of the fiber is compensated by erbium-doped fiber amplifiers spaced roughly 40 km apart; the transmission wavelength is at the gain wavelength of erbium, 1.54 μm. These technical advances in transoceanic transmission will influence the usage of terrestrial fibers, 10 million km of which are already in the ground. Terrestrial fibers have been designed to have zero group velocity dispersion at 1.3 μm. It was anticipated that all communication over the fibers would be accomplished at a wavelength of 1.3 μm, using laser diode amplifiers, even though the minimum loss of the fiber is at around 1.5 μm [32] (see Fig. 3.1). With zero dispersion, the pulses propagate with no distortion, except for the effects of third-order dispersion. It has turned out, however, that the erbium doped fiber amplifiers perform much better than the diode amplifiers. They have long gain relaxation times of the order of 1 ms and thus have no intersymbol crosstalk. Thus, it is likely, that most of the terrestrial network will also be ugraded to operate at 1.54 μm wavelength.

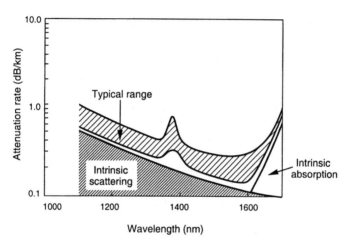

Fig. 3.1. The loss of a single-mode fiber as a function of wavelength (from [32])

3.1 Free-Space Propagation and Diffraction

In preparation for the study of optical-beam propagation, we solve Maxwell's equations in the paraxial limit, in the limit when all wave vectors composing the beam have small angles of inclination with respect to the axis of the beam. By solving for the vector potential along an axis transverse to the beam axis, a scalar equation is obtained. All three components of the electric and magnetic fields can be derived from the solution of this scalar equation. Optical Fabry–Pérot resonator fields can be constructed from these same solutions.

Propagation of optical beams is in everyone's daily experience. Sun rays passing through clouds delineate straight line designs in the sky. Thus, the dominant impression is that light propagates as rays, the foundation of the mathematical theory of ray optics. However, observation of light diffracted by a sharp edge or small holes (such as the weave of a parasol) is also a common experience, and these effects call for a refinement of ray optics by diffraction theory.

Maxwell's equations contain both ray optics and diffraction optics in certain limits. In diffraction optics, waves of different propagation vectors interfere with each other to produce collectively a beam. These beams do not maintain their cross section, they diffract. However, the diffraction may be small if the transverse dimension of the beam is many wavelengths. Since metals are too lossy at optical frequencies to provide efficient guidance of optical waves, as they do for microwaves, free-space beams are a convenient way to transmit power from one region of space to the other. The diffraction solutions of Maxwell's equations also provide the framework for the quantization of electromagnetic fields in free space. These are the reasons for the study of diffraction here.

Maxwell's equations are repeated here, as specialized to free space:

$$\nabla \times \boldsymbol{E} = -\mu_o \frac{\partial \boldsymbol{H}}{\partial t} \quad \text{(Faraday's law)} , \tag{3.1}$$

$$\nabla \times \boldsymbol{H} = \epsilon_o \frac{\partial \boldsymbol{E}}{\partial t} \quad \text{(Ampère's law)} , \tag{3.2}$$

$$\nabla \cdot \epsilon_o \boldsymbol{E} = 0 \quad \text{(Gauss's law)} , \tag{3.3}$$

$$\nabla \cdot \mu_o \boldsymbol{H} = 0 \quad \text{(Gauss's law)} . \tag{3.4}$$

From these equations one may derive the wave equation for the electric field

$$\nabla^2 \boldsymbol{E} = \epsilon_o \mu_o \frac{\partial^2 \boldsymbol{E}}{\partial t^2} \tag{3.5}$$

or an analogous relation for the magnetic field. We are interested in solutions that are plane-wave-like, but confined to a finite cross section that measures many wavelengths across. Under these conditions, one may make the paraxial wave approximation. It is more convenient to make this approximation in the wave equation for the vector potential than in the equation for the electric field, since then one may deal with a single-component vector field and a scalar wave equation, as we proceed to show [31].

The curl of the vector potential is defined by

$$\mu_o \boldsymbol{H} = \nabla \times \boldsymbol{A} . \tag{3.6}$$

In order to define a vector field completely, one needs to specify both its curl and its divergence. Equation (3.6) defines only the curl, in terms of the

H field. One may use this freedom to choose the divergence of the vector potential so as to obtain simple equations for the evolution of the vector potential. This is done by first noting that (3.1) and (3.6) give

$$E = -\frac{\partial A}{\partial t} - \nabla \Phi \,, \tag{3.7}$$

where Φ is an as yet unspecified scalar potential. Introducing (3.6) and (3.7) into (3.2), one finds

$$\nabla \times (\nabla \times A) = -\mu_o \epsilon_o \frac{\partial^2 A}{\partial t^2} - \mu_o \epsilon_o \frac{\partial}{\partial t} \nabla \Phi \,. \tag{3.8}$$

Using a well-known vector identitity, the curl of the curl of *A* can be written

$$\nabla \times (\nabla \times A) = \nabla \nabla \cdot A - \nabla^2 A \,. \tag{3.9}$$

Thus, if one chooses the so-called Lorentz gauge,

$$\nabla \cdot A + \mu_o \epsilon_o \frac{\partial \Phi}{\partial t} = 0 \,, \tag{3.10}$$

a simple wave equation is obtained for the vector potential:

$$\nabla^2 A - \mu_o \epsilon_o \frac{\partial^2 A}{\partial t^2} = 0 \,. \tag{3.11}$$

Gauss's law (3.3), in combination with (3.7), gives

$$\nabla \cdot \left(\frac{\partial}{\partial t} A + \nabla \Phi \right) = 0 \,. \tag{3.12}$$

When this relation is combined with the Lorentz gauge (3.10) one obtains the wave equation for the scalar potential Φ:

$$\nabla^2 \Phi - \mu_o \epsilon_o \frac{\partial^2 \Phi}{\partial t^2} = 0 \,. \tag{3.13}$$

Next we apply the wave equation obeyed by the vector potential, (3.11), to propagation of a beam in free space along the z direction of a Cartesian coordinate system. We assume a vector potential with a single component along the x axis. Substituting this ansatz into (3.11), we obtain a scalar wave equation for A_x:

$$\nabla^2 A_x = \epsilon_o \mu_o \frac{\partial^2 A_x}{\partial t^2} \,. \tag{3.14}$$

We now look for a solution of A_x in the form of a quasi-plane wave, i.e. we assume

$$A_x = \psi(x, y, z)e^{-i\omega t}e^{ikz} \tag{3.15}$$

and obtain the differential equation for the field envelope ψ:

$$\frac{\partial^2 \psi}{\partial x^2} + \frac{\partial^2 \psi}{\partial y^2} + 2ik\frac{\partial \psi}{\partial z} + \frac{\partial^2 \psi}{\partial z^2} = 0 , \tag{3.16}$$

where k is defined by $k = \omega\sqrt{\mu_o \epsilon_o}$, which is the dispersion relation of plane waves in vacuum. From the Lorentz gauge (3.10), one then obtains for the scalar potential Φ the following expression:

$$\Phi = -\frac{i}{\epsilon_o \mu_o \omega}\frac{\partial \psi}{\partial x}\exp(-i\omega t + ikz) ,$$

which manifestly satisfies the scalar wave equation (3.13).

If the beam has a cross section much larger than a wavelength, the z dependence of ψ is approximately given by e^{ikz}, and thus the correction to the z dependence, $d\psi/dz$, is relatively small. The second derivative of ψ with respect to z can be ignored, with the result

$$\frac{\partial \psi}{\partial x^2} + \frac{\partial^2 \psi}{\partial y^2} + 2ik\frac{\partial \psi}{\partial z} = 0 . \tag{3.17}$$

This is the paraxial wave equation. This equation also happens to be the Schrödinger equation of a free particle in two dimensions, if z is replaced by t. Equation (3.17) is of first order in z and thus describes waves that travel in the $+z$ direction only. A corresponding equation with k replaced by $-k$ gives waves traveling in the $-z$ direction. The simplest solution of the paraxial wave equation is a beam of Gaussian cross section

$$\psi(x, y, z) = A_o\frac{-ib}{z - ib}\exp\left(\frac{ik(x^2 + y^2)}{2(z - ib)}\right) , \tag{3.18}$$

where A_o and b are integration constants. The former is the amplitude at the beam center, $x = y = 0$, at $z = 0$; the latter is the so-called confocal parameter. This parameter determines the minimum diameter of the beam. In order to see this, we rewrite (3.18) by separating the real part and imaginary part of the exponent in the form

$$\psi(x, y, z) = \frac{A_o}{\sqrt{1 + z^2/b^2}}\exp\left(-\frac{x^2 + y^2}{w^2}\right)\exp\left(i\frac{k(x^2 + y^2)}{2R}\right)\exp(-i\phi) .$$
$$\tag{3.19}$$

Here the meaning of the parameters is easily identifiable: w is the radius at which the field amplitude is decreased from its peak value by $1/e$; R is the radius of curvature of the phase front surface defined by $k[(x^2+y^2)/2R]+kz = 0$; ϕ is a phase advance. All these parameters are related to the confocal parameter b. Indeed,

$$w^2 = \frac{2b}{k}\left(1 + \frac{z^2}{b^2}\right), \quad \frac{1}{R} = \frac{z}{z^2 + b^2}, \quad \text{and} \quad \phi = \tan^{-1}\frac{z}{b}. \tag{3.20}$$

The minimum beam diameter is

$$w_o = \sqrt{\frac{2b}{k}} = \sqrt{\frac{\lambda b}{\pi}}, \tag{3.21}$$

where $\lambda = 2\pi/k$ is the free-space wavelength. Equation (3.19) can also be written

$$\psi(x, y, z) = \frac{A_o}{w/w_o} \exp\left(-\frac{x^2 + y^2}{w^2}\right) \exp\left(i\frac{k(x^2 + y^2)}{2R}\right) \exp(-i\phi). \tag{3.19a}$$

The denominator w/w_o takes care of power conservation: the power flow density has to decrease with the square of the beam radius. The phase advance ϕ imparts to the beam a phase velocity larger than the speed of light. This is due to the the fact that the Gaussian beam is made up of a superposition of plane waves whose wave-vectors are inclined with respect to the z axis, and thus possess phase velocities as measured along the z axis that are larger than the speed of light. The group velocity is, of course, less than the speed of light (see Appendix A.1). One may say that vacuum is dispersive for a beam of any given beam radius w_o.

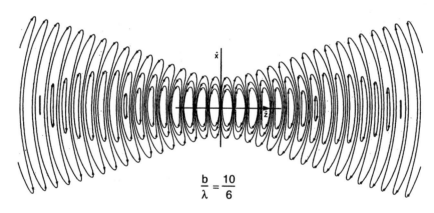

$$\frac{b}{\lambda} = \frac{10}{6}$$

Fig. 3.2. Electric field of Gaussian beam in x–z plane at one instant of time. The pattern moves to the right as a function of time; $b/\lambda = 10/6$

Figure 3.2 shows the electric field in the x–z plane of the fundamental Gaussian for a wave traveling in the $+z$ direction. The field has both z and x components, which are evaluated from (3.7) using the vector potential solution and the scalar potential associated with it according to (3.10). The electric field is found to be

$$E = \mathrm{i}\omega \left(i_x \psi + \frac{\mathrm{i}k}{\omega^2 \mu_o \epsilon_o} i_z \frac{\partial \psi}{\partial x} \right) \exp(-\mathrm{i}\omega t + \mathrm{i}kz) \,. \tag{3.22}$$

The derivation from the vector potential has paid off. We have found the total electric field from one single vector component of the vector potential. Had we set up the paraxial wave equation for the electric field, we would have had to solve three scalar wave equations separately and would have had to make the three solutions consistent with each other by setting the divergence of E equal to zero.

Fabry–Pérot resonators support Gaussian beams when formed from curved, spherical mirrors spaced at the appropriate distance so as to match the phase curvature of the Gaussian. The nodal surfaces of the modes fit the mirror surfaces, which may be thought to function as perfect conductors. One uses dielectric mirrors with periodic layers of dielectrics of different dielectric constant to construct highly reflecting surfaces at optical frequencies. These Fabry–Pérot resonators are the laser resonators for gas and many solid-state lasers, in which the medium cannot provide guidance of the optical wave.

In many cases it is possible to ignore the refractive properties of the medium and compute the electric field solely from the vacuum field. The laser medium supplies only the gain that balances the losses in the medium and the loss due to radiation passing through the partially transmitting end mirror used as the laser output mirror. The emitted laser beam outside the resonator does not experience vacuum dispersion, as we now discuss. We have pointed out that a Gaussian mode is supported between two curved mirrors of some given radius R. If we look at a symmetric resonator, with both mirrors of the same curvature R, spaced a distance d apart, then (3.20) yields a value for the b parameter

$$b = \sqrt{Rd/2 - (d/2)^2} \,. \tag{3.23}$$

The b parameter is fixed by the geometry; it is wavelength-independent. Hence, if many axial Gaussian modes of different frequencies are excited simultaneously within the laser by mode-locking the laser [31,33], a short pulse is produced within the laser. The different frequency components of the pulse all have the same b parameter, which means that they have different beam radii. If a group velocity is computed from the phase shift of the pulse in one pass, $2kd + 2\tan^{-1}(d/2b)$, one finds that it is equal to the speed of light c. The additional phase shift 2ϕ does not contribute to the group velocity since it is frequency independent. A pulse of this type emitted from the laser (if one of the mirrors is partially transmissive) does not experience "vacuum dispersion".

The paraxial wave equation has a complete set of solutions that are composed of products of Hermite Gaussians:

$$\varphi_m(\xi) \equiv H_m(\xi) \exp\left(-\frac{\xi^2}{2} \right) \,. \tag{3.24}$$

The solutions are

$$\psi_{mn}(x, y, z) = \frac{w_o}{w} \varphi_m\left(\frac{\sqrt{2}x}{w}\right) \varphi_n\left(\frac{\sqrt{2}y}{w}\right) \exp\left[\frac{ik}{2R}(x^2 + y^2)\right]$$

$$\times \exp[-i(m + n + 1)\phi] \,,$$

(3.25)

with w, R and ϕ given by (3.20). They have the same phase profile as the fundamental Gaussian, but different phase velocities. The fields experience a greater phase advance per unit distance of propagation the greater the order of the Hermite Gaussian. The reason for this is that the higher the order of the mode, the greater the inclination with respect to the z axis of the plane waves composing the mode. Thus, the higher order modes acquire phase velocities larger than the speed of light.

The Hermite Gaussians form a complete orthogonal set. The orthogonality could be proved by mathematical manipulation. However, there is a simple physical argument for the orthogonality. The power flow in the beam is formed from the integrals of complex-conjugate products of field profiles. A product of two mode patterns of different propagation constants has a z dependence. Since the time-averaged power flow must be z independent, such cross terms must be equal to zero.

An excitation described by a transverse electric-field distribution can be expanded in terms of this set. The radius w of the Hermite Gaussians is arbitrary, but should be chosen so that the number of terms in the expansion with appreciable amplitudes is minimized. For simple profiles of the excitation, the rule is to maximize the excitation of the fundamental Gaussian mode by proper choice of w. Some important relations among Hermite Gaussians are summarized in Appendix A.2.

3.2 Modes in a Cylindrical Piecewise Uniform Dielectric

A dielectric rod can guide microwaves. A rod of refractive index higher than that of the surrounding space confines the field in the rod and in its immediate vicinity. The eigenmode solutions for a dielectric rod are the same as those for an optical fiber of uniform core index. A fiber has a dielectric core of slightly higher index than that of the surrounding cladding. In ray optics parlance, optical radiation can be confined to the core and its periphery by total internal reflection if the rays constituting the mode have incidence angles greater than the critical angle. Figure 3.3 shows schematically a ray bouncing around in a dielectric cylinder with a step discontinuity in the index [34].

In terms of Maxwell's equations, guided modes appear as eigensolutions of the wave equation that decay exponentially towards infinity in the transverse plane. This analytic approach yields mode profiles and dispersion relations

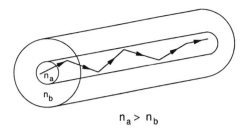

$n_a > n_b$

Fig. 3.3. Path of ray in single mode circular cylindrical step-index fiber

for the modes. We shall follow it here, using some of the results from [35]. The vector field $E(r, t)$ is assumed to be sinusoidally time-dependent (any general time-dependence can be built up by Fourier superposition):

$$E(r, t) = \text{Re}[E(r)\exp(-i\omega t)] \ . \tag{3.26}$$

From Maxwell's equations,

$$\nabla \times E = i\omega\mu_o H \ , \tag{3.27}$$

$$\nabla \times H = -i\omega\epsilon E \ , \tag{3.28}$$

$$\nabla \cdot \epsilon E = 0 \ , \tag{3.29}$$

$$\nabla \cdot \mu_o H = 0 \ , \tag{3.30}$$

one may derive the wave equation for the electric field if the dielectric is uniform. In a piecewise uniform dielectric this condition is obeyed separately in each region with a uniform medium:

$$\nabla^2 E(r) + \omega^2\mu_o\epsilon E(r) = 0 \ . \tag{3.31}$$

Similarly, a wave equation (or Helmholtz equation) can be derived for the magnetic field. If we consider a cylindrical waveguide of radius $\rho = a$ with index n_a, and an index n_b outside that radius (see Fig. 3.4), one may find solutions of (3.31) for the z component of the electric field. This equation, written in cylindrical coordinates, is

$$E_z = A(\omega)F(\rho)e^{im\phi}e^{i\beta z} \ , \tag{3.32}$$

in which the equation for F becomes

$$\frac{d^2F}{d\rho^2} + \frac{1}{\rho}\frac{dF}{d\rho} + \left(\kappa^2 - \frac{m^2}{\rho^2}\right)F = 0 \tag{3.33}$$

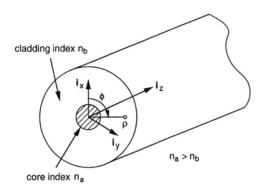

cladding index n_b

core index n_a

$n_a > n_b$

Fig. 3.4. Geometry of fiber with index step

with

$$\kappa^2 = n_a^2 k^2 - \beta^2 \; , \tag{3.34}$$

where $k = \omega\sqrt{\mu_o\epsilon_o}$. The solutions for F are

$$F = \begin{cases} J_m(\kappa\rho) \; , & \rho \leq a \\ K_m(\gamma\rho) \; , & \rho > a \end{cases} \tag{3.35}$$

with

$$\gamma^2 = \beta^2 - n_b^2 k^2 \; . \tag{3.36}$$

The J_ms are Bessel functions of order m and the K_ms are modified Bessel functions of order m. The modified Bessel functions K_m decay exponentially as $\rho \to \infty$ and are singular at the origin, but because they are not used to express the field at the origin the singularity does not occur in the field solution. The specific ρ–ϕ dependence of the z component of the electric field has associated with it a definite H field which is purely transverse. There is also an associated E field, which appears curl-free in the transverse plane, because $H_z = 0$. The solution thus obtained is a so called E wave. If the core were enclosed in a perfect conductor, the E wave could be made to satisfy all the boundary conditions. In an open structure, however, it is not possible to provide continuity of the tangential components of E and H at $\rho = a$, using only an E wave with two adjustable constants. Instead it is necessary to develop an analogous H wave solution of the same kind, with the same radial and ϕ dependence. The boundary conditions can be matched using a mixture of E and H waves. We do not present the details here, but refer the reader to the literature [35–37]. We simply state the determinantal equation that results from matching of the boundary conditions:

$$\left(\frac{J'_m(\kappa a)}{\kappa J_m(\kappa a)} + \frac{K'_m(\gamma a)}{\gamma K_m(\gamma a)}\right)\left(\frac{J'_m(\kappa a)}{\kappa J_m(\kappa a)} + \frac{n_b^2}{n_a^2}\frac{K'_m(\gamma a)}{\gamma K_m(\gamma a)}\right)$$

$$= \left(\frac{m\beta k(n_a^2 - n_b^2)}{a\kappa^2\gamma^2 n_a}\right)^2 \tag{3.37}$$

with

$$\kappa^2 + \gamma^2 = (n_a^2 - n_b^2)k^2 . \tag{3.38}$$

This is a rather complicated-looking determinantal equation. It is clear that it is the result of two-wave coupling, the two factors in parentheses representing some forms of limiting solutions in the limit $n_b \to n_a$. Of course, in this limit, no bound solution could in fact exist. Yet, the factors suggest that there may exist simpler, approximate determinantal equations related to either one of these factors. We shall show that this is indeed the case after some more discussion of the meaning of the determinantal equation.

As mentioned earlier, the modes along a fiber are mixtures of E waves and H waves, and hence it seems appropriate that they have been dubbed HE$_{mn}$ and EH$_{mn}$ modes. At any specific frequency only a finite number of these modes is guided. Below a certain frequency, the cutoff frequency of the first higher-order mode, only one mode propagates, the HE$_{11}$ mode. This is the dominant mode used in single-mode fiber propagation. It is, therefore, the most important mode and deserves further scrutiny. We shall derive its properties by the much simpler, approximate method of the next section.

3.3 Approximate Approach

The determinantal equation (3.37) is complicated because it expresses the interaction of E waves with H waves, coupled by the index discontinuity. One cannot arrive at normalized graphs that are independent of the ratio n_a/n_b, something possible with approximate analyses. If the index discontinuity is small, the coupling between E and H waves is weak, and either one or the other wave predominates. This is the reason that approximate approaches, which deal essentially with one type of wave, produce satisfactory answers. They arrive at graphs that are normalizable and universal (they do not depend on n_a/n_b) and give simple dispersion relations.

One of the approximate analyses is the approach that arrives at linearly polarized (LP) waves [38] by solving the wave equation for, say, an x directed field. It gives a scalar wave equation of the same type as the one solved for the z component of the electric field in the exact analysis. The electric field is exactly matched at the boundary, while the magnetic field is allowed to be slightly discontinuous.

The determinantal equation is

$$\frac{\kappa J_{m+1}(\kappa\rho)}{J_m(\kappa\rho)} = \frac{\gamma K_{m+1}(\gamma\rho)}{K_m(\gamma\rho)} . \tag{3.39}$$

The graphs are universal, they do not depend on the ratio n_a/n_b. The lowest-order, dominant mode is the one with the slowest transverse variation, with $m = 0$. The determinantal equation is

$$\frac{\kappa J_o'(\kappa\rho)}{J_o(\kappa\rho)} = \frac{\gamma K_o'(\gamma\rho)}{K_o(\gamma\rho)} , \tag{3.40}$$

where we have used the Bessel function recursion relation given in Appendix A.3.

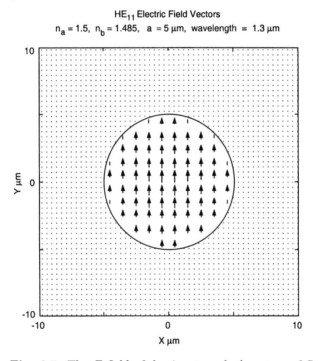

HE$_{11}$ Electric Field Vectors
n_a = 1.5, n_b = 1.485, a = 5 µm, wavelength = 1.3 µm

Fig. 3.5. The E field of dominant mode (courtesy of Sai-Tac Chu of Waterloo University). The lengths of the *arrows* indicate the magnitude of the electric field

The transverse field is illustrated in Fig. 3.5. In fact, the figure was obtained using the exact solution, but to the eye the difference is not noticeable. The fact that the dominant mode is identified with $m = 0$ in this approximate solution and with $m = 1$ in the exact approach is, at first, rather puzzling. In the exact analysis, Bessel functions of order $m = 1$ express the z component

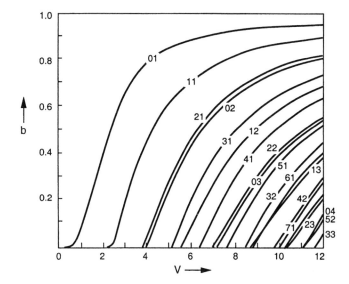

Fig. 3.6. Dispersion in normalized units for the propagation constant b and frequency V obtained from LP analysis (from [38])

of the field. The divergence relation connects the derivatives of the transverse field to $i\beta E_z$, and hence the transverse E field involves integrals of E_z. The integrals lead from J_1 to J_o, and K_1 to K_o.

The determination of the field is only one of the steps in the characterization of a fiber mode. Another important piece of information is the dispersion relation $\beta = \beta(\omega)$. Figure 3.6 shows the normalized propagation constant with n_a and n_b considered frequency-independent. The figure uses the normalized frequency

$$V = (n_a^2 - n_b^2)^{1/2} ka \qquad (3.41)$$

and the normalized propagation constant

$$b = (\beta/k - n_b)/(n_a - n_b) . \qquad (3.42)$$

At low frequencies, the mode extends far into the cladding and acquires a propagation constant characteristic of a plane wave in a medium of index n_b. At very high frequencies, the mode is very effectively reflected at the boundary between the two media and is essentially confined to the medium of index n_a. This explains the asymptotic behavior of the propagation constant for low and high frequencies. Note that a dispersion curve with zero group velocity dispersion (GVD) would be a horizontal, straight line in this graph, because V is proportional to k, and b is independent of k over the frequency range of zero GVD. Since the propagation constant is not a linear function of frequency, the inverse group velocity β' is a function of frequency. The system has GVD. Clearly, the geometry of the fiber imposes GVD. In practice, the situation is complicated by the fact that the index of silica is itself a function of frequency. It rises toward short wavelengths as the frequency gets closer to absorption bands in the ultraviolet. This greatly modifies the dispersion curve, the propagation constant as a function of frequency.

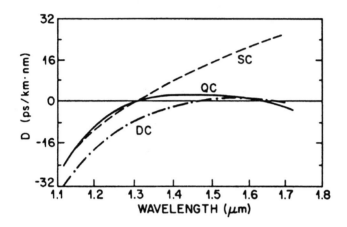

Fig. 3.7. Total dispersion D and relative contributions of material dispersion D_M and waveguide dispersion D_W for a conventional single-mode fiber. The zero-dispersion wavelength shifts to a higher value because of the waveguide contributions (from [39])

Figure 3.7 plots the parameter D_λ for plane wave propagation in silica [39, 40], where

$$D_\lambda \equiv -\frac{\lambda}{c}\frac{d^2n}{d\lambda^2}.$$ (3.43)

The parameter is derived from β'' by noting that the second derivative of β with respect to ω can be written in terms of the second derivative of n with respect to λ (whereas when written in terms of derivatives of n with respect to ω it would involve $dn/d\omega$ as well). The second derivative of β with respect to frequency is

$$\frac{d^2\beta}{d\omega^2} = \left(\frac{\lambda}{2\pi}\right)\left(\frac{\lambda}{c}\right)^2\left(\frac{d^2n}{d\lambda^2}\right)$$ (3.44)

and thus D_λ is proportional to β''. Figure 3.7 includes both the "waveguide dispersion" due to fiber geometry and the material dispersion.

3.4 Perturbation Theory

We are often interested in the change of the propagation constant caused by a small change of the index distribution. Thus, for example, the Kerr effect which changes the index as a function of electric field intensity, can change the propagation constant. The E field obeys the vector Helmholtz equation. If we separate the Laplacian into longitudinal and transverse components we obtain

$$\nabla_T^2 \boldsymbol{E} + \omega^2 \mu_o \epsilon \boldsymbol{E} - \beta^2 \boldsymbol{E} = 0 \ . \tag{3.45}$$

In this equation we treat ϵ as a continuous function of the transverse coordinates x and y. We suppose that ϵ changes by $\delta\epsilon$, $\boldsymbol{E}$ by $\delta\boldsymbol{E}$, and β by $\delta\beta$. These perturbations obey an equation that is derived from (3.45) by perturbing it to first order:

$$\nabla_T^2 \delta\boldsymbol{E} + \omega^2 \mu_o \epsilon \, \delta\boldsymbol{E} + \omega^2 \mu_o \, \delta\epsilon \, \boldsymbol{E} - \beta^2 \, \delta\boldsymbol{E} - 2\beta \, \delta\beta \, \boldsymbol{E} = 0 \ . \tag{3.46}$$

We dot-multiply (3.46) by $\boldsymbol{E}^*$ and the complex conjugate of (3.41) by $\delta\boldsymbol{E}$ and subtract, and integrate over the cross section. Solving for $\delta\beta$, we find

$$2\beta \, \delta\beta = \frac{\int_{\substack{\text{cross} \\ \text{section}}} \omega^2 \mu_o \, \delta\epsilon |\boldsymbol{E}|^2 \, dS}{\int_{\substack{\text{cross} \\ \text{section}}} |\boldsymbol{E}|^2 \, dS} \ , \tag{3.47}$$

where $dS = dx \, dy$ is an area element in a plane transverse to z. Note that $\delta\boldsymbol{E}$ has dropped out. We need not know the change of $\boldsymbol{E}$ to first order to be able to evaluate the change of the propagation constant. This is a very important finding that facilitates the introduction of perturbations into the propagation equations.

3.5 Propagation Along a Dispersive Fiber

Uniform waveguides propagate waves in both directions along the axis of the waveguide. We have had ample opportunity to study such modes in metallic waveguides. The propagation along dielectric guides and optical fibers is completely analogous. Here we develop the propagation equation for a traveling wave of a mode in a phenomenological way. Waves in metallic waveguides are just one special case in this more general approach. We consider a wave of an eigenmode in a lossless, uniform (with respect to z) wave-guiding structure, with the amplitude spectrum $A(\omega, z)$. Its z dependence is simply $e^{i\beta z}$. The wave obeys the differential equation

$$\frac{\partial}{\partial z} A(\omega, z) = i\beta(\omega) A(\omega, z) \ . \tag{3.48}$$

The propagation constant is a function of frequency. We concentrate on an investigation of a wave of narrow bandwidth within a frequency interval centered at the nominal carrier frequency ω_o. Carrying out an expansion to second order in the deviation $\Delta\omega$ from the carrier frequency, we obtain (see Fig. 3.8)

$$\beta(\omega) = \beta_o + \Delta\omega\,\beta' + \frac{1}{2}\Delta\omega^2\,\beta'' ,\tag{3.49}$$

where we use the following abbreviations:

$$\beta_o = \beta(\omega_o) ,\tag{3.50a}$$

$$\beta' = \frac{d\beta}{d\omega} = \frac{1}{v_g} \quad \text{(inverse group velocity)} ,\tag{3.50b}$$

$$\beta'' = \frac{d^2\beta}{d\omega^2} \quad \text{(group velocity dispersion)} .\tag{3.50c}$$

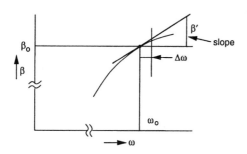

Fig. 3.8. Definition of parameters

In the next step, we take advantage of the narrowness of the spectrum. The spatial dependence at the carrier frequency is $\exp(i\beta_o z)$. The spatial dependence of the entire spectrum will deviate from this dependence, because the frequencies of the Fourier components differ from ω_o by $\Delta\omega$. We write

$$A(\omega, z) = a(\Delta\omega, z)\exp(i\beta_o z) .\tag{3.51}$$

When we introduce the ansatz (3.51) and the expansion (3.49) into (3.48), we obtain

$$\frac{\partial}{\partial z}a(\Delta\omega, z) = i\left(\Delta\omega\,\beta' + \frac{1}{2}\Delta\omega^2\,\beta''\right)a(\Delta\omega, z) .\tag{3.52}$$

We find that the spatial dependence of $a(\Delta\omega, z)$ is much slower than that of $A(\omega, z)$. Next we look at the temporal dependence of $A(\omega, z)$ by taking its inverse Fourier transform:

$$A(t, z) = \int_{-\infty}^{\infty} d\omega \, e^{-i\omega t} A(\omega, z)$$

$$= \exp(-i\omega_o t + i\beta_o z) \int_{-\infty}^{\infty} d\Delta\omega \exp(-i\,\Delta\omega\, t) a(\Delta\omega, z) \qquad (3.53)$$

$$= e^{-i\omega_o t + i\beta_o z} a(t, z) \, ,$$

where $a(t, z)$ is the inverse Fourier transform of $a(\Delta\omega, z)$. The fast space–time dependence of the wave amplitude is removed from $a(t, z)$, the so-called envelope of the wave. We further note the relation

$$\int_{-\infty}^{\infty} d\Delta\omega \exp(-i\,\Delta\omega\, t)(i\,\Delta\omega)^m a(\Delta\omega, z) = (-1)^m \frac{\partial^m}{\partial t^m} a(t, z) \, . \qquad (3.54)$$

Multiplication by $(i\,\Delta\omega)^m$ of the Fourier transform $a(\Delta\omega, z)$ produces $(-1)^m$ times the mth derivative of the inverse Fourier transform. Using this fact, we may inverse Fourier transform (3.52) to obtain

$$\frac{\partial}{\partial z} a + \frac{1}{v_g} \frac{\partial a}{\partial t} = -\frac{i}{2} \beta'' \frac{\partial^2 a}{\partial t^2} \, . \qquad (3.55)$$

If we introduce a new time variable that removes the time delay z/v_g,

$$\tau = t - \frac{z}{v_g} \, , \qquad (3.56a)$$

$$\xi = z \, , \qquad (3.56b)$$

we obtain the equation

$$\frac{\partial a}{\partial \xi} = -\frac{i}{2} \beta'' \frac{\partial^2 a}{\partial \tau^2} \, . \qquad (3.57)$$

This is the propagation equation for a mode in a fiber with group velocity dispersion. It also happens to be the Schrödinger equation of a free particle in one dimension.

3.6 Solution of the Dispersion Equation for a Gaussian Pulse

We shall now solve the group velocity dispersion equation. For simplicity and flexibility in notation we again denote the distance variable by z and the time variable by t, writing for (3.57)

$$\frac{\partial a}{\partial z} = -\frac{i}{2} \beta'' \frac{\partial^2 a}{\partial t^2} \, . \qquad (3.58)$$

This equation has the simple solution

$$a(t,z) = A_o \sqrt{\frac{ib}{z + ib}} \exp\left(-i\frac{t^2}{2\beta''(z + ib)}\right)$$

$$= A_o \sqrt{\frac{\tau_o}{\tau}} \exp\left(-\frac{t^2}{\tau^2}\right) \exp\left(-i\theta(t,z) + i\phi(z)\right),$$

(3.59)

where

$$\tau^2 = \tau_o^2\left(1 + \frac{z^2}{b^2}\right),$$

(3.60)

$$\theta(t,z) = \frac{z}{2\beta''(z^2 + b^2)}t^2,$$

(3.61)

$$\tau_o^2 = 2\beta''b,$$

(3.62)

and

$$\phi = \frac{1}{2}\tan^{-1}\left(\frac{z}{b}\right).$$

(3.63)

A pulse that is initially of constant phase at $z = 0$ acquires a time-dependent phase given by (3.61); it becomes chirped. Since $\partial\theta/\partial t$ can be identified with the instantaneous frequency, a Gaussian pulse propagating in a dispersive system acquires a time-dependent frequency (chirp). In doing so, it broadens (see (3.59)). The chirped pulse acquires a width that is greater than would be inferred from the width of the spectrum for an unchirped (transform-limited) pulse. The system being linear, the spectral width cannot change with propagation.

The propagation of a Gaussian pulse along a dispersive fiber bears a close analogy to the diffraction of a beam as discussed in Sect. 3.1. The paraxial wave equation (3.17) resembles the propagation equation along a dispersive fiber, except that the diffraction equation contains two second derivatives instead of one. If the diffraction equation is applied to a slab beam with one transverse dimension, the analogy becomes complete. Comparison of (3.18) and (3.59) shows the close resemblance. In two dimensions, the amplitude of the mode must decrease asymptotically linearly with $1/z$; in one dimension the amplitude must decrease asymptotically as $1/\sqrt{z}$. This fact accounts for the multiplier $-ib/(z - ib)$ in (3.18) and the multiplier $\sqrt{ib/(z + ib)}$ in (3.59).

The equation for dispersive propagation, analogously to the equation for diffraction of a one-dimensional slab beam, has a complete set of solutions. An initial excitation can be expressed as a superposition of these solutions. In analogy with the problem of a beam in two dimensions, with the solutions (3.25), the solutions of the equation of dispersive propagation are

$$\psi_m(t,z) = \sqrt{\frac{\tau_o}{\tau}} \phi_m\left(\frac{\sqrt{2}t}{\tau}\right) \exp[-i\theta(t,z)] \exp[i(m+n+1)\phi(z)] , \quad (3.64)$$

where we again denote the Hermite Gaussian of mth order by ϕ_m. The solutions are orthogonal, permitting the evaluation of the coefficients of the Hermite Gaussians for an input excitation $a(t,0)$ from the integrals $\int dt\, a(t,0)\phi_m(\sqrt{2}t/\tau_o)$. Identities that help in the evaluation are presented in Appendix A.2.

3.7 Propagation of a Polarized Wave in an Isotropic Kerr Medium

The simplest model of a Kerr medium is an isotropic medium in which the polarization is an instantaneous function of the cube of the electric field:

$$P(t) = \epsilon_o \chi^{(3)} E^2(t) E(t) , \quad (3.65)$$

where $\chi^{(3)}$ is the third-order susceptibility and the alignment of the polarization and field is implied by using scalars. The endpoints of P and E could follow complicated temporal curves, depending upon the temporal evolution of the E field. Suppose that at a particular instant the E field points in the (general) direction

$$\boldsymbol{E} = E_x \boldsymbol{i}_x + E_y \boldsymbol{i}_y + E_z \boldsymbol{i}_z . \quad (3.66)$$

The polarization points along the E field and is given by

$$P_x = \epsilon_o \chi^{(3)} E_x (E_x^2 + E_y^2 + E_z^2) , \quad (3.67)$$

$$P_y = \epsilon_o \chi^{(3)} E_y (E_x^2 + E_y^2 + E_z^2) , \quad (3.68)$$

and

$$P_z = \epsilon_o \chi^{(3)} E_z (E_x^2 + E_y^2 + E_z^2) . \quad (3.69)$$

Suppose next that the E field has one single frequency and lies in the x–y plane. Then

$$E_x(t) = \frac{1}{2}[E_x(\omega)e^{-i\omega t} + E_x^*(\omega)]e^{+i\omega t} , \quad (3.70)$$

where $E_x(\omega)$ is a shorthand for $|E_x(\omega)|e^{-i\phi_x}$, etc. When we introduce the above expression into (3.67)–(3.69) and retain only the terms with an $e^{-i\omega t}$ dependence, we obtain

$$\frac{1}{2}P_x(\omega) = \epsilon_o \frac{\chi^{(3)}}{8} [3|E_x(\omega)|^2 E_x(\omega) + 2|E_y(\omega)|^2 E_x(\omega)$$

$$+ E_y^2(\omega)E_x^*(\omega)] \ . \tag{3.71}$$

This expression consists of three types of term. There is the self-modulation term $\chi^{(3)} 3|E_x|^2 E_x$, which is the only term surviving when the field is polarized along x. Then there is the cross phase modulation term, which looks like a change of index produced by the E_y component and seen by the x component of the E field, namely $2|E_y|^2 E_x$. Finally, there is a "coherence" term which produces an x polarization due to E_x^* and depends on the phase of E_y. This is a term utilized in four-wave mixing. In a birefringent fiber, in which the two orthogonal polarizations have different propagation constants, with the slow axis along x and the fast axis along y, these effects will cancel on average, because they will contain spatial dependences like $\exp i(2k_y - k_x)z$, and the optical nonlinear effects take place, generally, over distances much larger than the period of intrinsic birefringence of even so-called nonbirefringent fibers.

Now let us relate this expression to the commonly employed Kerr nonlinearity in which the index is written

$$n = n_o + n_2 I \tag{3.72}$$

and I is the intensity (power per unit area) of the field. The polarization P is defined by

$$P = \epsilon_o(n^2 - 1)E \cong \left[\epsilon_o(n_o^2 - 1) + 2\epsilon_o n_o n_2 I\right] E \ , \tag{3.73}$$

where the last term is clearly the contribution of the nonlinearity. Thus

$$2\epsilon_o n_o n_2 I = 3\epsilon_o \frac{\chi^{(3)}}{4} |E_x|^2 \tag{3.74}$$

in the case of a linearly polarized field. Therefore, since the intensity I is given by

$$I = \frac{1}{2}\epsilon|E_x|^2 \frac{c}{n_o} = \frac{1}{2}\epsilon_o n_o c|E_x|^2 \ , \tag{3.75}$$

we have for n_2

$$n_2 = \frac{3}{4} \frac{\chi^{(3)}}{\epsilon_o n_o^2 c} \ . \tag{3.76}$$

In glass, the coefficient n_2 has the value [40–46]

$$n_2 = 2.2 \times 10^{-16} \ \text{cm}^2/\text{W} \ .$$

3.7.1 Circular Polarization

It turns out that it is convenient to introduce circularly polarized modes by means of

$$E_x = \frac{1}{\sqrt{2}}(E_+ + E_-) \,, \tag{3.77a}$$

$$E_y = \frac{1}{i\sqrt{2}}(E_+ - E_-) \,, \tag{3.77b}$$

so that

$$E_+ = \frac{1}{\sqrt{2}}(E_x + iE_y) \,, \tag{3.78a}$$

$$E_- = \frac{1}{\sqrt{2}}(E_x - iE_y) \,. \tag{3.78b}$$

If we then evaluate

$$\frac{1}{\sqrt{2}}(P_x \pm iP_y) = P_\pm \tag{3.79}$$

we find from (3.71) that

$$P_\pm = \epsilon_o \frac{\chi^{(3)}}{2}(|E_\pm|^2 + 2|E_\mp|^2)E_\pm \,. \tag{3.80}$$

The presence of a circular polarization of opposite sense of rotation affects the index twice as strongly as the original polarization. We find the very interesting result that circular polarization does not exhibit a "coherence" term that depends on the relative phase between the two polarizations of E, unlike the coherence term for linear polarization. There is a simple reason for this fact which it is well to remember. Consider a linear polarization in an isotropic medium. The linear polarization can be represented by two counterrotating circular polarizations of the same amplitude. Suppose that there were a coherence term in (3.80) involving E_+^* or E_-^*. Then the evolution of the polarization would depend upon the relative phase between E_+ and E_-. But this is not possible, because a change of the relative phase means rotation of the linear polarization, and we know that the evolution of the polarization cannot depend on the orientation of the linear polarization in an isotropic medium.

The analysis of the propagation of polarized light in a uniform medium can be applied directly to the propagation of the fundamental mode in a weakly guiding fiber. Indeed, the mode is essentially linearly polarized; two orthogonally polarized modes experience coupling very much like plane waves, except that the coupling coefficient must now include the mode profiles. The ratio of the coefficients of the self-phase modulation, cross phase modulation and coherence terms still remains 3 to 2 to 1.

3.8 Summary

In this chapter we have presented the analysis of Hermite Gaussian modes in free space. These are used to construct optical resonators and hence are basic to laser operation and to the quantization of optical fields in such resonators. We have presented an analysis of modes in optical fibers and discussed their dispersion. The dispersion is caused partly by the geometry of the index profile and partly by the material dispersion of glass. It is possible to manipulate the net dispersion by changes in the index profile. Whereas the zero-dispersion wavelength of a glass fiber with a step index profile is roughly 1.3 μm, it is possible to shift the zero-dispersion wavelength to 1.5 μm, the wavelength region of the erbium-doped fiber amplifier, by proper choice of the index profile of the fiber core.

An isotropic Kerr medium with an instantaneous response has a very specific response to signals with two orthogonal polarizations. The response contains a "coherence term" which is a function of the phase between the two signals. In the circular-polarization basis the response is much simpler, and no coherence term is present. Even though the analysis holds strictly only for plane waves, the formalism can be applied to modes in optical fibers, which are almost entirely linearly polarized. The change of propagation constant follows from the perturbation formula developed in Sect. 3.4. The Kerr effect is a nonlinear effect that affects long-distance fiber communications. It is either combatted by group velocity dispersion management (varying GVD along the fiber), in the so-called non-return-to-zero format of communications currently installed in repeaterless transoceanic cables, or used to balance the group velocity dispersion of fibers in long-distance soliton communications, as taken up in Chap. 10. The Kerr effect is also used to generate squeezed states of radiation, as discussed in Chaps. 12 and 13.

Problems

3.1 An optical wave passing through a thin convergent lens in the x–y plane acquires the phase profile

$$\phi(x, y) = \phi_o - \frac{k}{2f}(x^2 + y^2) , \tag{1}$$

where f is the focal length. This means that the complex wave amplitude is multiplied by

$$\exp\left\{i\left[\phi_o - \frac{k}{2f}(x^2 + y^2)\right]\right\} . \tag{2}$$

Prove this statement by considering the ray-optical picture of rays, normal to the phase front, heading for a focus.

3.2* The z dependence of a diffracting Gaussian beam is twofold.

i. The solution contains the factor $1/(z - ib)$. This multiplier gives a phase advance $\phi = \arctan(z/b)$, and a change in amplitude to compensate for the beam expansion. These parameters are of lesser interest than the next item.

ii. The solution also contains the exponential dependence $\exp[ik(x^2 + y^2)/2(z-ib)]$, which represents the changing beam diameter and phase profile.

The parameter $z - ib$ is the so-called q parameter and contains all the above information: $\mathrm{Re}(1/q) = 1/R$, where R is the radius of the phase front, and $\mathrm{Im}(1/q) = \lambda/\pi w^2$, where w is the beam diameter.

(a) As the beam passes through a set of lenses and free-space intervals, the q parameter transforms very simply. Propagation over a distance d yields $q' = q+d$; passage through a lens of focal length f gives $1/q' = 1/q - 1/f$. Prove this statement.

(b) For a beam $A_o \exp[-(x^2 + y^2)/w^2]$ passing through a lens of focal length f, find the position of the minimum beam diameter and its magnitude.

3.3* Show that, in the paraxial approximation, a mirror of radius R focuses a normally incident beam like a lens of focal length $R/2$.

3.4 A Fabry–Pérot resonator mode between two curved reflecting mirrors of radius R, a distance d apart, their concave sides facing each other, supports a mode with a minimum beam diameter w_o in the symmetry plane. The beam propagation can be broken down into a sequence of focusing lenses of focal length $R/2$.

(a) Evaluate the q-parameter transformation for propagation from the symmetry plane to the mirror, reflection by the mirror, and propagation back to the symmetry plane.

(b) Evaluate the q parameter that repeats itself under this transformation.

(c) Show that beyond a certain critical distance d, there are no Gaussian beam solutions.

3.5 The Gaussian solution for two-dimensional diffraction, such as for a slab beam, is

$$\frac{1}{\sqrt{z - ib}} \exp\left[\frac{ik(x^2 + y^2)}{2(z - ib)}\right] . \tag{1}$$

Two-dimensional diffraction is in one-to-one correspondence with dispersive propagation of a pulse of the form $A_o \exp(-t^2/\tau_o^2)$ along a fiber of dispersion β''. A filter that puts a phase profile $\exp[-i(t^2/2\tau_f^2)]$ onto the pulse affects the dispersive propagation similarly to the way a lens affects diffraction. The q parameter describes dispersive propagation equally well.

Describe how the pulse $A_o \exp(-t^2/\tau_o^2)$ propagates after passage through a filter that puts a phase profile $(t^2/2\tau_f^2)$ onto the pulse.

3.6 The electric field of a Gaussian beam (3.22) has an x component and a z component. The z component can be separated into a part that is in phase with ψ and one that is in quadrature with ψ. The in-phase component is responsible for the curvature of the field lines, which is equal to the curvature of the phase fronts. Prove this statement by evaluating $\mathrm{Re}(E_z/E_x)$ and noting that (see Fig. P3.6.1)

$$\tan\theta \cong \theta \cong \frac{x}{R} = -\mathrm{Re}\left(\frac{E_z}{E_x}\right).\tag{1}$$

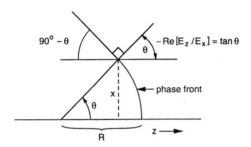

Fig. P3.6.1. The phase front and the definition of θ

3.7* Determine the dispersion parameter $d^2\beta/d\omega^2$ for the model of a dielectric developed in Prob. 1.6. Sketch $\omega_p c(d^2\beta/d\omega^2)$ versus ω/ω_p for $\omega_o/\omega_p = 0.5$.

3.8 Use the perturbation approach to evaluate the change of the propagation constant of the dominant-mode wave above cutoff in a square metallic waveguide of dimensions $a \times a$ caused by a dielectric rod of radius R and dielectric constant ε at the center of the waveguide. Assume $R \ll a$.

3.9 The major and minor axes of a polarization ellipse rotate under the influence of the Kerr effect. Find the ellipticity of the ellipse, $|E|_{\min}/|E|_{\max}$, for which the product of the rate of rotation and the transmission contrast $(T_{\max} - T_{\min})$ is maximized at a given power (assuming that the field is transverse to the direction of propagation).

3.10 This problem is relevant to so-called polarization mode dispersion in fibers. Consider the excitation column matrix

$$\begin{bmatrix} a_x \\ a_y \end{bmatrix} \equiv \boldsymbol{a}$$

containing the excitations of the x and y components of the E field of the mode. The output $\boldsymbol{b}$ is related to the input by a transfer matrix $\boldsymbol{T}$ obeying the losslessness condition.

Show that there are two orthogonal "principal" polarization state pairs $a^{(1)}(\omega) = a^{(1)}(\omega + \Delta\omega)$ and $a^{(2)}(\omega) = a^{(2)}(\omega + \Delta\omega)$ that transform into $b^{(I)}(\omega)$ and $b^{(I)}(\omega + \Delta\omega) = b^{(I)}(\omega) + (db^{(I)}/d\omega)\Delta\omega$, with $db^{(I)}/d\omega = \lambda^{(I)}b^{(I)}$, $I = 1, 2$, where λ is pure imaginary. Use the results of Probs. 2.6 and 2.10. The proof forms the basis of the analysis of pulse propagation in birefringent fibers. The two principal polarization states have distinct group delays. The energy trapped beween the two nodes of the wave packet remains trapped. Thus, the two principal polarization states have definable energy and group velocities.

Solutions

3.2

(a) The fact that $q' = q + d$ when the beam travels over a distance d follows from its definition. Next, consider the inverse of the q parameter $z - ib$. The imaginary part of the inverse of the q parameter gives the beam radius

$$\text{Im}\frac{1}{z - ib} = \frac{b}{z^2 + b^2} = \frac{\lambda}{\pi w^2} \tag{1}$$

and the inverse phase front radius is

$$\text{Re}\frac{1}{z - ib} = \frac{z}{z^2 + b^2} . \tag{2}$$

The lens transforms $1/q$ into $1/q'$, where

$$\frac{1}{q'} = \frac{z}{z^2 + b^2} - \frac{1}{f} + i\frac{b}{z^2 + b^2} = \frac{1}{q} - \frac{1}{f} . \tag{3}$$

(b) The minimum beam diameter is found where the q parameter becomes pure imaginary. The initial value of q is given by

$$\frac{1}{q} = i\frac{\lambda}{\pi w^2} . \tag{4}$$

After passing through the lens, the q parameter is

$$\frac{1}{q'} = i\frac{\lambda}{\pi w^2} - \frac{1}{f} . \tag{5}$$

After passing through a distance d, the new q parameter is

$$q' = \frac{1}{i(\lambda/\pi w^2) - 1/f} + d = -\frac{1/f + i(\lambda/\pi w^2)}{1/f^2 + (\lambda/\pi w^2)^2} + d$$

$$= -\frac{1/f + i(\lambda/\pi w^2) - d[1/f^2 + (\lambda/\pi w^2)^2]}{1/f^2 + (\lambda/\pi w^2)^2} . \tag{6}$$

The position of the minimum beam diameter is where q' is pure imaginary:

$$d = \frac{f}{[1 + (\lambda f/\pi w^2)^2]} \ . \tag{7}$$

The value of the minimum beam diameter is obtained from (6) for the value of d given by (7):

$$q' = -\frac{i(\lambda/\pi w^2)}{1/f^2 + (\lambda/\pi w^2)^2} = -i\frac{\pi w_{min}^2}{\lambda} \tag{8}$$

or

$$\pi w_{min}^2 = \lambda f \frac{(f\lambda/\pi w^2)}{1 + (\lambda f/\pi w^2)^2} \ . \tag{9}$$

3.3 Figure S3.3.1 shows two rays, one along the axis of the mirror, the other parallel to it. If the separation is small compared to R (paraxial approximation), it is easily seen that the two rays intersect at a distance $R/2$ in front of the mirror. This proves the fact that a spherical mirror acts as a lens of focal length $R/2$.

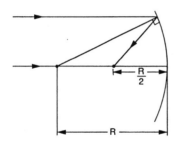

Fig. S3.3.1. Ray construction for focus of spherical mirror

3.7 The dielectric constant of the medium has been obtained in Sect. 1.7

$$\epsilon = \epsilon_o \left[1 + \frac{\omega_p^2}{\omega_o^2 - \omega^2} \right] , \tag{1}$$

where $\omega_p^2 = q^2 N/\epsilon_o m$ is the square of the so-called plasma frequency. The propagation constant is

$$\beta = \omega\sqrt{\mu_o \epsilon} = \omega\sqrt{\mu_o \epsilon_o \left(1 + \frac{\omega_p^2}{\omega_o^2 - \omega^2} \right)} \ . \tag{2}$$

Over part of the frequency range the propagation constant is pure imaginary. In this frequency regime no propagating waves exist. The second derivative is (Fig. S3.7.1)

$$\beta'' = \frac{1}{c}\frac{1}{\sqrt{\epsilon\epsilon_o}}\left[\frac{\partial\epsilon}{\partial\omega} + \frac{1}{2}\omega\frac{\partial^2\epsilon}{\partial\omega^2} - \frac{1}{4\epsilon}\omega\left(\frac{\partial\epsilon}{\partial\omega}\right)^2\right] . \tag{3}$$

The individual derivatives are

$$\frac{\partial\epsilon}{\partial\omega} = \epsilon_o\frac{2\omega\omega_p^2}{(\omega_o^2 - \omega^2)^2} ,$$

$$\frac{\partial^2\epsilon}{\partial\omega^2} = \epsilon_o\left\{\frac{2\omega_p^2}{(\omega_o^2 - \omega^2)^2} + \frac{8\omega^2\omega_p^2}{(\omega_o^2 - \omega^2)^3}\right\} . \tag{4}$$

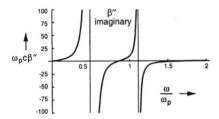

Fig. S3.7.1. A plot of $\omega_p c\beta''$ versus ω/ω_p

4. Shot Noise and Thermal Noise

It is well known that electronic amplifiers introduce noise. The noise can be heard in any radio receiver tuned between stations. Some of the noise comes from the environment, but most of the noise is generated internally in the amplifiers. One source of amplifier noise is the shot noise that accompanies a flow of electric current. Another source is thermal noise, emitted by any resistor at any given temperature. Amplification is a nonequilibrium process, and thus amplification involves noise sources other than thermal sources.

Shot noise was first analyzed by Schottky in 1918 [2]. He was studying the noise associated with the emission of electrons from a cathode in a vacuum tube and set himself the task of deriving a quantitative description of the effect. The name derives from the sound made by a fistful of gunshot dropped on the floor (*der Schrot Effekt*, in German) and not from an abbreviation of the name of its discoverer. In his paper, Schottky was asking the question as to whether there are fundamental limits to the signal-to-noise ratio set by the noise in vacuum tube amplifiers.

It is a fact that shot noise can be reduced by utilizing the mutual repulsion among the negatively charged electrons. An electron emitted from the cathode can inhibit the emission of electrons following it. This process is utilized to reduce the noise emission from cathodes in traveling-wave tubes [47]. On the other hand, if both the amplitude and the phase of an optical wave are to be detected in a heterodyne experiment (Chap. 8), one cannot rely on the repulsion effect if the amplitude changes of the wave are to be faithfully reproduced at frequencies as high as optical frequencies. In this case the full shot noise level has to be accepted. It turns out that shot noise is the fundamental noise process required to satisfy the uncertainty principle applied to a simultaneous measurement of the amplitude and phase of an optical field in heterodyne detection, as discussed in Chap. 8.

The power radiated by a "black body" at thermal equilibrium was derived by Planck. In order to arrive at a formula that agreed with Wien's law, he postulated the quantization of the electromagnetic energy. The classical limit of the Planck formula applied to a single mode of radiation gives the Nyquist formula [48]. The Nyquist noise is present in electronic circuits operating at or near room temperature. Electronic amplifiers are nonequilibrium devices and hence may be affected by other forms of noise in addition to shot noise and thermal noise. For the analysis of the signal-to-noise ratio of such amplifiers it is sufficient to know the mean square fluctuations of the amplitudes of the various noise sources expressed in terms of their spectral densities. This will be discussed in detail in Chap. 5.

The energy fluctuations of a mode at thermal equilibrium are predicted by the Bose–Einstein formula, which is derived at the end of this chapter. The Bose–Einstein fluctuations play an important role in the optical amplification of a digital bit stream (pulses and blanks), as discussed in detail in Chap. 9.

In this chapter we derive the spectrum of shot noise. Next we find the probability distribution of photoelectron emission from a thermionic cathode or the current in a p–n junction. We derive the power spectrum of the thermal noise associated with the waves of a uniform waveguide and the modes of a resonator from the equipartition theorem. We show that loss in a waveguide or a circuit calls for the introduction of noise sources if the circuit is to be at thermal equilibrium, and we derive the spectra of these so-called Langevin sources. We consider lossy multiports and identify the noise sources required for thermal equilibrium.

Finally, we derive the probability distribution of photons at thermal equilibrium, the so-called Bose–Einstein distribution, by maximization of the entropy. This is the energy, or power, distribution of thermal radiation. In the classical limit, the distribution becomes exponential. With a slight modification, the derivation can be used to show that a Gaussian amplitude distribution maximizes the entropy. It is also easily shown that the energy distribution of a Gaussian-distributed amplitude is exponential, the classical limit of the Bose–Einstein distribution.

4.1 The Spectrum of Shot Noise

Schottky assumed that the emission of the electrons was purely random. In deriving the shot noise formula, we shall adhere to the same assumption. We consider a diode consisting of a cathode and anode as shown in Fig. 4.1. The anode is a.c. short-circuited to the cathode. An electron emitted from the cathode induces a current in the short circuit that is a function of time, $h(t)$, extending from the time of emission to the time of collection, a time τ later, where τ is the transit time. The current in the short circuit within a time interval T is

$$i(t) = q \sum_r h(t - t_r) \,, \tag{4.1}$$

where $-q$ is the electron charge, t_r is the time of emission, and the summation is over all emission events within the time interval T.

The function $h(t)$ has area unity, $\int_{-\infty}^{+\infty} h(t)dt = 1$. The shape of the function depends on the velocity of the electron during transit. Figure 4.2a shows

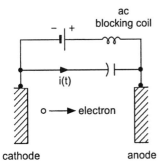

Fig. 4.1. Schematic illustration of diode emitting electrons

how one would evaluate the function. The point charge traveling between the two perfectly conducting plates of the cathode and anode induces image charges in the plates. The distance between the electrodes is assumed to be much smaller than the transverse dimensions of the electrodes. In order to satisfy the boundary conditions of zero tangential electric field on the electrodes, the charge and the image charges have to be repeated periodically along the x direction. The charge and its images are spatial unit impulse functions. These impulse functions can be represented by a Fourier series in the transverse dimensions y and z. The leading term in the Fourier expansion is a uniform surface charge density. All other Fourier components have zero net charge and do not contribute to the net charge. Hence, the net induced charge in the plates can be evaluated from the sheet charge model as shown in Fig. 4.2b. The E field is uniform on either side of the charge sheet, as shown in Fig. 4.2b, with a jump at the sheet:

$$\epsilon_o(E_{x+} - E_{x-})A = -q\,,\tag{4.2}$$

where A is the area of the electrodes (of transverse dimension much larger than their spacing). The fields on the two sides have to give zero net potential difference. Therefore

$$E_{x-}x = -E_{x+}(d-x)\,.\tag{4.3}$$

Solving for E_{x-}, one obtains from these two equations

$$\epsilon_o E_{x-}A = \frac{d-x}{d}q\,.\tag{4.4}$$

On the left-hand side is the net image charge in the cathode. Its time rate of change is given by the derivative and gives the current that passes from the anode to the cathode:

$$i(t) = -q\frac{v}{d}\,,\tag{4.5}$$

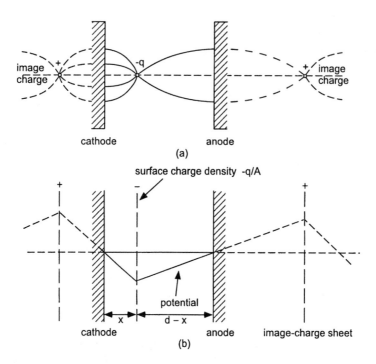

cathode   anode

(a)

surface charge density -q/A

cathode   anode   image-charge sheet

(b)

Fig. 4.2. (a) The charge and image charges in the space between the electrodes (in fact, there is an infinite number of image charges repeated periodically). (b) The set of sheet image charges

where v is the velocity of the electron. The integral over all time of the current is equal to $-q$, irrespective of the time dependence of the electron's velocity. The simplest case is when the velocity is a constant. Then the time dependence of the current is a square-wave function of duration τ, the transit time, and of unity area. The analysis applies equally well to the carrier flow in a p–n junction diode, either electrons or holes.

Next, we evaluate the autocorrelation function of the current induced by charges entering at times t_r. The spectrum of the current is then obtained by a Fourier transform of the autocorrelation function (Appendix A.4). The current is the superposition of the individual current pulses:

$$i(t) = q \sum_r h(t - t_r) , \qquad (4.6)$$

where $h(t)$ is the temporal dependence of the current induced by a charge, and the sum is extended over a long sample of duration T, ideally infinitely long. Figure 4.3 shows samples of filtered shot noise. The autocorrelation function is

$$\langle i(t)i(t-\tau)\rangle = q^2 \left\langle \sum_{r,r'} h(t-t_r)h(t'-t_{r'}) \right\rangle \quad \text{with} \quad t' = t - \tau, \quad (4.7)$$

where the angle brackets indicate a statistical average over an ensemble of sample functions. If the arrival times are random, then one must distinguish between product terms referring to the same event at t_r and different events that occur at different time instants, $r \neq r'$:

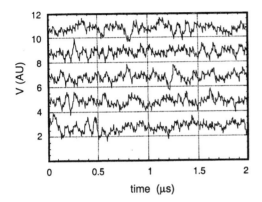

Fig. 4.3. Filtered shot noise as a function of time; filter center frequency 1 MHz, filter bandwidth 50 kHz

$$\langle i(t)i(t-\tau)\rangle = q^2 \left\langle \sum_{r=r'} h(t-t_r)h(t'-t_{r'}) \right\rangle$$

$$+q^2 \left\langle \sum_{r \neq r'} h(t-t_r)h(t'-t_{r'}) \right\rangle. \quad (4.8)$$

We look first at the case in which the probability of events is time-independent, a stationary process. The events occur at times randomly distributed over t_r. Within the infinitesimal time interval dt_r the probability of occurrence is $R\,dt_r$, where R is the average rate of occurrence. We have

$$\left\langle \sum_{r=r'} h(t-t_r)h(t'-t_{r'}) \right\rangle = R \int dt_r\, h(t-t_r)h(t'-t_r-\tau)$$

$$= R \int dt\, h(t)h(t-\tau). \quad (4.9)$$

The summation over different events calls for averaging of each of the factors, since the events are assumed to be statistically independent:

$$\left\langle \sum_{r \neq r'} h(t - t_r) h(t' - t_{r'}) \right\rangle \approx \left\langle \sum_{r} h(t - t_r) \right\rangle \left\langle \sum_{r'} h(t' - t_{r'}) \right\rangle$$

(4.10)

$$= \int R \, dt_r \, h(t - t_r) \int R \, dt_{r'} \, h(t' - t_{r'}) = R^2 .$$

We have used the approximation sign, since for N events in the time interval T, the double sum contains $N(N-1)$ terms, with the terms $r = r'$ omitted. If the samples are very long, as assumed, and $N \to \infty$, the approximation is a good one. Thus, we find for the autocorrelation function

$$\langle i(t) i(t - \tau) \rangle = q^2 \left[R \int dt \, h(t) h(t - \tau) + R^2 \right] .$$

(4.11)

The spectral density is the Fourier transform of the autocorrelation function:

$$\Phi_i(\omega) = \frac{1}{2\pi} \int d\tau \langle i(t) i(t - \tau) \rangle \exp(i\omega\tau) = \frac{q^2}{2\pi} [R|H(\omega)|^2 + 2\pi R^2 \delta(\omega)] ,$$

(4.12)

with

$$|H(\omega)|^2 = \int d\tau \int dt \, h(t) h(t - \tau) \exp(i\omega\tau)$$

$$= \int dt \, h(t) \exp(i\omega t) \int d(t - \tau) \, h(t - \tau) \exp[-i\omega(t - \tau)] \quad (4.13)$$

$$= H(\omega) H^*(\omega) ,$$

where $H(\omega)$ is the Fourier transform of $h(t)$. Note that $H(0) = 1$. The first term is the shot noise spectrum; the second term is the delta function at the origin expressing the deterministic part of the spectrum associated with the d.c. current. If the current pulses are short compared with the inverse bandwidth under consideration, the functions $h(t)$ can be approximated by delta functions and the noise spectrum becomes flat, i.e. "white":

$$\Phi_i(\omega) = \frac{q^2}{2\pi} [R + 2\pi R^2 \delta(\omega)] .$$

(4.14)

If the spectrum is measured by a spectrum analyzer with a filter of bandwidth $\Delta\omega$ centered at a frequency ω_o, both sides of the spectrum, corresponding to positive and negative frequencies, are accepted. The measured mean square current fluctuations are

$$2\Phi_i(\omega_o) \Delta\omega = \frac{q^2}{\pi} R \Delta\omega = 2q I_o B ,$$

(4.15)

with $I_o = qR$, the d.c. current. This is the famous shot noise formula, where the bandwidth B in Hz is $B = \Delta\omega/2\pi$.

If the current consists of a distribution of different response functions, all of unity area, that are independent of the time of the event, then the analysis changes very little. A second average has to be taken over the spectral response, so that $|H(\omega)|^2 \to \langle|H(\omega)|^2\rangle$.

If the rate R of the charge carrier flow is itself a function of time, the analysis can be modified to accommodate this time dependence. Consider the expectation value of the current

$$\langle i(t) \rangle = q \left\langle \sum_r h(t - t_r) \right\rangle = q \int dt_r\, R(t_r) h(t - t_r) . \tag{4.16}$$

Next, construct the autocorrelation function of the current. First, we evaluate the summation over the same events, $t_r = t'_r$:

$$q^2 \left\langle \sum_{r=r'} h(t - t_r) h(t - \tau - t_{r'}) \right\rangle = q^2 \int dt_r\, R(t_r) h(t - t_r) h(t - \tau - t_r) . \tag{4.17}$$

This is a convolution of the function $R(t)$ with the function $h(t)h(t - \tau)$. The summation over independent events at different times gives

$$q^2 \langle \sum_{r \neq r'} h(t - t_r) h(t - \tau - t_{r'}) \rangle$$
$$= q^2 \int dt_r\, R(t_r) h(t - t_r) \int dt_{r'}\, R(t_{r'}) h(t - \tau - t_{r'}) . \tag{4.18}$$

Therefore the correlation function becomes

$$\langle i(t)i(t - \tau) \rangle = q^2 \int dt_r\, R(t_r) h(t - t_r) h(t - \tau - t_r)$$
$$+ q^2 \int dt_r\, R(t_r) h(t - t_r) \int dt_{r'} R(t_{r'}) h(t - \tau - t_{r'}) . \tag{4.19}$$

The autocorrelation function depends not only on the time difference τ, but also on the time t, since the emission rate is time-dependent.

If the rate $R(t)$ is deterministic, then the second term in (4.19) can be recognized as the product of $\langle i(t) \rangle$ and $\langle i(t - \tau) \rangle$. The fluctuations of the current are obtained by subtraction of $\langle i(t) \rangle \langle i(t - \tau) \rangle$ from $\langle i(t)i(t - \tau) \rangle$:

$$\langle i(t)i(t - \tau) \rangle - \langle i(t) \rangle \langle i(t - \tau) \rangle = q^2 \int dt_r\, R(t_r) h(t - t_r) h(t - t_r - \tau) . \tag{4.20}$$

In the case where the emission rate itself is a stationary statistical function, an additional average over the ensemble of $R(t)$ renders the process stationary and makes the autocorrelation function time-independent [49]:

$$\langle i(t)i(t-\tau)\rangle = q^2\langle R(t)\rangle \int dt_r\, h(t-t_r)h(t-\tau-t_r)$$

$$+q^2\int dt_r \int dt_{r'}\, \langle R(t_r)R(t_{r'})\rangle h(t-t_r)h(t-\tau-t_{r'})\ . \tag{4.21}$$

In preparation for the evaluation of the spectrum of (4.21), we transform the second term by noting that $\langle R(t_r)R(t_{r'})\rangle$ is a function of the time difference $t_r - t_{r'} \equiv \tau'$ only, if the signal statistics are stationary. One then has

$$q^2\int dt_r \int dt_{r'}\, \langle R(t_r)R(t_{r'})\rangle h(t-t_r)h(t-\tau-t_{r'})$$

$$= q^2\int d\tau'\, \langle R(t_r)R(t_r-\tau')\rangle \int dt\, h(t)h(t-\tau+\tau')\ . \tag{4.22}$$

This term is the convolution of the autocorrelation functions of the rate function $R(t)$ and of the detector response $h(t)$. Its Fourier transform is the product of the Fourier transforms $\Phi_R(\omega)$ and $|H(\omega)|^2$. We obtain for the spectrum of the current, the Fourier transform of (4.21),

$$\Phi_i(\omega) = \frac{q^2}{2\pi}[\langle R(t)\rangle + 2\pi\Phi_R(\omega)]|H(\omega)|^2\ . \tag{4.23}$$

The first term is the shot noise contribution to the spectrum; the second term is the contribution of the signal. It is remarkable that the shot noise part of the spectrum still has the form for a process with a constant rate R, except that this rate is replaced by its average.

4.2 The Probability Distribution of Shot Noise Events

In the preceding section, we derived the spectrum of shot noise. This spectrum would be measured by a spectrum analyzer responding to the current fluctuations of the diode. There are other ways of interpreting the statistical process of the current, or charge, fluctuations. One may ask for the probability $p(n, \tau)$ that n charge carriers have been emitted from one of the electrodes if the rate of emission is R. This is obtained by deriving appropriate differential equations for the probabilities $p(m, \tau)$ for $m \leq n$ [50, 51]. Consider, first, a very short time interval $\Delta\tau$, in the limit $\Delta\tau \to 0$, and ask for the probability of emitting one electron in this time interval. This probability is

$$p(1, \Delta\tau) = R\Delta\tau\ . \tag{4.24}$$

The probability of emitting more than one electron is negligible, and thus the sum of the probabilities of emitting no electron, $P(0, \Delta\tau)$, and of emitting one electron, $P(1, \Delta\tau)$, must be equal to one:

$$p(0, \Delta\tau) + p(1, \Delta\tau) = 1\ . \tag{4.25}$$

Next, let us find the probability $p(0, \tau + \Delta\tau)$ of no emission in a total time interval $\tau + \Delta\tau$. Since the events in adjacent time slots are assumed to be independent, the probability is the product of the probabilities of no emission within τ and no emission within $\Delta\tau$:

$$p(0, \tau + \Delta\tau) = p(0, \tau)p(0, \Delta\tau) \, . \tag{4.26}$$

Substituting for $p(0, \Delta\tau)$ from (4.24) and (4.25), one finds

$$\frac{p(0, \tau + \Delta\tau) - p(0, \tau)}{\Delta\tau} = -Rp(0, \tau) \, . \tag{4.27}$$

In the limit $\Delta\tau \to 0$, this reduces to a differential equation, which can be solved to give

$$p(0, \tau) = \exp(-R\tau) \, , \tag{4.28}$$

where the following boundary condition has been used:

$$p(0, 0) = 1 \, . \tag{4.29}$$

Next, consider the probability that n electrons have been emitted in a time interval $\tau + \Delta\tau$. This is clearly

$$p(n, \tau + \Delta\tau) = p(n - 1, \tau)p(1, \Delta\tau) + p(n, \tau)p(0, \Delta\tau) \, . \tag{4.30}$$

Upon substituting from (4.24) and (4.25), we find in the limit $\Delta\tau \to 0$

$$\frac{dp(n, \tau)}{d\tau} + Rp(n, \tau) = Rp(n - 1, \tau) \, . \tag{4.31}$$

The solution of this equation gives a recursion formula

$$p(n, \tau) = \exp(-R\tau)R \int_0^\tau d\tau \exp(R\tau)p(n - 1, \tau) \, . \tag{4.32}$$

Evaluating $p(1, \tau)$ from the above, using the expression for $p(0, \tau)$, and continuing the process, we end up with

$$p(n, \tau) = \frac{(R\tau)^n}{n!} \exp(-R\tau) \, . \tag{4.33}$$

This is the Poisson probability distribution for a process with the average number $\langle n \rangle = R\tau$:

$$p_{\text{Poisson}}(n) = \frac{\langle n \rangle^n}{n!} e^{-\langle n \rangle} \, . \tag{4.34}$$

We shall encounter this distribution in the quantum analysis of coherent radiation. Figure 4.4 shows the Poisson distribution for different average numbers $\langle n \rangle$. One sees that the distribution becomes more and more symmetric around the average value $\langle n \rangle$ with increasing $\langle n \rangle$.

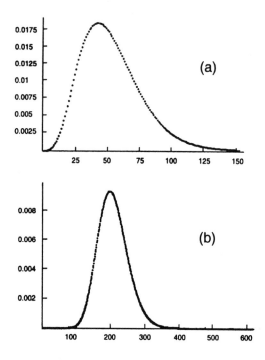

Fig. 4.4. Examples of Poisson distributions: (a) $\langle n \rangle = 50$; (b) $\langle n \rangle = 200$

4.3 Thermal Noise in Waveguides and Transmission Lines

In this section we arrive at the formula for the thermal noise in a bandwidth B in a system supporting single forward- and backward-propagating modes. The TEM mode of a transmission line is a good example, and so is an optical mode of one polarization in a single-mode fiber. The fundamental Gaussian beam of one polarization is another example. The derivation is the one-dimensional analog of the black-body radiation law that applies to radiation in a large, three-dimensional enclosure.

The derivation of the mean square fluctuations of thermal noise is based on the equipartition theorem [52]: every degree of freedom must have, on average, an energy of $\frac{1}{2}k\theta$ at the absolute temperature θ, where k is the Boltzmann constant. The simple interpretation of the equipartition theorem is that, at thermal equilibrium, all degrees of freedom have the same probability of excitation. We refer the reader to the literature [52] for the derivation of the equipartition theorem. Here we present a simple plausibility argument as to the validity of the theorem. A system containing N point particles has $3N$ degrees of freedom. If the particles are of finite size and have finite angular momenta, then the system has $6N$ degrees of freedom. If such a system is coupled to another one and the two systems are at thermal equilibrium, then both systems acquire the same temperature. A given temperature corresponds to an average energy of each of the component particles. This statement holds for any two macroscopic systems for which averages can be taken over all the particles. A degree of freedom is a microscopic concept that does not permit an average over all particles. However, it permits a time average. The energy associated with the degree of freedom can be averaged over arbitrarily long time intervals. At thermal equilibrium, this average must yield a value of energy that is consistent with the average energy of each of the degrees of freedom of each particle.

An electromagnetic mode of a resonator obeys a simple one-dimensional oscillator equation and thus has the same number of degrees of freedom as a one-dimensional oscillator, i.e. two. The equipartition theorem assigns an energy $k\theta$ to the mode in the low-frequency limit, and an energy $\hbar\omega/[\exp(\hbar\omega/k\theta) - 1]$ in the quantum limit. Since the thermal noise is caused by coupling to a thermal reservoir of many degrees of freedom, the central limit theorem [52] implies that the field amplitudes must have a Gaussian distribution. A Gaussian distribution is fully characterized by its mean square value, and thus the distribution is known when its mean square value is specified. At the end of this chapter we shall show that the classical electromagnetic field of a mode has a Gaussian distribution without appealing to the central limit theorem.

Consider a mode of amplitude A_n with propagation constant β_n of a single-mode waveguide (in a multimode waveguide the following analysis applies to each of the modes). The propagation constant is a function of frequency $\omega_n = \omega(\beta_n)$, and not necessarily a linear function of β_n if the waveguide is dispersive. The amplitudes A_n of the modes are so normalized that $|A_n|^2$ are the energies in the modes. We consider a ring waveguide closing on itself, of very long length L. The nth mode obeys the periodicity condition

$$\beta_n L = 2\pi n . \tag{4.35}$$

Each mode has two degrees of freedom, the electric field and the magnetic field. By the equipartition theorem, the statistical average of the square of the amplitude, which is equal to the expectation value of the energy, is the energy assigned to two degrees of freedom:

$$\langle |A_n|^2 \rangle = k\theta . \tag{4.36}$$

Stationarity of the process requires that the amplitudes of any two different modes are uncorrelated. Indeed, two modes of different β values β_n and β_m have different frequencies ω_n and ω_m and thus different time dependences. The statistical average of the energy would vary as $\cos[(\omega_n - \omega_m)t + \phi]$ unless

$$\langle A_n A_m^* \rangle = 0 , \tag{4.37}$$

and thus different modes of a stationary process must be uncorrelated. Equations (4.36) and (4.37) give full information on the thermal excitations of the modes of a ring resonator. The ring configuration was an artifice to relate the thermal excitations on a transmission line or waveguide to the excitations of a set of resonances. An open waveguide or transmission line also supports thermal excitations. However, in order to describe these excitations it is convenient to refer them not to a structure of length L, but rather to excitation amplitudes whose mean square expectation values are equal to the thermal energy per unit length propagating in the two directions along the guide or transmission line. We now proceed with the derivation of these mode amplitudes. This is done by noting that the energy of a mode of length L is converted into the energy per unit length by dividing it by L:

$$\left\langle \frac{A_n^* A_n}{L} \right\rangle = \text{energy per unit length in one mode} . \tag{4.38}$$

An increment of the propagation constant $\Delta\beta$ corresponds to a set of modes Δn, according to (4.35):

$$\Delta\beta L = 2\pi \Delta n . \tag{4.39}$$

The energy per unit length in the waveguide is given by the sum over all modes, an expression that can also be written as a double sum, using condition (4.37):

$$\text{energy per unit length } = \sum_n \left\langle \frac{A_n^* A_n}{L} \right\rangle = \sum_{n,m} \left\langle \frac{A_n^* A_m}{L} \right\rangle . \tag{4.40}$$

The double sum can be converted into a double integral of a differently defined mode amplitude. Note that the increment of integration $\Delta\beta = 2\pi/L$. The energy per unit length can be written

$$\sum_{n,m} \left\langle \frac{A_n^* A_m}{L} \right\rangle = \sum_{n,m} \frac{1}{L} \left(\frac{\Delta\beta L}{2\pi} \right)^2 \langle A_n^* A_m \rangle = \int d\beta \int d\beta' \langle a^*(\beta) a(\beta') \rangle , \tag{4.41}$$

with

$$a(\beta) \equiv \frac{\sqrt{L}}{2\pi} A_n . \tag{4.42}$$

The correlation conditions (4.36) and (4.37) can be summarized in the single equation

$$\langle a^*(\beta') a(\beta) \rangle = \frac{1}{2\pi} k\theta\delta(\beta - \beta') , \tag{4.43}$$

where $\delta(\beta - \beta')$ is a delta function of unity area and height $1/\Delta\beta = L/2\pi$. The power passing a filter of bandwidth $\Delta\Omega = (d\omega/d\beta)\Delta\beta$ is

$$\frac{d\omega}{d\beta} \times \text{ energy per unit length in bandwidth } \Delta\Omega$$

$$= \frac{d\omega}{d\beta} \int d\beta \int d\beta' \langle a^*(\beta')a(\beta)\rangle$$

$$= \frac{1}{2\pi}\frac{d\omega}{d\beta}k\theta \int d\beta \int d\beta'\delta(\beta - \beta') \qquad (4.44)$$

$$= \frac{1}{2\pi}\frac{d\omega}{d\beta}k\theta\Delta\beta = \frac{1}{2\pi}k\theta\Delta\Omega .$$

Note that we have considered modes labeled by their characteristic frequency ω, which was taken as positive. Thus, the spectrum (4.44) is specified only for positive frequencies. If both positive and negative frequencies are used, then (4.44) has to be reduced by a factor of $1/2$. The power within the frequency interval $\Delta\Omega$ is

$$\text{power in frequency interval } \Delta\Omega = \frac{1}{2\pi}k\theta\,\Delta\Omega . \qquad (4.45)$$

Equation (4.45) is the Nyquist formula [48] for the thermal power propagating in each mode in either of two directions within a bandwidth $B = \Delta\Omega/2\pi$. The spectral density of the thermal power is independent of frequency and thus the thermal power is infinite if extended over all frequencies. This is the ultraviolet catastrophe in a one-dimensional system. An analysis of modes in three dimensions would have led to the Rayleigh–Jeans law, with its even more pronounced ultraviolet catastrophe. In his effort to connect the Rayleigh–Jeans law to the experimentally observed Wien's law, Planck introduced the quantization of energy. We shall derive this generalized form of the Nyquist formula in Sect. 4.8 of this chapter.

It will be convenient to define mode amplitudes as a function of frequency rather than of propagation constant. We shall denote these by $a(\omega)$ and relate them to $a(\beta)$ by requiring that the statistical average of their square give the power flow

$$a(\omega) = \sqrt{\frac{d\beta}{d\omega}}a(\beta) . \qquad (4.46)$$

The power in a mode is given by the double integral over ω of $\langle a(\omega)a^*(\omega')\rangle$, where

$$\langle a(\omega)a^*(\omega')\rangle = \frac{1}{2\pi}k\theta\delta(\omega - \omega') . \qquad (4.47)$$

The different normalizations of the mode amplitudes are summarized in Appendix A.5.

4.4 The Noise of a Lossless Resonator

Thus far we have considered a uniform waveguide supporting a single mode propagating in both directions along the guide. The spectral density of the thermal noise power associated with the waves in both directions was white, according to classical theory. Reflections along the waveguide alter the spectral distribution. Reflectors placed at two cross sections of the waveguide form a resonator, open if one or both reflectors are only partially reflecting. The redistribution of the thermal noise spectrum in such a resonator is illustrated by analyzing the system of Fig. 4.5, a Fabry–Pérot resonator supporting a transverse mode and coupled to an incoming wave through a partially transmitting mirror. There are forward and backward waves in the resonator. We solve the problem in the limit of weak coupling to the waveguide, the case where perturbation theory is valid. In this case, the description of the resonator is particularly simple (see Chap. 2, Sect. 2.12). We describe the mode amplitude in the resonator by $U(t)$. The amplitude is so normalized that $|U(t)|^2$ represents the energy in the resonator. The natural time dependence of the mode in the closed resonator is that for when the partially transmitting mirror is made perfectly reflecting:

$$U(t) = U_o \exp(-i\omega_o t) , \tag{4.48}$$

where ω_o is the resonance frequency.

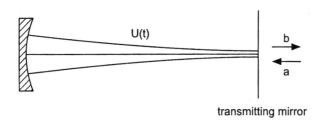

transmitting mirror

Fig. 4.5. A resonator with a single input port

An isolated resonance of a resonator is described by a second-order differential equation in time. Such a differential equation leads to two poles in the complex ω plane. If the resonator is lossless and uncoupled to the outside, the poles lie on the real axis at $\pm\omega_o$. A convenient equivalent circuit for the resonance is a parallel L–C circuit with $\omega_o = 1/\sqrt{LC}$. Coupling to the outside world moves the poles off the real axis, contributing imaginary parts to the location of the poles, indicating decay; $\pm\omega_o \rightarrow \pm\omega_o - i/\tau_e$. If the displacement is small, the Q of the resonance is high, and it is possible to ignore the coupling of positive frequencies associated with the pole at $+\omega_o - i\tau_e$ to negative-frequency excitations associated with the pole at

$-\omega_o - i\tau_e$. The equations of a resonator coupled to an input port reduce to first-order differential equations. We denote the decay rate of the amplitude due to the coupling of the resonator to the waveguide by $1/\tau_e$. The equation for the mode amplitude U is the first order differential equation

$$\frac{d}{dt}U = -i\omega_o U - \frac{1}{\tau_e}U + \sqrt{\frac{2}{\tau_e}}a \, , \tag{4.49}$$

where a is the wave incident upon the resonator from the input port. The incident and reflected waves in the port are related by

$$b = -a + \sqrt{\frac{2}{\tau_e}}U \, . \tag{4.50}$$

The steady state response of the resonator to an excitation at frequency ω is

$$U(\omega) = \frac{\sqrt{2/\tau_e}a(\omega)}{i(\omega_o - \omega) + 1/\tau_e} \, . \tag{4.51}$$

The energy in the resonator is then

$$\begin{aligned}
&\int d\omega \int d\omega' \langle U(\omega)U^*(\omega')\rangle \\
&= \int d\omega \int d\omega' \frac{(2/\tau_e)\langle a(\omega)a^*(\omega')\rangle}{[i(\omega_o - \omega) + 1/\tau_e][-i(\omega_o - \omega') + 1/\tau_e]} \, .
\end{aligned} \tag{4.52}$$

If we assume thermal equilibrium, then the incident wave must obey Nyquist's theorem. When one uses the expression for the cross-spectral density of the wave in the waveguide (4.47), one finds

$$\int d\omega \int d\omega' \langle U(\omega)U^*(\omega')\rangle = k\theta \, . \tag{4.53}$$

The energy storage *integrated over all frequencies* obeys the equipartition theorem. This is a generalization of the equipartition theorem which is, strictly, a statement about the energy of a resonator mode not "connected to the outside world."

It is interesting to ask what is the energy possessed by the waves within a resonator formed from a uniform waveguide with reflecting mirrors. In the absence of the mirrors, the waves would have a power spectral density (4.47) in both directions, independent of frequency. The energy spectrum in the resonator, $|U(\omega)|^2$, is made up of the energy spectra of the two waves traveling in opposite directions. The power in each of the waves is ((4.52) and (4.47))

$$\frac{(2/\tau_e)(v_g/2L)k\theta\Delta\omega/2\pi}{(\omega_o - \omega)^2 + 1/\tau_e^2} \, , \tag{4.54}$$

where L is the length of the resonator. At the resonance frequency $\omega = \omega_o$, the power within the frequency increment $\Delta\omega$ is

$$\frac{\tau_e v_g}{L} k\theta \Delta\omega/2\pi \ . \tag{4.55}$$

Now recall the meaning of $1/\tau_e$. It is the rate of decay of the amplitude of the resonant mode due to coupling to the outside waveguide. If there were no coupling mirror, there would be no resonant mode, and it is clear that a forward and backward wave occupying the segment of waveguide of length L would leave within a time $2L/v_g$. Hence, the multiplier in (4.55), $\tau_e v_g/L$, is much greater than unity. Thus, reflecting mirrors can greatly enhance the thermal power in the propagating waves in the forward and backward waves in a Fabry–Pérot-type resonator. When integrated over the resonance, they give an energy storage of $k\theta$ as dictated by the equipartition principle.

The reader may have noticed that the analysis of a resonator as described by (4.49) is not limited to an electromagnetic resonator. The same formalism can be applied to an acoustic resonator. The enhancement of the thermal radiation near the peak of the resonance is precisely the effect mentioned in the Preface, namely the "hearing of the ocean" when a large, hollow shell is held near one's ear.

We have found that the energy spectrum of the resonator excitation occupies a narrow frequency band. The integral of the spectrum gives $k\theta$. It is of interest to determine the spectrum of the wave reflected from the resonator. We have

$$b = -a + \sqrt{\frac{2}{\tau_e}} U \ . \tag{4.56}$$

Thus we find for the Fourier component

$$b(\omega) = -a(\omega) + \frac{2}{\tau_e} \frac{a(\omega)}{\mathrm{i}(\omega_o - \omega) + 1/\tau_e} = \frac{[-\mathrm{i}(\omega_o - \omega) + 1/\tau_e]a(\omega)}{\mathrm{i}(\omega_o - \omega) + 1/\tau_e} \ , \tag{4.57}$$

and the spectrum of $b(\omega)$ is

$$\langle b^*(\omega)b(\omega')\rangle = \langle a^*(\omega)a(\omega')\rangle = \frac{1}{2\pi}k\theta\delta(\omega - \omega') \ . \tag{4.58}$$

The spectrum is the same as that of the incident wave. This is indeed necessary, since the reflected wave travels along an open transmission line or waveguide, or is a freely propagating beam. As such, it has to have the thermal properties of a freely propagating wave. The thermal nature of the reflected wave is maintained through two processes. (i) The resonator radiates power within the frequency band of the resonance. This radiation is supplemented by (ii) the reflected radiation. Outside the band of the resonance, the b wave is solely due to reflection of the a wave.

4.5 The Noise of a Lossy Resonator

The analysis in the preceding section dealt with a resonator coupled to a connecting waveguide. The resonator itself was lossless. It is easy to include loss in the analysis, using the more general equation

$$\frac{dU}{dt} = -\left(i\omega_o + \frac{1}{\tau_o} + \frac{1}{\tau_e}\right)U + \sqrt{\frac{2}{\tau_o}}n_o(t) + \sqrt{\frac{2}{\tau_e}}a(t) . \tag{4.59}$$

The new decay rate $1/\tau_o$ calls for a noise source that compensates for the decay of the thermal radiation. It is easy to determine the spectrum of the source by analogy with the spectrum of the incident wave. Indeed, opening the resonator to the outside world introduced a decay rate $1/\tau_e$. The thermal excitation did not decay, since it was regenerated by the incident wave with the spectrum

$$\langle a^*(\omega)a(\omega')\rangle = \frac{1}{2\pi}k\theta\delta(\omega - \omega') . \tag{4.60}$$

Hence, by inspection, one sees that the noise source n_o must have the spectrum

$$\langle n_o^*(\omega)n_o(\omega')\rangle = \frac{1}{2\pi}k\theta\delta(\omega - \omega') , \tag{4.61}$$

so as to compensate for the new decay rate. The physical origin of the noise source is self-evident. Loss is due to the coupling of the radiation to the excitation of the charged particles in the lossy medium. These charged particles in turn are thermally excited. Their thermal excitation is represented by the noise source. Note that the spectra (4.60) and (4.61) are delta-function-correlated. This is the consequence of the stationary character of the thermal noise. Indeed, if components of different frequencies were correlated, the radiation would become time dependent, which is not permitted in a stationary process.

The Fourier transform of (4.61) gives the correlation function

$$\langle n_o^*(t)n_o(t')\rangle = k\theta\delta(t - t') . \tag{4.62}$$

The noise sources are delta-function-correlated in time as well, since the spectrum is frequency-independent (white).

It is of interest to derive the noise source correlation function directly from the conservation of the thermal excitation in the resonator. For this purpose one looks at the "stripped" model of the resonator, with no output, $1/\tau_e \to 0$. The equation is then

$$\frac{dU}{dt} = -\left(i\omega_o + \frac{1}{\tau_o}\right)U + \sqrt{\frac{2}{\tau_o}}n_o(t) . \tag{4.63}$$

The fluctuations at time t obey the differential equation

$$\frac{d\langle U^*U\rangle}{dt} = -\left(\frac{2}{\tau_o}\right)\langle U^*U\rangle + \sqrt{\frac{2}{\tau_o}}\langle U^*n_o + n_o^*U\rangle .$$

(4.64)

The first term on the right hand side gives the decay of the thermal radiation that must be compensated by the second term. One may suppose, at first sight, that the noise source is uncorrelated with U, and hence the second term should vanish. However, the delta function character of the correlation function means that the noise source "kicks" are very large. Within Δt, the excitation U acquires the average value $(1/2)\sqrt{2/\tau_o}\Delta t\, n_o$. Thus, the term in the brackets is

$$\sqrt{\frac{2}{\tau_o}}\langle U^*n_o + n_o^*U\rangle = \frac{2}{\tau_o}\langle n_o^*n_o\rangle\Delta t .$$

(4.65)

This contribution must cancel the decay, and thus

$$\frac{2}{\tau_o}\langle n_o^*n_o\rangle\Delta t = \frac{2}{\tau_o}k\theta .$$

(4.66)

Identifying the inverse of the short time interval Δt as the magnitude of the delta function divided by 2π, we derive (4.62). This is an independent derivation of the noise source in a way analogous to the approach used in the next section, which determines the noise sources for a distributed attenuator.

The question may be raised as to the spectrum of the noise source if the loss of the medium is itself frequency-dependent. This problem can be approached by a set of thought experiments. One may consider a large resonator, with many resonance frequencies, filled with the lossy medium. The decay rates will now be functions of frequency. For each of the resonator frequencies the noise source can be determined. If $2/\tau_o = 2/\tau_o(\omega)$ is a (slow) function of frequency, then the spectrum of the noise source $\sqrt{2/\tau_o}\,n_o$ will have the same frequency dependence. In the time domain, the correlation function will cease to be a delta function. However, the analysis of the resonator as outlined above does not change, since the delta function concept is a relative one. As long as the spectrum of the noise source can be considered white over the bandwidth of the resonator, the analysis can treat the associated correlation function as a delta function.

4.6 Langevin Sources in a Waveguide with Loss

We have derived the thermal noise power traveling in either direction in a uniform waveguide, i.e. the Nyquist formula (4.45). We are now ready to treat single-mode waveguides with loss at thermal equilibrium. If a lossy semi-infinite waveguide did not contain noise sources, then the thermal power

incident upon it from one side would be attenuated as it propagated along
the waveguide, leading to smaller and smaller fluctuations further and further
away from the input. But at thermal equilibrium the fluctuations of the modes
in the waveguide must maintain their equilibrium value. This is accomplished
by introducing noise sources into the linear equation for the wave propagation.
The derivation of these sources, called Langevin sources after the scientist
who first introduced them, is as follows. In a lossless waveguide, the mode
amplitude $a(\beta)$ obeys the differential equation

$$\frac{d}{dz}a(\beta) = i\beta a(\beta) . \tag{4.67}$$

If loss is present, the equation changes into

$$\frac{d}{dz}a(\beta) = i\beta a(\beta) - \alpha a(\beta) + s(\beta, z) , \tag{4.68}$$

where $-\alpha a(\beta)$ represents the loss per unit length and $s(z)$ is the source
required to maintain thermal equilibrium. Its expectation value follows from
the requirement that the noise spectrum be conserved at thermal equilibrium:

$$\frac{d}{dz}[\langle a(\beta)a^*(\beta')\rangle] = -2\alpha\langle a(\beta)a^*(\beta')\rangle + \langle s(\beta, z)a^*(\beta') + a(\beta)s^*(\beta', z)\rangle$$

$$= 0 . \tag{4.69}$$

The noise sources at different cross sections of the waveguide are uncorrelated, because each segment of the lossy guide is connected to its own
reservoir of charges. Now, one might think that the local noise source and
the mode amplitude traveling through it were uncorrelated as well, because
the noise is due to the reservoir responsible for the loss, and the amplitude
impinging upon it has come from statistically independent sources. However,
there is a contribution to $a(\beta)$ from the noise source $s(z)$ that grows from 0
to $\Delta z\, s(z)$ within the distance Δz. The average value is half the end value.
Thus, we have from (4.69)

$$-2\alpha\langle a(\beta)a^*(\beta')\rangle + \frac{1}{2}\Delta z\langle s(\beta, z)s^*(\beta', z) + s(\beta, z)s^*(\beta', z)\rangle = 0 . \tag{4.70}$$

Using (4.43), we conclude that the noise source term must be equal to

$$\langle s(\beta, z)s^*(\beta', z')\rangle = \frac{1}{2\pi}2\alpha k\theta\delta(z - z')\delta(\beta - \beta') . \tag{4.71}$$

The spatial delta function has amplitude $1/\Delta z$ within the increment of
distance Δz and is zero elsewhere. It expresses the fact that the noise sources
at two different points are uncorrelated. Because of the Gaussian character

of the noise processes, full information on the probability distribution of the noise source amplitude is contained in these equations.

If we replace the mode amplitude $a(\beta)$ with the mode amplitude introduced in (4.46), whose square is related to power flow, (4.68) remains unchanged in form:

$$\frac{d}{dz}a(\omega) = i\beta(\omega)a(\omega) - \alpha a(\omega) + s(\omega, z) \ . \tag{4.72}$$

The noise source correlation function becomes

$$\langle s(\omega, z)s^*(\omega', z')\rangle = \frac{1}{2\pi}2\alpha k\theta\delta(\omega - \omega')\delta(z - z') \ . \tag{4.73}$$

4.7 Lossy Linear Multiports at Thermal Equilibrium

In the preceding sections, we have treated the thermal noise in a lossless waveguide or transmission line, in a lossless resonator, in a resonator with loss, and in a lossy waveguide. Loss calls for the introduction of noise sources to maintain the thermal excitation. Such noise sources must be associated with any circuit that possesses loss. The simplest such circuit is a resistor. Figure 4.6 shows a resistor with an associated noise voltage generator E_s. The spectrum of this generator can be evaluated by a thought experiment in which the resistor R_s terminates a lossless transmission line of characteristic impedance $Z_o = R_s$. The power delivered by the resistor is given by

$$\text{power delivered by resistor} = \frac{\langle|E_s|^2\rangle}{4R_s} \ . \tag{4.74}$$

If the transmission line is to remain in thermal equilibrium, the power spectrum delivered by it to the resistor must be equal to the power spectrum delivered by the resistor and its source. Equating this power to the power absorbed by the material resistor from the wave impinging upon it (compare (4.47)), we have

$$\frac{\langle|E_s|^2\rangle}{4R_s} = \int d\omega \int d\omega' \langle a^*(\omega)a(\omega')\rangle = \frac{k\theta}{2\pi}\Delta\omega \ . \tag{4.75}$$

Fig. 4.6. Resistor terminating transmission line of characteristic impedance Z_o

If we use $B = \Delta\omega/2\pi$ for the bandwidth instead of a radian frequency interval, we have

$$\langle|E_s|^2\rangle = 4R_s k\theta B . \tag{4.76}$$

This is an alternative form of the Nyquist formula. It is left as an exercise for the reader to show that the noise source to be associated with a frequency-dependent impedance $Z_s(\omega)$ is

$$\langle|E_s|^2\rangle = 4\,\text{Re}[Z_s(\omega)]k\theta B . \tag{4.77}$$

When the series connection of the resistor R_s with its thermal noise source is replaced by its Norton equivalent, a conductance $G_s = 1/R_s$ in parallel with a noise current generator $I_s = E_s/R_s$, the mean square fluctuations of the current noise source are

$$\langle|I_s|^2\rangle = 4G_s k\theta B . \tag{4.76a}$$

An alternative derivation results if one defines the termination not as a resistor but as a reflector terminating the transmission line. Then the description of the termination is in terms of the wave formalism:

$$b = \Gamma a + s , \tag{4.78}$$

where a is the incident wave, b is the reflected wave, Γ is the reflection coefficient, and s is a noise wave source. A forward-wave noise source mounted at a point z' on a transmission line produces a traveling wave in the $+z$ direction for all $z > z'$ and no wave in the opposite direction, $z < z'$. A combination of a voltage source and a current source as shown in Fig. 4.7 can accomplish this. In the absence of the noise source, i.e. $s = 0$, the termination absorbs power within the bandwidth $B = \Delta\omega/2\pi$ due to the incident waves $a(\omega)$ that propagate on the transmission line at thermal equilibrium. This power is equal to

$$\left\langle \int d\omega \int d\omega'[a(\omega)a^*(\omega') - b(\omega)b^*(\omega')]\right\rangle$$

$$= (1 - |\Gamma|^2)\int d\omega \int d\omega'\langle a(\omega)a^*(\omega')\rangle \tag{4.79}$$

$$= (1 - |\Gamma|^2)k\theta B .$$

If thermal equilibrium is to be maintained, the internal noise source of the termination must reradiate the same power:

$$\int d\omega \int d\omega'\langle s(\omega)s^*(\omega')\rangle = (1 - |\Gamma|^2)k\theta B . \tag{4.80}$$

If the termination is matched to the line, $\Gamma = 0$ and the power radiated is $k\theta B$. If the source is reflecting, the reradiated power is less. Equation (4.80)

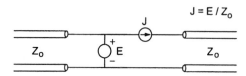

$$J = E / Z_0$$

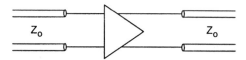

Symbol for Wave Generator:

Fig. 4.7. Wave noise source

is an alternative expression for the noise associated with a termination. Of course, this expression must be consistent with (4.76). In Appendix A.6 we show that this is indeed the case.

The double integrals become cumbersome after a while. For this reason it is customary to subsume the delta function correlation of all frequency components and use simpler symbols for the Fourier components of the excitations. Henceforth we shall make the replacement (see Appendix A.5)

$$\left\langle \int_{-\Delta\omega/2}^{\Delta\omega/2} d\omega \int_{-\Delta\omega/2}^{\Delta\omega/2} d\omega' a(\omega) a^*(\omega') \right\rangle \rightarrow \langle |a|^2 \rangle , \tag{4.81}$$

and analogously for all other excitation amplitudes. Note the change of units from $a(\omega)$ to a. One may consider a to be the amplitude of the forward wave within a narrow frequency band $\Delta\omega = 2\pi B$. Its mean square is equal to the power within the frequency increment $\Delta\omega$.

The wave formalism is easy to generalize to a multiport (Fig. 4.8). The multiport is characterized by its scattering matrix $\boldsymbol{S}$, the column matrix of the incident waves $\boldsymbol{a}$, and the column matrix of reflected waves $\boldsymbol{b}$. In analogy with (4.78),

$$\boldsymbol{b} = \boldsymbol{S}\boldsymbol{a} + \boldsymbol{s} , \tag{4.82}$$

where $\boldsymbol{s}$ is the column matrix of wave noise-source amplitudes. The correlation matrix $\langle \boldsymbol{s}\boldsymbol{s}^\dagger \rangle$, where the dagger superscript indicates a Hermitian transpose, can be evaluated by requiring that the expectation values of the products $\langle b_i b_j^* \rangle$ of the outgoing waves have the proper values corresponding to thermal equilibrium. From (4.82) and the fact that the noise sources are uncorrelated with the incident waves a_i, we have

$$\langle \boldsymbol{b}\boldsymbol{b}^\dagger \rangle = \boldsymbol{S} \langle \boldsymbol{a}\boldsymbol{a}^\dagger \rangle \boldsymbol{S}^\dagger + \langle \boldsymbol{s}\boldsymbol{s}^\dagger \rangle . \tag{4.83}$$

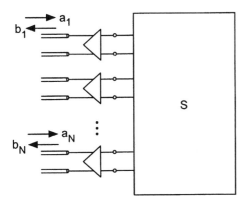

Fig. 4.8. Schematic of multiport with noise sources

Now, the incident waves in the different input ports are uncorrelated:

$$\langle aa^\dagger \rangle = k\theta B \mathbf{1} , \tag{4.84}$$

where $\mathbf{1}$ is the identity matrix. The outgoing waves have the same correlation matrix as the incoming waves. Using this fact, we obtain from (4.82)

$$\langle ss^\dagger \rangle = (\mathbf{1} - SS^\dagger)k\theta B . \tag{4.85}$$

This is the generalization of the Nyquist theorem to a multiport. An equivalent derivation was first given by Twiss [53]. Note that the noise source correlation matrix on the left hand side is positive definite or semidefinite. Hence the matrix $\langle \mathbf{1} - SS^\dagger \rangle$ must also be positive definite or semidefinite. This means that the network has to be dissipative, as shown in the next chapter. Indeed, only for such passive networks can thermal equilibrium be meaningfully defined. Active networks, by definition, cannot be at thermal equilibrium.

A lossless multiport does not require the introduction of noise sources. To prove this we check first the condition of losslessness. We must have

$$b^\dagger b = a^\dagger S^\dagger S a = a^\dagger a \tag{4.86}$$

or

$$a^\dagger (\mathbf{1} - S^\dagger S)a = 0 . \tag{4.87}$$

Since the excitation amplitudes are arbitrary, we find that the scattering matrix must be unitary:

$$S^\dagger S = \mathbf{1} \quad \text{or} \quad S^\dagger = S^{-1} . \tag{4.88}$$

A lossless network at thermal equilibrium does not contain internal noise sources. Indeed, if we substitute (4.88) into (4.85) we find that the noise correlation matrix vanishes.

4.8 The Probability Distribution of Photons at Thermal Equilibrium

Thus far we have studied thermal noise in lossy waveguides, resonators, transmission lines, and circuits using the Nyquist formula, which is based on the equipartition theorem. The noise spectral density in a single mode is then white and the power in a bandwidth B is $k\theta B$. If this relation were valid at all frequencies, the thermal power would be infinite. This leads to the so-called ultraviolet catastrophe, which is unphysical. Quantum theory removes the ultraviolet catastrophe by postulating that electromagnetic energy can only occur in quanta of energy $\hbar\omega$, where $\hbar$ is Planck's constant divided by 2π. At thermal equilibrium the photon distribution must be that of maximum randomness, i.e. maximum entropy. It can be shown that the equilibrium state of a system can depend only on the energies of the states [54]. The entropy of the system is [54]

$$S = -k \sum_i p_i \ln(p_i) , \tag{4.89}$$

where the p_is express the probabilities of the states with energy E_i. Thermal equilibrium is the state with maximum entropy. Denote the average energy by $\langle E \rangle$. We find the equilibrium state by maximizing (4.89) under the two constraints

$$\sum_i p_i = 1 \tag{4.90}$$

and

$$\sum_i E_i p_i = \langle E \rangle . \tag{4.91}$$

The task is to find the dependence of the p_is on the energies E_i. The maximization can be carried out with two Lagrange multipliers that take care of the two constraints. We extremize the function

$$f(p_i) = -k \left[\sum_i p_i \ln(p_i) + \lambda_1 \left(\sum_i p_i - 1 \right) + \lambda_2 \left(\sum_i E_i p_i - \langle E \rangle \right) \right] , \tag{4.92}$$

setting $\partial f/\partial p_i = (\partial f/\partial \lambda_1) = (\partial f/\partial \lambda_2) = 0$. From $\partial f/\partial p_i = 0$ we obtain the equation

$$1 + ln(p_i) + \lambda_1 + \lambda_2 E_i = 0 . \tag{4.93}$$

We find for p_i

$$p_i = \exp[-(\lambda_1 + 1)] \exp[-\lambda_2 E_i] . \tag{4.94}$$

We see that the Lagrange multiplier λ_1 fixes the normalization of the probability and the multiplier λ_2 gives the explicit dependence on energy. The probability must depend exponentially on the energy.

Next, we consider a harmonic oscillator of frequency ω_o, representing a mode in a resonator. We make Planck's assumption that the accessible energies occur in multiples of $\hbar\omega$, where $\hbar = h/2\pi$, and h is Planck's constant. This assumption was justified years later when the quantization of the harmonic oscillator was carried out according to the rules established by quantum mechanics. The quantization of the harmonic oscillator will be discussed in Sect. 6.1. Here we accept this ground rule and proceed to evaluate the probability distribution of the energy. We obtain from (4.94)

$$p_i \to p(n) = \exp-(\lambda_1 + 1)\exp-(\lambda_2 n\hbar\omega) , \qquad (4.95)$$

where n is the level of occupancy, or the photon number as used by Einstein in 1905 in the analysis of the photoelectric effect. The multiplier is set so that the probabilities add up to unity

$$\exp\left[-(\lambda_1 + 1)\right] = \frac{1}{\sum_{n=0}^{n=\infty} \exp\left[-(\lambda_2 n\hbar\omega)\right]} = 1 - \exp\left[-(\lambda_2 \hbar\omega)\right] . \qquad (4.96)$$

The average photon number is

$$\langle n \rangle = \sum_n np(n) = \frac{\sum_n n\exp\left[-n(\lambda_2\hbar\omega)\right]}{\sum_n \exp\left[-n(\lambda_2\hbar\omega)\right]} = \frac{\exp\left[-(\lambda_2\hbar\omega)\right]}{1 - \exp\left[-(\lambda_2\hbar\omega)\right]} . \qquad (4.97)$$

This equation defines λ_2 in terms of the average photon number. We find for the probability distribution

$$p(n) = \frac{1}{1 + \langle n \rangle}\left[\frac{\langle n \rangle}{1 + \langle n \rangle}\right]^n . \qquad (4.98)$$

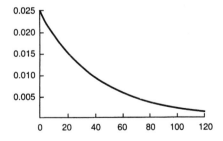

Fig. 4.9. Bose–Einstein probability distribution for $\langle n \rangle = 40$

Equation (4.98) is the so-called Bose–Einstein equilibrium distribution (see Fig. 4.9). To give further physical meaning to the Lagrange multiplier, consider the average energy $\langle n \rangle \hbar\omega$ in the classical limit as $\hbar \to 0$. In this limit the energy of an oscillator with two degrees of freedom has to be equal to $k\theta$. We find from (4.97)

$$\lim_{\hbar\omega \to 0} \langle n \rangle \hbar\omega = \hbar\omega \sum_n np(n) = \lim_{n \to 0} \frac{\hbar\omega \exp[-(\lambda_2 \hbar\omega)]}{1 - \exp[-(\lambda_2 \hbar\omega)]} = \frac{1}{\lambda_2} = k\theta . \quad (4.99)$$

The Langrange multiplier is proportional to the inverse temperature. The average photon number is thus in general

$$\langle n \rangle = \frac{1}{\exp(\hbar\omega/k\theta) - 1} . \quad (4.100)$$

The average energy is $\hbar\omega\langle n \rangle$. If we evaluate this average energy in the limit of low frequencies, $\hbar\omega \ll k\theta$, we find the value assigned by the equipartition theorem:

$$\lim_{\hbar\omega \ll k\theta} \hbar\omega\langle n \rangle = \lim_{\hbar\omega \ll k\theta} \frac{\hbar\omega}{\exp(\hbar\omega/k\theta) - 1} = k\theta . \quad (4.101)$$

Hence, all formulae involving the power at thermal equilibrium developed in the classical limit can be generalized to arbitrarily high frequencies by replacing $k\theta$ with $\hbar\omega/[\exp(\hbar\omega/k\theta) - 1]$ (see Fig. 4.9). The Bose–Einstein distribution applies to situations more general than thermal equilibrium. Amplified spontaneous emission is Bose–Einstein distributed, as we shall show in Chap. 9. Thus, the statistics at the output of an amplifier with no input mimic a hot thermal source.

4.9 Gaussian Amplitude Distribution of Thermal Excitations

We have mentioned earlier that the amplitude of a mode in a waveguide, or in a resonator, has a Gaussian distribution since the thermal excitation is due to coupling to a thermal reservoir with many degrees of freedom. The central limit theorem then requires the amplitude to have a Gaussian distribution. The Gaussian distribution can also be derived without appeal to the central limit theorem, but rather as the distribution that maximizes the entropy, a condition for thermal equilibrium. We may use the analysis in the preceding section almost unchanged, if we discretize a continuous amplitude distribution in such a way that the amplitude assumes only discrete values A_i with probabilities p_i. The constraints are

$$\sum_i p_i = 1 \quad (4.102)$$

and

$$\sum_i A_i^2 p_i = \langle E \rangle \, . \tag{4.103}$$

The function that is to be maximized is

$$f(p_i) = -k \left[\sum_i p_i \, \ln(p_i) + \lambda_1 \left(\sum_i p_i - 1 \right) + \lambda_2 \sum_i A_i^2 p_i - \langle E \rangle \right] \, . \tag{4.104}$$

The probability distribution is found completely analogously to the solution of (4.93):

$$p_i = \exp[-(\lambda_1 + 1)] \exp[-(\lambda_2 A_i^2)] \, . \tag{4.105}$$

This is a Gaussian distribution. A transition to a continuous distribution A, along with the normalization $\int_{-\infty}^{+\infty} dA \, p(A) = 1$, gives

$$p(A) = \frac{1}{\sqrt{2\pi\sigma^2}} \exp\left(-\frac{A^2}{2\sigma^2} \right) , \tag{4.106}$$

with $\sigma^2 = \langle E \rangle$. Thus, the Gaussian distribution maximizes the entropy under the constraint that the average energy is fixed.

Consider some further properties of a Gaussian-distributed electric field. It is clear that the description of a time-dependent Gaussian field calls for two components, an in-phase component and a quadrature component; one may also characterize them as a cosine component and a sine component. The energy, or power, is proportional to the sum of the squares of these amplitudes. We consider an electromagnetic wave with the cosine amplitude equal to A_c, and the sine amplitude equal to A_s. The square of the field is normalized to the energy w in a chosen time interval, equal to the sum of A_c^2 and A_s^2: $w = A_c^2 + A_s^2$. The expectation value of the energy w is equal to the sum of the mean square deviations $\sigma_c^2 + \sigma_s^2$. Let us determine the probability distribution of w. The combined probability distribution of A_c and A_s, when the two are statistically independent, is

$$\frac{1}{\sqrt{2\pi\sigma_c^2}} \exp\left(-\frac{A_c^2}{2\sigma_c^2} \right) dA_c \frac{1}{\sqrt{2\pi\sigma_s^2}} \exp\left(-\frac{A_s^2}{2\sigma_s^2} \right) dA_s$$

$$= \frac{1}{2\pi\sigma^2} \exp\left(-\frac{A_c^2 + A_s^2}{2\sigma^2} \right) dA_c \, dA_s \, , \tag{4.107}$$

where we have used the fact that the mean square deviations of the two fields are equal: $\sigma_c^2 = \sigma_s^2 \equiv \sigma^2$. This probability distribution can be written as a probability distribution for the energy w, if one integrates in the A_c–A_s plane around a circle of constant $w = A_c^2 + A_s^2$:

$$p(w)\, dw = \int_o^{2\pi} \sqrt{A_c^2 + A_s^2}\, d\phi \, \frac{1}{2\pi\sigma^2} \exp\left(-\frac{A_c^2 + A_s^2}{2\sigma^2}\right) d\sqrt{A_c^2 + A_s^2}$$

$$= \frac{1}{2\sigma^2} \exp\left(-\frac{A_c^2 + A_s^2}{2\sigma^2}\right) d\left(A_c^2 + A_s^2\right)$$

$$= \frac{1}{\langle w \rangle} \exp\left(-\frac{w}{\langle w \rangle}\right) dw .$$

$$(4.108)$$

The probability distribution of the energy is an exponential with the average value $\langle w \rangle = 2\sigma^2$. The mean square fluctuations of the power are

$$\int w^2 p(w)\, dw - \langle w \rangle^2 = \langle w \rangle^2 \int \frac{w^2}{\langle w \rangle^2} e^{-w/\langle w \rangle} d\left(\frac{w}{\langle w \rangle}\right) - \langle w \rangle^2 = \langle w \rangle^2 .$$

$$(4.109)$$

4.10 Summary

Shot noise is an important example of a random process that not only occurs in current flow through diodes and p–n junctions, but also plays an important role in optical detectors illuminated by a light source of constant intensity. We shall have ample opportunity to use the expressions for the shot noise spectrum and for the Poisson probability distribution. The power spectrum of an electromagnetic wave on a transmission line or in a waveguide was derived from the equipartition theorem. Note that we started with the modes in a ring resonator of assumed length L. The final expression for the power spectrum did not depend on the length, an important justification of the formalism, since dependence of physical quantities on such an artificial parameter would be unacceptable. Modes of resonators coupled to the outside world do not have a power spectral density independent of frequency. The spectral density of the mode energy peaks at the frequency of resonance. The integral over the resonance band yields an energy $k\theta$.

We found that linear lossy circuits call for the introduction of Langevin noise sources in order to maintain the thermal fluctuations against the power loss of the circuit. We derived the spectra for the Langevin sources in a lossy waveguide and in a multiport linear circuit at thermal equilibrium. Thermal noise of passive structures at thermodynamic equilibrium is another example of an important process which is in close analogy with the zero point fluctuations of quantum mechanics discussed later on. We derived the Bose–Einstein statistics of photon distributions at thermal equilibrium from the condition of maximum entropy. Finally, we obtained the Gaussian probability distribution of the amplitude of a mode at thermal equilibrium from maximization of entropy.

Problems

4.1* The formula for the current flowing in a diode is $i = I_o[\exp(qV/k\theta)-1]$. At equilibrium, this current can be thought of as consisting of two current flows in opposite directions of magnitude I_o, each and canceling when $V = 0$.

(a) What are the shot-noise current fluctuations at equilibrium?
(b) The conductance of the diode at equilibrium is di/dV. What are the short-circuit current fluctuations, from the Nyquist formula? Compare with (a).

4.2 A receiving microwave antenna has a bandwidth $B = 100$ MHz. If this antenna receives the cosmic background radiation of 2.75 K, what is the net power received?

4.3* In the text we evaluate the number of modes in a waveguide of length L in the frequency interval $\Delta\omega$ by setting $\beta L = 2\pi m$ (periodic boundary conditions) and determining $\Delta m = (d\beta/d\omega)L\,\Delta\omega/2\pi$. Had we used standing-wave boundary conditions we would have set $\beta L = \pi m$ and found $\Delta m = (d\beta/d\omega)L\,\Delta\omega/\pi$. This is twice the previous number. The two results are not in conflict, because the result with periodic boundary conditions includes only forward-traveling waves. Thus, the actual number of modes is the same in both cases.

In this problem you are asked to derive the number of standing wave modes in free space within a cubic box of side length L. Note that

$$\left(\frac{m\pi}{L}\right)^2 + \left(\frac{n\pi}{L}\right)^2 + \left(\frac{p\pi}{L}\right)^2 = k^2 = \frac{\omega^2}{c^2}$$

or

$$m^2 + n^2 + p^2 = \frac{\omega^2 L^2}{c^2\pi^2}.$$

One may think of each mode as a point in a space of dimensions $\omega L/c\pi$. Only positive mode numbers are to be included. The number of modes in one-eighth of a sphere of radius $\omega L/c\pi$ is equal to the volume $(\pi/6)(\omega L/c\pi)^3$. The number of modes in a shell of thickness $\Delta\omega L/c\pi$ is $(\pi/2)(L/c\pi)^3\omega^2\Delta\omega$. Noting that each mode has two polarizations, determine the electromagnetic energy per unit volume within the bandwidth $\Delta\omega$ at thermal equilibrium at temperature θ. You will have found the Rayleigh–Jeans law.

4.4 Sometimes one may prove relations derived from Maxwell's equations by referring to thermodynamic equilibrium considerations.

Consider an antenna with gain G. If it is thermally excited by a single mode waveguide, the power radiated into a narrow solid angle $\Delta\Omega$ in the direction Θ, ϕ is $k\theta G(\Theta, \phi)(\Delta\Omega/4\pi)$. By requiring that the antenna receive as much power as it transmits when in thermal equilibrium with its environment, prove that the receiving cross section A of the antenna is

$A(\Theta, \phi) = (\lambda^2/4\pi)G(\Theta, \phi)$. Note that, by the definition of receiving cross section, an antenna receives the power $P = A(\Theta, \phi)S$ if it is irradiated by a Poynting flux S traveling in the direction Θ, ϕ as expressed in spherical coordinates centered at the antenna.

4.5 Using the results of the previous problem, prove that the receiving cross section of a short dipole is $A = (3/2)(\lambda^2/4\pi)$.

4.6 In the preceding problem you have found that the receiving cross section of a dipole is independent of the length of the dipole. This is a surprising result until one realizes that the definition of receiving cross section assumes that the antenna is matched to its termination. Determine the matching impedance for a short dipole as a function of its length.

4.7* The Rayleigh–Jeans law exhibits the ultraviolet catastrophe. Planck's quantization removes the catastrophe. Derive Planck's law for the energy density per unit volume of electromagnetic radiation at thermal equilibrium at temperature θ.

4.8 Compare the short-circuit current fluctuations of the thermal noise of a 50 Ω resistor at room temperature with the shot noise of a current I_o flowing through the resistor. At what value of I_o is the latter equal to the former?

Solutions

4.1

(a) If the current in each direction is I_o, then the shot noise due to the two currents is $\langle i^2 \rangle = 4qI_oB$.

(b) The conductance at $V = 0$ is

$$di/dV = (qI_o/k\theta)\exp(qV/k\theta) = \frac{qI_o}{k\theta} = G .$$

The Nyquist formula gives $\langle i^2 \rangle = 4Gk\theta B = 4qI_oB$. The two results agree.

4.3 The electromagnetic energy is $k\theta$ per mode times the number of modes. The energy per unit volume and per unit solid angle within the bandwidth $\Delta\omega = 2\pi\Delta\nu$ is

$$\frac{1}{4\pi}k\theta \times 2 \times \frac{\pi}{2}\frac{\omega^2\Delta\omega}{(c\pi)^3} = 2\frac{\nu^2\Delta\nu}{c^3}k\theta .$$

4.7 At high frequencies, when the Planck formula replaces the equipartition theorem, the energy density per unit volume and unit solid angle becomes (compare Prob. 4.3):

$$\frac{\text{Energy density}}{\text{unit volume} \times \text{unit solid angle}} = 2\frac{\nu^2\Delta\nu}{c^3}\frac{h\nu}{\exp(h\nu/k\theta) - 1} .$$

This law does not diverge as the frequency goes to infinity.

5. Linear Noisy Multiports

Microwave and optical devices may all be described as multiports: signals are propagated into the device through input waveguides and emerge in output waveguides. If signal distortion is avoided, these devices are characterized as linear multiports. Of course, even linear multiports may distort a broadband signal by introducing frequency-dependent changes of the amplitudes and phases of the Fourier components of the signal. A linear multiport with loss does not only attenuate the signal, it also adds noise at thermal equilibrium. Linear multiports with gain amplify the signal, but also add noise in the process. In this chapter we study the basic noise properties of linear multiports. Linear multiports are described by an appropriate response matrix, which is a function of frequency, and a set of (Langevin) noise sources; there are N such sources for a multiport with N ports. Since the sources are generated by noise processes with a large number of degrees of freedom, they are usually Gaussian, according to the central limit theorem. Then, the correlation matrix of the noise sources, which is a function of frequency, is sufficient for their specification. In Sect. 4.7 we determined the noise sources for passive multiports at thermal equilibrium. Active multiports, such as amplifiers, contain noise sources that are determined by the physics of the amplifying process.

We shall start with the derivation of the characteristic noise matrix, which determines the stationary values of the power that can be extracted from a noisy multiport with variations in the loading of the network. We shall find that the stationary values of the power are given by the eigenvalues of a characteristic noise matrix. This thought experiment establishes a universal measure of "noisiness" of a network, which also underlies the noise performance of an active network used as an amplifier. Then we show how the characteristic noise matrix transforms from one network description to another network description. We show that its eigenvalues are invariant under such transformations. Finally, we express the characteristic noise matrix in the scattering-matrix notation, the notation most useful in optical systems terminology. The characteristic noise matrices of different matrix formulations relate to different thought experiments performed on the network. In the transfer matrix formulation, the characteristic noise matrix results from optimization of the noise performance of an amplifier.

Active two-ports are amplifiers. The purpose of signal amplification is to provide a signal level at the amplifier output so high that any further operations on the signal do not cause a significant deterioration of the signal-to-noise ratio. An amplifier raises the signal power level, at the expense of a decrease in signal-to-noise ratio. Clearly, the objective of good amplifier design is to achieve a minimum deterioration of the signal-to-noise ratio in the amplifying process.

In order to characterize the noise performance of an amplifier, one needs a measure of noise performance. The noise figure F, defined by Friis [55], is one such measure. It is defined by

$$F = \frac{\text{input signal-to-noise ratio}}{\text{output signal to noise ratio}} \ .$$

Since the signal-to-noise ratio deteriorates in passage through an amplifier, the noise figure F is greater than unity. It is, further, customary to define the noise at the input in terms of a thermal background at room temperature, $\theta_o = 290$ K. The signal level need not appear in the definition of noise figure, since the ratio of the signal levels at output and input is simply the gain G. One may write

$$F = \frac{\text{noise at the output}}{k\theta_o G} \ ,$$

where G is the available power gain of the amplifier (to be defined more precisely below). We concentrate here on the so-called spot noise figure, defined for bandwidths narrow enough that the amplifier characteristics do not vary over the chosen bandwidth. The definitions of noise figure (a) in terms of the input and output signal-to-noise ratios and (b) as applied to linear amplifiers in terms of the amplifier output noise were adopted by the Standards Committee of the Institute of Radio Engineers in 1959 [17]. The successor Institute of Electrical and Electronics Engineers adopted the same standard. Later, in Chap. 9, we shall discuss the definition of noise figure for optical amplifiers in current use and raise some important issues with regard to this usage. It suffices to state at this point that the noise figure is an adequate measure of amplifier noise performance only if the gain of the amplifier is large. Indeed, if one shorted the leads of a two-port amplifier from input to output, the noise figure of this modified arrangement would be unity, i.e. ideal. However, the gain of this structure is unity, and hence the whole purpose of amplification of a signal is vitiated. There must be a better way of measuring noise performance, namely with a measure that also includes the gain of the amplifier in such a way that an "amplifier" with unity gain does not appear to have a good noise performance. Confronted with this dilemma, Prof. R. B. Adler and the author constructed a measure of noise performance [56–61] which remains meaningful if the amplifier gain is not large. This so-called "noise measure" was defined by

$$M \equiv \frac{F - 1}{1 - 1/G} .$$

It is clear that this definition will not register an improvement in the noise performance when the two-port amplifier is shorted out. Indeed, when this happens, the "excess noise figure" $F - 1$ becomes zero, but so does the denominator. The noise measure becomes indeterminate, zero over zero. It does acquire a definite value if the limits are taken properly.

Further, the concept of available gain was generalized to allow for source or amplifier output impedances with negative real parts. In this chapter we shall address these issues in detail and arrive at unequivocal definitions of noise performance of linear amplifiers.

5.1 Available and Exchangeable Power from a Source

A source is a one-port, described by the voltage–current relation (see Fig. 5.1)

$$V = Z_s I + E_s , \tag{5.1}$$

where V is the voltage across the source, I is the current flowing into the source, and E_s is the open-circuit voltage across the source. The available power of the source is defined as the maximum power transferable from the source to a load, with adjustment of the load impedance. The power flowing into the load is

$$P_L = \frac{\langle |E_s|^2 \rangle \mathrm{Re}(Z_L)}{|Z_s + Z_L|^2} . \tag{5.2}$$

Here we use the notation $\langle \; \rangle$ for an ensemble average; if the noise is stationary, the ensemble average is equal to the time average. In the case of noise, we shall attach a very specific meaning to $\langle |E_s|^2 \rangle$: it will stand for the mean square voltage fluctuations in a bandwidth B. Thus, the open-circuit mean square fluctuations $\langle |E_s|^2 \rangle$ of a resistor at thermal equilibrium are $\langle |E_s|^2 \rangle = 4R_s k\theta B$.

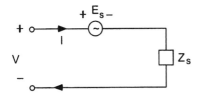

Fig. 5.1. The equivalent circuit of a source

The power flowing into the load is maximized when $Z_L = Z_s^*$. With this value of the load the maximum power is realized; P_L of (5.2) becomes the so-called "available power"

$$P_{av} = \frac{\langle |E_s|^2 \rangle}{2(Z_s + Z_s^*)} \, . \tag{5.3}$$

This relation assumes that the source impedance has a positive real part. This is not always the case. The impedance of a parametric amplifier (see Chap. 11) may have a negative real part. If the parametric amplifier is used as the first amplifier in a cascade of amplifiers, the combination of source and parametric amplifier may appear to the remainder of the cascade as a source with an internal impedance that has a negative real part. If the source impedance Z_s has a negative real part, a passive load $Z_L = -Z_s^*$ leads to a finite amount of power for $E_s = 0$ and an infinite amount of power if $E_s \neq 0$. In such a case one needs a generalization of the concept of available power, the *exchangeable power*. It is defined as the extremum of the power exchanged between source and load. For a source with a positive real part of its impedance the exchangeable power is the available power as discussed above; when the source impedance has a negative real part, it is the minimum (the extremum of the) power fed to the load. We shall use this extended definition of power from a source henceforth, so as to allow for the cascading of structures that may result in source impedances with negative real part:

$$P_{ex} = \frac{\langle |E_s|^2 \rangle}{2(Z_s + Z_s^*)} \, . \tag{5.4}$$

Note that the exchangeable power is negative when the real part of the source impedance is negative. Figure 5.2 shows the dependence of the power exchanged between source and load for the cases of a positive and a negative source resistance. The extrema occur for positive and negative load resistances, respectively.

5.2 The Stationary Values of the Power Delivered by a Noisy Multiport and the Characteristic Noise Matrix

In the preceding section we have studied the available power from a source of impedance Z_s and internal noise source E_s. When the source impedance had a negative real part, we generalized the concept of available power to that of exchangeable power, the value of the power that is stationary with respect to variation of the load impedance. In this section we generalize the concept of exchangeable power to multiports. Since the network has many terminals, one

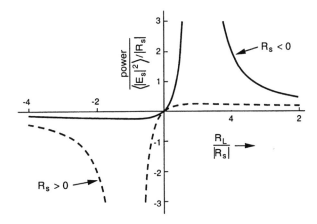

Fig. 5.2. The power flowing into the source $R_s = Re(Z_s)$ as a function of R_L

must be specific as to which terminal is being explored. However, varying the load on one terminal pair alone, leaving all other terminals open-circuited, does not allow for sufficient adjustment. For this reason, we analyze the more general case in which the network is embedded into an arbitrary lossless $2N$-port first, and then the load is varied on one of the terminals. This allows for a sufficiently wide range of adjustment. We shall find that the stationary values of the power are given by the eigenvalues of the "characteristic noise matrix" of the network.

At any particular frequency ω, a noisy multiport can be described by its impedance matrix expressing the terminal voltages in terms of the terminal currents (see Fig. 5.3):

$$V = ZI + E \ . \tag{5.5}$$

For a multiport of Nth order, the impedance matrix Z is a square matrix of Nth rank. The noise sources are arranged in a column vector E. They are specified in terms of the correlation matrix $\langle EE^\dagger \rangle$, whose ij element is $\langle E_i E_j^* \rangle$. As defined, the noise sources appear as voltage generators in series at the terminals of the multiport as shown in Fig. 5.3. We may now ask for the available or, more generally, the exchangeable power from one of the terminals of the N-port. With all ports open-circuited except the ith, the exchangeable power from the ith port is

$$P_{e,i} = \frac{1}{2} \frac{\langle E_i E_i^* \rangle}{Z_{ii} + Z_{ii}^*} = \frac{1}{2} \frac{\xi^\dagger \langle EE^\dagger \rangle \xi}{\xi^\dagger [Z + Z^\dagger] \xi} \ , \tag{5.6}$$

where the column matrix ξ consists of all zeros except for the ith row, which is a one ($\xi_j = 0, j \neq i; \xi_i = 1$). Equation (5.6) gives the stationary values of

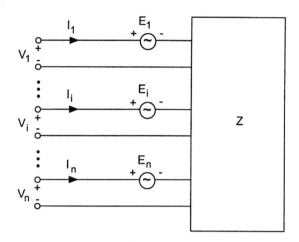

Fig. 5.3. Equivalent circuit of linear noisy N-port

the power with variation of the load on the ith terminal, but with no other adjustment of the network. More general is the case when the network of interest is embedded in a lossless, noise-free $2N$-terminal-pair network resulting in a new N-port, whose ith port is terminated in the complex conjugate of the open-circuit impedance of this new port. We shall now turn to the theory of embedding of an N-port in a lossless $2N$-port.

The impedance matrix of the lossless $2N$-port is subdivided into four impedance matrices of Nth rank (see Fig. 5.4):

$$Z_T = \begin{bmatrix} Z_{aa} & Z_{ab} \\ Z_{ba} & Z_{bb} \end{bmatrix} , \tag{5.7}$$

with the voltage–current relations

$$V_a = Z_{aa} I_a + Z_{ab} I_b , \tag{5.8}$$

$$V_b = Z_{ba} I_a + Z_{bb} I_b . \tag{5.9}$$

Since the embedding network is lossless, we must have

$$I^\dagger (Z_T + Z_T^\dagger) I = 0 , \tag{5.10}$$

for an arbitrary current excitation I, and thus

$$Z_T + Z_T^\dagger = 0 \quad \text{or} \quad \begin{bmatrix} Z_{aa} + Z_{aa}^\dagger & Z_{ab} + Z_{ba}^\dagger \\ Z_{ba} + Z_{ab}^\dagger & Z_{bb} + Z_{bb}^\dagger \end{bmatrix} = 0 . \tag{5.11}$$

The currents I of the original N-port are equal and opposite to the currents I_a fed into the embedding network; the voltages V are equal to the

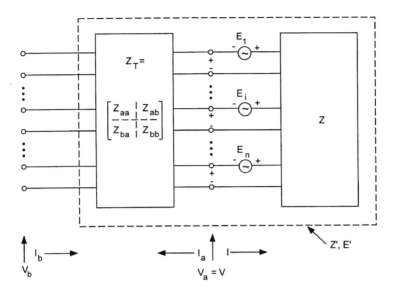

Fig. 5.4. A lossless $2N$-port with an embedded N-port

voltages $\boldsymbol{V}_a$ across the terminals of the embedding network. We obtain from (5.5) and (5.8)

$$\boldsymbol{I}_a = -(\boldsymbol{Z} + \boldsymbol{Z}_{aa})^{-1}\boldsymbol{Z}_{ab}\boldsymbol{I}_b + (\boldsymbol{Z} + \boldsymbol{Z}_{aa})^{-1}\boldsymbol{E} \ . \tag{5.12}$$

Using (5.9), the new network with terminal voltages $\boldsymbol{V}_b$ and terminal current $\boldsymbol{I}_b$ has the impedance matrix $\boldsymbol{Z}'$ and Langevin sources $\boldsymbol{E}'$:

$$\boldsymbol{V}_b = \boldsymbol{Z}'\boldsymbol{I}_b + \boldsymbol{E}' \tag{5.13}$$

with

$$\boldsymbol{Z}' = -\boldsymbol{Z}_{ba}(\boldsymbol{Z} + \boldsymbol{Z}_{aa})^{-1}\boldsymbol{Z}_{ab} + \boldsymbol{Z}_{bb} \tag{5.14}$$

and

$$\boldsymbol{E}' = \boldsymbol{Z}_{ba}(\boldsymbol{Z} + \boldsymbol{Z}_{aa})^{-1}\boldsymbol{E} \ . \tag{5.15}$$

The exchangeable power contains the matrices $\langle \boldsymbol{E}\boldsymbol{E}^\dagger \rangle$ and $\boldsymbol{Z} + \boldsymbol{Z}^\dagger$, in the numerator and denominator. Hence it is of interest to determine the transformation of these two matrices, using (5.14) and (5.15). Taking into account the condition (5.11) for losslessness of the embedding network, we obtain

$$\boldsymbol{Z}' + \boldsymbol{Z}'^\dagger = -\boldsymbol{Z}_{ba}(\boldsymbol{Z} + \boldsymbol{Z}_{aa})^{-1}\boldsymbol{Z}_{ab} + \boldsymbol{Z}_{bb} + \boldsymbol{Z}_{bb}^\dagger$$

$$-\boldsymbol{Z}_{ab}^\dagger(\boldsymbol{Z}^\dagger + \boldsymbol{Z}_{aa}^\dagger)^{-1}\boldsymbol{Z}_{ba}^\dagger$$

$$= \boldsymbol{Z}_{ba}[(\boldsymbol{Z} + \boldsymbol{Z}_{aa})^{-1} + (\boldsymbol{Z}^\dagger + \boldsymbol{Z}_{aa}^\dagger)^{-1}]\boldsymbol{Z}_{ba}^\dagger \ .$$

This expression can be transformed further:

$$Z' + Z'^\dagger = Z_{ba}(Z + Z_{aa})^{-1}[(Z^\dagger + Z_{aa}^\dagger) + (Z + Z_{aa})]$$

$$(Z^\dagger + Z_{aa}^\dagger)^{-1}Z_{ba}^\dagger \tag{5.16}$$

$$= D^\dagger(Z + Z^\dagger)D$$

with

$$D^\dagger \equiv Z_{ba}(Z + Z_{aa})^{-1} . \tag{5.17}$$

The transformation of the matrix $Z + Z^\dagger$ is a collinear transformation. The transformation of the noise correlation matrix follows from (5.15):

$$\langle E'E'^\dagger\rangle = D^\dagger\langle EE^\dagger\rangle D . \tag{5.18}$$

The same collinear transformation law is obeyed by both the correlation matrix and the impedance matrix plus its Hermitian conjugate. Note that both $Z + Z^\dagger$ and $\langle EE^\dagger\rangle$ are Hermitian matrices and that $\langle EE^\dagger\rangle$ is positive definite. It is possible to diagonalize both matrices with one and the same collinear transformation. To show this, suppose first that the positive definite Hermitian matrix $EE^\dagger$ is diagonalized by the unitary matrix U, a well-known operation. On the diagonal of the diagonalized matrix appear the *real* eigenvalues of the matrix. Next, we normalize the resulting matrix by a diagonal, real, normalizing matrix N, obtaining an identity matrix as the result:

$$N^\dagger U^\dagger\langle EE^\dagger\rangle UN = 1 .$$

Next, consider the matrix $Z + Z^\dagger$. We perform the same operations on this matrix and obtain a new matrix $N^\dagger U^\dagger(Z + Z^\dagger)UN$, which, of course, is not diagonal, in general, but is still Hermitian. Now we diagonalize the matrix with the unitary matrix V, so that $V^\dagger N^\dagger U^\dagger(Z + Z^\dagger)UNV$ is diagonal. Since we are looking for simultaneous diagonalization of $Z + Z^\dagger$ and $\langle EE^\dagger\rangle$, we must pre- and post-multiply $N^\dagger U^\dagger\langle EE^\dagger\rangle UN$ by $V^\dagger$ and V. But since the matrix $N^\dagger U^\dagger\langle EE^\dagger\rangle UN$ is the identity matrix, the operation leaves it unchanged. This proves the theorem that *two Hermitian matrices can be diagonalized simultaneously with a collinear transformation if one of them is positive definite.*

We can now ask for the exchangeable power from the ith terminal pair of the new network. It is

$$P'_{e,i} = \frac{1}{2}\frac{\xi^\dagger\langle E'E'^\dagger\rangle\xi}{\xi^\dagger[Z' + Z'^\dagger]\xi} . \tag{5.19}$$

When we use the transformation laws (5.16) and (5.18) to express the primed quantities in terms of the unprimed, original impedance and noise correlation matrices, we find

$$P'_{e,i} = \frac{1}{2} \frac{x^\dagger \langle EE^\dagger \rangle x}{x^\dagger [Z + Z^\dagger] x} , \qquad (5.20)$$

with

$$x^\dagger \equiv \xi^\dagger Z_{ba} (Z + Z_{aa})^{-1} = \xi^\dagger D^\dagger . \qquad (5.21)$$

Suppose next that we pick D so as to diagonalize simultaneously both $\langle EE^\dagger \rangle$ and $2(Z + Z^\dagger)$, with $\langle EE^\dagger \rangle$ transformed into the identity matrix. Denote the diagonal elements of the transformed matrix $2(Z' + Z'^\dagger)$ by $1/\lambda_i$. We then obtain for (5.20)

$$P'_{e,i} = \frac{|\xi_i|^2}{|\xi_i|^2 (1/\lambda_i)} = \lambda_i . \qquad (5.22)$$

It is obvious that λ_i is an extension of the concept of exchangeable power. Further, λ_i is one of the eigenvalues of the matrix

$$\frac{1}{2}(Z + Z^\dagger)^{-1} \langle EE^\dagger \rangle ,$$

which has undergone the similarity transformation

$$D^{-1} \frac{1}{2}(Z + Z^\dagger)^{-1} \langle EE^\dagger \rangle D ,$$

which rendered it diagonal. The same result can be obtained by an alternative route. Returning to (5.20), we note that the exchangeable power is the ratio of two scalars that are constructed from two Hermitian matrices $A = \frac{1}{2} \langle EE^\dagger \rangle$ and $B = Z + Z^\dagger$ by projection via the column matrix x. The extrema of this expression can be found by determining the stationary values of $x^\dagger Bx$ under the constraint $x^\dagger Ax = $ constant. With the Lagrange multiplier λ, and the recognition that $x^\dagger Bx$ and $x^\dagger Ax$ can be considered functions of either the x_i or the x_i^*, we find

$$\frac{\partial}{\partial x_i^*} \left(x_i^* B_{ij} x_j - \frac{1}{\lambda} x_i^* A_{ij} x_j \right) = 0$$

or

$$Ax - \lambda Bx = 0 . \qquad (5.23)$$

The values of λ are determined from the determinantal equation

$$\det(A - \lambda B) = 0 = \det(B^{-1}A - \lambda \mathbf{1}) . \qquad (5.24)$$

The eigenvalues of λ fix the extrema of the exchangeable power from terminal i. They are the eigenvalues of the matrix

$$B^{-1}A = \frac{1}{2}(Z + Z^\dagger)^{-1} \langle EE^\dagger \rangle . \qquad (5.25)$$

This matrix, or rather its negative, has been dubbed the "characteristic noise matrix", N_Z [61]. The choice of sign is motivated by the fact that the positive eigenvalues of the characteristic noise matrix so defined determine amplifier noise performance, as shown further on. This matrix is given by

$$N_Z \equiv -\frac{1}{2}(Z + Z^\dagger)^{-1}\langle EE^\dagger \rangle \,. \tag{5.26}$$

One feature of the characteristic noise matrix, when applied to a passive network at thermal equilibrium can be discerned right away. A passive network has open-circuit impedances with positive real parts only. At thermal equilibrium, the available power delivered to the matched load must be $k\theta B$. Thus all eigenvalues of the characteristic noise matrix must be equal to $-k\theta B$. We have

$$N_Z = -\frac{1}{2}(Z + Z^\dagger)^{-1}\langle EE^\dagger \rangle = -k\theta B\mathbf{1} \,, \tag{5.27}$$

where $\mathbf{1}$ is the identity matrix.

A few words about the sign of the eigenvalues of (5.26). The correlation matrix $\langle EE^\dagger \rangle$ is positive definite (or semidefinite in some limits). The matrix $(Z + Z^\dagger)^{-1}$ is positive definite if the network is passive, negative definite if it is totally active so that it cannot absorb power under any circumstances, or indefinite if the network can both generate and absorb power. The definiteness of a product of two matrices one of which is positive definite is that of the other matrix. Hence, the eigenvalues of (5.26) are all negative if the network is passive, all positive if the network is totally active, and both positive and negative if the network can both generate and absorb power.

5.3 The Characteristic Noise Matrix in the Admittance Representation Applied to a Field Effect Transistor

An analogous derivation can be carried out in the admittance matrix representation. The exchangeable power of a one-port is now

$$P_{ex} = \frac{\langle |J_s|^2 \rangle}{2(Y_s + Y_s^*)} \,, \tag{5.28}$$

where Y_s is the source admittance and J_s is the noise source. The generalization to a multiport, with all terminals short-circuited except the ith, is

$$P_{e,i} = \frac{1}{2} \frac{\xi^\dagger \langle JJ^\dagger \rangle \xi}{\xi^\dagger (Y + Y^\dagger)\xi} \,, \tag{5.29}$$

where the current–voltage relationship is

$$I = YV + J .$$ (5.30)

The embedding proceeds completely analogously. We find that the extrema of the exchangeable power are the eigenvalues of the characteristic noise matrix

$$N_Y = -\frac{1}{2}(Y + Y^\dagger)^{-1}\langle JJ^\dagger \rangle .$$ (5.31)

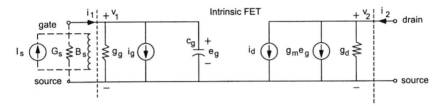

Fig. 5.5. Small-signal equivalent circuit of junction field effect transitor (JFET)

It is of interest to look at the simple example of a two-port amplifier. Figure 5.5 shows the small-signal equivalent circuit of a field effect transistor. The linearized equivalent circuit consists of a gate conductance g_g, a gate capacitance c_g, a drain conductance g_d, and a voltage-dependent voltage generator $g_m e_g$. The noise is represented by the two noise generators i_g and i_d. Because of the linearized form of the representation we use lower-case letters for all symbols. The current–voltage relations for this equivalent circuit are

$$i_1 = (g_g - i\omega c_g)v_1 + i_g ,$$ (5.32)

$$i_2 = g_m v_1 + g_d v_2 + i_d .$$ (5.33)

The matrix $\frac{1}{2}(Y + Y^\dagger)$ is

$$\frac{1}{2}(Y + Y^\dagger) = \begin{bmatrix} g_g & g_m/2 \\ g_m/2 & g_d \end{bmatrix} .$$ (5.34)

This matrix is not positive definite. Indeed, the determinant is

$$\det\left[\frac{1}{2}(Y + Y^\dagger)\right] = g_g g_d - \frac{g_m^2}{4} .$$ (5.35)

When $g_m^2/4 > g_g g_d$ the determinant is negative. This means that the network is capable of delivering net power, acting as an amplifier.

The inverse of the matrix $\frac{1}{2}(\boldsymbol{Y} + \boldsymbol{Y}^\dagger)$ is

$$\left[\frac{1}{2}(\boldsymbol{Y} + \boldsymbol{Y}^\dagger)\right]^{-1} = \frac{1}{g_g g_d - \frac{1}{4}g_m^2}\begin{bmatrix} g_d & -g_m/2 \\ -g_m/2 & g_g \end{bmatrix}. \tag{5.36}$$

The noise current source correlation function is

$$\langle \boldsymbol{J}\boldsymbol{J}^\dagger \rangle = \begin{bmatrix} \langle |i_g|^2 \rangle & \langle i_g i_d^* \rangle \\ \langle i_g^* i_d \rangle & \langle |i_d|^2 \rangle \end{bmatrix}, \tag{5.37}$$

and the characteristic noise matrix becomes

$$\boldsymbol{N}_Y = \frac{1}{2}\frac{1}{g_m^2/4 - g_g g_d}\begin{bmatrix} g_d\langle |i_g|^2 \rangle - \frac{1}{2}g_m\langle i_g^* i_d \rangle & g_d\langle i_g i_d^* \rangle - \frac{1}{2}g_m\langle |i_d|^2 \rangle \\ -\frac{1}{2}g_m\langle |i_g|^2 \rangle + g_g\langle i_g^* i_d \rangle & -\frac{1}{2}g_m\langle i_g i_d^* \rangle + g_g\langle |i_d|^2 \rangle \end{bmatrix}. \tag{5.38}$$

The eigenvalues of the characteristic noise matrix are

$$\lambda = \frac{1}{2}\frac{1}{g_m^2/4 - g_g g_d}\left[\frac{1}{2}(g_d\langle |i_g|^2 \rangle + g_g\langle |i_d|^2 \rangle - g_m\mathrm{Re}\langle i_g^* i_d \rangle)\right]$$

$$\pm\left[\frac{1}{4}(g_d\langle |i_g|^2 \rangle + g_g\langle |i_d|^2 \rangle - g_m Re\langle i_g^* i_d \rangle)^2 \right. \tag{5.39}$$

$$\left. + (g_m^2/4 - g_d g_g)(\langle |i_g|^2 \rangle\langle |i_d|^2 \rangle - |\langle i_g^* i_d \rangle|^2)\right]^{1/2}.$$

When the system has gain, the two eigenvalues are of opposite sign. With proper passive loading, the two-port can be made to oscillate in the absence of the internal sources and deliver an infinite amount of power in the presence of the internal sources. From the preceding analysis we know that the eigenvalues determine the exchangeable power from the two-port. The one with the negative sign gives the minimum power delivered by the network; the one with the positive sign gives the minimum power delivered to the two-port by active (negative-conductance) terminations. We shall later prove, in Sect. 5.7, that the positive eigenvalue determines the optimum noise measure of the amplifier.

5.4 Transformations of the Characteristic Noise Matrix

In Sect. 5.2, we evaluated the exchangeable power obtainable from a noisy N-port when the N-port is first embedded in a lossless network and then the ith port of the resulting network is terminated in a load, while all other ports are left open-circuited. This procedure provided a sufficient number of adjustable parameters for an arbitrary adjustment of the loading of the N-port. We arrived at a characteristic noise matrix of the original N-port in this manner. The same procedure can now be exploited to determine the change of the characteristic noise matrix following a lossless embedding. The original network has the following characteristic noise matrix (5.23):

$$N_Z = -\frac{1}{2}(Z + Z^\dagger)^{-1}\langle EE^\dagger\rangle \ . \tag{5.40}$$

A lossless embedding transforms $\langle EE^\dagger\rangle$ and $Z + Z^\dagger$ according to (5.16) and (5.18). Accordingly, the transformation of the characteristic noise matrix is

$$N'_Z = D^{-1}N_Z D \ . \tag{5.41}$$

The transformation is a similarity transformation, which leaves the eigenvalues of the characteristic noise matrix invariant! This finding will be exploited later to show that the optimum noise performance achievable with an amplifier is invariant under a lossless embedding, feedback being one special case of such an embedding.

Different forms of the characteristic noise matrix result from different matrix descriptions of a multiport, the impedance matrix description and the scattering-matrix description being two such examples. We shall show that these different forms of the characteristic noise matrix are also related by similarity transformations. Since the most important attribute of the characteristic noise matrix is its eigenvalues, and eigenvalues are invariant under similarity transformations, it is expedient to construct the characteristic noise matrix within the formalism used.

Suppose that the impedance matrix description of the network

$$V = ZI + E \tag{5.42}$$

is recast with new variables into a new formulation

$$v = Tu + \delta \ , \tag{5.43}$$

where the new variables are related to the voltage–current variables by the transformation

$$R\begin{bmatrix} v \\ u \end{bmatrix} = \begin{bmatrix} V \\ I \end{bmatrix} \ . \tag{5.44}$$

We shall now show that a characteristic noise matrix that is related to the characteristic noise matrix (5.40) by a similarity transformation emerges naturally in the new description of the network. To accomplish this most economically, we recast the terminal relations (5.42) and (5.43) into matrix format, introducing the column matrices (5.44), which are of twice the rank of either V or I. With these, we may rewrite the terminal relations (5.42) and (5.43)

$$[1 \quad - Z] \begin{bmatrix} V \\ I \end{bmatrix} = E \, , \tag{5.45}$$

$$[1 \quad - T] \begin{bmatrix} v \\ u \end{bmatrix} = \delta \, . \tag{5.46}$$

Now we introduce the transformation (5.44) into (5.45)

$$
\begin{aligned}
[1 \quad - Z] R R^{-1} \begin{bmatrix} V \\ I \end{bmatrix} &= [1 \quad - Z] R \begin{bmatrix} v \\ u \end{bmatrix} \\
&= [R_{11} - Z R_{21} \quad R_{12} - Z R_{22}] \begin{bmatrix} v \\ u \end{bmatrix} = E \, .
\end{aligned}
\tag{5.47}
$$

Multiplication of (5.47) by the matrix

$$M = (R_{11} - Z R_{21})^{-1} \tag{5.48}$$

puts (5.47) into the form of (5.46), where

$$[1 \quad - T] = M[1 \quad - Z] R \, , \tag{5.49}$$

$$\delta = M E \, . \tag{5.50}$$

This completes the transformation of the impedance matrix Z into the response matrix T, and the source vector E into the source vector δ.

Next we consider the representation of the power P flowing into the network in the absence of internal sources. Each matrix representation expresses the power in terms of the excitation variables via a specific matrix Q. Thus, consider the impedance representation, for which we write

$$P = \frac{1}{2} \langle V^\dagger I + I^\dagger V \rangle = \left\langle \begin{bmatrix} V \\ I \end{bmatrix}^\dagger Q_Z \begin{bmatrix} V \\ I \end{bmatrix} \right\rangle \, , \tag{5.51}$$

where Q_Z is the matrix

$$Q_Z = \frac{1}{2} \begin{bmatrix} 0 & 1 \\ 1 & 0 \end{bmatrix} \, . \tag{5.52}$$

Analogously, in the T representation, the power P is written

$$P = \left\langle \begin{bmatrix} v \\ u \end{bmatrix}^\dagger Q_T \begin{bmatrix} v \\ u \end{bmatrix} \right\rangle \, . \tag{5.53}$$

Since the power must be equal in the two descriptions, for all possible excitations, we must have

$$P = \left\langle \begin{bmatrix} v \\ u \end{bmatrix}^\dagger Q_T \begin{bmatrix} v \\ u \end{bmatrix} \right\rangle = \left\langle \begin{bmatrix} V \\ I \end{bmatrix}^\dagger Q_Z \begin{bmatrix} V \\ I \end{bmatrix} \right\rangle$$

$$= \left\langle \begin{bmatrix} v \\ u \end{bmatrix}^\dagger R^\dagger Q_Z R \begin{bmatrix} v \\ u \end{bmatrix} \right\rangle ,$$

$$(5.54)$$

and, thus,

$$Q_T = R^\dagger Q_Z R .$$

$$(5.55)$$

This is the law of transformation for the power matrix Q_T. Thus far, we have determined the transformations between two different matrix representations of a network involving the network matrices, the internal sources, and the transformation of the power flowing into the network in the absence of internal sources. Next we reformulate the characteristic noise matrix in the impedance formulation and determine its transformation into the T representation. The characteristic noise matrix (5.26), recast in terms of the reformulation (5.45), is seen to be

$$N_Z = \left\{ [1 \quad -Z] Q_Z^{-1} \begin{bmatrix} 1 \\ -Z^\dagger \end{bmatrix} \right\}^{-1} \langle EE^\dagger \rangle .$$

$$(5.56)$$

If we introduce the transformations (5.49) and (5.50) between the impedance matrix formulation and the T matrix formulation into (5.56) we find

$$N_Z = \left\{ M^{-1}[1 \quad -T] R^{-1} Q_Z^{-1} R^{\dagger-1} \begin{bmatrix} 1 \\ -T^\dagger \end{bmatrix} M^{\dagger-1} \right\}^{-1}$$

$$\times M^{-1} \langle \delta\delta^\dagger \rangle M^{\dagger-1}$$

$$(5.57)$$

$$= M^\dagger \left\{ [1 \quad -T] Q_T^{-1} \begin{bmatrix} 1 \\ T^\dagger \end{bmatrix} \right\}^{-1} \langle \delta\delta^\dagger \rangle M^{\dagger-1} .$$

We have derived a new characteristic noise matrix of the same generic form as that of N_Z in (5.56), namely

$$N_T = \left\{ [1 \quad -T] Q_T^{-1} \begin{bmatrix} 1 \\ -T^\dagger \end{bmatrix} \right\}^{-1} \langle \delta\delta^\dagger \rangle ,$$

$$(5.58)$$

and this matrix is related to N_Z by a similarity transformation.

We have studied the transformations among different matrix formulations of the same network. We have found that a new definition of the characteristic noise matrix emerges in every formulation. The different characteristic noise matrices are related by similarity transformations and thus possess the same eigenvalues. In each formulation, the eigenvalues of the characteristic noise matrix are equal to the stationary values of the power exchanged between the network and its terminations in a thought experiment in which the terminal conditions are varied. The stationary values of the exchangeable power in a thought experiment in which all terminals but one are open-circuited are given by the characteristic noise matrix in the impedance representation. A thought experiment that determines the stationary values of the exchangeable power when all but one terminal are short-circuited leads to the characteristic noise matrix in the admittance formulation. Since the two matrices are related by a similarity transformation, the stationary values in these two different thought experiments are, in fact, the same. In the next section we shall show that the characteristic noise matrix in the scattering-matrix representation gives the stationary values of the exchangeable power when all but one termination are matched.

5.5 Simplified Generic Forms of the Characteristic Noise Matrix

The matrix algebra in the preceding section was quite general, but it had to deal with manipulations of matrices of rank $2N$, a rank twice that of the network at hand. The expression for the characteristic noise matrix in any formalism can be simplified in many important cases, as indeed it was in the impedance matrix formulation, when we first encountered it by writing it in terms of matrices of rank N. For this purpose, two cases have to be distinguished.

(a) All ports of the network are equivalent. The response is in terms of input excitation variables (e.g. currents), defined at all ports in the same form, producing output excitation variables (e.g. voltages). Take, for example, the impedance description. We then have for Q_Z

$$Q_Z = \frac{1}{2} \begin{bmatrix} 0 & 1 \\ 1 & 0 \end{bmatrix} . \tag{5.59}$$

Another example is the admittance description. It has the same power matrix,

$$Q_Y = \frac{1}{2} \begin{bmatrix} 0 & 1 \\ 1 & 0 \end{bmatrix} . \tag{5.60}$$

If we use this definition and the fact that $T = Y$ for the admittance description, where

$$I = YV + J , \tag{5.61}$$

and J is a column matrix composed of noise current generators, we find for the characteristic noise matrix in the admittance formulation

$$N_Y = -\frac{1}{2}(Y + Y^\dagger)^{-1}\langle JJ^\dagger\rangle .\tag{5.62}$$

If we take the scattering matrix formulation as an example (compare Appendix A.6),

$$b = Sa + s ,\tag{5.63}$$

then

$$Q_S = \begin{bmatrix} -1 & 0 \\ 0 & 1 \end{bmatrix} ,\tag{5.64}$$

and the characteristic noise matrix with $T = S$ becomes

$$N_S = (SS^\dagger - 1)^{-1}\langle ss^\dagger\rangle .\tag{5.65}$$

(b) The network has an even number of ports, half of which are designated as input ports and half as output ports. The excitation variables at the output ports are expressed in terms of the excitation variables at the input ports. The two-port of Fig. 5.6 is an example. Port (2) is the "output" port, port (1) is the "input" port. The matrix T is the "transfer" or $ABCD$ matrix

$$\begin{bmatrix} V_1 \\ I_1 \end{bmatrix} = \begin{bmatrix} A & B \\ C & D \end{bmatrix}\begin{bmatrix} V_2 \\ I_2 \end{bmatrix} + \begin{bmatrix} E \\ J \end{bmatrix} .\tag{5.66}$$

Note the direction of positive current as defined in Fig. 5.6. In this case, the power matrix is of the form

$$Q_T = \begin{bmatrix} P_T & 0 \\ 0 & -P_T \end{bmatrix} ,\tag{5.67}$$

with

$$P_T = \frac{1}{2}\begin{bmatrix} 0 & 1 \\ 1 & 0 \end{bmatrix} ,\tag{5.68}$$

and the characteristic noise matrix assumes the form

$$N_T = (P_T^{-1} - TP_T^{-1}T^\dagger)^{-1}\langle\delta\delta^\dagger\rangle ,\tag{5.69}$$

with

$$T = \begin{bmatrix} A & B \\ C & D \end{bmatrix} .$$

If voltages and currents are not natural excitation variables, as is the case in the analysis of optical amplifiers, wave amplitudes can be used instead. If this is done for the transfer matrix formalism as represented by (5.66) in terms of voltages and currents, we obtain

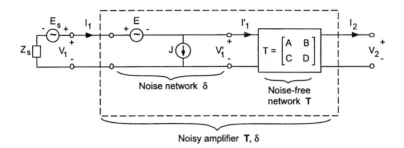

Fig. 5.6. The equivalent circuit of the two-port of (5.79)

$$\begin{bmatrix} a_1 \\ b_1 \end{bmatrix} = \begin{bmatrix} T_{aa} & T_{ab} \\ T_{ba} & T_{bb} \end{bmatrix} \begin{bmatrix} a_2 \\ b_2 \end{bmatrix} + \begin{bmatrix} \gamma_a \\ \gamma_b \end{bmatrix} . \tag{5.70}$$

The power matrix is

$$\boldsymbol{Q}_T = \begin{bmatrix} \boldsymbol{P}_T & 0 \\ 0 & \boldsymbol{P}_T \end{bmatrix} , \tag{5.71}$$

with

$$\boldsymbol{P}_T = \begin{bmatrix} 1 & 0 \\ 0 & -1 \end{bmatrix} .$$

In this formalism, the characteristic noise matrix is of the same form as in (5.69), the only change being the new interpretations of $\boldsymbol{P}_T$, the transfer matrix, and the noise source column matrix.

The characteristic noise matrix arises naturally in the scattering-matrix formulation when the question is asked about the stationary values of the power delivered to a load connected to the ith port, with all other ports *matched*, i.e. $a_j = 0$ for $j \neq i$. We proceed to prove this assertion. With all ports except the ith, matched, the equation of the ith port is

$$b_i = S_{ii} a_i + s_i . \tag{5.72}$$

The available power or, more generally, the exchangeable power, is realized when the termination impedance of the ith port is the complex conjugate of the internal impedance presented by the N-port. This means that the reflection coefficient of the termination is the complex conjugate of S_{ii}:

$$\text{reflection coefficient} = \frac{a_i}{b_i} = S_{ii}^* . \tag{5.73}$$

The power flowing into the load is

$$P_{ex,i} = |b_i|^2 - |a_i|^2 = \frac{\langle |s_i|^2 \rangle}{1 - |S_{ii}|^2} . \tag{5.74}$$

This can also be written in matrix form, with the column vector $\boldsymbol{\xi}$ such that $\xi_j = 0$ for $j \neq i$, and $\xi_i = 1$:

$$P_{ex,i} = \frac{\boldsymbol{\xi}^\dagger \langle \boldsymbol{ss}^\dagger \rangle \boldsymbol{\xi}}{\boldsymbol{\xi}^\dagger (1 - \boldsymbol{SS}^\dagger) \boldsymbol{\xi}} \ . \tag{5.75}$$

We can again consider a lossless embedding of the network that transforms the source correlation matrix and the matrix in the denominator. Lossless embeddings have been studied in the impedance formulation. We do not need to rederive the transformations in the scattering-matrix formulation, since we may transform both matrices into the impedance form. A lossless embedding transforms the resulting matrices by a similarity transformation. Transforming back into the scattering-matrix formulation, we obtain

$$P_{ex,i} = \frac{\boldsymbol{x}^\dagger \langle \boldsymbol{ss}^\dagger \rangle \boldsymbol{x}}{\boldsymbol{x}^\dagger (1 - \boldsymbol{SS}^\dagger) \boldsymbol{x}} \quad \text{with} \quad \boldsymbol{x} = \boldsymbol{M}^{\dagger-1} \boldsymbol{D} \boldsymbol{M}^\dagger \boldsymbol{\xi} \ . \tag{5.76}$$

Now the column matrix $\boldsymbol{x}$ is arbitrarily adjustable and can be varied for extremization. The eigenvalues of the characteristic noise matrix $\boldsymbol{N}_S = (\boldsymbol{SS}^\dagger - 1)^{-1} \langle \boldsymbol{ss}^\dagger \rangle$ now yield the extrema of the exchangeable power at the ith port, with all other ports matched.

The characteristic noise matrix in the transfer matrix formalism (5.66) is the basis of a thought experiment in which the so-called "noise measure" is extremized with adjustment of the source and load impedances, as we shall show in the next section.

5.6 Noise Measure of an Amplifier

Within a narrow frequency band, a linear amplifier is described completely by its scattering matrix and the correlation matrix of its noise sources. Amplifiers are not always connected to transmission lines in which the definition of incident and reflected waves is unequivocal. Hence, the scattering matrix formalism is not best suited for the study of amplifier noise performance at both low frequencies and microwave frequencies. The voltage–current description is more appropriate for this purpose. We shall start with this formalism and express only the final results in the scattering matrix terminology, which is natural for the description of optical amplifiers, for which the voltage–current description lacks specificity (equivalent circuits for optical structures are not unique).

5.6.1 Exchangeable Power

To facilitate the evaluation of the available gain, or the more general concept of exchangeable gain, it is expedient to derive the available or exchangeable power by matrix manipulation. The equation for the source connected to the amplifier (see Fig. 5.6) can be written

$$x^\dagger v = E_s \quad \text{with} \quad x \equiv \begin{bmatrix} 1 \\ Z_s^* \end{bmatrix} \quad \text{and} \quad v \equiv \begin{bmatrix} V_1 \\ I_1 \end{bmatrix} . \tag{5.77}$$

The exchangeable power is

$$P_{ex} = \frac{\langle |E_s|^2 \rangle}{x^\dagger P_T^{-1} x} , \tag{5.78}$$

where

$$P_T \equiv \frac{1}{2} \begin{bmatrix} 0 & 1 \\ 1 & 0 \end{bmatrix}$$

is the power matrix as defined earlier in (5.71).

5.6.2 Noise Figure

A matrix description best suited for the analysis of amplifiers in cascade is the "transfer" or $ABCD$ matrix expressing the input voltage and current in terms of the output voltage and current (see Fig. 5.6):

$$\begin{bmatrix} V_1 \\ I_1 \end{bmatrix} = \begin{bmatrix} A & B \\ C & D \end{bmatrix} \begin{bmatrix} V_2 \\ I_2 \end{bmatrix} + \begin{bmatrix} E \\ J \end{bmatrix} \tag{5.79}$$

or, in abbreviated matrix notation,

$$v = Tu + \delta \tag{5.80}$$

with

$$v = \begin{bmatrix} V_1 \\ I_1 \end{bmatrix} ,$$

$$u = \begin{bmatrix} V_2 \\ I_2 \end{bmatrix} ,$$

$$\delta = \begin{bmatrix} E \\ J \end{bmatrix} , \tag{5.81}$$

$$T = \begin{bmatrix} A & B \\ C & D \end{bmatrix} .$$

Here E and J are the internal voltage and current noise sources at the input of the amplifier as shown in Fig. 5.6. In this representation, the noise figure is already completely determined by E and J; no details of the $ABCD$

transfer matrix enter into its evaluation. Indeed, after incorporation of the noise sources in front of the amplifier into the voltage generators of the signal and the noise of the input source impedance, the signal-to-noise ratio does not change when the signal and added noise pass through the noise-free remainder of the equivalent circuit [62] (see Fig. 5.6). We compute the noise figure as defined in the introduction to this chapter:

$$F = \frac{\text{input signal-to-noise ratio}}{\text{output signal-to-noise ratio}} . \tag{5.82}$$

We may evaluate the noise figure as the ratio of the mean square noise voltage at the primed terminal pair of Fig. 5.6, divided by the mean square noise voltage in the absence of the amplifier noise sources; the input source is at thermal equilibrium at temperature θ_o:

$$F = \frac{\langle |E_s|^2 \rangle + \langle |E + Z_s J|^2 \rangle}{\langle |E_s|^2 \rangle} = 1 + \frac{\langle E + Z_s J|^2 \rangle}{4\text{Re}(Z_s)k\theta_o B} . \tag{5.83}$$

It is helpful to cast the noise figure expression into matrix notation. Using the column vectors x and δ as defined in (5.77) and (5.81), the excess noise figure can be written in the form

$$F - 1 = \frac{x^\dagger \langle \delta \delta^\dagger \rangle x}{k\theta_o B x^\dagger P_T^{-1} x} . \tag{5.84}$$

Note that the excess noise figure is equal to the exchangeable power at the input of the amplifier with the noise sources of the amplifier assigned to the source, divided by $k\theta_o B$. If the source impedance has a positive real part, as it always does at the input to the first amplifier in a cascade, then the excess noise figure is equal to the available power of the noise sources of the amplifier assigned to the source, divided by $k\theta_o B$:

$$F - 1 = \frac{P_{av,1}}{k\theta_o B} . \tag{5.85}$$

5.6.3 Exchangeable Power Gain

Next we determine the exchangeable power gain, defined as the ratio of output exchangeable power to input exchangeable power of an amplifier connected to an input source impedance Z_s and a signal voltage source E_s with no internal noise sources (see Fig. 5.7). The exchangeable power of the source is given by (compare (5.78))

$$P_{ex,1} = \frac{\langle |E_s|^2 \rangle}{x^\dagger P_T^{-1} x} , \tag{5.86}$$

where

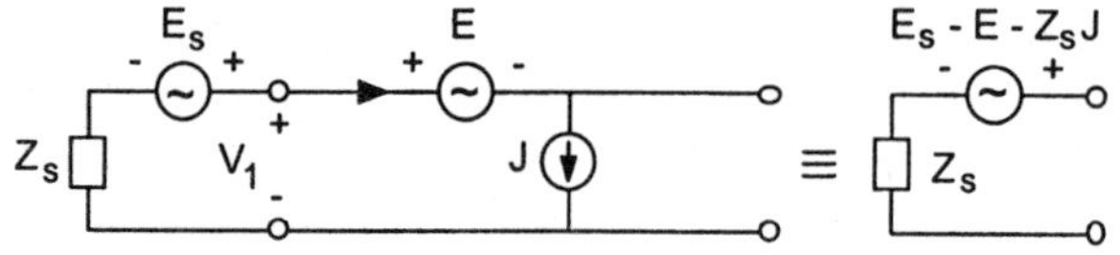

Fig. 5.7. The equivalent input noise source

$$x = \begin{bmatrix} 1 \\ Z_s^* \end{bmatrix} . \tag{5.87}$$

It is instructive to note that the expression for the exchangeable power (5.86) is constructed from the power matrix P_T and the components of the voltage–current relation of the source

$$x^\dagger v = E_s , \tag{5.88}$$

an expression that, written out explicitly, reads

$$V_1 + Z_s I_1 = E_s . \tag{5.89}$$

The exchangeable power at the amplifier output can be constructed similarly, if we note that (5.79), written in terms of the output voltage and current of the amplifier, assumes the form

$$x^\dagger T u = E_s \tag{5.90}$$

or, written out explicitly,

$$(A + Z_s C)V_2 + (B + Z_s D)I_2 = E_s . \tag{5.91}$$

We cast (5.91) into the form of (5.89) by multiplying it by $\alpha = 1/(A + Z_s C)$, obtaining the expression

$$V_2 + \frac{(B + Z_s D)}{(A + Z_s C)} I_2 = \frac{E_s}{A + Z_s C} \tag{5.92}$$

or, written more succinctly,

$$\alpha x^\dagger T u = \alpha E_s . \tag{5.93}$$

The exchangeable power at the amplifier output is thus, comparing (5.89), (5.92), and (5.93),

$$P_{\text{ex},2} = \frac{|\alpha|^2 \langle |E_s|^2 \rangle}{\alpha x^\dagger T P_T^{-1} T^\dagger x \alpha^*} = \frac{\langle |E_s|^2 \rangle}{x^\dagger T P_T^{-1} T^\dagger x} . \tag{5.94}$$

The exchangeable power gain is thus

$$G = \frac{P_{ex,2}}{P_{ex,1}} = \frac{x^\dagger P_T^{-1} x}{x^\dagger T P_T^{-1} T^\dagger x} \ . \tag{5.95}$$

The reader will remember that the exchangeable gain reduces to the available power gain when the output impedance of the amplifier connected to the source has a positive real part. This is the desirable situation, since amplifiers with output impedances of negative real part are prone to oscillate. In fact, whenever a case arises in which the output impedance has a negative real part, the circuit is usually modified via a circulator, so that the system looks matched as seen from the output port (see Fig. 5.8). In this case the exchangeable gain reduces to the available gain.

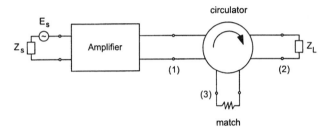

Fig. 5.8. Use of circulator to eliminate effect of negative output resistance of amplifier

5.6.4 The Noise Measure and Its Optimum Value

We are now ready to evaluate the noise measure

$$M \equiv \frac{F - 1}{1 - 1/G} \ , \tag{5.96}$$

with the gain G interpreted as the exchangeable gain. Combining (5.84) and (5.95) we find

$$M \equiv \frac{F - 1}{1 - 1/G} = \frac{x^\dagger \langle \delta\delta^\dagger \rangle x}{k\theta_o B x^\dagger (P_T^{-1} - T P_T^{-1} T^\dagger) x} \ . \tag{5.97}$$

Note that the noise measure becomes equal to the excess noise figure $F - 1$ when the exchangeable gain is large. Further, note that the noise measure is negative when the gain is less than unity. Hence, we are only interested in positive values of an amplifier noise measure.

The noise measure is the ratio of two scalars that are constructed from two Hermitian matrices $A = \langle \delta\delta^\dagger \rangle$ and $B = P_T^{-1} - T P_T^{-1} T^\dagger$ by projection via the column matrix x (see Sect. 5.2). The eigenvalues λ fix the extrema of the noise measure. They are the eigenvalues of the matrix

$$N_T = B^{-1}A = (P_T^{-1} - TP_T^{-1}T^\dagger)^{-1}\langle \delta\delta^\dagger \rangle , \qquad (5.98)$$

which is the characteristic noise matrix (5.69) in the transfer matrix notation. We have arrived at a new and interesting insight. We have studied earlier the transformation of the characteristic noise matrix as the consequence of a change in the network description. The eigenvalues of the characteristic noise matrix remained invariant under such a transformation. We showed that every new form of the characteristic noise matrix is associated with a thought experiment of exchangeable-power extremization via changes of the network loading. In the impedance matrix description, this corresponded to finding the extremum of the power delivered to a load on one terminal pair, with all other terminals open-circuited, after embedding of the network in a lossless network. In the scattering-matrix notation, it was the extremum of the power into a conjugately matched load at one terminal pair, with all other pairs terminated in matched transmission lines so that no waves were reflected from them. In the case of the $ABCD$ matrix of a two-port, we found that the eigenvalues of the characteristic noise matrix give the extremum of the noise measure of the two-port.

A characteristic noise matrix of second rank has two eigenvalues. These can be positive as well as negative. As pointed out earlier, only positive eigenvalues are of interest, since they are associated with gain. The smallest positive eigenvalue determines the lowest achievable noise measure, or excess noise figure at large gain. We have shown that the characteristic noise matrix of a passive network has only negative eigenvalues, a totally active network has only positive eigenvalues, and one that can both absorb and generate power has both negative and positive eigenvalues. The most common amplifiers are both active and passive for good reasons.

(a) Amplification is only possible if the network is capable of generating power.

(b) To prevent undesirable feedback effects due to reflections of the load at the output port of the amplifier it is desirable that the amplifier appear matched at its output. This is only possible if there is absorption of a wave incident upon the output port.

The range of values of the noise measure is illustrated in Fig. 5.9. The eigenvalues determine the extrema; the noise measure ranges from the positive eigenvalue to plus infinity and up from minus infinity to the negative eigenvalue, when the characteristic noise matrix is indefinite, and between the two positive eigenvalues when the characteristic noise matrix is positive definite (as mentioned earlier, the less common case).

Lossless embeddings leave the eigenvalues of the matrix (5.98) invariant as well. A special case of a lossless embedding is feedback produced by connecting reactive impedances between the input and output of an amplifier two-port. Thus the noise measure achievable with such feedback is subject to the same limits.

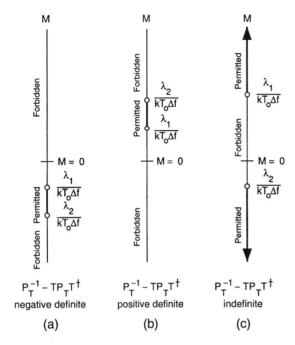

Fig. 5.9. Range of values of noise measure

5.7 The Noise Measure in Terms of Incident and Reflected Waves

In Sect. 5.4 we studied the transformation of the characteristic noise matrix from one formalism to another and showed that the characteristic noise matrices that emerge in different formalisms are related by similarity transformations. We also saw that each formalism is associated with a different thought experiment of extremization. Thus, one may ask for the extremization of the noise measure with respect to the source reflection coefficient, and write the resulting noise measure in terms of incident and reflected waves rather than terminal voltages and currents. This kind of description is particularly appropriate in the discussion of the noise measure of optical amplifiers, since their response is expressed naturally in terms of incident and reflected waves. In this section, we shall study this transformation in detail. In the next section we shall use it to simplify the algebra incurred when analyzing the noise measure of a field effect transistor.

The description of a two-port in the transfer matrix formulation is given by (5.80) and the noise measure was presented in (5.97), as repeated below:

$$M = \frac{\boldsymbol{x}^\dagger \langle \delta\delta^\dagger \rangle \boldsymbol{x}}{k\theta_o B \boldsymbol{x}^\dagger (\boldsymbol{P}_T^{-1} - \boldsymbol{T} \boldsymbol{P}_T^{-1} \boldsymbol{T}^\dagger) \boldsymbol{x}} , \tag{5.99}$$

with the vector (column matrix) $\boldsymbol{x}$ given in terms of the source impedance Z_s:

$$\boldsymbol{x} = \begin{bmatrix} 1 \\ Z_s^* \end{bmatrix} .$$

(5.100)

When the optimum noise measure is achieved, $\boldsymbol{x}$ is an eigenvector of the eigenvalue equation

$$\boldsymbol{N}_T \boldsymbol{x} = \lambda \boldsymbol{x} ,$$

(5.101)

where λ is the least positive eigenvalue of the characteristic noise matrix $\boldsymbol{N}_T$. Now, suppose we use wave variables instead of voltage–current variables to describe the excitation of the two-port. In the wave formalism, the source generator and the two-port noise generators become wave generators. Indeed, the equation of the source in the wave formalism is

$$a + \Gamma_s b = s .$$

(5.102)

The equivalent circuit with the wave generator is shown in Fig. 5.10a. The reader will recall that a wave generator is a combination of a voltage generator in series and a current generator in parallel. A signal wave passing through the wave generator is unaffected. The transformation from the voltage–current variables to the wave variables is based on the following two relations; the transformation from V, I to a, b:

$$a = \frac{1}{2}(\sqrt{Y_o}V + \sqrt{Z_o}I), \quad b = \frac{1}{2}(\sqrt{Y_o}V - \sqrt{Z_o}I) .$$

(5.103)

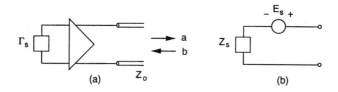

(a) Z_o (b)

Fig. 5.10. Equivalent source circuits: (a) the wave representation; (b) the voltage–current representation

The relation between the voltage source and wave source is

$$s = \frac{\sqrt{Y_o}E_s}{1 + Z_s/Z_o} .$$

(5.104)

The relation between the reflection coefficient and impedance is

$$\Gamma_s = \frac{1 - Z_s/Z_o}{1 + Z_s/Z_o} \; . \tag{5.105}$$

Again, the description in the wave formalism can be written concisely:

$$\boldsymbol{x}^\dagger \boldsymbol{v} = s \; , \quad \text{where} \quad \boldsymbol{x} = \begin{bmatrix} 1 \\ \Gamma_s^* \end{bmatrix} \quad \text{and} \quad \boldsymbol{v} = \begin{bmatrix} a \\ b \end{bmatrix} \; . \tag{5.106}$$

The equation for the noisy two-port has the standard form

$$\begin{bmatrix} a_1 \\ b_1 \end{bmatrix} = \begin{bmatrix} T_{aa} & T_{ba} \\ T_{ab} & T_{bb} \end{bmatrix} \begin{bmatrix} a_2 \\ b_2 \end{bmatrix} + \begin{bmatrix} \gamma_a \\ \gamma_b \end{bmatrix} , \quad \text{or}$$

$$\boldsymbol{v} = \boldsymbol{T}\boldsymbol{u} + \boldsymbol{\delta} \quad \text{with} \quad \boldsymbol{v} = \begin{bmatrix} a_1 \\ b_1 \end{bmatrix}, \; \boldsymbol{u} = \begin{bmatrix} a_2 \\ b_2 \end{bmatrix} , \quad \text{and} \quad \boldsymbol{\delta} = \begin{bmatrix} \gamma_a \\ \gamma_b \end{bmatrix} ,$$
$$\tag{5.107}$$

where γ_a and γ_b are noise wave generators (see Fig. 5.10b).

5.7.1 The Exchangeable Power Gain

The exchangeable power of the source can be evaluated directly from (5.78) by substituting (5.104) and (5.105):

$$P_{ex,1} = \frac{\langle |s|^2 \rangle}{1 - |\Gamma_s|^2} = \frac{\langle |s|^2 \rangle}{\boldsymbol{x}^\dagger \boldsymbol{P}_T^{-1} \boldsymbol{x}} \; . \tag{5.108}$$

This expression looks very much like (5.78), with

$$\boldsymbol{P}_T = \begin{bmatrix} 1 & 0 \\ 0 & -1 \end{bmatrix} \; . \tag{5.109}$$

The exchangeable output power in the general matrix notation is of the same form as in (5.94):

$$P_{ex,2} = \frac{\langle |s|^2 \rangle}{\boldsymbol{x}^\dagger \boldsymbol{T} \boldsymbol{P}_T^{-1} \boldsymbol{T}^\dagger \boldsymbol{x}} \; . \tag{5.110}$$

Finally, we obtain for the exchangeable power gain G

$$G = \frac{\boldsymbol{x}^\dagger \boldsymbol{P}_T^{-1} \boldsymbol{x}}{\boldsymbol{x}^\dagger \boldsymbol{T} \boldsymbol{P}_T^{-1} \boldsymbol{T}^\dagger \boldsymbol{x}} \; . \tag{5.111}$$

This expression is identical in form to (5.95), except for the fact that the matrices have all been redefined.

5.7.2 Excess Noise Figure

The reader will have sensed the drift of the derivation. If one asks for the excess noise figure, one again finds complete parallelism with the derivation in Sect. 5.6. The noise is now described by two wave generators as shown in Fig. 5.11. The evaluation of the noise figure can ignore the presence of the noise-free structure following the wave generators. We note that the equivalent noise source at port (1) due to the internal noise of the amplifer is

$$\gamma_a + \Gamma_s \gamma_b = x^\dagger \delta \ . \tag{5.112}$$

The exchangeable power of this source is

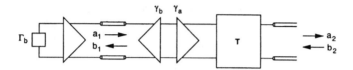

Fig. 5.11. The noisy two-port in the wave representation

$$P_{\text{ex,noise}} = \frac{x^\dagger \langle \delta \delta^\dagger \rangle x}{x^\dagger P_T^{-1} x} \ , \tag{5.113}$$

and the noise measure is

$$\frac{F-1}{1-1/G} = \frac{x^\dagger \langle \delta \delta^\dagger \rangle x}{k\theta_o B x^\dagger (P_T^{-1} - TP_T^{-1}T^\dagger) x} \ . \tag{5.114}$$

The noise measure has the same appearance as in the voltage–current formulation. The optimum noise measure is given by the lowest positive eigenvalue of the characteristic noise matrix

$$N_T = (P_T^{-1} - TP_T^{-1}T^\dagger)^{-1} \langle \delta \delta^\dagger \rangle \ . \tag{5.115}$$

The fact that we obtained the same formal expressions for the noise measure using the wave formalism as with the voltage–current formalism may appear surprising. Have we not noted that every new formulation of the matrix equations for the linear multiport corresponds to the extremization of a different thought experiment? In the current-voltage formalism the termination of one of the ports is varied while all the other ports remain open-circuited. In the wave formalism, as one of the terminations is varied, all other ports are matched. The reason why the two formalisms do not differ in the case of the noise measure rests on the fact that the noise measure is independent of the termination of the output port.

One issue we have not addressed thus far is the question of the optimum source impedance for minimization of the noise measure. In the wave formalism one asks for the optimum reflection coefficient of the source. This is obtained as follows. The optimization is achieved when the following equation holds:

$$N_{21}x_1 + N_{22}x_2 = \lambda x_2 , \tag{5.116}$$

where we have dropped the subscript T on the characteristic noise matrix N. The eigenvector x involves the reflection coefficient of the source for optimum noise performance (compare (5.106)). From (5.106) and (5.116) we find

$$\frac{x_2}{x_1} = \Gamma_s^* = \frac{N_{21}}{\lambda - N_{22}} . \tag{5.117}$$

The positive eigenvalue is given by

$$\lambda = \frac{N_{11} + N_{22}}{2} + \sqrt{\left(\frac{N_{11} - N_{22}}{2}\right)^2 + N_{12}N_{21}} , \tag{5.118}$$

and thus

$$\Gamma_s^* = \frac{N_{21}}{(N_{11} - N_{22})/2 + \sqrt{[(N_{11} - N_{22})/2]^2 + N_{12}N_{21}}} . \tag{5.119}$$

This termination is physically realizable if one finds that $|\Gamma_s| < 1$. If this condition is not met, then the optimization cannot be achieved with a passive source. In this case, in order to achieve optimization a lossless embedding of the two-port may be required.

5.8 Realization of Optimum Noise Performance

A simple example should serve to illustrate the general theory presented thus far. In particular, it will be shown how the optimum noise performance is achieved in one particular case. We shall look at the optimum noise performance of a junction field effect transistor (JFET) operating at microwave frequencies, with the equivalent circuit shown in Fig. 5.5 [63]. We ask the question as to how the optimum noise performance is realized: can it be achieved simply by choice of a proper source impedance or does the optimization entail a lossless embedding?

The answer to this question is best obtained if the network of Fig. 5.5 is properly normalized. We first connect an inductive admittance in parallel to make the input admittance purely real. This additional admittance may be thought to be associated with the source admittance. Next we connect transformers to the input and output to transform the admittances Y_{11} and

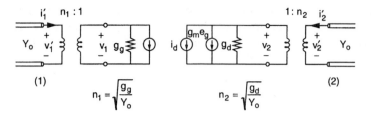

$$n_1 = \sqrt{\frac{g_g}{Y_o}} \qquad n_2 = \sqrt{\frac{g_d}{Y_o}}$$

Fig. 5.12. The equivalent circuit of the JFET with transformers and transmission lines

Y_{22}, which are now real, to be equal to the characteristic admittances of two transmission lines connected to the input and output (see Fig. 5.12). This is done in preparation for a wave-excitation formulation of the problem, a formulation that simplifies the ensuing algebra. The equations of the network become

$$i_1' = Y_o v_1' + \sqrt{\frac{Y_o}{g_g}} i_g , \tag{5.120}$$

$$i_2' = \frac{g_m}{\sqrt{g_g g_d}} Y_o v_1' + Y_o v_2' + \sqrt{\frac{Y_o}{g_d}} i_d . \tag{5.121}$$

Next we write the equations in the transfer matrix form in which port (1) is the output port and port (2) is the input port in accordance with the $ABCD$ matrix representation of (5.79):

$$v_1' = -\frac{1}{\mu}[v_2' + Z_o(-i_2')] - \frac{Z_o}{\mu}\sqrt{\frac{Y_o}{g_d}} i_d , \tag{5.122}$$

$$i_1' = -\frac{Y_o}{\mu}[v_2' + Z_o(-i_2')] - \frac{1}{\mu}\sqrt{\frac{Y_o}{g_d}} i_d + \sqrt{\frac{Y_o}{g_g}} i_g , \tag{5.123}$$

with

$$\mu \equiv \frac{g_m}{\sqrt{g_g g_d}} . \tag{5.124}$$

Finally, we introduce wave amplitudes as in Sect. 5.7. The equations assume a particularly simple form:

$$a_1 = -\frac{2}{\mu} a_2 + \gamma_a , \tag{5.125}$$

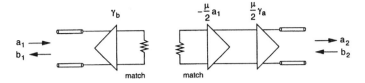

Fig. 5.13. The wave generator equivalent circuit of an FET

$$b_1 = \gamma_b , \tag{5.126}$$

with

$$\gamma_a = -\frac{1}{\mu\sqrt{g_d}}i_d + \frac{1}{2\sqrt{g_g}}i_g , \tag{5.127}$$

$$\gamma_b = -\frac{1}{2\sqrt{g_g}}i_g . \tag{5.128}$$

Figure 5.13 shows the equivalent circuit of the FET in the normalized wave representation. We have omitted the transformers. One must remember that the gate resistance and drain resistance are greatly different, the former being much larger than the latter. Hence, the characteristic impedance of the transmission lines at the input and output in Fig. 5.13 are very different. The representation has been expressed in terms of the wave formalism of Sect. 5.7. The T matrix is of the simple form (compare (5.107))

$$T = \begin{bmatrix} -2/\mu & 0 \\ 0 & 0 \end{bmatrix} , \tag{5.129}$$

and leads to a simple expression for the characteristic noise matrix (5.115). The characteristic noise matrix is now

$$N_T = \frac{1}{1 - 4/\mu^2} \begin{bmatrix} |\gamma_a|^2 & \gamma_a\gamma_b^* \\ -(1 - 4/\mu^2)\gamma_a^*\gamma_b & -(1 - 4/\mu^2)|\gamma_b|^2 \end{bmatrix} . \tag{5.130}$$

The positive eigenvalue of the characteristic noise matrix is

$$\lambda_+ = \frac{1}{1 - 4/\mu^2} \left\{ \frac{1}{2}\left[\langle|\gamma_a|^2\rangle - (1 - 4/\mu^2)\langle|\gamma_b|^2\rangle \right] \right.$$

$$\left. + \sqrt{\frac{1}{4}\left[\langle|\gamma_a|^2\rangle + \left(1 - \frac{4}{\mu^2}\right)\langle|\gamma_b|^2\rangle \right]^2 + \left(1 - \frac{4}{\mu^2}\right)|\langle\gamma_a^*\gamma_b\rangle|^2} \right\}$$

$$\tag{5.131}$$

and is, of course, equal to the positive eigenvalue of (5.39), but now looks much simpler. The eigenvector x for the optimum noise performance is given by (5.116), so that

$$|\Gamma_s| = \left|\frac{x_2}{x_1}\right| = \left(1 - \frac{4}{\mu^2}\right)|\langle\gamma_a^*\gamma_b\rangle|/\left\{\frac{1}{2}\left[|\gamma_a|^2 + \left(1 - \frac{4}{\mu^2}\right)|\gamma_b|^2\right]\right.$$

$$\left. + \sqrt{\frac{1}{4}\left[\langle|\gamma_a|^2\rangle + \left(1 - \frac{4}{\mu^2}\right)\langle|\gamma_b|^2\rangle\right]^2 + \left(1 - \frac{4}{\mu^2}\right)|\langle\gamma_a^*\gamma_b\rangle|^2}\right\}. \tag{5.132}$$

When the noise wave generators at the input and output are uncorrelated, the optimum source impedance is a matching resistance. Then the noise of the wave generator at the input escapes into the source, and the noise at the output is determined solely by the drain noise wave generator. When the two noise sources are correlated, then there is an advantage in mismatching the source impedance to partially cancel the effect of the drain wave generator. Figure 5.14 shows the magnitude of the reflection coefficient for different correlation coefficients. The phase of the reflection coefficient is equal to the phase of $\langle\gamma_a\gamma_b^*\rangle$. Figure 5.15 shows the normalized optimum noise figure. The correlation can be used to improve the noise figure by proper choice of the source reflection coefficient. But even for a correlation coefficient as high as 0.8, the improvement is small. It is worth noting that the optimum noise measure goes to zero when $\langle|\gamma_a|^2\rangle$ goes to zero. This is self-evident from Fig. 5.15. With no source at the output port (2) and a match at the input port (1), no noise is fed to the output.

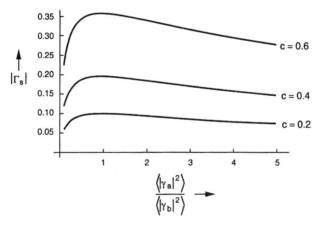

Fig. 5.14. The source reflection coefficient for optimum noise performance as a function of $\sqrt{\langle|\gamma_a^2|\rangle/\langle|\gamma_b^2|\rangle}$ for varying correlation coefficient $c = |\langle\gamma_a\gamma_b^*\rangle|/\sqrt{\langle|\gamma_a^2|\rangle\langle|\gamma_b^2|\rangle}$; $\mu = 10$

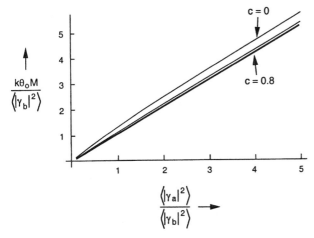

c = 0.8

Fig. 5.15. The normalized optimum noise measure $k\theta_o M/\langle|\gamma_b^2|\rangle$ as a function of $\sqrt{\langle|\gamma_a^2|\rangle/\langle|\gamma_b^2|\rangle}$ for varying correlation coefficient $c = |\langle\gamma_a\gamma_b^*\rangle|/\sqrt{\langle|\gamma_a^2|\rangle\langle|\gamma_b^2|\rangle}$; $\mu = 10$

5.9 Cascading of Amplifiers

When the gain of an amplifier is not large, the noise measure is larger than the excess noise figure, an indication of the inadequacy of the gain. In order to appreciate the role of gain, it is useful to consider amplifiers in cascade as shown in Fig. 5.16. Indeed, if the first amplifier does not raise the signal level sufficiently, then it is necessary to follow it with another one, whose noise performance cannot be ignored.

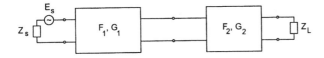

Fig. 5.16. Cascade of two noisy two-ports

The noise figure, as originally defined, is the ratio of the signal-to-noise ratio at the input of an amplifier to the signal to-noise ratio at the output. The noise at the input is the available power at a standard temperature θ_o. Since the ratio of the signal powers at the input and output involves only the available gain, the signal powers can be eliminated and the noise figure becomes the ratio of the available noise power at the output of the amplifier to the power that would be available at the output if the amplifier were noise-free. In our treatment we found it necessary to generalize the concept of available power to exchangeable power. With this generalization we have

$$F = \frac{k\theta_o BG + P_{ex}}{k\theta_o BG} = 1 + \frac{P_{ex}}{k\theta_o BG} , \tag{5.133}$$

where P_{ex} is the exchangeable noise power at the output due to the internal noise sources of the amplifier, and G is the exchangeable gain.

The definition (5.133) is the starting point for the evaluation of the noise figure of a cascade of amplifiers [55]. Note that the exchangeable gain of two amplifiers in cascade is the product of the exchangeable gains of the individual amplifiers. Attaching subscripts 1 and 2 to the quantities pertaining to amplifiers (1) and (2), we have for the excess noise figure $F - 1$ of the cascade

$$F - 1 = \frac{G_2 P_{ex,1} + P_{ex,2}}{k\theta_o BG_1 G_2} = \frac{P_{ex,1}}{k\theta_o BG_1} + \frac{P_{ex,2}}{k\theta_o BG_1 G_2}$$

$$= F_1 - 1 + \frac{F_2 - 1}{G_1} . \tag{5.134}$$

The exchangeable power at the output of the second amplifier can be negative, if the output impedance of the first amplifier has a negative real part. However, then G_1 is also negative and $F - 1$ of (5.134) is a sum of positive noise contributions, as one would expect from a proper definition of noise performance. Equation (5.134), in conjunction with the definition of noise measure, gives the cascading formula for the noise measure:

$$M = \frac{F - 1}{1 - 1/G} = \frac{F_1 - 1 + (F_2 - 1/G_1)}{1 - 1/G_1 G_2} = M_1 + (M_2 - M_1)\frac{G_2 - 1}{G - 1} . \tag{5.135}$$

If the two amplifiers have the same noise measure, the noise measure of the cascade is the same as that of the individual amplifiers. This is one of the invariance properties of the noise measure. This equation also shows that the amplifier with the lowest noise measure should be placed first in the cascade.

5.10 Summary

This chapter studied linear noisy multiports as basic components of any microwave or optical system. Optics has not been emphasized at this point, because in order to understand the noise at optical frequencies, it will be necessary to quantize the electromagnetic field. Yet all of the observations made about the general properties of microwave systems will be applicable in the domain of optics. We established general properties of linear noisy multiports as determined in a narrow frequency band B. If the performance of such networks over a broad bandwidth is of interest, the characteristic noise matrix must be treated as a function of frequency. We started this chapter with the study of the available power or, more generally, exchangeable power, obtainable from a linear noisy N-port, within a narrow bandwidth centered at one frequency. We embedded the N-port in a lossless, noise-free $2N$-port and then evaluated the stationary values of the power flowing out of the jth port, with all other ports open-circuited. The stationary values were found to be the eigenvalues of the characteristic noise matrix. The analysis was carried out in the impedance formulation of the network. We then showed that a transformation into the scattering-matrix formulation defined a new characteristic noise matrix in terms of the new network parameters. The new characteristic noise matrix was related to the original one by a similarity transformation, and, thus, possessed the same eigenvalues. We also found that a lossless embedding of the network resulted in a similarity transformation of the characteristic noise matrix. Further, we showed that the characteristic noise matrix in the scattering-matrix formulation related to a thought experiment involving the extremization of power flow that differed from the one in the impedance formulation.

Then we studied the noise performance of linear two-ports used as amplifiers. The noise figure of an amplifier measures the deterioration of the signal-to-noise ratio caused by the amplifying process. However, the noise figure by itself is not sufficient to characterize the noise performance since it does not discriminate against low gain of the amplifier. The noise measure is a better measure of noise performance. In particular, the optimum achievable noise measure is the lowest positive eigenvalue of the characteristic noise matrix. This property of the noise measure endows it with fundamental significance.

We studied one equivalent circuit of a junction field effect transistor and determined its optimum noise performance. We showed that the optimum value of the noise measure could be realized simply by adjustment of the source impedance, i.e. with an appropriate impedance transformer between source and amplifier.

The optimum noise measure of a two-port amplifier is a measure of the quality of the amplifier. When a particular gain and noise figure are realized with the amplifier, the noise measure achieved can be compared with the optimum value in order to decide whether further changes in the design of the system are worthwhile, or whether the noise measure achieved is close enough to the ultimate limit. In order to determine the ultimate limit it is, of course, necessary to determine it from measurements on the two-port. There is standard equipment to measure the scattering matrix of a two-port. With the two-port matched on both sides, the measurement of the noise power escaping from the two-port in a bandwidth B determines $\langle |s_1^2| \rangle$ and $\langle |s_2^2| \rangle$. With one side of the two-port shorted out, the cross correlation $\langle s_1 s_2^* \rangle$ can be determined from a measurement of the output noise power at the other port and the knowledge of the scattering matrix.

The work described in this chapter was carried out by the author and his colleague Richard B. Adler in the 1950s. Only the example of the junction field transistor is of later vintage. The noise measure was proposed as the appropriate measure of noise performance of an amplifier. This proposal did not catch on, mainly because commercial amplifiers at RF and microwave frequencies tended to have large gains, and thus the difference between the excess noise figure and the noise measure was not significant. Today, when optical doped-fiber amplifiers are used, often with gains less than 10 dB, it may be appropriate to reconsider the use of the noise measure instead of the noise figure.

The work in the 1950s on microwave amplifiers led the author to the conclusion that there was no fundamental classical limit on the noise measure of an amplifier, in particular if cooling of the device to low temperatures was an option. For a parametric amplifier, well known in those days, and discussed in detail in Chap. 11, classical theory predicted an arbitrarily low noise measure if the amplifier was cooled to absolute zero temperature. With the advent of the laser in the early 1960s amplifiers became available whose noise measure could not be made arbitrarily low, since quantum effects could not be ignored. This led the author into the study of noise in lasers and to many issues described further on in the text.

Problems

5.1* In the text, we found the extremum of an expression of the form

$$x^\dagger A x - \lambda x^\dagger B x$$

by differentiating the expression with respect to $x^\dagger$, treating x as a constant. A and B are Hermitian matrices. In fact, the extremization is with respect to the amplitudes $|x_i|$ and phases ϕ_i, which are contained in both x and $x^\dagger$. It is the purpose of this problem to show that differentiation with respect to the amplitudes $|x_i|$ and phases ϕ_i as independent variables is equivalent to differentiation with respect to $x^\dagger$ keeping x fixed.

(a) Express $x^\dagger A x$ and $x^\dagger B x$ in terms of $|x_i|$ and $\exp(i\phi_i)$.
(b) Differentiate with respect to $|x_i|$.
(c) Differentiate with respect to ϕ_i.
(d) Combine the two sets of equations in such a way as to arrive at

$$A x - \lambda B x = 0 .$$

5.2 An amplifier can be constructed from a negative conductance gY_o ($g < 0$) connected to port (2) of a circulator (Fig. 5.8). The negative conductance has an associated noise source.

(a) Derive the scattering matrix for the two-port between reference ports (1) and (3).
(b) Determine the characteristic noise matrix and its eigenvalues.
(c) By direct evaluation determine the noise measure of the amplifier.
(d) How is the optimum noise measure achieved?

5.3* Consider the nonreciprocal circuit of Fig. 5.5. Under what conditions could this circuit represent a network at thermal equilibrium?

5.4* Determine the noise figure and noise measure of the circuit of Fig. 5.5 by direct evaluation.

5.5 The scattering matrix of a two-port can be measured with standard equipment. Show how the noise correlation matrix could be obtained experimentally by placing appropriate shorts into the input and output.

5.6 A reactive feedback admittance Y is connected between the gate and the drain of the FET of Fig. 5.5. Show that the characteristic noise matrix remains unchanged.

5.7 Assume that the two noise sources in the equivalent circuit of Fig. 5.12 are uncorrelated. How does the noise figure vary as a function of the turns ratio of the transformer at the input of the amplifier?

5.8 The cascading formula is valid even if there is mismatch from stage to stage, because the definition of the noise figure takes this mismatch properly into account.

Determine the noise figure of a cascade of two FETs using the results of Prob. 5.7. The noise sources within each amplifier are assumed uncorrelated.

5.9 Assume that the two amplifiers in the preceding problem are identical and that transformers are placed between the source and the first amplifier and between the two amplifiers, of turns ratios n_1 and n_2, respectively. Derive the noise figure as a function of the transformer turns ratio.

5.10* It is well known that a Hermitian matrix has real eigenvalues. The characteristic noise matrix is not Hermitian but is composed of the product of two Hermitian matrices, one of which is positive definite. Prove that such matrices have real eigenvalues. Hint: use the fact that the factor matrices can be diagonalized by the same similarity transformation.

Solutions

5.1

(a) The products, written out in terms of magnitudes and phases, are

$$x^\dagger A x = A_{ij}|x_i||x_j|\exp\left[i(\phi_j - \phi_i)\right],$$

$$x^\dagger B x = B_{ij}|x_i||x_j|\exp[i(\phi_j - \phi_i)].$$

$$\frac{\partial}{\partial |x_k|} A_{ij}|x_i||x_j|\exp[\mathrm{i}(\phi_j - \phi_i)]$$

$$= A_{kj}|x_j|\exp[\mathrm{i}(\phi_j - \phi_k)] + A_{ik}|x_i|\exp[\mathrm{i}(\phi_k - \phi_i)]$$

$$= A_{kj}|x_j|\exp[\mathrm{i}(\phi_j - \phi_k)] + A_{ki}^*|x_i|\exp[-\mathrm{i}(\phi_i - \phi_k)]$$

$$= 2\mathrm{Re}[e^{-\mathrm{i}\phi_k}\boldsymbol{A}\boldsymbol{x}] ,$$

since $\boldsymbol{A}$ is a Hermitian matrix.

(b)

$$\frac{\partial}{\partial \phi_k} A_{ij}|x_i||x_j|\exp[\mathrm{i}(\phi_j - \phi_i)] = \mathrm{i}|x_k|e^{-\mathrm{i}\phi_k}\sum_j A_{kj}|x_j|\exp(\mathrm{i}\phi_j)$$

$$-\mathrm{i}|x_k|e^{\mathrm{i}\phi_k}\sum A_{jk}|x_j|\exp(-\mathrm{i}\phi_j) = 2|x_k|\,\mathrm{Im}(e^{-\mathrm{i}\phi_k}\boldsymbol{A}\boldsymbol{x}) .$$

(c) We obtain

$$2\,\mathrm{Re}\,[e^{-\mathrm{i}\phi_k}(\boldsymbol{A}\boldsymbol{x} - \lambda\boldsymbol{B}\boldsymbol{x})] = 0 ,$$

$$2|x_k|\,\mathrm{Im}[e^{-\mathrm{i}\phi_k}(\boldsymbol{A}\boldsymbol{x} - \lambda\boldsymbol{B}\boldsymbol{x})] = 0 .$$

It is clear that these two sets of equations imply

$$\boldsymbol{A}\boldsymbol{x} - \lambda\boldsymbol{B}\boldsymbol{x} = 0 .$$

5.3 The equations of the system are

$$i_1 = (g_g - \mathrm{i}\omega c_g)v_1 + i_g ,$$

$$i_2 = g_m v_1 + g_d v_2 + i_d .$$

From the admittance matrix

$$\frac{1}{2}(\boldsymbol{Y} + \boldsymbol{Y}^\dagger) = \begin{bmatrix} g_g & \frac{1}{2}g_m \\ \frac{1}{2}g_m & g_d \end{bmatrix} ,$$

we may judge whether the network is passive or active. If all determinants and subdeterminants are positive definite, the network is passive. The determinant is

$$\det\frac{1}{2}(\boldsymbol{Y} + \boldsymbol{Y}^\dagger) = g_g g_d - \frac{1}{4}g_m^2 .$$

The network is passive if $g_m^2 < 4g_g g_d$. Then this network can be at thermal equilibrium. Such a network has a characteristic noise matrix that is proportional to the identity matrix. From this fact we may derive the noise current correlation matrix

$$
\begin{bmatrix} \langle|i_g^2|\rangle & \langle i_g i_d^*\rangle \\ \langle i_g^* i_d\rangle & \langle|i_d^2|\rangle \end{bmatrix} = 4k\theta B \begin{bmatrix} g_s & \frac{1}{2}g_m \\ \frac{1}{2}g_m & g_d \end{bmatrix} .
$$

5.4 In order to compute the available gain, we must compute the available input power from the source $P_{s,\mathrm{av}}$ and the available output power $P_{\mathrm{out,av}}$. Suppose that the admittance of the source is G_s, and the source current generator is I_s.

$$
P_{s,\mathrm{av}} = |i_s|^2/4G_s; \quad P_{\mathrm{out,av}} = |g_m e_g|^2/4g_d ,
$$

where

$$
e_g = i_s/(G_s + g_g) .
$$

The available gain G is

$$
G = |e_g g_m|^2 G_s/|i_s|^2 g_d) = \frac{g_m^2}{(G_s + g_g)^2} \frac{G_s}{g_d} .
$$

The available output noise power is

$$
\langle|g_m e_g + i_d|^2\rangle/4g_d = \langle|g_m \frac{i_g}{G_s + g_g} + i_d|^2\rangle/4g_d
$$

$$
= \frac{g_m^2}{4g_d(G_s + g_g)^2}\langle|i_g^2|\rangle + \frac{\langle|i_d^2|\rangle}{4g_d} + \frac{2\,\mathrm{Re}(\langle i_g i_d^*\rangle)g_m}{4g_d(G_s + g_d)} .
$$

The noise measure is

$$
M = \frac{1}{k\theta B} \frac{\{[g_m^2/4g_d(G_s + g_g)^2]\langle|i_g^2|\rangle + \langle|i_d^2|\rangle/4g_d}{([g_m^2/(G_s + g_g)^2](G_s/g_d) - 1)}
$$

$$
+ \frac{2\,\mathrm{Re}(\langle i_g i_d^*\rangle)g_m/4g_d(G_s + g_d)\}}{([g_m^2/(G_s + g_g)^2](G_s/g_d) - 1)} .
$$

5.10 The eigenvalue problem of a Hermitian matrix H is

$$
Hx = \lambda x .
$$

The eigenvalues are all real. The eigenvalue equation to be considered is

$$
Ax = B\lambda x ;
$$

the A matrix is Hermitian and the B matrix is positive definite and Hermitian. It is possible to diagonalize and reduce to the identity the matrix B with premultiplication by D and postmultiplication $D^\dagger$. We obtain

$$
DAD^\dagger D^{\dagger -1}x = \lambda D^{\dagger -1}x .
$$

The matrix $DAD^\dagger$ is Hermitian. Thus, we have reduced the problem to a Hermitian eigenvalue problem.

6. Quantum Theory of Waveguides and Resonators

Thus far we have studied shot noise and thermal noise. In the case of thermal noise we extended the analysis by including Planck's quantum postulate of energy occurring in quanta. This led to the Bose–Einstein distribution of photons, a distribution of thermal equilibrium. Then we studied the noise of classical linear systems, and determined the optimum noise performance of a linear two-port amplifier with specified internal noise. In this chapter, we begin the study of quantum noise as governed by the quantized equations of motion.

The Schrödinger equation for the wave function of a quantum system is a linear equation of motion. On the other hand, the world around us is "nonlinear", yet the nonlinear behavior of a system is not easily perceived from the Schrödinger representation. It is also well known that it requires a certain effort to derive from the Schrödinger formalism the correspondence principle, which shows the connection with classical equations of motion. The Heisenberg equations of motion of operators, on the other hand, contain the correspondence principle in their very appearance. The observables, representable by c numbers classically, are replaced by operators that obey the classical equations of motion, provided, of course, that these observables have classical interpretations. (The spin operators are examples of operators that do not have a classical counterpart.) If the classical equations of motion are nonlinear, this nonlinearity carries over into the Heisenberg equations of motion. It is for this reason that we use the Heisenberg formalism for the representation of mode propagation in optical waveguides or optical fibers. The operator formalism ensures that quantum fluctuations (such as amplified spontaneous emission) are properly treated.

Quantization of the electromagnetic field is accomplished by treating the eigenmodes of the electromagnetic-field system as harmonic oscillators. The harmonic oscillators are quantized in the standard way. Section 6.1 reviews this quantization procedure. Section 6.2 looks in greater detail at creation and annihilation operators, and Sect. 6.3 studies the eigenstates of the annihilation operator, so-called coherent states. Section 6.4 describes the close connection between the uncertainty principle and noise. With this background, we can address the problem of noise in a laser below threshold. We show the need for the introduction of Langevin sources and determine their commutators from the requirement that commutators must be preserved in the evolution of a resonator mode. Then we investigate the case of waveguides with loss and gain. We determine the amplified spontaneous emission generated by an amplifier as required by the conservation of commutator brackets. We describe an experiment that determines the amplified spontaneous emission and thus the noise figure or noise measure of an optical amplifier. Finally, we study the quantum noise in a laser resonator below threshold.

6.1 Quantum Theory of the Harmonic Oscillator

The simplest approach to the quantization of electromagnetic fields takes advantage of the fact that waveguide modes and resonator modes obey the equations of harmonic oscillators. Quantization of the modes is in one-to-one correspondence with the quantization of the harmonic oscillator. Interactions between modes are taken into account by coupling Hamiltonians added to the Hamiltonian of the oscillators. Hence, an understanding of the quantum theory of the harmonic oscillator is basic to the understanding of the quantization of electromagnetic fields.

A classical harmonic oscillator of mass m and spring constant k obeys the equation of motion

$$\frac{dp}{dt} = -kq , \tag{6.1}$$

where p is the momentum and q is the displacement. The momentum is related to the displacement by

$$m\frac{dq}{dt} = p . \tag{6.2}$$

Elimination of p leads to the following second-order differential equation for q:

$$\frac{d^2}{dt^2}q + \omega_o^2 q = 0 , \quad \text{where} \quad \omega_o^2 = \frac{k}{m} . \tag{6.3}$$

These equations of motion can be obtained more formally from the Hamiltonian, which is the sum of the kinetic and potential energies:

$$H = \frac{1}{2}\left(\frac{p^2}{m} + kq^2\right) . \tag{6.4}$$

Equations (6.1) and (6.3) follow from the use of Poisson brackets on the Hamiltonian. The Poisson bracket is defined by a difference of derivative pairs:

$$\{u, v\}_{q,p} = \frac{\partial u}{\partial q}\frac{\partial v}{\partial p} - \frac{\partial v}{\partial q}\frac{\partial u}{\partial p} . \tag{6.5}$$

The time derivative of q is equal to the Poisson bracket in which u is identified with q and v is identified with the Hamiltonian H:

$$\frac{dq}{dt} = \{q, H\}_{q,p} = \frac{\partial q}{\partial q}\frac{\partial H}{\partial p} - \frac{\partial H}{\partial q}\frac{\partial q}{\partial p} = \frac{\partial H}{\partial p} = \frac{p}{m} .$$ (6.6)

The time derivative of p is equal to the Poisson bracket in which u is identified with p and v is identified with the Hamiltonian H:

$$\frac{dp}{dt} = \{p, H\}_{q,p} = \frac{\partial p}{\partial q}\frac{\partial H}{\partial p} - \frac{\partial H}{\partial q}\frac{\partial p}{\partial p} = -\frac{\partial H}{\partial q} = -kq .$$ (6.7)

These are indeed the correct equations of motion.

The harmonic oscillator is quantized by representing the observables p and q by operators $\hat{p}$ and $\hat{q}$ and by replacement of the Poisson brackets with the commutator brackets divided by $i\hbar$. The Hamiltonian becomes

$$\hat{H} = \frac{1}{2}\left(\frac{\hat{p}^2}{m} + k\hat{q}^2\right) .$$ (6.8)

Note that in the classical regime, the Poisson bracket of the position and momentum is

$$\{q, p\}_{q,p} = \frac{\partial q}{\partial q}\frac{\partial p}{\partial p} - \frac{\partial p}{\partial q}\frac{\partial q}{\partial p} = 1 .$$ (6.9)

If the commutator bracket divided by $i\hbar$ is to yield unity, then the momentum operator in the q representation must be identified with

$$\hat{p} = -i\hbar\frac{\partial}{\partial\hat{q}} .$$ (6.10)

Indeed, the commutator of $\hat{q}$ and $\hat{p}$ is then

$$[\hat{q}, \hat{p}] = \hat{q}\hat{p} - \hat{p}\hat{q} = -i\hbar\left(\hat{q}\frac{\partial}{\partial\hat{q}} - \frac{\partial}{\partial\hat{q}}\hat{q}\right) = i\hbar .$$ (6.11)

Several considerations enter into the operations carried out in the above expression. First of all, the products $\hat{q}\hat{p}$ and $\hat{p}\hat{q}$ are operators intended to operate on a function of $\hat{q}$. Secondly, derivatives with respect to $\hat{q}$ of powers of $\hat{q}$ in, say, a Taylor expansion of a function of $\hat{q}$ behave like derivatives with respect to a classical (c number) variable, since every operator commutes with itself.

Application of the same rules gives for the equations of motion

$$\frac{d\hat{q}}{dt} = -\frac{i}{\hbar}[\hat{q}, \hat{H}] = \frac{1}{m}\hat{p}$$ (6.12)

and

$$\frac{d\hat{p}}{dt} = -\frac{i}{\hbar}[\hat{p}, \hat{H}] = -k\hat{q} .\tag{6.13}$$

These are the Heisenberg equations of motion. They are in one-to-one correspondence with their classical counterparts. This is one form of the correspondence principle, which requires that the quantum mechanical description of physical processes merge with the classical description in the limit when the energy of excitation of the oscillator is large compared with $\hbar\omega_o$.

It proves convenient to introduce normalized variables

$$\hat{Q} = \sqrt{\frac{k}{\hbar\omega_o}}\hat{q} \quad \text{and} \quad \hat{P} = -i\frac{\partial}{\partial Q} .\tag{6.14}$$

The commutator of $\hat{Q}$ and $\hat{P}$ is

$$[\hat{Q}, \hat{P}] = \left[\hat{Q}, -i\frac{\partial}{\partial \hat{Q}}\right] = i .\tag{6.15}$$

In terms of these variables the Hamiltonian simplifies to

$$\hat{H} = \frac{\hbar\omega_o}{2}(\hat{Q}^2 + \hat{P}^2) = \frac{\hbar\omega_o}{2}\left(\hat{Q}^2 - \frac{\partial^2}{\partial \hat{Q}^2}\right) .\tag{6.16}$$

The Heisenberg equations of motion for the operators give

$$\frac{d}{dt}\hat{P} = -\frac{i}{\hbar}[\hat{P}, \hat{H}] = -\omega_o\hat{Q} ,\tag{6.17}$$

$$\frac{d}{dt}\hat{Q} = -\frac{i}{\hbar}[\hat{Q}, \hat{H}] = \omega_o\hat{P} .\tag{6.18}$$

Elimination of one of the two observables from (6.17) and (6.18) leads to a second order differential equation. Instead, one may introduce "canonical" variables that lead to two uncoupled first-order differential equations. These canonical variables are denoted by $\hat{A}$ and $\hat{A}^\dagger$:

$$\hat{A} = \frac{1}{\sqrt{2}}(\hat{Q} + i\hat{P}) \quad \text{and} \quad \hat{A}^\dagger = \frac{1}{\sqrt{2}}(\hat{Q} - i\hat{P}) .\tag{6.19}$$

The Heisenberg equations of motion of these two operators are obtained by addition and subtraction of (6.17) and (6.18), appropriately multiplied by i:

$$\frac{d\hat{A}}{dt} = -i\omega_o\hat{A} ,\tag{6.20}$$

$$\frac{d\hat{A}^\dagger}{dt} = i\omega_o\hat{A}^\dagger .\tag{6.21}$$

The operators $\hat{A}$ and $\hat{A}^\dagger$ obey the commutation relation

$$[\hat{A}, \hat{A}^\dagger] = 1 . \tag{6.22}$$

The Hamiltonian, written in terms of the operators $\hat{A}$ and $\hat{A}^\dagger$, has the form

$$\hat{H} = \hbar\omega_o \left(\hat{A}^\dagger \hat{A} + \frac{1}{2} \right) . \tag{6.23}$$

The energy eigenstates of the harmonic oscillator are obtained from the Schrödinger equation for quantum states

$$i\hbar \frac{d}{dt}\psi = \hat{H}\psi . \tag{6.24}$$

The state of constant energy E obeys the equation

$$i\hbar \frac{d}{dt}\psi = \hat{H}\psi = E\psi . \tag{6.25}$$

In the Q representation, this equation leads to the differential equation

$$\frac{\hbar\omega_o}{2} \left(\hat{Q}^2 - \frac{\partial^2}{\partial\hat{Q}^2} \right) \psi(\hat{Q}) = E\psi(\hat{Q}) . \tag{6.26}$$

Since this equation involves only the operator $\hat{Q}$, the operator can be treated as a c number. The solutions of this equation are Hermite Gaussians

$$\psi_n(Q) = H_n(Q) \exp(-Q^2/2) , \tag{6.27}$$

with the eigenvalues

$$E_n = \hbar\omega_o(n + 1/2) . \tag{6.28}$$

Figure A2.1 of Appendix A.2 shows some of the lowest-order Hermite Gaussians. Further details are given in Appendix A.2. Any general state $\varphi(Q)$ can be represented as a superposition of energy eigenstates:

$$\varphi(Q) = \sum_{n=0}^{\infty} c_n \psi_n(Q) . \tag{6.29}$$

Next we show that the operation of $\hat{A} = (1/\sqrt{2})(\hat{Q} + i\hat{P})$ on an eigenfunction $\psi_n(Q)$ produces the eigenfunction $\psi_{n-1}(Q)$, i.e. the operation annihilates a quantum of energy $\hbar\omega_o$. This is why the operator $\hat{A}$ is called the annihilation operator. For the proof, we note that

$$Q^2 - \frac{\partial^2}{\partial Q^2} = \left(Q - \frac{\partial}{\partial Q} \right) \left(Q + \frac{\partial}{\partial Q} \right) + 1$$

$$= \left(Q + \frac{\partial}{\partial Q} \right) \left(Q - \frac{\partial}{\partial Q} \right) - 1 . \tag{6.30}$$

We obtain from (6.26) and (6.30)

$$\frac{1}{\sqrt{2}}\left(Q + \frac{\partial}{\partial Q}\right)\frac{1}{2}\left(Q^2 - \frac{\partial^2}{\partial Q^2}\right)\psi_n(Q)$$

$$= \left[\frac{1}{2}\left(Q^2 - \frac{\partial^2}{\partial Q^2}\right)\frac{1}{\sqrt{2}}\left(Q + \frac{\partial}{\partial Q}\right) + 1\right]\psi_n(Q) \tag{6.31}$$

$$= \frac{E_n}{\hbar\omega_o}\frac{1}{\sqrt{2}}\left(Q + \frac{\partial}{\partial Q}\right)\psi_n(Q) \, .$$

This leads to the result

$$\left[\frac{1}{2}\left(Q^2 - \frac{\partial^2}{\partial Q^2}\right)\frac{1}{\sqrt{2}}\left(Q + \frac{\partial}{\partial Q}\right)\right]\psi_n(Q)$$

$$= \left(\frac{E_n}{\hbar\omega_o} - 1\right)\frac{1}{\sqrt{2}}\left(Q + \frac{\partial}{\partial Q}\right)\psi_n(Q) \, . \tag{6.32}$$

Hence we have proven that $(1/\sqrt{2})(Q + \partial/\partial Q)\psi_n(Q) = \hat{A}\psi_n(Q)$ is an energy eigenstate with an energy lowered by $\hbar\omega_o$ from the energy of $\psi_n(Q)$. The multiplier produced by the operation is gleaned from the matrix element $\langle\psi_n|\hat{A}^\dagger\hat{A}|\psi_n\rangle = \langle\psi_n|n|\psi_n\rangle = n$. We find

$$\hat{A}\psi_n(Q) = \sqrt{n}\,\psi_{n-1}(Q) \, , \tag{6.33}$$

within a phase factor that can be set to unity. Similarly, we find

$$\hat{A}^\dagger\psi_n(Q) = \sqrt{n+1}\,\psi_{n+1}(Q) \, . \tag{6.34}$$

Operation on the eigenfunction ψ_n by $\hat{A}^\dagger$ produces the eigenfunction of a higher-lying state on the next rung on the ladder of energy states. It is for this reason that $\hat{A}^\dagger$ is called the creation operator, because it creates one energy quantum.

The next issue to be taken up is the relation of the quantized harmonic oscillator to the quantization of electromagnetic fields. Equations (2.154) and (2.155) of Chap. 2 for the source-free cavity are

$$\epsilon\frac{dV_\nu}{dt} = p_\nu I_\nu \tag{6.35}$$

and

$$\mu\frac{dI_\nu}{dt} = -p_\nu V_\nu \, . \tag{6.36}$$

These equations establish the analogy between the displacement q of the harmonic oscillator and the amplitude V_ν of the electric field of the νth resonator mode, and between the momentum p of a harmonic oscillator and the amplitude I_ν of the magnetic field. If the field patterns are normalized as in (2.145) and (2.148) of Chap. 2, the Hamiltonian is

$$H = \frac{1}{2\mathcal{V}}(\epsilon V_\nu^2 + \mu I_\nu^2) \ . \tag{6.37}$$

The field is quantized by replacing $\sqrt{\epsilon/\mathcal{V}}V_\nu$ with the operator $\sqrt{\hbar\omega_\nu}\hat{Q}_\nu$, and $\sqrt{\mu/\mathcal{V}}I_\nu$ with the operator $\sqrt{\hbar\omega_\nu}\hat{P}_\nu$, a pair of operators for every mode. From here on, the analysis follows the harmonic oscillator analysis. Each operator has a set of energy eigenstates $|\psi(Q_\nu)\rangle$. Each mode has assigned to it creation and annihilation operators $\hat{A}_\nu^\dagger$ and $\hat{A}_\nu$.

6.2 Annihilation and Creation Operators

The quantization of electromagnetic waves proceeds by representing them as modes of a system with periodic boundary conditions, such as the modes of an optical fiber ring. This step relates the waves directly to the modes of a harmonic oscillator. Consider a mode on a fiber ring of length L. The choice of the length L depends on the physical situation under consideration, and in particular on the choice of the measurement apparatus. As another example one may consider a free-space Hermite Gaussian mode, repeatedly refocused by a periodic sequence of lenses.

The complex amplitude of the mode is expressed *classically* by $A_m(t)$, and this amplitude obeys the differential equation:

$$\frac{dA_m}{dt} = -\mathrm{i}\omega_m A_m \ , \tag{6.38}$$

with the solution

$$A_m(t) = A_m^o \exp(-\mathrm{i}\omega_m t) \ . \tag{6.39}$$

The mode, of frequency ω_m, has a propagation constant β_m. The mode obeys the boundary condition $\beta_m = (2\pi/L)m$, with m an integer. The amplitude can be normalized so that the energy w in the mode is given by

$$w = |A_m|^2 \ . \tag{6.40}$$

The energy w is interpreted as the total energy of the mode, composed of the energy in the electromagnetic field as well as that in the medium. If the medium is dispersive, the expressions developed in Chap. 2 apply. The presence of a dispersive medium affects the dispersion of the waveguide, the dependence on the propagation constant of the frequency w, i.e. $\omega = \omega(\beta)$. The amplitude $A_m(t)$ is a complex function of time, whereas the electric field is a real function of time. The electric field amplitude is proportional to

$$E(t) \propto \frac{1}{2}[A_m(t) + A_m^*(t)] \equiv A_m^{(1)}(t) , \tag{6.41}$$

where we shall call $A_m^{(1)}(t)$ the "in-phase component" of the electric field. One may construct a component in quadrature to the electric field as

$$\frac{1}{2i}[A_m(t) - A_m^*(t)] \equiv A_m^{(2)}(t) , \tag{6.42}$$

which is also a real function of time.

Quantization is accomplished when the modes of the electromagnetic field are identified with the modes of harmonic oscillators, one oscillator per mode. Comparison of (6.38) and (6.20) shows that the operator representing the complex field amplitude $A_m(t)$ is an annihilation operator $\hat{A}_m(t)$. The energy of the mth harmonic oscillator in the state ψ_m is, according to (6.28),

$$w = \hbar\omega_m\left(n_m + \frac{1}{2}\right) . \tag{6.43}$$

It is convenient to introduce the Dirac ket and bra notation for the energy eigenstates of the harmonic oscillator and their Hermitian adjoints: ψ_m of the mth harmonic oscillator is written $|n_m\rangle$. Thus, (6.33) and (6.34), rewritten in this notation for the mth harmonic oscillator, assume the form

$$\hat{A}_m|n_m\rangle = \sqrt{n_m}|n_m - 1\rangle \tag{6.44}$$

and

$$\hat{A}_m^\dagger|n_m\rangle = \sqrt{n_m + 1}|n_m + 1\rangle . \tag{6.45}$$

The operators of different harmonic oscillators commute, so that (6.22) can be generalized to

$$[\hat{A}_m, \hat{A}_n^\dagger] = \delta_{mn} . \tag{6.46}$$

The operator $\hat{A}_m^\dagger \hat{A}_m$ is the photon number operator of the mth mode. Indeed, combining (6.44) and (6.45), we see that operation of this operator on a number state gives

$$\hat{A}_m^\dagger \hat{A}_m|n_m\rangle = n_m|n_m\rangle . \tag{6.47}$$

The Hamiltonian includes the energy of all harmonic oscillators:

$$\hat{H} = \sum_m \hbar\omega_m\left(\hat{A}_m^\dagger \hat{A}_m + \frac{1}{2}\right) . \tag{6.48}$$

The Heisenberg equation for the evolution of the operator $\hat{A}_m$,

$$\frac{d}{dt}\hat{A}_m = -\frac{i}{\hbar}[\hat{A}_m, \hat{H}] \, ,$$
(6.49)

and the commutation relation (6.46) lead to

$$\frac{d}{dt}\hat{A}_m = -i\omega_m\hat{A}_m \, .$$
(6.50)

This is the same equation of motion as for the classical complex field amplitude. Note that the addend $1/2$ to the energy due to the zero-point fluctuations does not contribute to the equation of motion, since it is a c number and commutes with the operator $\hat{A}_m$. Further details on wave functions and operators are presented in Appendix A.7.

The annihilation and creation operators are not Hermitian. This means they do not represent observables, since observables must be represented by Hermitian operators. On the other hand, the operators representing the in-phase and quadrature components of the electric field as defined classically in (6.41) and (6.42) are Hermitian:

$$\hat{A}_m^{(1)} = \frac{1}{2}(\hat{A}_m + \hat{A}_m^\dagger) \, ,$$
(6.51)

$$\hat{A}_m^{(2)} = \frac{1}{2i}(\hat{A}_m - \hat{A}_m^\dagger) \, .$$
(6.52)

The expectation value of an operator is evaluated by "projection" with the state of the system. In Dirac notation, the expectation value of the photon number operator $\hat{A}_m^\dagger\hat{A}_m$ when the system is in a number state $|n_m\rangle$ is

$$\langle n_m|\hat{A}_m^\dagger\hat{A}_m|n_m\rangle = n_m \, .$$
(6.53)

This result follows from (6.44) and (6.45) and the normalization of the eigenstates

$$\langle n_m|n_p\rangle = \delta_{mp} \, .$$
(6.54)

The higher order moments of the number operator are

$$\langle n_m|(\hat{A}_m^\dagger\hat{A}_m)^N|n_m\rangle = n_m^N \, .$$
(6.55)

This shows that the photon number of a number state has the definite value n_m as expected.

6.3 Coherent States of the Electric Field

Quantum states of the electromagnetic field may exhibit nonclassical behavior. Some examples of such behavior will be discussed later on in this chapter and in Chap. 7. The quantum states that are closest in behavior to classical fields are the so-called "coherent states" [64–66]. The coherent state of the mth mode is represented by the ket $|\alpha_m\rangle$. In order to simplify the notation we shall drop the subscript m on the mode when a single mode is considered. Whenever a superposition of modes is treated, we shall restore the subscripts. The coherent state $|\alpha\rangle$ can be written in terms of the energy eigenstates, the so-called photon number states [66, 67]

$$|\alpha\rangle = e^{-|\alpha|^2/2} \sum_n \frac{\alpha^n}{\sqrt{n!}} |n\rangle , \tag{6.56}$$

where α is a complex number. One may confirm easily that the coherent state is an eigenstate of the annihilation operator, using (6.44):

$$\hat{A}|\alpha\rangle = \alpha|\alpha\rangle \tag{6.57}$$

Similarly, the Hermitian conjugate operation leads to

$$\langle\alpha|\hat{A}^\dagger = \alpha^*\langle\alpha| . \tag{6.58}$$

The coherent state has Poissonian photon statistics. Let us evaluate the Mth moment of the photon number in a coherent state $|\alpha\rangle$:

$$\langle n^M \rangle = \langle\alpha|(\hat{A}^\dagger \hat{A})^M|\alpha\rangle = e^{-|\alpha|^2} \sum_{m,n} \frac{\alpha^{*m}\alpha^n}{\sqrt{m!n!}} \langle m|n^M|n\rangle$$

$$= e^{-|\alpha|^2} \sum_n \frac{|\alpha|^{2n}}{n!} n^M . \tag{6.59}$$

The expectation value of n^M is the probability-weighted sum:

$$\langle n^M \rangle = \sum_n p(n) n^M . \tag{6.60}$$

Comparing (6.59) and (6.60), we find that the probability distribution is

$$p(n) = e^{-|\alpha|^2} \frac{|\alpha|^{2n}}{n!} . \tag{6.61}$$

The average photon number $\langle n \rangle$ is

$$\sum_n p(n)n = \sum_n e^{-|\alpha|^2} \frac{|\alpha|^{2n}}{n!} n = |\alpha|^2 \sum_n e^{-|\alpha|^2} \frac{|\alpha|^{2n}}{n!} = |\alpha|^2 . \tag{6.62}$$

Thus, we may write the probability (6.61) in terms of the average photon number $\langle n \rangle = |\alpha|^2$:

$$p(n) = e^{-\langle n \rangle} \frac{\langle n \rangle^n}{n!} ,\tag{6.63}$$

which is the Poisson distribution. The expectation value of the photon number operator, $\langle |\hat{A}^\dagger \hat{A}| \rangle$, gives the photon number in the mode in a ring of length L. Appendix A.7 discusses further properties of wave functions in the photon number state representation.

The expectation values of the in-phase and quadrature fields for a coherent state are

$$\langle \alpha | \hat{A}^{(1)} | \alpha \rangle = \frac{1}{2} \langle \alpha | (\hat{A} + \hat{A}^\dagger) | \alpha \rangle = \frac{1}{2}(\alpha + \alpha^*)\tag{6.64}$$

and

$$\langle \alpha | \hat{A}^{(2)} | \alpha \rangle = \frac{1}{2i} \langle \alpha | (\hat{A} - \hat{A}^\dagger) | \alpha \rangle = \frac{1}{2i}(\alpha - \alpha^*) .\tag{6.65}$$

We find that the complex parameter α represents the expectation value of the electric field in the complex phasor plane.

The commutator is closely related to the mean square "vacuum" fluctuations of the field. We start by asking for the number of photons in a waveguide mode in the ground state, the state $|0\rangle$. It is clear that

$$\langle 0 | \hat{A}^\dagger \hat{A} | 0 \rangle = 0 .\tag{6.66}$$

On the other hand, using the commutation relation, one finds that

$$\langle 0 | \hat{A} \hat{A}^\dagger | 0 \rangle = 1 .\tag{6.67}$$

The expectation value of the field in the ground state is zero:

$$\langle 0 | \hat{A}^{(1)} | 0 \rangle = \langle 0 | \frac{1}{2}(\hat{A}^\dagger + \hat{A}) | 0 \rangle = 0 .\tag{6.68}$$

The mean square fluctuations of the field are

$$\langle 0 | (\hat{A}^{(1)})^2 | 0 \rangle = \langle 0 | \frac{1}{4}(\hat{A}^\dagger \hat{A} + \hat{A} \hat{A} + \hat{A}^\dagger \hat{A}^\dagger + \hat{A} \hat{A}^\dagger) | 0 \rangle = \frac{1}{4} .\tag{6.69}$$

The mean square field fluctuations are due to the operator $\hat{A} \hat{A}^\dagger$, which is in reverse order to the photon number operator. Thus, even in the ground state, there is a contribution to the mean square field. These are the so-called zero-point fluctuations or vacuum fluctuations of the field. Similarly,

$$\langle 0 | (\hat{A}^{(2)})^2 | 0 \rangle = \frac{1}{4} .\tag{6.70}$$

The fluctuations of the in-phase and quadrature components contribute to the Hamiltonian of (6.48) the term $1/2$. The mean square fluctuations of a coherent state are the same as those of the vacuum state. They are evaluated from the following expression

$$\langle\alpha|\Delta\hat{A}^{(1)^2}|\alpha\rangle = \frac{1}{4}\langle\alpha|\hat{A}\hat{A} + \hat{A}^\dagger\hat{A}^\dagger + \hat{A}\hat{A}^\dagger + \hat{A}^\dagger\hat{A}|\alpha\rangle$$

$$-\frac{1}{4}\langle\alpha|\hat{A} + \hat{A}^\dagger|\alpha\rangle^2 .$$

(6.71)

Expressions such as (6.71) are easily evaluated if one notes that, for any coherent state $|\alpha\rangle$, the following theorem holds:

$$\langle\alpha|(\hat{A}^\dagger)^m\hat{A}^n|\alpha\rangle = \alpha^{*m}\alpha^n = \langle\alpha|\hat{A}^\dagger|\alpha\rangle^m\langle\alpha|\hat{A}|\alpha\rangle^n .$$

(6.72)

In words: the expectation value of the product of the creation operator $\hat{A}^\dagger$ to the mth power and the annihilation operator $\hat{A}$ to the nth power is equal to the product of the mth power of the expectation value of the operator $\hat{A}^\dagger$ and the nth power of the expectation value of the operator $\hat{A}$. This statement is true when the product of the operators is written in *normal order*, the creation operators precede the annihilation operators.

The theorem is useful in the evaluation of expressions like (6.71). If one casts the sum of operators in the first expression into normal order, then all terms of second order in the operators cancel against the square of the expectation value. Left over is a term due to the commutator, which is introduced in reversing the order of the term not in normal order. In this way one finds simply

$$\langle\alpha|\Delta\hat{A}^{(1)^2}|\alpha\rangle = \frac{1}{4}\langle\alpha|\hat{A}\hat{A} + \hat{A}^\dagger\hat{A}^\dagger + 2\hat{A}^\dagger\hat{A} + 1|\alpha\rangle$$

$$-\frac{1}{4}\langle\alpha|\hat{A}\hat{A}^\dagger|\alpha\rangle^2 = \frac{1}{4} .$$

(6.73)

The rearrangement into normal order saves a great deal of algebra when evaluating mean square fluctuations of coherent states. Equation (6.73) shows that the in-phase component of a coherent state has the same mean square fluctuations as the vacuum state. Figure 6.1a displays the electric field of a coherent state in the complex plane. The complex parameter α gives the phasor in phase and amplitude. The endpoint of the phasor lies at the center of a circle that shows the half-locus of the probability distribution (the locus outside of which the probability of finding a member of the ensemble is less than $\exp(-1/2)$). We shall prove in Sect. 7.6 that the distribution of endpoints is Gaussian. The ground state, or state of the vacuum at absolute zero, is illustrated in Fig. 6.1b. The distribution of the field amplitude is symmetric around the origin.

One may ask the question as to the physical meaning of graphs like the ones shown in Fig. 6.1a,b. They were obtained by asking for the expectation

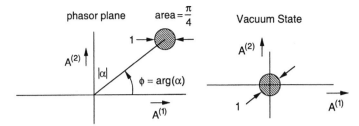

Fig. 6.1. (a) Representation of coherent state in complex phasor plane; (b) representation of vacuum state in complex phasor plane

value of the complex field and for its mean square fluctuations. Quantum theory describes the world probabilistically. Quantum theory does not give information about one physical system, but only about an ensemble of identically prepared systems. If ideal measurements that do not perturb the value of the observable are performed on such an ensemble of systems (we shall have the opportunity of studying some of such measurements), then quantum theory predicts the probability distribution of the outcomes of measurements on such an ensemble. In the practical world, in the absence of the availability of an ensemble of systems, one may proceed with an ensemble of measurements on the same system, making sure that the system starts in each case from the same initial state. The rule is that the expectation value (or average) of an observable, represented by a Hermitian operator, is obtained by the projection of the operator via the bra and ket of the state, in the case of a coherent state $\langle \alpha |$ and $| \alpha \rangle$, respectively. Squares of observables are, of course, also observables. Using this rule for evaluating expectation values, one may determine all the moments as well as the mean square deviations of observables.

6.4 Commutator Brackets, Heisenberg's Uncertainty Principle and Noise

Quantum theory treats as harmonic oscillators the "unbounded" modes on an open waveguide or transmission line, free-space Hermite Gaussian modes, and the modes on a fiber. The excitation of a harmonic oscillator of frequency ω is described by its position and momentum $\hat{q}$ and $\hat{p}$, respectively, or the annihilation and creation operators $\hat{A}$ and $\hat{A}^\dagger$, respectively. In the description of the electromagnetic field, $\hat{A}$ is analogous to the classical complex field amplitude. The position and momentum operators $\hat{q}$ and $\hat{p}$ play the role of in-phase and quadrature components of the electric field as referred to a phase reference of, say, a classical oscillator of fixed phase and frequency ω. If written as in-phase and quadrature components, their dimensions become identical. Whereas there is no particular significance attached to the elliptic phase diagram of the motion of $\hat{q}$ and $\hat{p}$ of a harmonic oscillator, the phase diagram of the motion of the in-phase and quadrature components must be a circle. We shall come back to the deep significance of the phase diagrams of the in-phase and quadrature components.

The creation and annihilation operators of the modes m and n obey the commutator relation (6.46)

$$[\hat{A}_m, \hat{A}_n^\dagger] = \delta_{mn} . \tag{6.74}$$

The commutators are an inalienable property of unbounded modes. They are also intimately related to their fluctuations and thus to fundamental quantum noise. Indeed, Heisenberg's uncertainty principle states that the root mean square deviations of the expectation values of two Hermitian operators $\hat{F}$ and $\hat{G}$ are proportional to their commutator C, if the commutator is a c number, as it is in the harmonic-oscillator cases of interest. Consider two operators $\hat{F}$ and $\hat{G}$ with the commutator

$$[\hat{F}, \hat{G}] = iC . \tag{6.75}$$

Then one may show through the use of Schwarz's inequality (see Appendix A.8) that the product of their mean square deviations obeys the inequality

$$(\langle \hat{F}^2 \rangle - \langle \hat{F} \rangle^2)(\langle \hat{G}^2 \rangle - \langle \hat{G} \rangle^2) \geq \frac{1}{4} C^2 . \tag{6.76}$$

But, mean square deviations are noise. Thus, the commutators determine the noise of electromagnetic modes. At the very least they establish a lower limit on the noise.

In (6.76) we looked at two general Hermitian operators. The in-phase and quadrature components of the electromagnetic field of a mode, $\hat{A}^{(1)} = (1/2)(\hat{A} + \hat{A}^\dagger)$ and $\hat{A}^{(2)} = (1/2i)(\hat{A} - \hat{A}^\dagger)$, are Hermitian operators and obey the commutator relation, a consequence of (6.74),

$$[\hat{A}^{(1)}, \hat{A}^{(2)}] = \frac{i}{2} . \tag{6.77}$$

Thus, the product of their mean square fluctuations must be greater than or equal to $1/16$.

Here we should remind ourselves that the in-phase and quadrature components referred to a time dependence $\cos(\omega t)$ are sines and cosines. If the noise is stationary, then the sine and cosine components must be uncorrelated and equal in the mean-square sense. Hence, one may conclude immediately that for a stationary process

$$\langle [\hat{A}^{(1)}]^2 \rangle - \langle \hat{A}^{(1)} \rangle^2 \geq \frac{1}{4} \quad \text{and} \quad \langle [\hat{A}^{(2)}]^2 \rangle - \langle \hat{A}^{(2)} \rangle^2 \geq \frac{1}{4} . \tag{6.78}$$

This establishes the minimum amount of quantum noise associated with the in-phase and quadrature components.

6.5 Quantum Theory of an Open Resonator

In Sect. 6.2 we discussed the quantization of modes in ring resonators. In this section, we shall study the equations of motion of the closed and the open resonator in greater detail. The present approach should be compared with the analysis in Sect. 2.12. We shall concentrate on a single resonance and drop the subscript m on the mode. We denote the creation and annihilation operators of the complex amplitude of the resonator mode by $\hat{U}^\dagger$ and $\hat{U}$. The Hamiltonian of the closed resonator is

$$\hat{H} = \hbar\omega_o\left(\hat{U}^\dagger\hat{U} + \frac{1}{2}\right). \tag{6.79}$$

The commutation relation

$$[\hat{U}, \hat{U}^\dagger] = 1, \tag{6.80}$$

employed in the Heisenberg equation of motion, leads to the differential equation

$$\frac{d\hat{U}}{dt} = -i\omega_o\hat{U}. \tag{6.81}$$

This is the description of the closed resonator. The equations for the open resonator are more subtle. In Sect. 6.4, we discussed the classical description of a resonator coupled to a waveguide. The coupling introduced a decay of the mode due to leakage into the coupled waveguide. A decay has no simple quantum description, since it "smacks" of irreversibility, and the equations of quantum mechanics are reversible. Now, it is well known that decay can be simulated in a quantum system by coupling it to an infinite set of modes. This is a very fundamental concept, and hence it is of interest to arrive at the quantum formulation of the classical equation (4.49) using this approach.

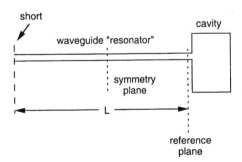

Fig. 6.2. Resonator coupled to long waveguide ($L \to \infty$)

Consider a resonator with one single resonant mode of interest, of frequency ω_o, coupled to a very long waveguide, which in turn may be modeled as a resonator with infinitesimally closely spaced resonance frequencies (see Fig. 6.2). The Hamiltonian of the total system, in terms of the creation and annihilation operators of the two subsystems, is

$$\hat{H} = \hbar\omega_o\left(\hat{U}^\dagger\hat{U} + \frac{1}{2}\right) + \sum_j \hbar\omega_j\left(\hat{V}_j^\dagger\hat{V}_j + \frac{1}{2}\right)$$

$$+ \hbar\sum_j\left(K_j\hat{U}^\dagger\hat{V}_j + K_j^*\hat{V}_j^\dagger\hat{U}\right),$$

(6.82)

where the ω_j are the frequencies of the waveguide modes, ω_o is the frequency of the resonator mode, and the K_j are the coupling coefficients of the waveguide modes to the resonator modes and vice versa. The Heisenberg equations of motion are

$$\frac{d\hat{U}}{dt} = -\mathrm{i}\omega_o\hat{U} - \mathrm{i}\sum_j K_j\hat{V}_j,$$

(6.83)

$$\frac{d\hat{V}_j}{dt} = -\mathrm{i}\omega_j\hat{V}_j - \mathrm{i}K_j^*\hat{U}.$$

(6.84)

Note that the coefficient of coupling of $\hat{V}_j$ to $\hat{U}$ is the complex conjugate of the coefficient of coupling of $\hat{U}$ to $\hat{V}_j$. This is the consequence of the fact that the equations are derived from a Hamiltonian and thus conserve energy.

At this point, it is of interest to ask about the nature of the modes associated with the operators $\hat{V}_j$ in the long waveguide "resonator" attached to the resonator under study, called simply the "cavity". The classical picture of a resonator mode radiating into an output waveguide can be used as a guide. If an initial excitation in the cavity starts to radiate into the external guide at $t = 0$, the electromagnetic field may be constructed from a sequence of impulses traveling into the guide in Fig. 6.2 from right to left. How is this phenomenon represented by a superposition of modes in the shorted waveguide "resonator" of length L?

The shorted-waveguide "resonator" has standing-wave solutions that are symmetric and antisymmetric with respect to the central symmetry plane between the two end shorting planes. These modes by themselves cannot couple directly through the shorting plane into the resonator. Coupling is achieved by placing surface currents at the reference plane, surface currents that represent the cavity field at the reference plane, as discussed in Chap. 2. Here we do not need to be concerned with the details of the current distribution, since the coupling is represented by the coefficients K_j in the Hamiltonian (6.82). Traveling waves may be constructed from a superposition of the standing waves. When the cavity is excited at $t = 0$, such traveling waves emerge from the cavity traveling to the left. The fact that the waveguide resonator is terminated in an electric short at the far left end does not affect the solution until the wave hits this termination. With $L \to \infty$ this does not happen within a finite time.

The operator amplitudes $\hat{V}_j$ of the waveguide modes obey the commutation relations

$$[\hat{V}_j, \hat{V}_k^\dagger] = \delta_{jk} \ . \tag{6.85}$$

These commutators are inherent attributes of the modes. The coupling of the resonator to the waveguide alters the modes in that the $\hat{V}_j$ acquire a contribution from the resonator mode leaking into the waveguide. This contribution is from the coupling of $\hat{U}$ in (6.84) and consists of waves traveling away from the resonator. With an assumed time dependence of $\hat{U}$ of the form $\exp(-i\omega t)$, we find for the part of $\hat{V}_j$ affected by $\hat{U}$

$$\hat{V}_j^{(U)} = \frac{K_j^* \hat{U}}{\omega - \omega_j} \ . \tag{6.86}$$

When this expression is substituted back into (6.83), we find the determinantal equation for ω:

$$\omega - \omega_o = \sum_j \frac{|K_j|^2}{\omega - \omega_j} \ . \tag{6.87}$$

The summation over the closely spaced resonances can be replaced by an integration. Assuming that the coupling coefficients do not vary with frequency over the frequency interval of interest, setting $|K_j^2| = \kappa^2 \Delta \beta_j = \kappa^2 (\Delta \beta / \Delta \omega) \Delta \omega_j = (\kappa^2 / v_g) \Delta \omega_j$, and using the fact that the integral passes around the pole in a semicircle, we obtain $\sum_j |K_j|^2/(\omega - \omega_j) \to -(\kappa^2/v_g)\pi i$. The determinantal equation (6.87) becomes

$$-i(\omega - \omega_o) = -\frac{\pi}{v_g}\kappa^2 \equiv -\frac{1}{\tau_e} \ . \tag{6.88}$$

We have found a decay rate $1/\tau_e$ due to the coupling to the waveguide. Equation (6.83) has acquired a decay and is modified to read

$$\frac{d\hat{U}}{dt} = -i\left(\omega_o - \frac{i}{\tau_e}\right)\hat{U} \ . \tag{6.89}$$

However, this is only half of the story. The decay of $\hat{U}$ was found by first evaluating the excitation of the $\hat{V}_j$ by the resonator mode, using (6.84) and then reintroducing these excitations into (6.83). An equation has been obtained that leads to the erroneous conclusion that the commutator of $\hat{U}$ and $\hat{U}^\dagger$ decays at the rate $2/\tau_e$. Indeed, we find from (6.89)

$$\frac{d}{dt}[\hat{U}, \hat{U}^\dagger] = \left[\frac{d\hat{U}}{dt}, \hat{U}^\dagger\right] + \left[\hat{U}, \frac{d\hat{U}^\dagger}{dt}\right] = -\frac{2}{\tau_e}[\hat{U}, \hat{U}^\dagger] \ .$$

What has happened is that we have ignored the excitation of $\hat{U}$ by the $\hat{V}_j$ as evidenced by (6.83). Hence, we write instead of (6.89)

$$\frac{d\hat{U}}{dt} = -i\left(\omega_o - \frac{i}{\tau_e}\right)\hat{U} - i\sum_j K_j \hat{V}_j \ . \tag{6.90}$$

Here the $\hat{V}_j$ are sources driving $\hat{U}$. The contribution is of waves traveling in the direction of the resonator and hence unaffected by $\hat{U}$. The situation has become analogous to the one encountered in a cavity at thermal-equilibrium. There, a decay of the mode called for the introduction of a Langevin source so as to maintain the energy in the resonator at the thermal-equilibrium value. On the other hand, the appearance of the sources in (6.90) is the natural consequence of the Hamiltonian description of the resonator modes and the modes in the output waveguide. In order to show that the sources are precisely the ones necessary to maintain the commutator time-independent, we solve for $\hat{U}$, noting that the modes $\hat{V}_j$ drive $\hat{U}$ at their respective frequencies ω_j:

$$\hat{U} = \frac{-i\sum_j K_j \hat{V}_j}{i(\omega_o - \omega_j) + 1/\tau_e} \ . \tag{6.91}$$

The commutator of $\hat{U}$ is given by

$$[\hat{U}, \hat{U}^\dagger] = \sum_{j,k} K_j K_k^* [\hat{V}_j, \hat{V}_k^\dagger]\frac{1}{[i(\omega_o - \omega_j) + 1/\tau_e][-i(\omega_o - \omega_k) + 1/\tau_e]}$$

$$= \sum_j \frac{|K_j|^2}{(\omega_o - \omega_j)^2 + 1/\tau_e^2} \rightarrow \int \frac{(\kappa^2/v_g)d\omega_j}{(\omega_o - \omega_j)^2 + 1/\tau_e^2} = \pi\tau_e \frac{\kappa^2}{v_g} = 1 \ . \tag{6.92}$$

The commutator is unity. The sources due to the coupling to the waveguide compensate for the decay of the commutator.

The preceding analysis demonstrates a very general principle. The commutator of an observable is a physical attribute of the observable. This attribute must be conserved, the commutator must not change with time. Loss causes decay of an excitation. In quantum theory, such a decay is modeled by coupling of the system to a reservoir with a very large number of modes. This coupling does not only cause the decay, it also introduces sources that keep the commutator of the system invariant with time. These sources are the quantum counterpart to the Langevin sources required to maintain the thermal fluctuations in a lossy system at thermal equilibrium.

6.6 Quantization of Excitations on a Single-Mode Waveguide

The operators in Sect. 6.2 can be renormalized in a way analogous to the renormalization of the mode amplitudes in Chap. 4 when dealing with thermal noise. There are subtle differences in the renormalization, however, which arise from the nature of the quantum description of physical processes. Classically, one analyzes steady-state excitations in waveguides and transmission lines as evolutions in space. The classical approach ends up naturally with a Fourier decomposition in the frequency domain, namely spectra of signals and noise.

The concept of a steady state evolving in space at a set of frequencies is foreign to quantum theory, since it describes the evolution of operators in time, in terms of the Heisenberg equation of motion. This fact manifests itself in the effort one must expend to arrive at quantum descriptions of processes that would have been denoted as a steady state in the classical domain. A good example is the propagation of waves along a single mode waveguide. One selects a forward-wave "packet" occupying a length L, and one follows its propagation in time. This wave travels forward at the group velocity and occupies different spatial regions as it proceeds. If the wave packet hits an obstacle, it is partially reflected and partially transmitted. Eventually, a wave propagating in the reverse direction appears on one side of the obstacle, and a transmitted wave appears on the other side. The Heisenberg equation of motion describes this evolution of the wavepacket in terms of a scattering event. A steady state analogous to the classical steady state is established when many wavepackets follow each other and their statistics are stationary in time.

Because quantum theory describes evolution in time, a Fourier decomposition in the frequency of the excitation is not natural. It is more natural to look at modes of a given propagation constant and study their evolution in time. The operator A_m represents a mode excitation on a waveguide of length L. It is so normalized that $\langle|\hat{A}_m^\dagger \hat{A}_m|\rangle$ is equal to the photon number.

The modes obey periodic boundary conditions:

$$\beta_m = \frac{2\pi}{L}m .$$ (6.93)

We now introduce a new normalization of the creation and annihilation operators, which then permits us to treat the excitation of modes as a continuum in the sense of a Fourier integral rather than a Fourier series in the limit $L \to \infty$. Compare Appendix A.5 and (4.42). This normalization is

$$\hat{a}(\beta) = (\sqrt{L}/2\pi)\hat{A}_m .$$ (6.94)

In the limit $L \to \infty$, renormalization changes the commutation relations (6.46) into

$$[\hat{a}(\beta), \hat{a}^\dagger(\beta')] = \frac{1}{2\pi}\delta(\beta - \beta') \ . \tag{6.95}$$

Operation of the creation operator on a photon number state of a mode of propagation constant β gives

$$\hat{a}^\dagger(\beta)|n\rangle = \frac{\sqrt{L}}{2\pi}\sqrt{n+1}|n+1\rangle \ , \tag{6.96}$$

with an analogous relation for the annihilation operator

$$\hat{a}(\beta)|n\rangle = \frac{\sqrt{L}}{2\pi}\sqrt{n}|n-1\rangle \ . \tag{6.97}$$

The Hamiltonian of the mode becomes

$$\hat{H} = 2\pi\hbar \int d\beta\, \omega(\beta)\hat{a}^\dagger(\beta)\hat{a}(\beta) \ . \tag{6.98}$$

In the Hamiltonian we have omitted the contribution of the zero-point fluctuations. If the mode spectrum extends to infinity, this contribution becomes infinite as well. It does not contribute to the Heisenberg equations of motion, and thus is conveniently suppressed. The integral over propagation constants in (6.98) has to be interpreted carefully. A dispersion-free waveguide, such as a structure supporting a TEM mode, propagates both forward and backward waves. The forward waves have positive propagation constants β, and the backward waves have negative propagation constants. A forward-propagating pulse is composed only of waves with positive propagation constants. Hence the integral in the Hamiltonian (6.98) describing a pulse involves only positive propagation constants, clustered around a "carrier" propagation constant β_o.

The operation of the annihilation operator on a coherent state $|\alpha(\beta)\rangle$ gives

$$\hat{a}(\beta)|\alpha(\beta)\rangle = \frac{\sqrt{L}}{2\pi}\alpha(\beta)|\alpha(\beta)\rangle \ . \tag{6.99}$$

The Heisenberg equation of motion for the new operators follows via the use of the commutation relation (6.95) with the Hamiltonian (6.98):

$$\frac{d}{dt}\hat{a}(\beta) = -i\omega(\beta)\hat{a}(\beta) \ . \tag{6.100}$$

Note that the frequency of the mode is now treated as a function of the propagation constant. The mode may be dispersive if the frequency is not a linear function of β, as given by the dispersion relation of the waveguide. Since, quantum mechanically, $\hbar\beta$ is the momentum of the mode, the dispersion relation is now the relation between the energy $\hbar\omega$ and momentum $\hbar\beta$. Note that the simple formalism presented here addresses waveguides with relatively small dispersion so that the criterion for the propagation direction of a wave is simple.

In a way analogous to the definitions of the Hermitian in-phase and quadrature operators, one may define renormalized versions of these two operators:

$$\hat{a}^{(1)}(\beta) = \frac{1}{2}[\hat{a}(\beta) + \hat{a}^\dagger(\beta)] \tag{6.101}$$

and

$$\hat{a}^{(2)}(\beta) = \frac{1}{2i}[\hat{a}(\beta) - \hat{a}^\dagger(\beta)] \ . \tag{6.102}$$

6.7 Quantum Theory of Waveguides with Loss

In this section, we consider the quantum description of a waveguide with loss. We focus on the evolution of the operator $\hat{a}(\beta)$ of propagation constant β. If we remove the natural time dependence $\exp(-i\omega t)$ by replacing $\hat{a}(\beta)$ by $\hat{a}(\beta)\exp(-i\omega t)$, we obtain from (6.100), in the case of zero loss, the equation of motion

$$\frac{d}{dt}\hat{a}(\beta) = 0 \ . \tag{6.103}$$

When the waveguide is lossy, the operator $a(\beta)$ decays as it propagates. In order to preserve commutator brackets, we need to introduce operator noise sources [16]. Denote the decay rate by $\sigma(\beta)$. We obtain the equation of motion

$$\frac{d}{dt}\hat{a}(\beta) = -\sigma(\beta)\hat{a}(\beta) + \hat{s}(\beta) \ , \tag{6.104}$$

where $\hat{s}(\beta)$ represents operator sources due to the coupling to loss reservoirs. The loss reservoirs can be represented by distributions of resonators coupled to the waveguide at every cross section. This is analogous to the representation of decay and commutator conservation in the case of the open resonator in Sect. 6.5, which introduced sources into the resonator equation representing the mode excitations of the output waveguide. A similar model could be used for the determination of the loss and the noise sources for the modes of a waveguide. The mode of the waveguide of propagation constant β_m could be coupled to a continuum of modes. The coupling leads to temporal decay and the appearance of noise sources. The sources maintain the commutator of the mode annihilation and creation operators, integrated over the bandwidth of the resonance. The modes with propagation constants β_{m-1} and β_{m+1} decay in the same way and possess analogous noise sources. Here we need not go through a detailed model of such couplings. Instead, we can derive the properties of the noise operators simply from the requirement of conservation of commutator brackets. The rate of change of the commutator follows from (6.104):

$$\frac{d}{dt}[\hat{a}(\beta), \hat{a}^{\dagger}(\beta')] = -[\sigma(\beta) + \sigma(\beta')][\hat{a}(\beta), \hat{a}^{\dagger}(\beta')]$$

$$+[\hat{s}(\beta), \hat{a}^{\dagger}(\beta')] + [\hat{a}(\beta), \hat{s}^{\dagger}(\beta')] \qquad (6.105)$$

$$= 0 .$$

The decay rate must be equal to zero since the commutator is an intrinsic property of the operator. One would expect that the noise source operators and the mode amplitude operators would commute. However, in the short time interval Δt, the mode amplitude acquires a contribution from the source, just as discussed in Chap. 4 in connection with the evaluation of the thermal noise source. Thus $[\hat{s}(\beta), \hat{a}^{\dagger}(\beta')] = (1/2)[\hat{s}(\beta), \hat{s}^{\dagger}(\beta')]\Delta t$ and, since $[\hat{a}(\beta), \hat{a}^{\dagger}(\beta')] = (1/2\pi)\delta(\beta - \beta')$ according to (6.95), we find

$$[\hat{s}(\beta), \hat{s}^{\dagger}(\beta')] = \frac{1}{2\pi}2\sigma(\beta)\delta(\beta - \beta')\delta(t - t') . \qquad (6.106)$$

We see that the commutator behaves in a way similar to the correlation spectrum of the thermal noise sources. The commutator referring to different times and different propagation constants is zero. The operators $\hat{s}(\beta)$ and $\hat{s}^{\dagger}(\beta)$ have the characteristics of annihilation and creation operators, respectively, since $\sigma(\beta) > 0$. They create or destroy photons of the optical mode through interaction with the loss reservoirs.

Some remarks are in order with regard to the integration of a linear differential equation involving operators of the form of (6.104). Because the equation is linear in the operators, integration of the equation never encounters products of the operators and hence never need consider commutation relations. For this reason, the integration proceeds in the same way as if the operators were c numbers. The operator $\hat{a}(\beta, T)$ at the time T is found from the initial conditions by integration of (6.104):

$$\hat{a}(\beta, T) = \exp(-\sigma T)\hat{a}(\beta, 0) + \exp(-\sigma T) \int_0^T dt \exp(\sigma t)\hat{s}(\beta) . \qquad (6.107)$$

Equation (6.107), and the equation for the creation operator, the Hermitian conjugate of (6.107), can be used to evaluate expectation values of the operators and their moments when the input excitation is specified. We shall concentrate here on coherent-state excitations of the waveguide input and ground state excitations of the noise reservoir. The system is in a product state $|\alpha(\beta)\rangle|0\rangle$. This product state is a generalized coherent state of the system as seen when it is operated upon by the annihilation operator $a(\beta, T)$. We find

$$\hat{a}(\beta, T)|\alpha(\beta)\rangle|0\rangle = \frac{\sqrt{L}}{2\pi}\exp(-\sigma T)\alpha(\beta)|\alpha(\beta)\rangle|0\rangle . \qquad (6.108)$$

The state is an eigenstate of the annihilation operator. Using (6.108), we find for the expectation value of the photon number

$$\langle n \rangle = 2\pi \langle 0 | \langle \alpha(\beta) | \int d\beta \, \hat{a}^\dagger(\beta, T) \hat{a}(\beta, T) | \alpha(\beta) \rangle | 0 \rangle$$

$$= \exp(-2\sigma T) |\alpha(\beta)|^2 \ . \tag{6.109}$$

The photon number has been reduced by an attenuation factor of the amplitude squared. For the expectation value of the operators in antinormal order, using the commutator (6.95), we find

$$\langle 0 | \langle \alpha(\beta) | \hat{a}(\beta, T) \hat{a}^\dagger(\beta', T) | \alpha(\beta) \rangle | 0 \rangle$$

$$= \langle 0 | \langle \alpha(\beta) | \hat{a}^\dagger(\beta, T) \hat{a}(\beta', T) | \alpha(\beta) \rangle | 0 \rangle + \frac{1}{2\pi} \delta(\beta - \beta') \tag{6.110}$$

$$= \frac{L}{(2\pi)^2} \exp(-2\sigma T) |\alpha(\beta)|^2 + \frac{1}{2\pi} \delta(\beta - \beta') \ .$$

One may evaluate the mean square fluctuations of the in-phase and quadrature operators in the same way. The algebra is rather cumbersome if done routinely. Instead, it is better to take advantage of the fact that a coherent state is an eigenstate of the annihilation operator. As shown in Sect. 6.2, one may write $\hat{a}^{(1)}(\beta)\hat{a}^{(1)}(\beta') = \frac{1}{4}[\hat{a}(\beta) + \hat{a}^\dagger(\beta)][\hat{a}(\beta') + \hat{a}^\dagger(\beta')]$ in normal order. When $\langle \hat{a}^{(1)}(\beta) \rangle \langle \hat{a}^{(1)}(\beta') \rangle$ is subtracted from $\langle \hat{a}^{(1)}(\beta)\hat{a}^{(1)}(\beta') \rangle$, only the contribution of the commutator remains, so that

$$\langle 0 | \langle \alpha(\beta) | \hat{a}^{(1)}(\beta, T) \hat{a}^{(1)}(\beta', T) | \alpha(\beta) \rangle | 0 \rangle$$

$$- \langle 0 | \langle \alpha(\beta) | \hat{a}^{(1)}(\beta, T) | \alpha(\beta) \rangle | 0 \rangle \langle 0 | \langle \alpha(\beta) | \hat{a}^{(1)}(\beta', T) | \alpha(\beta) \rangle | 0 \rangle \tag{6.111}$$

$$= \frac{1}{4} \frac{1}{2\pi} \delta(\beta - \beta') \ .$$

In the same way we find

$$\langle 0 | \langle \alpha(\beta) | \hat{a}^{(2)}(\beta, T) \hat{a}^{(2)}(\beta', T) | \alpha(\beta) \rangle | 0 \rangle$$

$$- \langle 0 | \langle \alpha(\beta) | \hat{a}^{(2)}(\beta, T) | \alpha(\beta) \rangle | 0 \rangle \langle 0 | \langle \alpha(\beta) | \hat{a}^{(2)}(\beta', T) | \alpha(\beta) \rangle | 0 \rangle \tag{6.112}$$

$$= \frac{1}{4} \frac{1}{2\pi} \delta(\beta - \beta') \ .$$

The spectrum of the fluctuations expressed as a function of propagation constant is β-independent, analogously to the frequency spectrum of thermal noise, which is ω-independent.

The quantum theory of a waveguide with loss bears a close analogy to the classical analysis of the same waveguide at thermal equilibrium. The thermal fluctuations of the electromagnetic field would decay in the waveguide, were it not for the Langevin noise sources that reexcite the modes and maintain the thermal fluctuations. In the quantum-theoretical treatment Langevin operator sources are required to maintain the commutator relations. Maintenance of the commutator relations then ensures maintenance of the mean square field fluctuations of the waveguide field in its ground state. The zero-point fluctuations bear a close resemblance to thermal fluctuations.

We have computed the fluctuations produced by the noise sources under the assumption that they are in the ground state. The question may be asked whether this is a severely restricting assumption. Generally, the loss reservoirs would be thermally excited. If the temperature of the reservoir is of the order of room temperature $\theta_o = 290$ K, the contribution of its excitation is negligible compared with the contribution of the zero-point fluctuations, since $k\theta_o \ll \hbar\omega$ for an optical frequency ω. The ratio $\hbar\omega/k\theta_o$ is typically of the order of 40.

The introduction of the noise source for conservation of the commutator bracket may seem ad hoc. In the next chapter, we shall show models of loss that are based on a Hamiltonian description of the system. The loss will be due to output ports that are not explicitly included in the description of the output, and the noise sources will be shown to arise from ports of the network not accessed by the signal.

6.8 The Quantum Noise of an Amplifier with a Perfectly Inverted Medium

Quantum theory permits a generalization to active devices not possible in the classical physics of thermal equilibrium, which is only applicable to passive systems. Indeed, if the system has gain, then $\sigma(\beta) < 0$, and the right-hand side of (6.106) becomes negative. The solution for the output operator $\hat{a}(\beta, T)$ is of the same form as (6.107),

$$\hat{a}(\beta, T) = \exp(|\sigma|T)\hat{a}(\beta, 0) + \exp(|\sigma|T) \int_{-\infty}^{T} dt \exp(-|\sigma|t)\hat{s}(\beta) . \quad (6.113)$$

The product state $|\alpha(\beta)\rangle|0\rangle$ is not any more an eigenstate of the operator $\hat{a}(\beta, T)$ which, in turn, ceases to act as an annihilation operator. In the analysis it is necessary to treat the two operators on the right hand side of (6.113) separately, the first one as an annihilation operator, the second one as a creation operator. The consequence of this reversal is that photons appear at the output of the amplifier even if no photons are fed into its input. Consider the product state $|\alpha(\beta)\rangle|0\rangle$, i.e. the situation when a coherent state is fed into the amplifier and the reservoirs of the noise sources are in the ground state. This is the case when the population of a laser medium is in the upper level and is equilibrated at the temperature of the host medium. Since this temperature is of the order of magnitude of room temperature $\theta_o = 290$ K, and $\hbar\omega(\beta) \gg k\theta_o$, one may approximate the equilibration temperature as equal to zero, with the states of the upper level in the ground state. We have for the photon number at the output of an amplifier of length ℓ, traversed within the time $T = \ell/v_g$,

$$\langle n \rangle = 2\pi \langle 0| \langle \alpha(\beta)| \int d\beta \, \hat{a}^\dagger(\beta, T)\hat{a}(\beta, T)|\alpha(\beta)\rangle |0\rangle$$

$$= 2\pi \exp(2|\sigma|T)\langle \alpha(\beta)| \int d\beta \, \hat{a}^\dagger(\beta, 0)\hat{a}(\beta, 0)|\alpha(\beta)\rangle$$

$$+2\pi \exp(2|\sigma|T)\langle 0| \int d\beta \int_0^T dt \int_0^T dt' \, \hat{s}^\dagger(\beta, t)\hat{s}(\beta, t')$$

$$\times \exp(-|\sigma|(t + t'))|0\rangle \, .$$

(6.114)

The first term is the amplified input signal and gives the contribution $\langle n \rangle_{\text{signal}} = G|\alpha(\beta)|^2$, where $G = \exp(2|\sigma|T)$. The second term follows from the commutator (6.106). This commutator is negative, indicating that $\hat{s}(\beta, t)$ is a creation operator, and its Hermitian conjugate an annihilation operator. The expectation value of the operator product $\hat{s}^\dagger(\beta, t)\hat{s}(\beta', t')$ is

$$\langle 0|\hat{s}^\dagger(\beta, t)\hat{s}(\beta', t')|0\rangle = \frac{1}{2\pi} 2|\sigma(\beta)|\delta(\beta - \beta')\delta(t - t') \, .$$

(6.115)

The double integral over time gives

$$\int_0^T dt \int_0^T dt' \exp[-|\sigma|(t + t')]\langle 0|\hat{s}^\dagger(\beta, t)\hat{s}(\beta', t')|0\rangle$$

$$= \frac{1}{2\pi}[1 - \exp(-2|\sigma|T)]\delta(\beta - \beta') \, .$$

(6.116)

The height of the delta function is $L/2\pi$, and it vanishes outside the interval of propagation constant $\Delta\beta = 2\pi L$. This interval is set by the bandwidth of the amplifier system. Thus consider a coherent signal state that extends over a length L, covering the time L/v_g. A time-varying signal of bandwidth $B = v_g/L$ is represented by a succession of coherent states, each occupying a time interval L/v_g. If the amplifier is followed by a filter of bandwidth equal to the signal bandwidth, the noise passed by the filter occupies the same bandwidth. Using this result, we find for the second term in (6.114), the term caused by the noise source, the amplified spontaneous emission (ASE),

$$\langle n \rangle_{\text{ASE}} = G - 1 \, .$$

(6.117)

Amplification of signal photons by the factor G entails the addition of $G - 1$ noise photons to the signal, provided the signal bandwidth and noise bandwidth are the same. Each increment $\Delta\beta$ carries $G - 1$ ASE photons. This is the well-known amplified-spontaneous-emission term of an ideal amplifier with perfect inversion. If the inversion is not perfect, if the lower laser level is occupied, some of the generated photons are reabsorbed by excitation of the lower level into the upper level. This requires the inclusion of an absorption term into the evolution equation (6.114), as treated in the next section.

Each increment $\Delta\beta$ corresponds to a frequency bandwidth $\Delta\omega = v_g \Delta\beta$. If an optical filter of bandwidth $\Delta\Omega = N\Delta\omega$ is inserted at the amplifier output, the rate of ASE photons passing through the filter is

$$\text{rate of ASE photons} = Nv_g \frac{\langle n \rangle_{\text{ASE}}}{L} = Nv_g \frac{2\pi}{L} \frac{G-1}{2\pi}$$

$$= (G-1)\frac{\Delta\Omega}{2\pi} = (G-1)B \ , \tag{6.118}$$

where B is the filter bandwidth in Hz.

Next, we compute the fluctuations of the in-phase and quadrature components when a coherent state $|\alpha(\beta)\rangle|0\rangle$ is specified as the initial condition. In computing the expectation value of the fluctuations of $\hat{a}^{(1)}(\beta, T)\hat{a}^{(1)}(\beta', T)$, we write

$$\hat{a}^{(1)}(\beta, T)\hat{a}^{(1)}(\beta', T) = \frac{1}{4}[\hat{a}(\beta, T) + \hat{a}^\dagger(\beta, T)][\hat{a}(\beta', T) + \hat{a}^\dagger(\beta', T)] \ , \tag{6.119}$$

and we cast the annihilation operators and the creation operators into normal order. This must be done separately for the waveguide mode operator $\hat{a}(\beta, T)$ and the noise source operator $\hat{a}(\beta)$. When this is done, we find for the in-phase fluctuations

$$\text{in-phase fluctuations} = \langle 0|\langle \alpha(\beta)|\hat{a}^{(1)}(\beta, T)\hat{a}^{(1)}(\beta', T)|\alpha(\beta')\rangle|0\rangle$$

$$- [\langle 0|\langle \alpha(\beta)|\hat{a}^{(1)}(\beta, T)|\alpha(\beta)\rangle|0\rangle]^2 \tag{6.120}$$

$$= \left[\frac{1}{2}(G-1) + \frac{1}{4}\right]\frac{1}{2\pi}\delta(\beta - \beta') \ .$$

The same result is found for the quadrature fluctuations:

$$\text{quadrature fluctuations} = \langle 0|\langle \alpha(\beta)|\hat{a}^{(2)}(\beta, T)\hat{a}^{(2)}(\beta', T)|\alpha(\beta')\rangle|0\rangle$$

$$- [\langle 0|\langle \alpha(\beta)|\hat{a}^{(2)}(\beta, T)|\alpha(\beta)\rangle|0\rangle]^2$$

$$= \left[\frac{1}{2}(G-1) + \frac{1}{4}\right]\frac{1}{2\pi}\delta(\beta - \beta') \ . \tag{6.121}$$

When the gain is large, the fluctuations are twice as large as the input fluctuations amplified by G. This finding has a deep significance in the context of quantum measurements of two noncommuting observables, as we shall see in Chap. 7.

6.9 The Quantum Noise
of an Imperfectly Inverted Amplifier Medium

An imperfectly inverted medium has a nonzero population in the lower energy level. This population acts as an absorber. One may analyze the amplifier as an active waveguide of gain coefficient σ_u due to the population in the upper laser level, interspersed with a passive one of loss coefficient σ_ℓ, with $\sigma_\ell < \sigma_u$, due to the population in the lower laser level. The equation for the mode propagation is

$$\frac{d}{dt}\hat{a}(\beta) = (\sigma_u - \sigma_\ell)\hat{s}(\beta) + \hat{s}_u(\beta) + \hat{s}_\ell(\beta) , \tag{6.122}$$

where $\hat{s}_u$ and $\hat{s}_\ell$ are the associated noise sources, with the commutation spectral densities

$$[\hat{s}_u(\beta), \hat{s}_u^\dagger(\beta')] = -2\sigma_u \frac{1}{2\pi}\delta(\beta - \beta')\delta(t - t') , \tag{6.123}$$

$$[\hat{s}_\ell(\beta), \hat{s}_\ell^\dagger(\beta')] = 2\sigma_\ell \frac{1}{2\pi}\delta(\beta - \beta')\delta(t - t') . \tag{6.124}$$

The integral of (6.122) is

$$\hat{a}(\beta, T) = \exp[(\sigma_u - \sigma_\ell)T]\hat{a}(\beta, 0) + \exp[(\sigma_u - \sigma_\ell)T]$$
$$\times \left| \int_0^T dt \exp[-(\sigma_u - \sigma_\ell)T][\hat{s}_u(\beta) + \hat{s}_\ell(\beta)] \right. . \tag{6.125}$$

The output operator $\hat{a}(\beta, T)$ consists of the amplified input operator, $\sqrt{G}\hat{a}(\beta, 0)$, with $\sqrt{G} \equiv \exp(\sigma_u - \sigma_\ell)T$, and two noise sources

$$\hat{n}_u(\beta) = \exp[(\sigma_u - \sigma_\ell)T] \int_0^T dt \exp[-(\sigma_u - \sigma_\ell)t]\hat{s}_u(\beta, t) \tag{6.126a}$$

and

$$\hat{n}_\ell(\beta) = \exp[(\sigma_u - \sigma_\ell)T] \int_0^T dt \exp[-(\sigma_u - \sigma_\ell)t]\hat{s}_\ell(\beta, t) . \tag{6.126b}$$

The commutators of these noise sources are

$$[\hat{n}_u(\beta), \hat{n}_u^\dagger(\beta')] = \exp[2(\sigma_u - \sigma_\ell)T]$$

$$\times \int_0^T dt \int_0^T dt' \exp[-(\sigma_u - \sigma_\ell)(t + t')][\hat{s}_u(\beta, t), \hat{s}_u^\dagger(\beta', t)] \qquad (6.127a)$$

$$= \frac{1}{2\pi} \delta(\beta - \beta') \frac{\sigma_u}{\sigma_u - \sigma_\ell}(1 - G)$$

and

$$[\hat{n}_\ell(\beta), \hat{n}_\ell^\dagger(\beta')] = \exp[2(\sigma_u - \sigma_\ell)T]$$

$$\times \int_0^T dt \int_0^T dt' \exp[-(\sigma_u - \sigma_\ell)(t + t')][\hat{s}_\ell(\beta, t), \hat{s}_\ell^\dagger(\beta', t)] \qquad (6.127b)$$

$$= \frac{1}{2\pi} \delta(\beta - \beta') \frac{\sigma_\ell}{\sigma_u - \sigma_\ell}(G - 1) .$$

According to the sign of the commutator, $\hat{n}_u(\beta)$ can be identified as a creation operator and $\hat{n}_\ell(\beta)$ as an annihilation operator. The first one is contributed by the upper level of the gain medium, the second by the lower level. The analysis can be simplified if we assume the presence of filters that accommodate a signal occupying a spatial slot of length L, corresponding to a bandwidth $B = v_g/L$. We may then revert to the original operators $\hat{A}$ introduced in Sects. 6.1 and 6.2 and related to $\hat{a}(\beta)$ by (6.94). Denoting the output and input by $(2\pi/\sqrt{L})\hat{a}(\beta, T) \equiv \hat{B}$ and $(2\pi/\sqrt{L})\hat{a}(\beta, T) \equiv \hat{A}$, respectively, we have from (6.125)

$$\hat{B} = \sqrt{G}\hat{A} + \hat{N}_u + \hat{N}_\ell . \qquad (6.128)$$

The creation operator $\hat{N}_u$ is responsible for the ASE and has the commutator

$$[\hat{N}_u, \hat{N}_u^\dagger] = \chi(1 - G) , \qquad (6.129)$$

where

$$\chi \equiv \frac{\sigma_u}{\sigma_u - \sigma_\ell} . \qquad (6.130)$$

The parameter χ is the so-called inversion parameter. It is equal to unity when the medium is perfectly inverted, and becomes greater than unity for a partially inverted gain medium.

The annihilation operator $\hat{N}_\ell$ represents the noise introduced by the lower level and has the commutator

$$[\hat{N}_\ell, \hat{N}_\ell^\dagger] = (\chi - 1)(G - 1) . \qquad (6.131)$$

This is a compact form of the amplifier description which will be of use in the evaluation of the probability distribution of the field in the next chapter.

The simple expression (6.128), along with (6.129) and (6.131), can be used to answer the question as to when optical amplifiers behave like classical amplifiers with additive noise. Thus, we may compute the mean square fluctuations of $\hat{B}^{(1)}$ and $\hat{B}^{(2)}$ for a coherent input state $|\alpha\rangle$. The noise sources are in the ground state, indicated by a single factor $|0\rangle$ for simplicity

$$\langle 0|\langle \alpha|(\Delta\hat{B}^{(1)})^2 + (\Delta\hat{B}^{(2)})^2|\alpha\rangle|0\rangle$$

$$= \langle 0|\langle \alpha|(\hat{B}^{(1)})^2 + (\hat{B}^{(2)})^2|\alpha\rangle|0\rangle - \langle 0|\langle \alpha|\hat{B}^{(1)} + B^{(2)}|\alpha\rangle|0\rangle^2 . \tag{6.132}$$

When the in-phase and quadrature operators are expressed in terms of creation and annihilation operators and are put into normal order, only the contributions of the commutators remain. We find

$$\langle 0|\langle \alpha|(\Delta\hat{B}^{(1)})^2 + (\Delta\hat{B}^{(2)})^2|\alpha\rangle|0\rangle$$

$$= \frac{1}{2}G + \frac{1}{2}\langle 0|\hat{N}_u^\dagger\hat{N}_u + \hat{N}_\ell\hat{N}_\ell^\dagger|0\rangle^2 \tag{6.133}$$

$$= \frac{1}{2}G + \frac{1}{2}(2\chi - 1)(G - 1) .$$

The first term comes from the amplified zero-point fluctuations of the signal, and the second term comes from the noise contributions of the upper and lower level. In order to cast this expression in terms of signal power and additive noise power, we transform the above into a flow in units of power by converting the net mean square fluctuations to unit distance through division by L and through multiplication by the group velocity v_g and the photon energy $\hbar\omega$:

$$\frac{1}{L}\hbar\omega v_g\langle 0|\langle \alpha|(\Delta\hat{B}^{(1)})^2 + (\Delta\hat{B}^{(2)})^2|\alpha\rangle|0\rangle$$

$$= \frac{1}{L}\hbar\omega v_g\left[\frac{1}{2} + \chi(G - 1)\right] . \tag{6.134}$$

Now, $2\pi/L$ is the mode separation $\Delta\beta$, corresponding to a bandwidth $\Delta\omega = (d\omega/d\beta)\Delta\beta = v_g\Delta\beta = 2\pi B$. Thus

$$\frac{1}{L}\hbar\omega v_g\langle 0|\langle \alpha|(\Delta\hat{B}^{(1)})^2 + (\Delta\hat{B}^{(2)})^2|\alpha\rangle|0\rangle = \hbar\omega B\left[\frac{1}{2} + \chi(G - 1)\right] . \tag{6.135}$$

On the other hand, the power flow of the amplified spontaneous emission is

$$\text{ASE power} = \frac{1}{L}\hbar\omega v_g\langle 0|\hat{N}_u^\dagger\hat{N}_u|0\rangle = \hbar\omega B\chi(G - 1) . \tag{6.136}$$

If we reasoned classically, we would assign half of the power flow to the mean square fluctuations of the in-phase and quadrature components in power flow units. Comparison with (6.146) shows that this gives the right answer in the limit of large gain G.

The expression for the amplified spontaneous emission was known in the years of the invention of the laser [68]. In the early days of the laser it was not easy to separate out the fundamental Gaussian mode from a pumped crystal emitting into a large solid angle. With the advent of optical waveguides and single-mode fibers, this presents no problem, and it is easy to verify (6.136) experimentally. Figure 6.3 shows the experimental arrangement. An erbium-doped fiber laser is pumped by a laser diode operating at 980 nm wavelength and emits at 1.54 μm wavelength. No signal is applied to the amplifier. An optical filter of bandwidth $\Delta\Omega = 2\pi B$, much less than the amplifier bandwidth, is put in front of the power detector. The detector can be a bolometer, measuring power by a temperature rise, or a photodiode calibrated in power units.

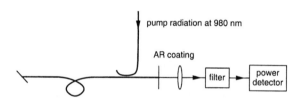

Fig. 6.3. Experimental arrangement for measurement of amplified spontaneous emission

6.10 Noise in a Fiber with Loss Compensated by Gain

We have emphasized several times that the quantum noise of optical components used in communications can be thought of as additive, in the same sense as thermal noise can be viewed as additive to a classical signal. The mean square fluctuations of the signal amplitude after passage through the component can be evaluated from the sum of the signal fluctuations and the fluctuations of the added noise. In this section, we develop this semiclassical picture of quantum noise in the case of an optical waveguide (fiber) whose loss is compensated by distributed gain. This is the simplest model for long-distance fiber communications in which distributed amplification compensates the fiber loss. We shall use the results in Chap. 10 in the derivation of the timing jitter of soliton propagation.

We consider one segment of length Δz of the waveguide composed of a loss section of loss $\mathcal{L}(<1)$ followed by a gain section with gain G. The gain section is described by (6.126b), where $\hat{A}$ is the input operator and $\hat{B}$ is the output operator

$$\hat{B} = \sqrt{G}\hat{A} + \hat{N}_u + \hat{N}_\ell .$$ (6.137)

The commutators of the noise sources are

$$[\hat{N}_u, \hat{N}_u^\dagger] = \chi(1 - G)$$ (6.138)

and

$$[\hat{N}_\ell, \hat{N}_\ell^\dagger] = (\chi - 1)(G - 1) .$$ (6.139)

The loss section by itself obeys the relation

$$\hat{B} = \sqrt{\mathcal{L}}\hat{A} + \hat{N}_\mathcal{L} ,$$ (6.140)

where the noise source has the commutator

$$[\hat{N}_\mathcal{L}, \hat{N}_\mathcal{L}^\dagger] = 1 - \mathcal{L} .$$ (6.141)

We shall assume that the loss and gain are very small, i.e. $1 - \mathcal{L} \ll 1$, $G - 1 \ll 1$. Further, we assume that the loss and gain balance, so that $\mathcal{L}G = 1$. Under these conditions, the cascade of the two segments, with the output of the loss section being the input of the gain section, has the overall response

$$\hat{B} = \sqrt{G}(\sqrt{\mathcal{L}}\hat{A} + \hat{N}_\mathcal{L}) + \hat{N}_u + \hat{N}_\ell \simeq \hat{A} + \hat{N}_\mathcal{L} + \hat{N}_u + \hat{N}_\ell .$$ (6.142)

The signal remains unchanged and quantum noise sources have been added to it. We find for the expectation value of the in-phase component of the signal

$$\langle \hat{B}^{(1)} \rangle = \langle A^{(1)} \rangle .$$ (6.143)

The mean square fluctuations are evaluated as usual:

$$\langle (\Delta B^{(1)})^2 \rangle = \langle (B^{(1)})^2 \rangle - \langle B^{(1)} \rangle^2 .$$ (6.144)

If the signal is in a coherent state, the operator products can be put into normal order and the mean square fluctuations result solely from the commutators. The term $G - 1$ can be expressed in terms of the gain per unit length. From the gain within a time interval T,

$$G = \exp[2(\sigma_u - \sigma_\ell)T] ,$$ (6.145)

we may construct $G - 1$ when $G - 1 \ll 1$

$$G - 1 = 2(\sigma_u - \sigma_\ell)T = 2\alpha \, \Delta z ,$$ (6.146)

where Δz is the distance traveled by the signal within the time T, and α is the gain per unit length. Since the loss is equal to the gain, we also have

$$\mathcal{L} - 1 = 2\alpha \, \Delta z \,. \tag{6.147}$$

The mean square fluctuations of the excitation after passage through one segment of length Δz are

$$\langle (\Delta B^{(2)})^2 \rangle = \frac{1}{4}G + \frac{1}{4}G(1-L) + \frac{1}{4}\chi G(1-L) + \frac{1}{4}(\chi - 1)G \,. \tag{6.148}$$

The first term is the zero-point fluctuation of the signal at the input that has passed through the gain and loss; the second term is the contribution of the noise source associated with the loss; the third term is the contribution of the upper level of the gain medium; and the last term is the contribution of the lower level of the gain medium. When account is taken of the fact that G differs little from unity and that $G = 1 + 2\alpha \, \Delta z$ and $\mathcal{L} = 1 - 2\alpha \, \Delta z$, the above expression becomes

$$\langle (\Delta B^{(1)})^2 \rangle = \frac{1}{4}[1 + \chi(2\alpha \, \Delta z)] \,. \tag{6.149}$$

The first term is the zero-point fluctuation accompanying the signal; the second term is the added noise due to gain and loss. The quadrature component has the same fluctuations. In the semiclassical picture, the fluctuations are additive to a noise-free signal. To bring this picture into correspondence with the picture of signal and additive thermal noise, we transform the above into a flow in units of power by converting the net mean square fluctuations to the value for unit distance through division by L and by multiplication by the group velocity and $\hbar\omega$:

$$\frac{\hbar\omega}{L}v_g \langle (\Delta \hat{B}^{(1)})^2 + (\Delta B^{(2)})^2 \rangle = \frac{\hbar\omega}{L}v_g \left(\frac{1}{2} + \chi(2\alpha \, \Delta z) \right) \,. \tag{6.150}$$

Now, $2\pi/L$ is the mode separation $\Delta\beta$ corresponding to a bandwidth $\Delta\omega = (d\omega/d\beta)\Delta\beta = v_g \, \Delta\beta = 2\pi B$. Thus

$$\frac{\hbar\omega}{L}v_g \langle (\Delta \hat{B}^{(1)})^2 + (\Delta \hat{B}^{(2)})^2 \rangle = \hbar\omega B \left(\frac{1}{2} + \chi(2\alpha \, \Delta z) \right) \,. \tag{6.151}$$

This formula shows that the mean square fluctuations are proportional to the bandwidth and grow linearly with distance along a fiber whose gain is balanced by the loss. Note that the added mean square fluctuations correspond to the added ASE power. Indeed, this power is

$$\Delta(\text{ASE power}) = \hbar\omega B\chi(2\alpha \, \Delta z) \,. \tag{6.152}$$

The process can be described by propagation of a classical amplitude $a(\omega)$ in the presence of a noise source

$$\frac{d}{dz}a(\omega) = s(\omega, z) \,, \tag{6.153}$$

with

$$\langle s(\omega, z)s^*(\omega', z)\rangle = \frac{1}{2\pi}\chi 2\alpha \,\delta(\omega - \omega')\delta(z - z') \,. \tag{6.154}$$

We shall use this semiclassical formula in Chap. 10 for the evaluation of the noise accompanying soliton propagation. It should be emphasized that the results are correct quantum mechanically if applied to a signal in a coherent state. The noise is additive. Further, one may note that the noise is composed of a contribution from the gain and one from the loss.

6.11 The Lossy Resonator and the Laser Below Threshold

In Sect. 6.5 we derived the commutator conservation of an open resonator from a Hamiltonian description. The decay of the commutator of the resonator mode due to radiation into the connecting waveguide was compensated by the coupling to the commutators of the waveguide acting as a reservoir. In the subsequent sections we treated the waveguide modes from several points of view. Using the formalism developed thus far, we may treat the open resonator problem in a different way, starting from the classical equations of the open-resonator and quantizing them by replacing the complex amplitudes with operators. We have from (2.221)

$$\frac{d\hat{U}}{dt} = -(i\omega_o + 1/\tau_e)\hat{U} + \sqrt{\frac{2}{\tau_e}}\hat{a} \,. \tag{6.155}$$

In the transition to the quantum description attention has to be paid to the meaning of the amplitudes. In the classical formalism, $|U|^2$ represents the energy in the resonator. This suggests that $\hat{U}^\dagger\hat{U}$ should be interpreted as the photon number operator, as has already been done in Sect. 6.5. In the classical description, $|a(t)|^2$ is the power flow of the mode incident upon the resonator. Therefore, in the quantum formulation, the operator $\hat{a}^\dagger(t)\hat{a}(t)$ must give the photon flow rate in the time domain. In the propagation constant description, $2\pi \int d\beta\, \hat{a}^\dagger(\beta)\hat{a}(\beta)$ is the photon number operator $\hat{n}$ assigned to a wavepacket of length L. The photon flow rate is $v_g\hat{n}/L = (d\omega/d\beta)\hat{n}/L$. Hence the photon flow operator is given by

$$\text{photon flow operator} = \frac{2\pi}{L}\int d\omega\, \hat{a}^\dagger[\beta(\omega)]\hat{a}[\beta(\omega)] \,. \tag{6.156}$$

If the photon flow rate is finite, the integral must go to infinity as L goes to infinity. Division by L gives a finite result. Further, if the process is stationary, operators with different frequencies must be uncorrelated. Hence it makes sense to write the photon flow operator of a stationary process as a double integral (note $2\pi/L = \Delta\beta = (d\beta/d\omega)\Delta\omega$):

$$\text{photon flow operator} = \int d\omega \int d\omega' \frac{d\beta}{d\omega} \hat{a}^\dagger[\beta(\omega)]\hat{a}[\beta'(\omega')]$$

$$= \int d\omega \int d\omega' \, \hat{a}^\dagger(\omega)\hat{a}(\omega') \, . \tag{6.157}$$

The new operator $\hat{a}(\omega)$ obeys the commutation relation

$$[\hat{a}(\omega), \hat{a}^\dagger(\omega')] = \frac{d\beta}{d\omega}[\hat{a}(\beta), \hat{a}^\dagger(\beta')]$$

$$= \frac{d\beta}{d\omega}\frac{1}{2\pi}\delta(\beta - \beta') = \frac{1}{2\pi}\delta(\omega - \omega') \, . \tag{6.158}$$

This operator is related to the Fourier transform of $\hat{a}(t)$, which gives the photon flow rate in the time domain as $\hat{a}^\dagger(t)\hat{a}(t)$. The Fourier transform pair is

$$\hat{a}(t) = \int d\omega \, \hat{a}(\omega)\exp(-i\omega t); \quad \hat{a}(\omega) = \frac{1}{2\pi}\int dt \, \hat{a}(t)\exp(i\omega t) \, . \tag{6.159}$$

We have for the operator $\hat{a}^\dagger(t)\hat{a}(t)$

$$\hat{a}^\dagger(t)\hat{a}(t) = \int d\omega \int d\omega' \, \hat{a}^\dagger(\omega')\hat{a}(\omega)\exp[i(\omega' - \omega)t] \, . \tag{6.160}$$

We see from (6.160) that the expectation value of $\hat{a}^\dagger(t)\hat{a}(t)$ is given by the expectation value of the photon flow operator (6.156) when the operators at different frequencies are uncorrelated.

A few remarks as to the meaning of the Fourier transform pair are appropriate. The operator $\hat{a}(\omega)$ has the time dependence $\exp(-i\omega t)$, where the frequency ω is positive. The Hermitian conjugate creation operator has the time dependence $\exp(i\omega t)$. Fourier transforms are normally defined as relations between functions of time and functions of frequency which extend over the entire frequency range from minus infinity to plus infinity. The quantum operator $\hat{a}(\omega)$ is defined only for positive frequencies. As long as the spectrum of ω_o is clustered around a carrier frequency ω_o, the analysis is self-consistent. We shall discuss this issue in more detail in Chap. 12.

The equation for the excitation of the reflected wave is

$$\hat{b} = -\hat{a} + \sqrt{\frac{2}{\tau_e}}\hat{U} \, . \tag{6.161}$$

Appendix A.9 connects the reservoir analysis of Sect. 6.5 with (6.155) and (6.161).

Equation (6.155) is a linear operator equation. In solving a linear differential equation no commutators of the operators appear. For this reason the solution of a linear operator equation is indistinguishable from the solution of its classical, c-number counterpart. We solve (6.155) in the Fourier transform domain by assuming a time dependence of the form $\exp(-i\omega t)$:

$$\hat{U}(\omega) = \frac{\sqrt{2/\tau_e}\,\hat{a}(\omega)}{i(\omega_o - \omega) + 1/\tau_e} \; . \tag{6.162}$$

The commutator of the resonator excitation is

$$[\hat{U}(\omega), \hat{U}^\dagger(\omega')] = \frac{(2/\tau_e)[\hat{a}(\omega), \hat{a}^\dagger(\omega')]}{[i(\omega_o - \omega) + 1/\tau_e][-i(\omega_o - \omega) + 1/\tau_e]} \tag{6.163}$$

$$= \frac{2/\tau_e}{[(\omega_o - \omega)^2 + 1/\tau_e^2]} \frac{1}{2\pi}\delta(\omega - \omega') \; ,$$

where we have used the commutator relation (6.158). The resonator commutator has become a function of frequency. This is a consequence of the boundary conditions imposed on the resonator mode. Commutators of excitations within enclosures do not have unchanging universal properties. Thus, for example, if one introduced partially transmitting irises into a uniform waveguide to form a transmission resonator, the commutator spectrum of the excitations internal to the resonator would change. This is analogous to the change of the thermal excitations in equilibrium when partially transmitting irises are introduced into a uniform waveguide. The thermal excitations peak around the resonance frequencies, and are much smaller in the frequency regimes between the resonances.

The double integral over frequency of the resonator mode commutator gives unity:

$$\int d\omega \int d\omega' [\hat{U}(\omega), \hat{U}^\dagger(\omega')]$$

$$= \int d\omega \int d\omega' \frac{2/\tau_e}{[(\omega_o - \omega)^2 + 1/\tau_e^2]} \frac{1}{2\pi}\delta(\omega - \omega') = 1 \; . \tag{6.164}$$

Since the reflected wave is generated by the incident wave via interaction with the resonator, it is not obvious that the commutator of the reflected wave has remained unchanged. On the other hand, we have emphasized that the commutator of an excitation amplitude of a wave in an open waveguide is a property of the wave and should not change under any circumstances. Hence, conservation of this commutator serves as a check on the self-consistency of the theory. Let us check the value of this commutator. We find for $\hat{b}(\omega)$, using (6.161),

$$\hat{b}(\omega) = \frac{i(\omega - \omega_o) + 1/\tau_e}{-i(\omega - \omega_o) + 1/\tau_e}\hat{a}(\omega) \; . \tag{6.165}$$

It thus follows that

$$[\hat{b}(\omega),\hat{b}^\dagger(\omega')] = [\hat{a}(\omega),\hat{a}^\dagger(\omega')] = \frac{1}{2\pi}\delta(\omega-\omega') . \tag{6.166}$$

Thus, the commutator of the reflected wave has indeed retained its proper value.

We can now show how the introduction of loss can be handled smoothly with the present formalism. If loss is introduced, the equation of motion (6.155) has to be modified in two ways: (a) a decay rate $1/\tau_o$ has to be introduced; (b) in order to conserve commutators, a noise source must appear. Thus (6.155) changes into

$$\frac{d\hat{U}}{dt} = -(\mathrm{i}\omega_o + 1/\tau_e + 1/\tau_o)\hat{U} + \sqrt{\frac{2}{\tau_o}}n_o + \sqrt{\frac{2}{\tau_e}}\hat{a} . \tag{6.167}$$

In the present perturbational approach one may turn on one perturbation at a time and check for self-consistency. Thus we may ignore the coupling to the outside waveguide and look at the truncated equation

$$\frac{d\hat{U}}{dt} = -(\mathrm{i}\omega_o + 1/\tau_o)\hat{U} + \sqrt{\frac{2}{\tau_o}}n_o . \tag{6.168}$$

The noise source must maintain the commutator of the resonator excitation, which in the absence of the noise source would decay at the rate $2/\tau_o$. From (6.167) we find

$$\frac{d}{dt}[\hat{U},\hat{U}^\dagger] = \left[\frac{d\hat{U}}{dt},\hat{U}^\dagger\right] + \left[\hat{U},\frac{d\hat{U}^\dagger}{dt}\right] \tag{6.169}$$

$$= -\frac{2}{\tau_o}[\hat{U},\hat{U}^\dagger] + \sqrt{\frac{2}{\tau_o}}\{[\hat{n}_o,\hat{U}^\dagger] + [\hat{U},\hat{n}_o^\dagger]\} .$$

Since the loss is frequency independent, the noise source has to be delta-function-correlated in time. Whereas one might expect that the resonator excitation and the noise source commute, since they are independent, this fact does not reduce the second term in (6.169) to zero. Indeed, within the time interval Δt, the resonator amplitude acquires the contribution $(1/2)\sqrt{2/\tau_o}\Delta t\,\hat{n}$ from the noise source, so that the right hand side of (6.169) becomes

$$-\frac{2}{\tau_o}[\hat{U},\hat{U}^\dagger] + \sqrt{\frac{2}{\tau_o}}\{[\hat{n}_o,\hat{U}^\dagger] + \hat{U},\hat{n}_o^\dagger]\} \tag{6.170}$$

$$= -\frac{2}{\tau_o}[\hat{U},\hat{U}^\dagger] + \frac{2}{\tau_o}\Delta t[\hat{n}_o,\hat{n}_o^\dagger] .$$

Thus, conservation of commutator brackets is ensured for a noise source with the commutator

$$[\hat{n}_o(t), \hat{n}_o(t')] = \delta(t - t') .$$ (6.171)

In the Fourier transform domain, the commutator is

$$[\hat{n}_o(\omega), \hat{n}_o(\omega')] = \frac{1}{2\pi}\delta(\omega - \omega') .$$ (6.172)

The approach taken here has some resemblance to the introduction of noise sources in the classical analysis of linear systems at thermal equilibrium. Note that we could have started with (6.155), treating $\hat{a}$ as an undetermined noise source required to maintain commutator conservation in the presence of the decay rate $1/\tau_e$. An analysis identical to the determination of the commutator of $\hat{n}_o$ would have led us to find (6.158) for the commutator $[\hat{a}(\omega), \hat{a}^\dagger(\omega')]$. Thus, the conservation-of-commutator principle can replace a detailed analysis of loss induced by coupling to a reservoir.

The next step we undertake is to introduce gain into (6.155). The fact that one may make statements about the nature of the noise source in this nonequilibrium case has no classical thermodynamic analog. Again, we look at the truncated equation for the resonator with nothing but gain, represented by the growth rate $1/\tau_g$:

$$\frac{d\hat{U}}{dt} = -(i\omega_o - 1/\tau_g)\hat{U} + \sqrt{\frac{2}{\tau_g}}\hat{n}_g .$$ (6.173)

The analysis is completely analogous that carried through in the case with loss, with the result that the commutator of the noise source is now

$$[\hat{n}_g(t), \hat{n}_g^\dagger(t')] = -\delta(t - t') ,$$ (6.174)

or, Fourier transformed,

$$[\hat{n}_g(\omega), \hat{n}_g^\dagger(\omega')] = -\frac{1}{2\pi}\delta(\omega - \omega') .$$ (6.175)

Note the appearance of the minus sign. This means that the roles of the creation and annihilation operators have been reversed. We should note further that (6.174) and (6.175) do not require that $\hat{n}_g(t)$ be a pure creation operator; it could be composed of a sum of a creation operator and an annihilation operator that commute with each other. The only requirement is that the commutation relation of the sum operator and its Hermitian conjugate obeys (6.174) or (6.175). The physical meaning is that the gain mechanism consists of two opposing processes, one with gain, the other with loss. Gain is provided by a two-level system with inversion, in which the occupation of the upper level is higher than that of the lower level. The upper level experiences induced emission, in which a photon causes a transition to the lower level; it also experiences spontaneous emission, in which a laser particle spontaneously decays to the lower level, emitting a photon that is uncorrelated with the induced photons. The lower level can absorb photons as particles in the lower level make transitions to the upper level. In this case the gain mechanism must be represented by a sum of a creation operator $\hat{n}_u$, representing the excitation of the upper level, and an annihilation operator $\hat{n}_\ell$, representing the excitation of the lower level. The upper level causes a growth rate $1/\tau_u$, the lower level a decay rate $1/\tau_\ell$. The net growth rate is

$$\frac{1}{\tau_g} = \frac{1}{\tau_u} - \frac{1}{\tau_\ell} , \tag{6.176}$$

and the commutators of the two noise sources are

$$[\hat{n}_u(t), \hat{n}_u^\dagger(t')] = -\delta(t - t') \quad \text{and} \quad [\hat{n}_\ell(t), \hat{n}_\ell^\dagger(t')] = \delta(t - t') . \tag{6.177}$$

This description of an incompletely inverted medium is indistinguishable from the case of a perfectly inverted medium in a resonator with a loss rate $1/\tau_o = 1/\tau_\ell$.

We may now assemble all three physical mechanisms studied thus far in one single equation for the resonator mode:

$$\frac{d\hat{U}}{dt} = -(\mathrm{i}\omega_o + 1/\tau_o - 1/\tau_g + 1/\tau_e)\hat{U} + \sqrt{\frac{2}{\tau_o}}\hat{n}_o + \sqrt{\frac{2}{\tau_g}}\hat{n}_g + \sqrt{\frac{2}{\tau_e}}\hat{a} . \tag{6.178}$$

The equation for the excitation of the reflected wave remains unchanged. It is easily checked that the commutator of the reflected wave, (6.161), is preserved, as it should be, in the presence of all three noise sources, which are all mutually uncorrelated and commute.

Equation (6.178) can be used to evaluate the photon number inside the resonator. These photons represent amplified spontaneous emission if no signal is fed into the resonator. At this point we must decide on the states of the different noise sources, or rather the reservoirs they represent. If there were thermal excitation, it would be near room temperature θ_o. The energy levels under consideration are optical levels, with energies of the order of 40 times the value of $k\theta_o$. Thus, one may assume that the noise sources are all unexcited; they are in the ground state. This means that $\langle \hat{a}^\dagger(\omega)\hat{a}(\omega')\rangle = \langle \hat{n}_o^\dagger(\omega)\hat{n}_o(\omega')\rangle = 0$. On the other hand, we have identified the operator $\hat{n}_g$ as a creation operator and its Hermitian conjugate as an annihilation operator. Since

$$\langle [\hat{n}_g(\omega), \hat{n}_g^\dagger(\omega')]\rangle = \langle \hat{n}_g(\omega)\hat{n}_g^\dagger(\omega')\rangle - \langle \hat{n}_g^\dagger(\omega)\hat{n}_g(\omega')\rangle = -\frac{1}{2\pi}\delta(\omega - \omega') ,$$

we must conclude that

$$\langle \hat{n}_g^\dagger(\omega)\hat{n}_g(\omega')\rangle = \frac{1}{2\pi}\delta(\omega - \omega') .$$

We have

$$\langle \hat{U}^\dagger(\omega)\hat{U}(\omega')\rangle \Delta\omega\,\Delta\omega'$$

$$= \frac{(2/\tau_g)\langle \hat{n}_g^\dagger(\omega)\hat{n}_g(\omega')\rangle \Delta\omega\,\Delta\omega'}{[-\mathrm{i}(\omega_o - \omega) + 1/\tau_o - 1/\tau_g + 1/\tau_e][\mathrm{i}(\omega_o - \omega') + 1/\tau_o - 1/\tau_g + 1/\tau_e]}$$

$$= \frac{1}{2\pi} \frac{2/\tau_g}{[-\mathrm{i}(\omega_o - \omega) + 1/\tau_o - 1/\tau_g + 1/\tau_e][\mathrm{i}(\omega_o - \omega') + 1/\tau_o - 1/\tau_g + 1/\tau_e]}$$

$$\times \Delta\omega\,\Delta\omega'\,\delta(\omega - \omega')\ .$$

$$(6.179)$$

Since the photon flow rate is expressed as a double integral over frequency, the photon number in the resonator is also obtained from a double integral:

$$\int d\omega \int d\omega'\,\langle \hat{U}^\dagger(\omega)\hat{U}(\omega')\rangle$$

$$= \int d\omega \frac{1}{2\pi} \frac{2/\tau_g}{[(\omega_o - \omega)^2 + (1/\tau_o - 1/\tau_g + 1/\tau_e)^2]} \qquad (6.180)$$

$$= \frac{1/\tau_g}{1/\tau_o - 1/\tau_g + 1/\tau_e}\ .$$

Next, we consider the photon flow $\langle \hat{b}^\dagger(\omega)\hat{b}(\omega')\rangle \Delta\omega\,\Delta\omega'$ from the resonator with gain. We limit ourselves to a perfectly inverted medium. We find

$$\langle \hat{b}^\dagger(\omega)\hat{b}(\omega')\rangle \Delta\omega\,\Delta\omega'$$

$$= \frac{2}{\tau_e} \frac{(2/\tau_g)\langle \hat{n}_g^\dagger(\omega)\hat{n}_g(\omega')\rangle \Delta\omega\,\Delta\omega'}{[-\mathrm{i}(\omega_o - \omega) + 1/\tau_o - 1/\tau_g + 1/\tau_e][\mathrm{i}(\omega_o - \omega') + 1/\tau_o - 1/\tau_g + 1/\tau_e]}$$

$$= \frac{4/\tau_e\tau_g}{(\omega_o - \omega)^2 + (1/\tau_o - 1/\tau_g + 1/\tau_e)^2} \frac{\Delta\omega}{2\pi}\delta(\omega - \omega')\ .$$

$$(6.181)$$

The net photon flow is

$$\int d\omega \int d\omega'\,\langle \hat{b}^\dagger(\omega)\hat{b}(\omega')\rangle$$

$$= \int d\omega \frac{1}{2\pi} \frac{4/\tau_e\tau_g}{(\omega_o - \omega)^2 + (1/\tau_o - 1/\tau_g + 1/\tau_e)^2} \qquad (6.182)$$

$$= \frac{2/(\tau_e\tau_g)}{1/\tau_o - 1/\tau_g + 1/\tau_e}\ .$$

This is the photon flow of the amplified spontaneous emission for a fully inverted medium in the presence of resonator loss. The flow of photons goes to infinity as the threshold is reached, as $1/\tau_g \to 1/\tau_o + 1/\tau_e$. It is clear that this trend cannot persist when the threshold is passed. In Chap. 11 we look at the "lasing" operation of the resonator, the operation above threshold.

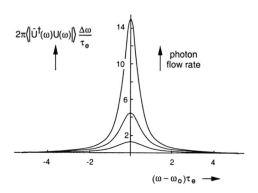

Fig. 6.4. The spectrum $2\pi\langle|\hat{U}^\dagger(\omega)\hat{U}(\omega)|\rangle\Delta\omega/\tau_e$ for $\tau_e/\tau_g = 0.2, 0.4$ and 0.6

Figure 6.4 shows the spectrum of the photon flow for three different values of τ_e/τ_g, with no loss in the resonator, i.e. $1/\tau_o = 0$. Next we study the in-phase and quadrature components of the wave emitted by the resonator. The in-phase and quadrature components are Hermitian operators that have to be constructed from the sum and difference of $\hat{b}$ and $\hat{b}^\dagger$. We have

$$\hat{b}^{(1)} = \hat{a}^{(1)} - \frac{1}{2}\left[\sqrt{2/\tau_e}\frac{\sqrt{2/\tau_o}\hat{n}_o(\omega) + \sqrt{2/\tau_g}\hat{n}_g(\omega) + \sqrt{2/\tau_e}\hat{a}(\omega)}{i(\omega - \omega_o) + 1/\tau_o - 1/\tau_g + 1/\tau_e} + \text{H.c.}\right]$$

$$= \hat{a}^{(1)} - \left(\frac{1}{\tau_o} - \frac{1}{\tau_g} + \frac{1}{\tau_e}\right)$$

$$\times \sqrt{\frac{2}{\tau_e}}\frac{\sqrt{2/\tau_o}\hat{n}_o^{(1)}(\omega) + \sqrt{2/\tau_g}\hat{n}_g^{(1)}(\omega) + \sqrt{2/\tau_e}\hat{a}^{(1)}(\omega)}{(\omega - \omega_o)^2 + (1/\tau_o - 1/\tau_g + 1/\tau_e)^2}$$

$$- (\omega - \omega_o)\sqrt{\frac{2}{\tau_e}}\frac{\sqrt{2/\tau_o}\hat{n}_o^{(2)}(\omega) + \sqrt{2/\tau_g}\hat{n}_g^{(2)}(\omega) + \sqrt{2/\tau_e}\hat{a}^{(2)}(\omega)}{(\omega - \omega_o)^2 + (1/\tau_o - 1/\tau_g + 1/\tau_e)^2},$$

$$(6.183)$$

where "H.c." stands for "Hermitian conjugate". It is of interest to note that the quadrature component couples to the in-phase component off resonance. This phase-to-amplitude coupling is, in fact, characterisic of all resonant structures excited off their resonance frequency. FM detectors are constructed on this principle.

From (6.183) and the mean square fluctuations of the noise sources, one may construct the mean square fluctuations of the wave emitted from the resonator:

$$\langle |\hat{b}^{(1)}|^2 \rangle$$

$$= \frac{1}{4} \left\{ \left[1 - \left(\frac{1}{\tau_o} - \frac{1}{\tau_g} + \frac{1}{\tau_e} \right) \frac{2/\tau_e}{(\omega - \omega_o)^2 + (1/\tau_o - 1/\tau_g + 1/\tau_e)^2} \right]^2 \right.$$

$$+ \left(\frac{1}{\tau_o} - \frac{1}{\tau_g} + \frac{1}{\tau_e} \right)^2 \frac{2}{\tau_e} \frac{2/\tau_o + 2/\tau_g}{[(\omega - \omega_o)^2 + (1/\tau_o - 1/\tau_g + 1/\tau_e)^2]^2}$$

$$+ (\omega - \omega_o)^2 \frac{2}{\tau_e} \frac{2/\tau_o + 2/\tau_g + 2/\tau_e}{[(\omega - \omega_o)^2 + (1/\tau_o - 1/\tau_g + 1/\tau_e)^2]^2} \left. \right\} \frac{1}{2\pi} \delta(\omega - \omega')$$

$$= \frac{1}{4} \left[1 + \frac{8/(\tau_e \tau_g)}{(\omega - \omega_o)^2 + (1/\tau_o - 1/\tau_g + 1/\tau_e)^2} \right] \frac{1}{2\pi} \delta(\omega - \omega') .$$

$$(6.184)$$

The same result is obtained for $\langle |\hat{b}^{(2)}|^2 \rangle$. The physical significance of the result is plainly evident. If there is no gain in the resonator, the exterior fluctuations are zero-point fluctuations. If there is gain, the fluctuations at and near the resonance frequency are enhanced. Away from resonance, they revert to simple zero-point fluctuations.

6.12 Summary

We started with a review of the classical Hamiltonian mechanics of the harmonic oscillator and reviewed its quantization. The quantization of electromagnetic fields uses the fact that electromagnetic modes obey harmonic-oscillator equations. It was Planck who arrived at this quantization procedure with great ingenuity, long before the quantum formalism was developed. The excitation of the waveguide is described by creation and annihilation operators that are in one-to-one correspondence with the classical complex amplitudes of the electric field of the mode. These operators obey commutation relations that are intrinsic to their nature.

We developed the quantum formalism for a cavity coupled to an external waveguide. The decay of the cavity mode was derived from the coupling of the cavity mode to an infinite number of modes in the coupling waveguide, assumed to be so long that the period of the beats, associated with the coupling of two lossless modes was extended to infinity and the effect of the coupling appeared as a simple decay of the resonator mode. The analysis introduced automatically an operator source that ensured conservation of the commutator brackets of the resonator mode operators.

Next we addressed loss in a waveguide. Having learned that coupling to a reservoir of modes that introduces decay calls for an operator noise source in order to ensure commutator conservation, we introduced such a source and determined its commutator. It was then possible to evaluate the mean square fluctuations of the mode field under the assumption that the reservoir modes were in their ground (vacuum) state. The analysis was carried through analogously for a waveguide with gain. Here, it was possible to show that an amplifier must emit photons even in the absence of an input signal, namely the photons of amplified spontaneous emission.

Finally, we quantized the classical equations of a resonator with loss and gain coupled to an external waveguide. This description was in full agreement with the Hamiltonian description of a resonator coupled to a reservoir of the modes of a long waveguide developed in Sect. 6.5. The simplicity of the formalism permitted us to obtain answers to a number of questions as to the photon flow emitted by such a structure and the mean square fluctuations of the in-phase and quadrature components of the emitted field.

Problems

6.1 Show that the states of the harmonic oscillator are orthogonal.

6.2 Show that different Hermite Gaussians are orthogonal.

6.3

(a) Determine the enhancement of the ASE associated with incomplete inversion as described in Sect. 6.6, by taking advantage of the equations with perfect inversion, but with an additional loss rate. This additional loss rate may be identified as being due to occupation of the lower lasing level.

(b) Derive the photon flow for a resonator of zero resonator loss containing an incompletely inverted medium.

6.4* Derive the mean square quantum fluctuations of a coherent-state wave transmitted through a transmission resonator with coupling rates $1/\tau_{e1}$ and $1/\tau_{e2}$ and an internal loss rate $1/\tau_o$.

6.5* A wave incident from port (1) onto a beam splitter with the scattering matrix

$$S = \begin{bmatrix} r & i\sqrt{1-r^2} \\ i\sqrt{1-r^2} & r \end{bmatrix}$$

exits partly in port (3) and partly in port (4). As viewed from ports (1) and (3), the system appears as a lossy system. Conservation of the commutator brackets requires the addition of noise. Show that the amplitude operator entering through port (2), with part of it emerging in port (3), fully accounts for a "source" that preserves commutators.

6.6 Show that the formalism of Sect. 6.5 that arrives at the equations of an open resonator and the associated noise sources by coupling to the modes of an external waveguide can be used to derive the equations for a classical resonator at thermal equilibrium. Compare Appendix A.9.

6.7 Show that the transmission of power through a transmission resonator for a coherent state $|\alpha\rangle$ of frequency ω incident from port (1) is in one-to-one correspondence with the transmission of power through a classical resonator.

6.8* Show that for any pair of operators $\hat{A}$ and $\hat{B}$ the following relationship holds:

$$\hat{A}\exp(\hat{B})\hat{A}^{-1} = \exp(\hat{A}\hat{B}\hat{A}^{-1}) \ .$$

Solutions

6.4 This solution uses a generalization to the two-port resonator of Prob. 2.8. Otherwise it follows closely the derivation of Sect. 6.11. The equation of the resonator mode is

$$\frac{d\hat{U}}{dt} = -\left(i\omega_o + \frac{1}{\tau_{e1}} + \frac{1}{\tau_{e2}} + \frac{1}{\tau_o}\right)\hat{U} + \sqrt{\frac{2}{\tau_{e1}}}\hat{a}_1 + \sqrt{\frac{2}{\tau_{e2}}}\hat{a}_2 + \sqrt{\frac{2}{\tau_o}}\hat{n}_o \ . \quad (1)$$

With an assumed time dependence $\exp(-i\omega t)$, we find

$$\hat{U}(\omega) = \frac{\sqrt{2/\tau_{e1}}\hat{a}_1 + \sqrt{2/\tau_{e2}}\hat{a}_2 + \sqrt{2/\tau_o}\hat{n}_o}{i(\omega_o - \omega) + 1/\tau_{e1} + 1/\tau_{e2} + 1/\tau_o} \ . \quad (2)$$

The excitation of port (2) is

$$\hat{b}_2 = -\hat{a}_2 + \sqrt{\frac{2}{\tau_{e2}}}\hat{U} \quad (3)$$

and therefore

$$\hat{b}_2(\omega) = -\hat{a}_2(\omega) + \sqrt{\frac{2}{\tau_{e2}}}\frac{\sqrt{2/\tau_{e1}}\hat{a}_1(\omega) + \sqrt{2/\tau_{e2}}\hat{a}_2(\omega) + \sqrt{2/\tau_o}\hat{n}_o(\omega)}{i(\omega_o - \omega) + 1/\tau_{e1} + 1/\tau_{e2} + 1/\tau_o} \ . \quad (4)$$

The mean square output for the two phases is

$$\langle \hat{b}_2^{(i)}(\omega)\hat{b}_2^{(i)}(\omega')\rangle \ , \quad \text{where} \quad i = 1, 2 \ . \quad (5)$$

Since the input is in a coherent state, all operator products should be put into normal order. Then, only the contribution of the commutator remains:

$$\langle \hat{b}_2^{(i)}(\omega) \hat{b}_2^{(i)}(\omega') \rangle - |\langle \hat{b}_2^{(i)}(\omega) \rangle|^2$$

$$= \frac{1}{4} \frac{1}{2\pi} \delta(\omega - \omega') \left[1 - \frac{2}{\tau_{e2}} \frac{2/\tau}{(\omega - \omega_o)^2 + 1/\tau^2} + \frac{2}{\tau_{e2}} \frac{2/\tau}{(\omega - \omega_o)^2 + 1/\tau^2} \right]$$

$$= \frac{1}{4} \frac{1}{2\pi} \delta(\omega - \omega') ,$$

where

$$\frac{1}{\tau} = \frac{1}{\tau_{e1}} + \frac{1}{\tau_{e2}} + \frac{1}{\tau_o} .$$

We get standard zero-point fluctuations. This is as expected, since in the absence of laser action, a wave in an open waveguide must exhibit standard fluctuations.

6.5 The equation for the output in terms of the input is

$$\hat{B}_3 = r\hat{A}_1 + i\sqrt{1 - r^2}\hat{A}_2 . \tag{1}$$

The commutator of the output wave is

$$[\hat{B}_3, \hat{B}_3^\dagger] = r^2 [\hat{A}_1, \hat{A}_1^\dagger] + (1 - r^2)[\hat{A}_2, \hat{A}_2^\dagger] = 1 . \tag{2}$$

Thus, (1) written as

$$\hat{B}_3 = r\hat{A}_1 + \hat{N} \tag{3}$$

has acquired a noise source with the proper commutator:

$$[\hat{N}, \hat{N}^\dagger] = 1 - r^2 . \tag{4}$$

6.8 The identity is proven by expanding the exponential into a power series

$$\hat{A}\exp(\hat{B})\hat{A}^{-1} = \hat{A}\sum \frac{\hat{B}^n}{n!}\hat{A}^{-1} .$$

Consider one term in the expansion. We have

$$\hat{A}\hat{B}^n A^{-1} = \hat{A}\hat{B}A^{-1} \cdot \hat{A}\hat{B}A^{-1} \cdot \hat{A}\hat{B}A^{-1} \ldots n \text{ times} .$$

By introducing this identity into the series, we prove the assertion.

7. Classical and Quantum Analysis of Phase-Insensitive Systems

In Chap. 6 we investigated the quantization of open resonators and of waves on transmission lines. We treated one example of a simple linear system, namely a resonator coupled to a waveguide. Practical electromagnetic systems consist of RLC circuits, resonators, waveguide junctions, fibers, beam splitters, and, of course, amplifiers, to name only a few. Such systems, if linear, are described classically by impedance matrices or scattering matrices (Chap. 2) that are functions of frequency. This formalism is well developed in the classical domain. In this chapter, we review the classical formalism and its generalization to quantum theory. We define Hamiltonians which, via the Heisenberg equations of motion, lead to equations that are in direct correspondence with the classical circuit equations. If the multiports are lossy or exhibit gain, they must contain noise sources in order to conserve commutator brackets from input to output. The commutator brackets determine the minimum amount of noise added to the signal as it passes through the network. Hence one may determine the optimum noise measure achievable in a quantum circuit directly from these relations.

Amplifiers with high gain provide a signal level at their output that is "classical", which, for example, can be viewed on a scope without any ambivalence as to what is being observed. Two observables whose operators do not commute cannot be measured simultaneously. Yet, in a classical display one may view observables whose operators do not commute, such as the in-phase and quadrature components of the field amplitude. We shall show that these can be observed simultaneously, but that the simultaneous measurement of both observables is accompanied by a penalty of additional noise.

The Heisenberg equations of motion of the field operators have a close correspondence with the classical equations of motion of the complex field amplitudes. This is the correspondence principle that requires the emergence of classical equations of motion for observables when quantum effects can be neglected. The Schrödinger formalism, which expresses the time evolution of the states rather than of the operators, does not display the correspondence principle directly, since quantum states have no classical counterpart. Conversely it is also true that the Heisenberg equations of motion do not directly display quantum behavior, such as that contained in so-called entangled states. Entangled states are a wellspring of paradoxes associated with quantum measurements. In preparation for their discussion in Chap. 14, we study the peculiar properties of entangled states, using the Schrödinger formalism.

7.1 Renormalization of the Creation and Annihilation Operators

In the analysis of waveguides, we found it convenient to use the operators $\hat{a}(\beta)$ and $\hat{a}^\dagger(\beta)$ for the mode amplitudes, essentially a spectral representation in β space. In this chapter, we analyze multiports excited by several waveguides that may have different dispersions: modes of the same frequency have different β values in the different waveguides. In linear multiports, modes of the same frequency in different input waveguides couple to each other, and modes of different frequencies do not. Hence, the modes entering from the different waveguides must be identified by frequency, not propagation constant. There is a further problem. The quantization in a waveguide was done for modes occupying a length L. In the excitation of a multiport from different waveguides within a narrow band of wavelengths and/or frequencies, the excitations from the different waveguides enter the multiport, interact, and leave. They do so moving at their own group velocities. The lengths L in the different waveguides must be in the inverse ratio of their group velocities to be properly synchronized. For this reason it is appropriate to use operators that do not depend on these length assignments. The operators are redefined as follows. Remember that the photon number within the length L was given by

$$\hat{A}_m^\dagger \hat{A}_m \,, \tag{7.1}$$

with the commutation relation

$$[\hat{A}_m, \hat{A}_n^\dagger] = \delta_{nm} \,. \tag{7.2}$$

The photon number can be converted into a photon number flow by division by L and multiplication by the group velocity

$$\text{photon flow} = \frac{v_g}{L} \hat{A}_m^\dagger \hat{A}_m = \frac{1}{2\pi} v_g \Delta\beta \, \hat{A}_m^\dagger \hat{A}_m = \hat{A}_m^\dagger \hat{A}_m \frac{\Delta\omega_q}{2\pi} \,, \tag{7.3}$$

where $\Delta\omega_q = v_g \Delta\beta$ is the interval of quantization. We introduce the new notation

$$\hat{a} \equiv \hat{A}_m \sqrt{\frac{\Delta\omega_q}{2\pi}} \,. \tag{7.4}$$

These new operators, assigned to a frequency ω and the frequency interval $\Delta\omega_q$, obey the commutation relation

$$[\hat{a}, \hat{a}^\dagger] = \frac{\Delta\omega_q}{2\pi} \tag{7.5a}$$

when $\hat{a}$ and $\hat{a}^\dagger$ have the same frequency, and

$$[\hat{a}, \hat{a}^\dagger] = 0 \tag{7.5b}$$

for $\hat{a}$'s of different frequencies.

Operation on a coherent state by $\hat{a}$, still an eigenstate of $\hat{a}$, gives the result

$$\hat{a}|\alpha\rangle = \sqrt{\frac{\Delta\omega_q}{2\pi}}\,\alpha|\alpha\rangle\ , \tag{7.6}$$

so that the expectation value $\langle\alpha|\hat{a}^\dagger\hat{a}|\alpha\rangle$ is

$$\langle\alpha|\hat{a}^\dagger\hat{a}|\alpha\rangle = \frac{v_g}{L}|\alpha|^2\ , \tag{7.7}$$

i.e. the photon flow of the coherent state. Strictly, a coherent state has a bandwidth. The state is defined over a length L, and hence a time interval $T = L/v_g$. Outside this time interval another coherent state is defined, and hence the duration of the coherent state is $T = 2\pi/\Delta\omega_q$. If communication is performed with a sequence of coherent states, and the noise accompanying the signal is to be properly filtered, a filter bandwidth $\Delta\omega$ must be chosen. In the subsequent analysis it will be assumed that the signal bandwidth and noise bandwidth are made equal, and we shall drop the subscript "q" on $\Delta\omega$.

7.2 Linear Lossless Multiports in the Classical and Quantum Domains

Consider the excitation of a linear multiport from N waveguides as shown in Fig. 7.1. The excitations of the waveguides at one frequency may be written in terms of the N incident waves a_j and the N reflected waves b_j. Because the circuit is linear, the b_i are related linearly to the a_j and no other frequency components are generated by the excitation of the circuit. We form column matrices of the excitation amplitudes a_i and b_j. The multiport is described by the $N \times N$ scattering matrix S and the following relation holds:

$$b = Sa + s\ , \tag{7.8}$$

where s contains the noise sources. A passive multiport at thermal equilibrium requires such noise sources in order to conserve the thermal radiation from input to output. The spectra of these noise sources were evaluated in Chap. 4.

In this section, we look first at some lossless multiports. Lossless multiports have no internal noise sources and the scattering equation simplifies to

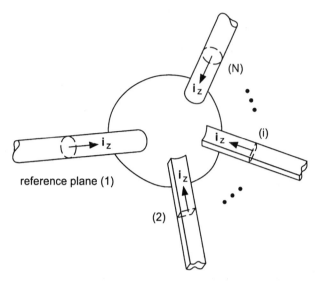

Fig. 7.1. A linear multiport excited by incident waves a_j

$$b = Sa \,, \tag{7.9}$$

where S is a unitary matrix. One of the simplest two-ports is a lossless, partially transmitting, mirror with amplitude reflection r (see Fig. 7.2). Its scattering matrix must be unitary, as proven in Chap. 2. Such a mirror must also be reciprocal since it is described by the reciprocal Maxwell's equations. The reciprocity condition implies symmetry of the scattering matrix (Chap. 2). The unitarity condition for a complex matrix of second rank leads to two real equations and one complex equation, four real equations in toto. Thus, the eight real parameters of a complex matrix of second rank are reduced to six by symmetry of the matrix, and further reduced to three free parameters by the unitarity condition. Returning to the lossless mirror, we find that we may choose arbitrarily the reference planes for the incident and reflected waves, which disposes of two free parameters. Thus, a lossless two-port has only one real free parameter. In the present case, it is the reflectivity r of the mirror. Thus, the scattering matrix of a mirror is

$$S = \begin{bmatrix} r & -i\sqrt{1-r^2} \\ -i\sqrt{1-r^2} & r \end{bmatrix} \,. \tag{7.10}$$

How does one describe a mirror quantum mechanically? The wave amplitude operators in the incoming and outgoing ports are the quantities $\hat{a}$ of the preceding section. The operator is assigned to one mode. Reflection from the mirror constitutes a scattering event. Incident waves are transformed into reflected waves. The transformation is described by an integral of the

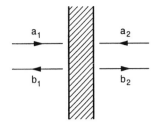

Fig. 7.2. A partially transmitting mirror

Heisenberg equation of motion:

$$\frac{d}{dt}\hat{a}_j = \frac{i}{\hbar}[\hat{H}, \hat{a}_j] . \tag{7.11}$$

Since the interaction is linear, the Hamiltonian must be a quadratic expression in the $\hat{a}_j$. (Remember the commutator removes one operator factor; therefore the commutator of an operator with a quadratic Hamiltonian is linear in the operator(s).) If we suppress the natural time dependence $\exp(-i\omega t)$ of the operators, we may assume a Hamiltonian of the form

$$\hat{H} = \hbar(M_{12}\hat{a}_1^\dagger\hat{a}_2 + M_{21}\hat{a}_2^\dagger\hat{a}_1) , \tag{7.12}$$

where $M_{12} = M_{21}^*$, because the Hamiltonian is Hermitian and thus $M_{12} = Me^{i\theta}, M_{21} = Me^{-i\theta}$ with M real. In (7.12) we have omitted the contribution of zero-point fluctuations, since it does not affect the equations of motion.

The equations of motion are

$$\frac{d}{dt}\hat{a}_1 = -iM_{12}\hat{a}_2 , \tag{7.13}$$

$$\frac{d}{dt}\hat{a}_2 = -iM_{21}\hat{a}_1 . \tag{7.14}$$

The solutions of these equations are the functions $\exp(-iMT)$. The meaning of this exponential with a matrix as its argument is extracted from the Taylor expansion. Its first-order term is

$$-i\begin{bmatrix} 0 & M_{12} \\ M_{21} & 0 \end{bmatrix} T = -i\begin{bmatrix} 0 & e^{i\theta} \\ e^{-i\theta} & 0 \end{bmatrix} MT .$$

Its second-order term is

$$-\frac{1}{2}\begin{bmatrix} 0 & M_{12} \\ M_{21} & 0 \end{bmatrix}\begin{bmatrix} 0 & M_{12} \\ M_{21} & 0 \end{bmatrix} T^2 = -\frac{1}{2}\begin{bmatrix} M_{12}M_{21} & 0 \\ 0 & M_{21}M_{12} \end{bmatrix} T^2$$

$$= -\frac{1}{2}\begin{bmatrix} 1 & 0 \\ 0 & 1 \end{bmatrix} M^2 T^2 .$$

It easy to see that the Taylor series gives the solutions

$$\exp(-\mathrm{i}MT) = \begin{bmatrix} \cos(MT) & -\mathrm{i}e^{\mathrm{i}\theta}\sin(MT) \\ -\mathrm{i}e^{-\mathrm{i}\theta}\sin(MT) & \cos(MT) \end{bmatrix}, \tag{7.15}$$

and thus

$$\begin{bmatrix} \hat{a}_1(T) \\ \hat{a}_2(T) \end{bmatrix} = \begin{bmatrix} \cos(MT) & -\mathrm{i}e^{\mathrm{i}\theta}\sin(MT) \\ -\mathrm{i}e^{-\mathrm{i}\theta}\sin(MT) & \cos(MT) \end{bmatrix} \begin{bmatrix} \hat{a}_1(0) \\ \hat{a}_2(0) \end{bmatrix}. \tag{7.16}$$

The excitations after evolution over the time T must be interpreted as the outgoing waves $\hat{b}_1$ and $\hat{b}_2$. Thus we find correspondence with the classical scattering matrix of the mirror, with $r = \cos(MT)$ and $\theta = 0$. The quantum analysis implies losslessness, but not necessarily reciprocity, and thus it ends up with an arbitrary phase angle, which can be removed from the classical scattering matrix on the basis of reciprocity.

After this simple example of a two-port we may turn to the analysis of a general lossless multiport of N ports by considering the Hamiltonian

$$\hat{H} = \hbar \hat{a}^\dagger M \hat{a}, \tag{7.17}$$

where M is an $N \times N$ matrix. Here we have arranged the operator excitations into column matrices. The dagger indicates the Hermitian conjugate of the operator as well as the transpose of the column matrix. The Heisenberg equation of motion becomes

$$\frac{d}{dt}\hat{a} = -\mathrm{i}M\hat{a}. \tag{7.18}$$

Integration of the equation over a time T gives the scattered waves $\hat{b}$ in terms of the incident waves,

$$\hat{b} = S\hat{a}, \tag{7.19}$$

with the scattering matrix

$$S = \exp(-\mathrm{i}MT). \tag{7.20}$$

It should be noted that the input and output excitations in the quantum case refer to photon packets, whereas classically the excitations are traveling waves. If the group velocities in the different waveguides are different, the packets occupy different lengths, the lengths being in the ratio of the respective group velocities.

Let us look at an important example of a lossless four-port, a beam splitter (see Fig. 7.3). This schematic shows which excitation from each port makes it to some other port. An input excitation in port (1) exits from ports (2) and (3), an excitation in port (2) exits from ports (1) and (4), etc. Only two numbers, the reflection r and a phase θ, describe the whole operation, because

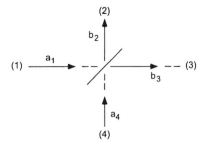

Fig. 7.3. Lossless beam splitter

the four-port has to obey the conditions of reciprocity (i.e. have a symmetric scattering matrix) and power conservation (it must be unitary). If arbitrary phases are removed by proper choice of the positions of the reference planes, then the beam splitter is described by the following scattering relation with a symmetric unitary matrix of fourth rank:

$$
\begin{bmatrix} \hat{b}_1 \\ \hat{b}_2 \\ \hat{b}_3 \\ \hat{b}_4 \end{bmatrix}
=
\begin{bmatrix}
0 & r & -\mathrm{i}\sqrt{1-r^2}e^{-\mathrm{i}\theta} & 0 \\
r & 0 & 0 & -\mathrm{i}\sqrt{1-r^2}e^{\mathrm{i}\theta} \\
-\mathrm{i}\sqrt{1-r^2}e^{-\mathrm{i}\theta} & 0 & 0 & r \\
0 & -\mathrm{i}\sqrt{1-r^2}e^{\mathrm{i}\theta} & r & 0
\end{bmatrix}
\begin{bmatrix} \hat{a}_1 \\ \hat{a}_2 \\ \hat{a}_3 \\ \hat{a}_4 \end{bmatrix},
$$

$$(7.21)$$

both classically and quantum mechanically. In the latter case, the amplitudes become annihilation operators. The scattering matrix of the beam splitter (7.21) applies equally well to a waveguide coupler propagating forward and backward waves, as shown in Fig. 7.4. The waveguide coupler is a lossless four-port. A forward wave couples gradually to a copropagating wave in the adjacent waveguide without coupling to the backward-propagating waves. Backward-propagating waves couple in a similar manner to each other. Using the fact that forward- and backward-propagating waves do not couple to each other, and the constraints imposed by losslessness and reciprocity, we arrive at the scattering matrix of (7.21).

The appearance of the operator evolution (7.19), in which an operator is premultiplied by a unitary matrix, is a bit surprising to those of us who know that, in the Heisenberg representation, the time evolution of an operator is described by pre- and post-multiplication of the operator by a unitary matrix. In the next section, we show that the two approaches are consistent when applied to linear lossless multiports.

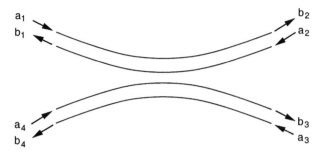

Fig. 7.4. Lossless waveguide coupler

7.3 Comparison of the Schrödinger and Heisenberg Formulations of Lossless Linear Multiports

Thus far we have quantized guided waves and resonant modes using the Heisenberg representation. The Heisenberg representation is in strong correspondence with classical field theory. If the system is linear, the mode annihilation operators evolve in time in the same way as the classical complex field amplitudes. Wave functions describing the state of the system are used only when expectation values of the operators are evaluated. Further, the wave functions used to find the expectation values are those of the initial states of the operators. The time evolution of the system is contained fully in the time evolution of the operators. This description has the advantage that the correspondence principle is rendered self-evident. It has the disadvantage that it does not display explicitly effects that are inherently quantum mechanical, such as the strange behavior of entangled states. Of course, such quantum effects are still contained in the theory and can be extracted from the expectation values of the field operators. When these peculiar quantum effects are present, then the expectation values of the moments of the field operators cannot be predicted from classical probability considerations.

In the Schrödinger representation, the operators are time-independent; the wave functions evolve in time. The correspondence principle is not self-evident, since wave functions have no place in classical physical theory. On the other hand, entangled states, which are a wellspring of paradoxes associated with quantum mechanics, emerge clearly in this representation. In fact, we present the Schrödinger formalism with the intent to use it in Chap. 14 to elucidate the behavior of entangled states and to present a resolution of the Schrödinger cat paradox.

The evolution of the wave function in the Schrödinger representation is given by

$$|\psi(t)\rangle = \hat{U}(t)|\psi(0)\rangle , \tag{7.22}$$

where $\hat{U}$ is a unitary operator related to the Hamiltonian by

$$\hat{U}(t) = \exp\left(-\frac{i}{\hbar}\hat{H}t\right) . \tag{7.23}$$

The transition between the Schrödinger and Heisenberg representations follows from the expression for the expectation value of an operator. Taking the annihilation operator $\hat{A}(t)$ as an example, we evaluate its expectation value from the wave function $|\psi(0)\rangle$ in the Heisenberg representation:

$$\langle\psi(0)|\hat{A}(t)|\psi(0)\rangle = \langle\psi(0)|\hat{U}^\dagger(t)\hat{A}(0)\hat{U}(t)|\psi(0)\rangle = \langle\psi(t)|\hat{A}(0)|\psi(t)\rangle . \tag{7.24}$$

This equation shows that either one may use the Schrödinger evolution of the wave function as in (7.22), keeping the operator of the observable $\hat{A}$ at its initial value $\hat{A}(0)$, or one may vary the operator according to the law

$$\hat{A}(t) = \hat{U}^\dagger(t)\hat{A}(0)\hat{U}(t) . \tag{7.25}$$

This approach can be extended to column matrices of observables. To indicate the transition to column matrices we write the operators in bold type:

$$\begin{bmatrix} \hat{A}_1 \\ \hat{A}_2 \\ \ldots \\ \hat{A}_N \end{bmatrix} = \hat{\boldsymbol{A}} . \tag{7.26}$$

The unitary operator $\hat{U}$ involves the Hamiltonian of the entire system:

$$\hat{H} = \hbar \sum_{j,k=1}^{N} \left(M_{jk}\hat{A}_j^\dagger\hat{A}_k + \frac{1}{2}\right) . \tag{7.27}$$

The unitary evolution matrix has the form of (7.23) and remains a scalar, rather than becoming a column matrix. The input state $|\psi\rangle$ is now a product state:

$$|\psi(0)\rangle = |\psi_1(0)\rangle \otimes |\psi_2(0)\rangle \otimes \ldots \otimes |\psi_N(0)\rangle = \prod_{j=1}^{j=N} |\psi_j(0)\rangle . \tag{7.28}$$

In the Schrödinger formalism, the state evolves according to the law

$$|\psi(t)\rangle = \hat{U}(t)|\psi(0)\rangle . \tag{7.29}$$

In the Heisenberg formalism, the column matrix operator $\hat{\boldsymbol{A}}$ evolves according to the law

$$\hat{\boldsymbol{A}}(t) = \hat{U}^\dagger(t)\hat{\boldsymbol{A}}(0)\hat{U}(t) . \tag{7.30}$$

If the time evolution extends over a time interval T, the operator $\hat{\boldsymbol{A}}(T)$ is

$$\hat{A}(T) = \exp[i\hat{A}^\dagger(0)M\hat{A}(0)T]\hat{A}(0)\exp[-i\hat{A}^\dagger(0)M\hat{A}(0)T] . \qquad (7.31)$$

From now on we denote by the operator $\hat{a}$ its value at $t = 0$. In the notation of (7.19), $\hat{A}(T) = \hat{B}$, and hence the pre- and post-multiplication by the unitary operator ought to be equivalent to premultiplication by the scattering matrix. This equivalence is not obvious at first glance, but we now proceed to prove it. For this we need an operator identity.

Consider the following function of the c number ξ containing the operators $\hat{Q}$ and $\hat{R}$:

$$\hat{f}(\xi) = \exp(\xi\hat{R})\hat{Q}\exp(-\xi\hat{R}) . \qquad (7.32)$$

We may expand this function of ξ into a Taylor series in ξ. For this purpose, evaluate $d\hat{f}/d\xi$:

$$\frac{d\hat{f}}{d\xi} = \hat{R}\exp(\xi\hat{R})\hat{Q}\exp(-\xi\hat{R}) - \exp(\xi\hat{R})\hat{Q}\exp(-\xi\hat{R})\hat{R} = [\hat{R}, \hat{f}(\xi)] . \qquad (7.33)$$

Repetition of this procedure gives the Taylor expansion

$$\hat{f}(\xi) = \hat{f}(0) + \frac{\xi}{1!}\frac{d\hat{f}}{d\xi} + \frac{\xi^2}{2!}\frac{d^2\hat{f}}{d\xi^2} + \cdots \qquad (7.34)$$

$$= \hat{Q} + \frac{\xi}{1!}[\hat{R}, \hat{Q}] + \frac{\xi^2}{2!}[\hat{R}, [\hat{R}, \hat{Q}]] + \cdots .$$

Note that the right hand side of (7.31) is of the form of $\hat{f}(\xi)$, with $\hat{R} = \hat{A}^\dagger M\hat{A}$, $\hat{Q} = A$, and $\xi = iT$. Thus, using the result just obtained and noting that

$$[\hat{R}, \hat{Q}] = [\hat{A}^\dagger M\hat{A}, \hat{A}] = [\hat{A}_i^\dagger M_{ij}\hat{A}_j, \hat{A}_k] = M_{ij}(\hat{A}_i^\dagger \hat{A}_j \hat{A}_k - \hat{A}_k \hat{A}_i^\dagger \hat{A}_j)$$

$$= -M_{ij}\delta_{ik}\hat{A}_j = -M_{kj}\hat{A}_j , \qquad (7.35)$$

we find that the commutator is equal to $-M\hat{A}$. Repeating the same algebra, we find

$$[\hat{R}, [\hat{R}, \hat{Q}]] = M^2\hat{A} , \quad [\hat{R}, [\hat{R}, [\hat{R}, \hat{Q}]]] = -M^3\hat{A} , \qquad (7.36)$$

and so forth. Hence

$$\hat{A}(T) = \hat{A} - (i/1!)MT\hat{A} - (1/2!)(MT)^2\hat{A} + (i/3!)(MT)^3\hat{A} + \cdots$$

$$= \exp(-iMT)\hat{A} . \qquad (7.37)$$

Thus, we have recovered (7.19). The present exercise confirms the legitimacy of multiplying an excitation operator from one side with a unitary operator to describe its time evolution in a linear system, whereas when the Heisenberg formulation is first encountered, the time evolution of an operator is described by pre- and post-multiplication of the operator by a unitary matrix and its Hermitian conjugate, respectively.

7.4 The Schrödinger Formulation and Entangled States

Thus far, we have used the Heisenberg representation to describe the effects of optical elements on the annihilation operators, stand-ins for the classical complex field amplitudes. The Heisenberg representation of optical phenomena takes the form of classical equations of evolution of the observables represented by the operators. In the Schrödinger representation, the wave functions change, and not the operators. It is of interest to compare the two approaches. We may follow the change of the wave function through a phase shifter or beam splitter, just as we have followed the change of the annihilation operator through a phase shifter or a beam splitter.

Let us consider first the action of a phase shifter. Since we shall analyze operations on number states, it is convenient to revert to the annihilation and creation operators $\hat{A}$ and $\hat{A}^\dagger$, which have the simple properties of (6.44) and (6.45) when operating on a number state. We shall omit the subscript m. In the Heisenberg representation, an excitation described by the annihilation operator $\hat{A}$, when passed through a phase shifter producing a phase shift θ, is described by multiplying $\hat{A}$ by $\exp(i\theta)$. We have seen that this operation is equivalent to a pre- and post-multiplication of the operator by $\exp(-i\theta\hat{A}^\dagger\hat{A})$ and $\exp(i\theta\hat{A}^\dagger\hat{A})$, respectively. This means that, in the Schrödinger representation, the wave function is multiplied by $\exp(i\theta\hat{A}^\dagger\hat{A})$. Consider first the case of a coherent state $|\alpha\rangle$. Detailed evaluation gives

$$
\exp(i\theta\hat{A}^\dagger\hat{A})|\alpha\rangle = e^{-i|\alpha|^2/2}\exp(i\theta\hat{A}^\dagger\hat{A})\sum_{n=0}\frac{\alpha^n}{\sqrt{n!}}|n\rangle
$$

$$
= e^{-i|\alpha|^2/2}\sum_{n=0}\exp(i\theta n)\frac{\alpha^n}{\sqrt{n!}}|n\rangle \tag{7.38}
$$

$$
= e^{-i|\alpha|^2/2}\sum_{n=0}\frac{(e^{i\theta}\alpha)^n}{\sqrt{n!}}|n\rangle = |e^{i\theta}\alpha\rangle \ .
$$

Thus, the passage of a coherent state through the phase shifter transforms the state $|\alpha\rangle$ into the state $|e^{i\theta}\alpha\rangle$. The result is simplicity itself. Indeed, the complex parameter α of the coherent state describes the endpoint of the phasor in the complex plane, in one-to-one correspondence with the complex amplitude of the electric field. This amplitude behaves classically.

Next, consider the transformation of a photon state $|1\rangle$ or $|0\rangle$ by a phase shifter. We have

$$\exp(i\theta\hat{A}^\dagger\hat{A})|1\rangle = \exp(i\theta)|1\rangle . \tag{7.39}$$

Even though the photon state $|1\rangle$ does not have a well-defined phase, passage of the photon state through a phase shifter does impart a phase shift. Interference of the photon state with its phase shifted version can lead to interference fringes. In a similar way we have

$$\exp(i\theta\hat{A}^\dagger\hat{A})|0\rangle = |0\rangle . \tag{7.40}$$

The ground state remains unchanged. This shows that the ground state is unaffected by a phase shift, because it cannot lead to interference with itself.

Next, we take up the operation of a beam splitter. A beam splitter is described by the Hamiltonian

$$\hat{H} = \hbar(M\hat{A}^\dagger\hat{B} + M^*\hat{B}^\dagger\hat{A}) . \tag{7.41}$$

Integration of the Schrödinger equation of motion

$$\frac{d|\psi\rangle}{dt} = -\frac{i}{\hbar}\hat{H}|\psi\rangle \tag{7.42}$$

gives

$$|\psi(T)\rangle = \exp\left(-\frac{i}{\hbar}\hat{H}T\right)|\psi(0)\rangle . \tag{7.43}$$

For convenience we choose M real and positive. In order to simplify the notation, we write

$$\frac{\hat{H}}{\hbar}T = \phi(\hat{A}^\dagger\hat{B} + \hat{B}^\dagger\hat{A}) , \tag{7.44}$$

where $\phi = MT$. Let us start with a single photon in port (1) and vacuum fed into port (2). Then, the input state is

$$|\psi(0)\rangle = |1\rangle \otimes |0\rangle . \tag{7.45}$$

The output is obtained by expanding the exponential into a Taylor series

$$|\psi(T)\rangle = \sum_{n=0}^{n=\infty} \frac{1}{n!}\left(-\frac{i}{\hbar}\hat{H}T\right)^n |1\rangle \otimes |0\rangle$$

$$= \sum_{n=0}^{n=\infty} \frac{(-i\phi)^n}{n!}(\hat{A}^\dagger\hat{B} + \hat{B}^\dagger\hat{A})^n |1\rangle \otimes |0\rangle . \tag{7.46}$$

Now consider the effect of the operation $(\hat{A}^\dagger\hat{B} + \hat{B}^\dagger\hat{A})^n$ on the product state $|1\rangle \otimes |0\rangle$, where the operators $\hat{A}$ and $\hat{A}^\dagger$ operate on $|1\rangle$ and the operators $\hat{B}$ and $\hat{B}^\dagger$ are applied to $|0\rangle$. We have

$$(\hat{A}^\dagger\hat{B} + \hat{B}^\dagger\hat{A})|1\rangle \otimes |0\rangle = |0\rangle \otimes |1\rangle . \tag{7.47}$$

In a similar way we find

$$(\hat{A}^\dagger\hat{B} + \hat{B}^\dagger\hat{A})^2|1\rangle \otimes |0\rangle = |1\rangle \otimes |0\rangle . \tag{7.48}$$

In this way we find

$$
\begin{aligned}
|\psi(T)\rangle &= \sum_{n=0}^{n=\infty} \frac{(-\mathrm{i}\phi)^n}{n!}(\hat{A}^\dagger\hat{B} + \hat{B}^\dagger\hat{A})^n|1\rangle \otimes |0\rangle \\
&= \sum_{n=\text{even}}^{n=\infty} \frac{(-\mathrm{i}\phi)^n}{n!}|1\rangle \otimes |0\rangle + \sum_{n=\text{odd}}^{n=\infty} \frac{(-\mathrm{i}\phi)^n}{n!}|0\rangle \otimes |1\rangle \\
&= \cos\phi|1\rangle \otimes |0\rangle - \mathrm{i}\sin\phi|0\rangle \otimes |1\rangle .
\end{aligned}
\tag{7.49}
$$

The output wave function is a coherent superposition of two states, a simple example of an entangled state. Entangled states have no classical analog. Let us look at this state in greater detail. Although photon states are not classical in their nature either, classical language can be applied to many processes that transform photons. The input is in a product state $|1\rangle \otimes |0\rangle$. The density matrix ρ (Appendix A.10) at the input is the product of two diagonal density matrices:

$$\rho(0) = |1\rangle \otimes |0\rangle\langle 0| \otimes \langle 1| = |1\rangle\langle 1| \otimes |0\rangle\langle 0| . \tag{7.50}$$

The probability of finding one photon in the input port (1) is the value of the diagonal element $|1\rangle\langle 1|$, which is unity. Similarly, the probability of finding zero photons in the input port (2) is the value of the diagonal element $|0\rangle\langle 0|$, also equal to unity. At the output of the beam splitter, the density matrix is

$$\rho(T) = \left\{ \begin{aligned} &\cos^2\phi|1\rangle\langle 1| \otimes |0\rangle\langle 0| + \mathrm{i}\sin\phi\cos\phi|1\rangle\langle 0| \otimes |0\rangle\langle 1| \\ &-\mathrm{i}\sin\phi\cos\phi|0\rangle\langle 1| \otimes |1\rangle\langle 0| + \sin^2\phi|0\rangle\langle 0| \otimes \langle 1|\langle 1| \end{aligned} \right\} . \tag{7.51}$$

The density matrix is not diagonal, it is made up of a sum that contains off-diagonal elements $|1\rangle\langle 0| \otimes |0\rangle\langle 1|$ and $|0\rangle\langle 1| \otimes |1\rangle\langle 0|$. This is the density matrix of a so-called "entangled state". Measurements on the system can yield outcomes with no classical interpretation, because the off-diagonal elements of the density matrix may contribute terms to the expectation values that "interfere", thus preventing a classical interpretation in terms of the probabilities of photons exiting in ports (3) and (4). Appendix A.11 looks at some further operations of the beam splitter.

The case of coherent inputs into ports (1) and (2) described by

$$|\psi(0)\rangle = |\alpha\rangle \otimes |\beta\rangle , \tag{7.52}$$

could be analyzed in the same way. However, since the states $|\alpha\rangle$ and $|\beta\rangle$ are not eigenstates of the creation operators $\hat{a}^\dagger$ and $\hat{b}^\dagger$, the analysis gets quickly out of hand. There is a better way to approach the problem, as is done in the next section. Suffice it to state here that the operation of the beam splitter leads to the output wave function

$$|\psi(T)\rangle = |\cos\phi\alpha - i\sin\phi\beta\rangle \otimes |-i\sin\phi\alpha + \cos\phi\beta\rangle . \tag{7.53}$$

The state remains a product state; the complex amplitudes of the coherent states add like classical complex field amplitudes. No entanglement occurs.

7.5 Transformation of Coherent States

The Heisenberg representation of linear systems transforms incident-wave operators into outgoing wave operators in a way described by a simple scattering process and bears a close analogy to the classical description. This is one of the advantages of the Heisenberg representation, since it is one of the manifestations of the correspondence principle: when the observables are expressed in terms of operators, the equations of the operators assume the form of classical equations of motion. The correspondence principle is not obtained as easily in the Schrödinger representation. Yet, it is of interest to derive it in this representation as well, since then we can show what input states bear the closest analogy with classical physics. In his seminal paper, Glauber [66] introduced expansions in terms of coherent states with the intention to clarify the correspondence between classical optics and its quantum description. "Such expansions have the property that whenever the field possesses a classical limit, they render that limit evident while at the same time preserving an intrinsically quantum-mechanical description of the field." [66]

Consider a linear network represented by the following Hamiltonian in normal order:

$$\hat{H} = \hbar M_{jk}\hat{A}_j^\dagger \hat{A}_k , \tag{7.54}$$

where M is a Hermitian matrix, and we use the Einstein summation convention. Suppose that the input state $|\psi\rangle$ is a product state of coherent states

$$|\psi\rangle = \prod_j |\alpha_j\rangle . \tag{7.55}$$

Schrödinger's equation of evolution leads to the differential equation

$$\frac{\partial}{\partial t}|\psi\rangle = -\frac{i}{\hbar}\hat{H}|\psi\rangle . \tag{7.56}$$

The following manipulations are greatly simplified if we introduce the renormalized states of Glauber [66]

$$||\alpha_j\rangle \equiv e^{|\alpha_j|^2/2}|\alpha_j\rangle = \sum_{n_j} \frac{\alpha_j^{n_j}}{\sqrt{n_j!}}|n_j\rangle . \tag{7.57}$$

These functions have the remarkable property that operation by the creation operator is equivalent to taking a derivative with respect to α [66]:

$$\hat{A}_j^\dagger ||\alpha_j\rangle = \frac{\partial}{\partial\alpha_j}||\alpha_j\rangle , \tag{7.58}$$

as can be easily confirmed using the properties of the creation operator operating on a photon state. We now assume that a coherent product state maintains its product character as it evolves according to (7.56). We shall then show that this assumption is correct and leads to a simple solution of the Schrödinger equation. The state $|\psi\rangle$ of (7.55) can be written

$$|\psi\rangle = e^{-\sum_j |\alpha_j|^2/2} \prod_\ell ||\alpha_\ell\rangle . \tag{7.59}$$

Still following the assumption that the solution can be represented as a product state of coherent states, we take into account that energy conservation ensures time independence of the sum over the squares of the $|\alpha_j|$. The time derivative of (7.59) is thus

$$\frac{\partial}{\partial t}|\psi\rangle = e^{-\sum_j |\alpha_j|^2/2} \left(\frac{\partial}{\partial t}\alpha_j\right)\frac{\partial}{\partial\alpha_j}\prod_\ell ||\alpha_\ell\rangle . \tag{7.60}$$

Next, consider what form the same equation takes when operation by the Hamiltonian operator is carried out according to the Schrödinger equation (7.56). When the ansatz (7.58) is introduced into (7.56) we obtain

$$\frac{\partial}{\partial t}|\psi\rangle = -\mathrm{i}M_{jk}\hat{A}_j^\dagger\hat{A}_k e^{-\sum_j |\alpha_j|^2/2}\prod_\ell ||\alpha_\ell\rangle$$

$$= -\mathrm{i}e^{-\sum_j |\alpha_j|^2/2}M_{jk}\alpha_k\frac{\partial}{\partial\alpha_j}\prod_\ell ||\alpha_\ell\rangle . \tag{7.61}$$

We find that (7.60) and (7.61) are consistent when

$$\frac{\partial}{\partial t}\alpha_j = -\mathrm{i}M_{jk}\alpha_k . \tag{7.62}$$

We have found that the complex amplitudes of the coherent states of the different modes obey linear equations of the same form as the annihilation operators of the modes. These are the classical equations of motion of the mode amplitudes. Our analysis has accomplished several objectives. First of all, we have found classical equations of motion for the amplitudes of the coherent states. In this way we have established a correspondence principle in the Schrödinger picture. Secondly, we have proven that a state constructed as a product of coherent states remains such a product state as it evolves in a linear system. Thirdly, since coherent states have Poissonian photon statistics, we have proven that Poissonian statistics are preserved in the scattering process of a linear system. We shall confirm this result in Chap. 9 using a different approach.

7.6 Characteristic Functions and Probability Distributions

In the analysis of linear circuits, such as discussed in Chap. 5, one deals with amplitudes of the electric field. Hence, in the context of linear circuits one is interested in the probability distribution of the field. It is clear that the so-called "characteristic function", defined by

$$C(\xi) = \langle \exp(i\xi \hat{E}) \rangle , \tag{7.63}$$

contains all the moments of the electric field. Indeed, expansion of the exponential gives

$$C(\xi) = \sum_{n=0}^{\infty} \frac{i\xi^n}{n!} \langle \hat{E}^n \rangle . \tag{7.64}$$

We now turn to the evaluation of the characteristic function of the electric field, using the creation and annihilation operators:

$$C(\xi) = \left\langle \exp\left[i\xi \frac{1}{2}(\hat{A}^\dagger + \hat{A}) \right] \right\rangle . \tag{7.65}$$

An expansion of the exponential involves products of the annihilation operators in various orders. This is an inconvenient form of the expansion. The analysis is greatly simplified through the use of the Baker–Hausdorff identity [66], which puts exponentials of sums of noncommuting operators into normal order. If $\hat{A}$ and $\hat{B}$ are operators, and their commutator $[\hat{A}, \hat{B}]$ is a c number, then the Baker–Hausdorff theorem states (see Appendix A.12)

$$\exp\left[i\xi(\hat{A} + \hat{B}) \right] = \exp(i\xi \hat{A}) \exp(i\xi \hat{B}) \exp\left(\frac{\xi^2}{2} \left[\hat{A}, \hat{B} \right] \right) . \tag{7.66}$$

7.6.1 Coherent State

When this theorem is applied to a coherent state $|\alpha\rangle$ we find

$$C(\xi) = \exp\left(-\frac{\xi^2}{8}\right) \langle\alpha| \exp\left(i\xi\hat{A}^\dagger/2\right) \exp\left(i\xi\hat{A}/2\right) |\alpha\rangle$$

$$= \exp\left(-\frac{\xi^2}{8}\right) \exp(i\xi\alpha^*/2) \exp(i\xi\alpha/2) \tag{7.67}$$

$$= \exp\left(-\frac{\xi^2}{8}\right) \exp[i\xi(\alpha^* + \alpha)/2] .$$

This is the characteristic function of a Gaussian for an E field centered at $E_o = (\alpha^* + \alpha)/2$. Indeed, using the classical interpretation of the characteristic function of a field E with the probability distribution $p(E)$, we obtain

$$C(\xi) = \int_{-\infty}^{\infty} dE\, p(E) \exp(i\xi E)$$

$$= \int_{-\infty}^{\infty} dE \frac{1}{\sqrt{2\pi\sigma^2}} \exp\left(-\frac{(E - E_o)^2}{2\sigma^2}\right) \exp(i\xi E) \tag{7.68}$$

$$= \exp\left(-\frac{\xi^2\sigma^2}{2}\right) \exp(i\xi E_o) ,$$

where we have used a Gaussian distribution of mean square deviation σ, centered around E_o. We find indeed that the characteristic function of a coherent state is equal to the characteristic function of a Gaussian with $\sigma = 1/2$ and centered at $E_o = (\alpha^* + \alpha)/2$. The characteristic function for the quadrature field $(1/2i)(\hat{A} - \hat{A}^\dagger)$ gives the same kind of expression, with $E_o = (\alpha - \alpha^*)/(2i)$. Let us look at some further properties of the characteristic function in the classical interpretation. If we expand $C(\xi)$ we have

$$C(\xi) = \int_{-\infty}^{\infty} dE\, p(E) \exp(i\xi E) = \int_{-\infty}^{\infty} dE\, p(E) \sum_m \frac{(i\xi)^m E^m}{m!}$$

$$= \sum_m \frac{(i\xi)^m}{m!} \langle E^m \rangle . \tag{7.69}$$

The characteristic function contains the moments of the field as the coefficients of the expansion.

The expansion of the characteristic function of a Gaussian distribution with zero average field ($E_o = 0$) gives

$$\exp\left(-\frac{\xi^2\sigma^2}{2}\right) = \sum_m (-1)^m \frac{1}{m!} \left(\frac{\xi^2}{2}\right)^m \sigma^{2m} . \tag{7.70}$$

Comparison of (7.69) and (7.70) gives zero for all odd-order moments, and for the even-order moments

$$\langle |E|^{2m} \rangle = \frac{(2m)!}{m!2^m} \sigma^{2m} .$$ (7.71)

Hence, all moments of a Gaussian distribution with zero average field can be expressed in terms of σ^2. This is why a Gaussian distribution is fully described by its mean square deviation.

7.6.2 Bose–Einstein Distribution

Next we evaluate the characteristic function of the in-phase field component for a Bose–Einstein distribution:

$$|\psi\rangle = \sum_n c_n |n\rangle ,$$ (7.72)

with

$$\langle c_n c_m^* \rangle = \delta_{nm} p_{B-E}(n) .$$ (7.73)

In analogy with (7.67) we find

$$C(\xi) = \exp\left(-\frac{\xi^2}{8}\right) \langle\psi| \exp(i\xi\hat{A}^\dagger/2) \exp(i\xi\hat{A}/2)|\psi\rangle$$

$$= \exp\left(-\frac{\xi^2}{8}\right) \langle\psi| \sum_r \frac{(i\xi\hat{A}^\dagger/2)^r}{r!} \sum_q \frac{(i\xi\hat{A}/2)^q}{q!} |\psi\rangle$$

$$= \exp\left(-\frac{\xi^2}{8}\right) \sum_r (-1)^r \frac{(\xi/2)^{2r}}{(r!)^2} \sum_n p_{B-E}(n) n(n-1)(n-r+1) .$$ (7.74)

The characteristic function contains the falling factorial moments F_r of the Bose–Einstein distribution. These will be derived in Sect. 9.1. Here we use the result (9.13) of Chap. 9. We find the following simple answer for (7.74):

$$C(\xi) = \exp\left(-\frac{\xi^2}{8}\right) \sum_r (-1)^r \frac{(\xi/2)^{2r}}{(r!)^2} r! \langle n\rangle^r$$

$$= \exp\left(-\frac{\xi^2}{8}\right) \sum_r (-1)^r \frac{(\xi^2\langle n\rangle/4)^r}{(r!)}$$ (7.75)

$$= \exp\left(-\frac{\xi^2(1+2\langle n\rangle)}{8}\right) .$$

According to (7.68), this is the characteristic function of a Gaussian distribution of zero average field with the mean square deviation $\sigma^2 = 1/4 + \langle n\rangle/2$. The quadrature field of a Gaussian distribution has the same fluctuations.

7.7 Two-Dimensional Characteristic Functions and the Wigner Distribution

In the preceding section we looked at the characteristic function of the in-phase and quadrature components of the electric field. The Fourier transform gave us Gaussian probability distributions of these components. The Fourier transformation of the characteristic function of a single observable always leads to a positive definite function that can be interpreted as a probability distribution.

Two classical random variables x_1 and x_2 are described by the joint probability distribution $p(x_1, x_2)$, which is the Fourier transform of the characteristic function

$$C(\xi_1, \xi_2) = \langle \exp(i\xi_1 x_1 + i\xi_2 x_2) \rangle . \tag{7.76}$$

Indeed, let us evaluate (7.76) with the aid of the joint probability distribution $p(x_1, x_2)$:

$$C(\xi_1, \xi_2) = \int dx_1 \int dx_2\, p(x_1, x_2) \exp(i\xi_1 x_1 + i\xi_2 x_2) . \tag{7.77}$$

The Fourier transformation of $C(\xi_1, \xi_2)$ gives

$$\frac{1}{(2\pi)^2} \int d\xi_1 \int d\xi_2 \exp(-i\xi_1 X_1 - i\xi_2 X_2) C(\xi_1, \xi_2)$$

$$= \frac{1}{(2\pi)^2} \int dx_1 \int dx_2 \int d\xi_1 \int d\xi_2 \exp[-i\xi_1(x_1 - X_1)$$

$$-i\xi_2(x_2 - X_2)]p(x_1, x_2) \tag{7.78}$$

$$= \int dx_1 \int dx_2\, \delta(x_1 - X_1)\delta(x_2 - X_2)p(x_1, x_2)$$

$$= p(X_1, X_2) .$$

Thus, the Fourier transform of the characteristic function of two random variables gives the joint probability distribution.

The characteristic function of two quantum observables $\hat{x}_1$ and $\hat{x}_2$ is well defined as

$$C(\xi_1, \xi_2) = \langle \exp(i\xi_1 \hat{x}_1 + i\xi_2 \hat{x}_2) \rangle . \tag{7.79}$$

An expansion of the exponential in powers of ξ_1 and ξ_2 contains terms like $\xi_1^n \xi_2^m \langle \hat{x}_1^n \hat{x}_2^m \rangle$. Thus the characteristic function gives full information about the moments of the observables. However, the Fourier transform of the characteristic function of two noncommuting observables is not always positive definite. Even if positive definite, it cannot always be interpreted as a probability distribution in the classical sense. This is shown by the following example.

We consider the entangled state produced by a 50/50 beam splitter as derived in Sect. 7.4. There we found that a beam splitter produces the wave function

$$|\psi\rangle = (\cos\phi|1\rangle_1 \otimes |0\rangle_2 - \mathrm{i}\sin\phi|0\rangle_1 \otimes |1\rangle_2) \ . \tag{7.80}$$

We have added subscripts as a reminder of the fact that the $\hat{A}$ operator operates only on the wave function with the subscript 1, and the $\hat{B}$ operator only on the wave function with the subscript 2. For a 50/50 beam splitter $\phi = \pi/4$, and

$$|\psi\rangle = \frac{1}{\sqrt{2}}(|1\rangle_1 \otimes |0\rangle_2 - \mathrm{i}|0\rangle_1 \otimes |1\rangle_2) \ . \tag{7.81}$$

The characteristic function of the photon number at the two outputs is

$$C(\xi_1, \xi_2) = \langle \exp[\mathrm{i}(\xi_i\hat{A}^\dagger\hat{A} + \xi_2\hat{B}^\dagger\hat{B})]\rangle$$

$$= \frac{1}{2}(_2\langle 0| \otimes_1 \langle 1| + \mathrm{i}_2\langle 1| \otimes_1 \langle 0|) \exp[\mathrm{i}(\xi_1\hat{A}^\dagger\hat{A} + \xi_2\hat{B}^\dagger\hat{B})] \tag{7.82}$$

$$\times (|1\rangle_1 \otimes |0\rangle_2 - \mathrm{i}|0\rangle_1 \otimes |1\rangle_2) \ .$$

Thus, we obtain

$$C(\xi_1, \xi_2) = \frac{1}{2}[\exp(\mathrm{i}\xi_1) + \exp(\mathrm{i}\xi_2)] \ . \tag{7.83}$$

The inverse Fourier transform gives for the probability of the photon numbers n_1 and n_2

$$p(n_1, n_2) = \frac{1}{2} \text{ for } n_1 = 1, n_2 = 0, \quad \text{and } \frac{1}{2} \text{ for } n_2 = 1, n_1 = 0 \ . \tag{7.84}$$

This result seems very "classical": if a photon enters the beam splitter from the input port (a), it ends up with probability 1/2 in either of the two output ports. This classical interpretation is, however, misleading. To see this, pass the output of the beam splitter through another 50/50 beam splitter. The classical interpretation would say that we pass the photon that ended up with probability 1/2 in one of the output ports through the second beam splitter and again it would end up with probability 1/2 in either of the two output ports. We do the same for the events when the photon ended up in the second port of the first beam splitter. Again the photon ends up with equal probabilities in either of the two output ports of the second beam splitter. Thus the answer is that we see a photon in either of the output ports with probability $1/2$. The quantum problem arrives at a completely different answer. Instead of $\phi = \pi/4$ in (7.80), the two beam splitters in cascade are described by $\phi = \pi/2$. The input photon to the two beam splitters ends up with certainty in output port (a). This is a consequence of the coherence in the wave function at the output of the first beam splitter. Wave functions add, not probabilities!

The quantum nature of a situation emerges when one deals with the characteristic function of two noncommuting observables. The characteristic function itself is well defined, since it deals with the weighted moments of an observable that is the linear combination of the two observables $\xi_1 \hat{x}_1 + \xi_2 \hat{x}_2$. The Fourier transform of the characteristic function is the Wigner function

$$W(x_1, x_2) = \left(\frac{1}{2\pi}\right)^2 \int d\xi_1 \int d\xi_2 \, C(\xi_1, \xi_2) \exp(-i\xi_1 x_1 - i\xi_2 x_2) \, . \quad (7.85)$$

The Wigner function integrated over one of the two variables is positive and can be interpreted as a probability

$$p(x_1) = \int dx_2 \, W(x_1, x_2) \, . \quad (7.86)$$

However, if one attempts to interpret the Wigner function as the joint probability of both observables, one may run into negative values of the function. The experimental measurement of the Wigner function of particle diffraction through a double slit has actually been carried out, in which $\hat{x}_1$ is the the position $\hat{q}$ and $\hat{x}_2$ is the momentum $\hat{p}$ [69]. Appendix A.13 evaluates the Wigner function for the position and momentum of a particle.

On the other hand, the Wigner function of a coherent state shows no idiosyncrasies, a fact which reinforces the picture of coherent states as quantum states with a classical character. Let us evaluate this Wigner function for a coherent state. It is convenient to recast the characteristic function in terms of creation and annihilation operators. The characteristic function can be written as

$$C(\xi_1, \xi_2) = \langle \exp[i(\xi_j \hat{A}_j)]\rangle = \langle \exp(\eta^* \hat{A} - \eta \hat{A}^\dagger)\rangle \, , \quad (7.87)$$

where $\eta = (1/2i)(\xi_1 + i\xi_2)$, and where we use the Einstein summation convention. The Fourier transform gives the Wigner function:

$$W(A_1, A_2) = \left(\frac{1}{2\pi}\right)^2 \int d\xi_1 \int d\xi_2 \langle \exp[i\xi_j(\hat{A}_j - A_j)]\rangle$$

$$= \left(\frac{1}{\pi}\right)^2 \int d^2\eta \, \langle \exp[\eta^*(\hat{A} - A) - \eta(\hat{A}^\dagger - A^*)]\rangle \quad (7.88)$$

$$\equiv W(A) \, .$$

The two expressions for the Wigner function are equivalent. The first expression is written in terms of the real coordinates A_1 and A_2 and the Wigner function is a function in the A_1–A_2 plane. The second expression uses complex notation and creation and annihilation operators. The integral is a function in the A_1–iA_2 complex plane. If one is in doubt about how to carry out the double integral in the complex plane, one may always resort to the two-dimensional Fourier integral with real variables.

Note that the integral of the Wigner function over A_1 and A_2 is unity. Indeed,

$$\int dA_1 \int dA_2\, W(A_1, A_2)$$

$$= \left(\frac{1}{2\pi}\right)^2 \int dA_1 \int dA_2 \int d\xi_1 \int d\xi_2 \langle \exp[i\xi_j(\hat{A}_j - A_j)] \rangle$$

$$= \left(\frac{1}{2\pi}\right)^2 \int dA_1 \int dA_2 \int d\xi_1 \int d\xi_2 \exp(i\xi_j A_j) \langle \exp(i\xi_j \hat{A}_j) \rangle \tag{7.89}$$

$$= \int d\xi_1 \int d\xi_2\, \delta(\xi_1)\delta(\xi_2) \langle \exp(i\xi_j \hat{A}_j) \rangle = 1 \ .$$

In this respect, the Wigner function satisfies a condition of a probability distribution. Let us now evaluate the Wigner function for a coherent state $|\alpha\rangle$. For this purpose we put the kernel of the integrand in (7.88) into normal order using the Baker–Hausdorff theorem:

$$\exp[\eta^*(\hat{A} - A) - \eta(\hat{A}^\dagger - A^*)]$$

$$= \exp[-\eta(\hat{A}^\dagger - A^*)] \exp[\eta^*(\hat{A} - A)] \exp\left(-\frac{|\eta|^2}{2}\right) \ . \tag{7.90}$$

We evaluate next the expectation value of the kernel:

$$\langle \alpha | \exp[-\eta(\hat{A}^\dagger - A^*)] \exp[\eta^*(\hat{A} - A)] \exp\left(-\frac{|\eta|^2}{2}\right) | \alpha \rangle$$

$$= \exp[-\eta(\alpha^* - A^*)] \exp[\eta^*(\alpha - A)] \exp\left(-\frac{|\eta|^2}{2}\right) \ . \tag{7.91}$$

The arguments in the exponentials can be written in terms of the original coordinates ξ_1 and ξ_2 and in terms of the in-phase components A_1 and A_2. When this is done and the Fourier transformation (7.85) is carried out, we obtain for the Wigner function

$$W(A_1, A_2) = \frac{1}{2\pi} \exp\{-2[A_1 - \mathrm{Re}(\alpha)]^2\} \exp\{-2[A_2 - \mathrm{Im}(\alpha)]^2\} \ . \tag{7.92}$$

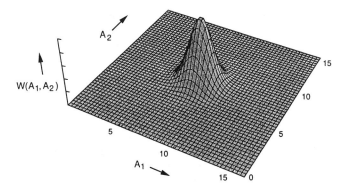

Fig. 7.5. The Wigner function of a coherent state

This is the Gaussian shown in Fig. 7.5 in the A_1–A_2 plane. This Gaussian was illustrated by the shaded circle in Fig. 6.1. The probability distribution is in perfect correspondence with that of a classical signal amplitude with additive Gaussian noise.

7.8 The Schrödinger Cat State and Its Wigner Distribution

In this book we are mainly interested in the quantum noise of electromagnetic fields at the optical frequencies that are used in optical communications. In all practical situations, these fields are relatively intense, in that they carry many photons per mode. The quantum noise of such fields bears a close resemblance to classical fields in the presence of additive thermal noise and thus permits simple interpretations. In this context one does not encounter the strange behavior exhibited by optical fields with only a few photons. However, optical fields of higher intensities may also exhibit strange behavior if they are prepared by a nonlinear system sensitive to the presence or absence of a photon. Quantum states of this kind are called Schrödinger cat states [70]. The name derives from Schrödinger's thought experiment concerning the prediction of the state of a cat whose life or death is determined by the outcome of a quantum measurement. How such states can be generated in principle will be discussed in Chap. 12. The Schrödinger cat thought experiment itself will be discussed in more detail in Chap. 14, where we shall attempt to show that the seeming paradoxes associated with this thought experiment can be removed by a proper definition of the experiment that determines the fate of the cat. At this point we consider Schrödinger cat states of a photon field in order to show the strangeness of the associated Wigner function.

An example of a pure state is a photon state $|n\rangle$ or a coherent state α. An example of an entangled state was considered in the preceding section, of a photon state passing through a beam splitter. Entangled states need not involve two observables, such as the photons in each of the ports of the beam splitter; they can be constructed in the Hilbert space of one observable. Thus, the state $|\psi\rangle$ formed from the superposition of two coherent states

$$|\psi\rangle = N(e^{-i\chi}|\alpha\rangle + e^{i\chi}|\beta\rangle), \tag{7.93}$$

is an entangled state. Here N is a normalizing factor to ensure a unity magnitude of $\langle\psi|\psi\rangle$. In order to evaluate N we need to know $\langle\alpha|\beta\rangle$. This projection is found easily using the photon state representation of a coherent state:

$$\langle\alpha|\beta\rangle = e^{-|\alpha|^2/2}e^{-|\beta|^2/2}\sum_{n,m}\langle n|m\rangle\frac{\alpha^{*n}\beta^m}{\sqrt{n!m!}}$$

$$= e^{-|\alpha|^2/2}e^{-|\beta|^2/2}\sum_n\frac{\alpha^{*n}\beta^n}{n!} \tag{7.94}$$

$$= \exp[-(|\alpha|^2 + |\beta|^2)/2]\exp(\alpha^*\beta).$$

In this way one finds for N

$$\frac{1}{|N|^2} = 2\left[1 + \cos(2\chi + \phi)\exp\left(-\frac{(|\alpha| - |\beta|)^2}{2}\right)\right], \tag{7.95}$$

where $\phi = \arg(\alpha^*\beta)$. Let us consider the state (7.93) with

$$|\psi\rangle = N(e^{i\pi/4}|\alpha\rangle + e^{-i\pi/4}|-\alpha\rangle) \tag{7.96}$$

and evaluate its Wigner function. We must evaluate projections with the bras and kets of α and $-\alpha$. The four expectation values in the kernel of (7.88) are the following.

Self-term $\langle\alpha||\alpha\rangle$:

$$\langle\alpha|\exp[-\eta(\hat{A}^\dagger - A^*)]\exp[\eta^*(\hat{A} - A)]\exp\left(-\frac{|\eta|^2}{2}\right)|\alpha\rangle \tag{7.97a}$$

$$= \exp[-\eta(\alpha^* - A^*)]\exp[\eta^*(\alpha - A)]\exp\left(-\frac{|\eta|^2}{2}\right).$$

Self-term $\langle-\alpha||-\alpha\rangle$:

$$\langle-\alpha|\exp[-\eta(\hat{A}^\dagger - A^*)]\exp[\eta^*(\hat{A} - A)]\exp\left(-\frac{|\eta|^2}{2}\right)|-\alpha\rangle \tag{7.97b}$$

$$= \exp[\eta(\alpha^* + A^*)]\exp[-\eta^*(\alpha + A)]\exp\left(-\frac{|\eta|^2}{2}\right).$$

Cross term $\langle-\alpha||\alpha\rangle$:

$$\langle -\alpha | \exp[-\eta(\hat{A}^\dagger - A^*)] \exp[\eta^*(\hat{A} - A)] \exp\left(-\frac{|\eta|^2}{2} \right) |\alpha\rangle \qquad (7.97c)$$

$$= \exp[\eta(\alpha^* + A^*)] \exp[\eta^*(\alpha - A)] \exp\left(-\frac{|\eta|^2}{2} \right) \exp(-2|\alpha|^2) \ .$$

Cross term $\langle \alpha || - \alpha\rangle$:

$$\langle \alpha | \exp[-\eta(\hat{A}^\dagger - A^*)] \exp[-\eta^*(\hat{A} - A)] \exp\left(-\frac{|\eta|^2}{2} \right) | - \alpha\rangle \qquad (7.97d)$$

$$= \exp[-\eta(\alpha^* - A^*)] \exp[-\eta^*(\alpha + A)] \exp\left(-\frac{|\eta|^2}{2} \right) \exp(-2|\alpha|^2) \ .$$

With these four terms we construct the kernel in the Fourier transform (7.85) that leads to the Wigner function. The arguments in the exponentials can be written in terms of the original coordinates ξ_1 and ξ_2 and in terms of the in-phase components A_1 and A_2. We assume for simplicity that α is real. Then

$$W(A_1, A_2) = \left(\frac{1}{2\pi}\right) \frac{1}{|N|^2} \exp(-2A_2^2)$$

$$\times \{\exp[-2(A_1 - \alpha)^2] + \exp[-2(A_1 + \alpha)^2] \qquad (7.98)$$

$$+ 2\sin(4\alpha A_2) \exp(-2A_1^2)\} \ .$$

If the state were an incoherent superposition of the two coherent states $|\alpha\rangle$ and $| - \alpha\rangle$, the Wigner function would consist of two Gaussian peaks at $A_1 = \pm\alpha, A_2 = 0$. We shall denote these terms the "self-terms". The quantum character of the Schrödinger cat state is expressed by the coherence beat at the origin at $A_1 = 0, A_2 = 0$, which we shall call the "cross term." The cross term depends on the phase of the superposition of the states $|\alpha\rangle$ and $| - \alpha\rangle$. Had we used the state $|\psi\rangle \propto |\alpha\rangle + | - \alpha\rangle$, the beat term would be a cosine, rather than a sine. The cross term is not positive definite, indicating that the Wigner function does not allow an interpretation in terms of a classical probability distribution. The Wigner function is shown in Fig. 7.6. We also see that one may not define a probability of the field being either in the state $|\alpha\rangle$ or $| - \alpha\rangle$.

The Schrödinger cat state illustrates the peculiar nature of entangled states. Clearly, the Wigner function of Fig. 7.6 does not permit an interpretation in terms of a signal amplitude with additive noise. This is in spite of the fact that the parameter α could be large, the photon number of the state could be large, i.e. the state could have an intensity that is "macroscopic". Entangled states do not occur in optical communication systems operating with large photon numbers, since these states are extremely fragile, as we now proceed to show. The quantum beat, the cross term, is destroyed by the loss of a very few photons. To see this, let us pass the Schrödinger cat state through a beam splitter. The state after the beam splitter is easily

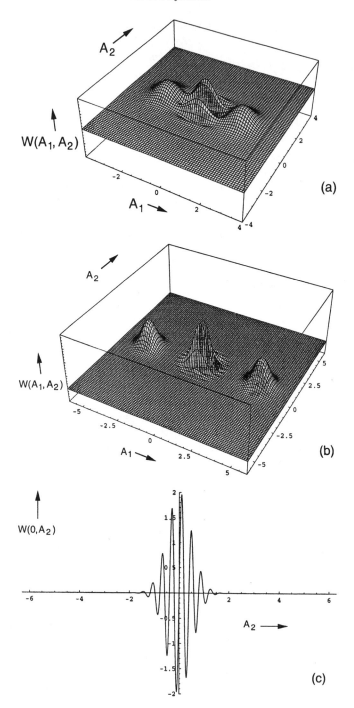

Fig. 7.6. The Wigner function of a Schrödinger cat state: (a) $\alpha = 2$; (b) $\alpha = 4$; (c) $W(0, A_2)$; $\alpha = 4$

evaluated for each of the two components of the product state entering the beam splitter, the second port being unexcited. The state is a superposition of products of coherent states, which remain products of coherent states after passage through the beam splitter. The output state is

$$|\psi\rangle = N(e^{i\pi/4}|\cos\phi\,\alpha\rangle \otimes |-i\sin\phi\,\alpha\rangle + e^{-i\pi/4}|\cos\phi\,\alpha\rangle \otimes |i\sin\phi\,\alpha\rangle)\,,$$
(7.99)

where N is a new normalization constant. The Wigner function of the cat state now involves four new projections with pairwise combinations of the product states. If the beam splitter lets most of the Schrödinger cat state through, $\phi \ll 1$, the self-terms remain unchanged when $\cos\phi$ is replaced by unity. The cross terms involve the projection

$$\langle -i\sin\phi\,\alpha|i\sin\phi\,\alpha\rangle \approx \langle -i\phi\alpha|i\phi\alpha\rangle = \exp(-2|\phi\alpha|^2)\,.$$
(7.100)

The term in the exponent is twice the number of photons siphoned off by the beam splitter. The cross term is decreased exponentially with the number of photons lost by the Schrödinger cat state. Once the cross term is removed, the Schrödinger cat state becomes an incoherent superposition of two coherent α states, a state with classical probabilistic character.

7.9 Passive and Active Multiports

In the preceding three sections we have studied the probability distributions of the in-phase and quadrature components of the field of a coherent state, and the Wigner function of a coherent state. We evaluated the Wigner function of a Schrödinger cat state in order to show the peculiar behavior of such special quantum states. However, as mentioned earlier, our main interest is in the simpler case of linear systems with quantum noise that is additive to the signal amplitude, in analogy with linear noisy classical networks. Our objective is to compare the behavior of linear quantum multiports with their classical counterparts.

The classical description of a passive or active linear multiport is

$$\boldsymbol{b} = \boldsymbol{S}\boldsymbol{a} + \boldsymbol{s}\,,$$
(7.101)

where $\boldsymbol{a}$ is the column matrix containing the signal waves, $\boldsymbol{b}$ is the matrix containing output waves, and $\boldsymbol{s}$ contains the noise sources. If the multiport is passive and at thermal equilibrium, the noise sources are evaluated as shown in Chap. 4. Active multiports contain media that provide gain. The media are never strictly linear in the sense that excitation at one frequency produces a response whose amplitude is proportional to the amplitude of the excitation field and that no mixing of different frequency components occurs. Since every medium with gain saturates, nonlinear frequency mixing occurs.

Thus, active systems can be described as linear multiports only after certain linearizing approximations have been made. Further, active multiports cannot be at thermal equilibrium, and their noise sources must be determined by the physics of the gain mechanism.

The quantization of lossless systems has led us to scattering-matrix descriptions of the output in terms of the input. Systems with loss or gain are physically more sophisticated. We have seen how loss can be treated as coupling to a reservoir of oscillators. Gain can be treated analogously, where the oscillators must now be able to supply energy to the system. The complexity of this analysis can be avoided if we adhere to the principle of commutator conservation and introduce sources that will ensure such conservation. Thus let us look at a multiport with loss or gain. This multiport is described by the operator analog of (7.101),

$$\hat{b} = S\hat{a} + \hat{s} .\tag{7.102}$$

The incoming waves contained in the column matrix $\hat{a}$ and the outgoing waves contained in $\hat{b}$ are waves on open waveguides with commutators that are their fundamental physical characteristics. In the normalization introduced in Sect. 7.1 (compare (7.5a) and (7.5b) with $\Delta\omega_q \to \Delta\omega$),

$$[\hat{b}, \hat{b}^\dagger] = [\hat{a}, \hat{a}^\dagger] = \frac{\Delta\omega}{2\pi}\mathbf{1} .\tag{7.103}$$

Using the equation for the multiport (7.102), the implication of (7.103) is best evaluated using subscript notation:

$$[\hat{b}_i, \hat{b}_j^\dagger] = [S_{ik}\hat{a}_k + \hat{s}_i, S_{j\ell}^*\hat{a}_\ell^\dagger + \hat{s}_j^\dagger] = [S_{ik}\hat{a}_k, S_{j\ell}^*\hat{a}_\ell^\dagger] + [\hat{s}_i, \hat{s}_j^\dagger]$$

$$= S_{i\ell}S_{j\ell}^* + [\hat{s}_i, \hat{s}_j^\dagger] = \frac{\Delta\omega}{2\pi}\delta_{ij} ,\tag{7.104}$$

since the mode amplitude operators and noise source operators commute. We find for the commutators of the noise sources

$$[\hat{s}_i, \hat{s}_j^\dagger] = \frac{\Delta\omega}{2\pi}(\delta_{ij} - S_{i\ell}S_{j\ell}^*)\tag{7.105}$$

or, in matrix notation,

$$[\hat{s}, \hat{s}^\dagger] = \frac{\Delta\omega}{2\pi}(\mathbf{1} - SS^\dagger) .\tag{7.105a}$$

Let us apply this relation to a section of a lossy waveguide with a power loss $\mathcal{L}(< 1)$ and a scattering matrix relating the input wave to the output wave

$$S = \begin{bmatrix} 1 & \sqrt{\mathcal{L}} \\ \sqrt{\mathcal{L}} & 1 \end{bmatrix} .\tag{7.106}$$

We find from (7.106)

$$[\hat{s}_1, \hat{s}_1^\dagger] = [\hat{s}_2, \hat{s}_2^\dagger] = (1 - \mathcal{L})\frac{\Delta\omega}{2\pi} \quad \text{and} \quad [\hat{s}_1, \hat{s}_2^\dagger] = 0 \, . \tag{7.107}$$

It is possible to justify physically the use of an operator noise source for conservation of the commutator. Consider the case of a lossy waveguide. The output power is less than the input power, because some of the power is lost on the way. This situation is represented equivalently by a waveguide coupler as shown in Fig. 7.4, where a_1 produces the outputs b_3 and b_2. If we did not look at waveguide (3), we would conclude that the waveguide was lossy; part of the power has been lost. Starting with the equation $\hat{b}_2 = S_{21}\hat{a}_1 + S_{24}\hat{a}_4$, suppressing subscripts in the spirit that we are looking only at an incident wave $\hat{a}$ and transmitted wave $\hat{b}$, we would write $\hat{b} = \sqrt{\mathcal{L}}\hat{a} + \hat{s}$, with $\sqrt{\mathcal{L}} = S_{21}$ and $\hat{s} = S_{24}\hat{a}_4$. Clearly, a noise source has appeared, which in the present case of a coupler can be identified with the input to the second guide (4). If the guide is unexcited, the noise is due to zero-point fluctuations. The commutator of the noise source is

$$[\hat{s}, \hat{s}^\dagger] = |S_{24}|^2[\hat{a}_4, \hat{a}_4^\dagger] = |S_{24}|^2 = 1 - |S_{21}|^2 = 1 - \mathcal{L} \, . \tag{7.108}$$

Thus, we have recovered the commutator of the noise source that accounts for the conservation of the commutator bracket. The noise comes from the unexcited port, which is fed only zero-point fluctuations. In this way we have justified the model of a loss element, starting with a fully reversible system. The irreversibility is introduced by suppressing the accounting for the outputs in the other waveguides.

The noise source operators associated with a lossy segment of waveguide have the usual interpretation of annihilation ($\hat{s}_i$) and creation ($\hat{s}_i^\dagger$) operators, since $\mathcal{L} < 1$. If the reservoir modes associated with the noise source are in the ground state, then the expectation value of the operator product $\hat{s}_i^\dagger\hat{s}_j$ is zero,

$$\langle 0|\hat{s}_i^\dagger\hat{s}_j|0\rangle = 0 \, . \tag{7.109}$$

On the other hand, because of the commutation relation (7.107), the expectation value of the operators in reverse order is

$$\langle 0|\hat{s}_i\hat{s}_j^\dagger|0\rangle = \delta_{ij}(1 - \mathcal{L})\frac{\Delta\omega}{2\pi} \, . \tag{7.110}$$

If the waveguide has power gain $G \, (> 1)$, we obtain

$$S = \begin{bmatrix} 1 & \sqrt{G} \\ \sqrt{G} & 1 \end{bmatrix} , \tag{7.111}$$

and

$$[\hat{s}_1, \hat{s}_1^\dagger] = [\hat{s}_2, \hat{s}_2^\dagger] = (1 - G)\frac{\Delta\omega}{2\pi} \quad \text{and} \quad [\hat{s}_1, \hat{s}_2^\dagger] = 0 . \tag{7.112}$$

In the case of a segment of waveguide with gain, the roles of creation and annihilation operators are reversed, since $G > 1$. This has profound consequences, already partially explored in Chap. 6. With the gain reservoir modes in the ground state, and the operator $\hat{s}_i$ interpreted as a creation operator and $\hat{s}_j^\dagger$ as an annihilation operator, we have the relations

$$\langle 0|\hat{s}_i\hat{s}_j^\dagger|0\rangle = 0 \quad \text{and} \quad \langle 0|\hat{s}_i^\dagger\hat{s}_j|0\rangle = \delta_{ij}(G - 1)\frac{\Delta\omega}{2\pi} . \tag{7.113}$$

In Chap. 9 we shall use the commutators to derive the full probability distribution of the output photons from an amplifier. Here, let us evaluate the photon number flow and the field fluctuations for a passive structure of loss $\mathcal{L}$ and an active structure of gain G. We assume that the reservoir of the noise sources is at absolute zero, in the ground state (since at optical frequencies with $\hbar\omega \gg k\theta$, room temperature can be approximated by zero temperature for all practical purposes). We assume that the signal is in a coherent state α. Thus, the input state is the product state $|\alpha\rangle|0\rangle$. We have for the photon number at the output

$$\langle 0|\langle\alpha|\hat{b}_2^\dagger\hat{b}_2|\alpha\rangle|0\rangle = \langle 0|\langle\alpha|[(\sqrt{\mathcal{L}}a_1^\dagger + \hat{s}_1^\dagger)(\sqrt{\mathcal{L}}\hat{a}_1 + \hat{s}_1)]|\alpha\rangle|0\rangle$$

$$= \frac{v_g}{L}\mathcal{L}|\alpha|^2 . \tag{7.114}$$

The output photon flow is the input photon flow reduced by a factor of $\mathcal{L}$. The mean square fluctuations of the in-phase field component at the input port (1), indicated by the superscript (1), are:

$$\langle|\hat{E}_1^{(1)2}|\rangle - \langle|\hat{E}_1^{(1)}|\rangle^2 = \frac{1}{4}[\langle|(\hat{a}_1^\dagger + \hat{a}_1)^2|\rangle - \langle|(\hat{a}_1^\dagger + \hat{a}_1)\rangle^2]$$

$$= \frac{1}{4}[\langle|\hat{a}_1^{\dagger 2} + \hat{a}_1^\dagger\hat{a}_1 + \hat{a}_1\hat{a}^\dagger + \hat{a}_1^2|\rangle - \frac{v_g}{L}(\alpha^* + \alpha)^2] \tag{7.115}$$

$$= \frac{1}{4}\frac{v_g}{L}[\alpha^{*2} + 2\alpha^*\alpha + \alpha^2 + 1 - (\alpha^* + \alpha)^2] = \frac{1}{4}\frac{\Delta\omega}{2\pi} .$$

These are the standard zero-point fluctuations. If we repeat the calculation (7.114) with the b operators replacing the a operators and use the commutation relations for the noise sources, we find the fluctuations at the output port (2) to be

$$\langle|\hat{E}_2^{(1)2}|\rangle - \langle|\hat{E}_2^{(1)}|\rangle^2 = \frac{1}{4}[\langle|(\hat{b}_2^\dagger + \hat{b}_2)^2|\rangle - \langle|(\hat{b}_2^\dagger + \hat{b}_2)|\rangle^2]$$

$$= \frac{1}{4}[\langle|[(\sqrt{\mathcal{L}}\hat{a}_1^\dagger + \hat{s}_1^\dagger) + (\sqrt{\mathcal{L}}\hat{a}_1 + \hat{s}_1)]^2|\rangle] - \frac{1}{4}\frac{v_g}{L}(\sqrt{\mathcal{L}}\alpha^* + \sqrt{\mathcal{L}}\alpha)^2$$

$$= \frac{1}{4}\frac{v_g}{L}[\mathcal{L}(\alpha^{*2} + 2\alpha^*\alpha + \alpha^2) + \mathcal{L} + \langle|\hat{s}_1^\dagger\hat{s}_1 + \hat{s}_1\hat{s}_1^\dagger|\rangle - (\sqrt{\mathcal{L}}\alpha^* + \sqrt{\mathcal{L}}\alpha)^2]$$

$$= \frac{1}{4}\frac{\Delta\omega}{2\pi} .$$

$$(7.116)$$

The field at the output experiences the same zero-point fluctuations as the input. The situation is very different for an amplifier. We have for the expectation value of the output photon number flow

$$\langle 0|\langle\alpha|\hat{b}_2^\dagger\hat{b}_2|\alpha\rangle|0\rangle = \langle 0|\langle\alpha|[(\sqrt{G}\hat{a}_1^\dagger + \hat{s}_1^\dagger)(\sqrt{G}\hat{a}_1 + \hat{s}_1)]|\alpha\rangle|0\rangle$$

$$(7.117)$$

$$= \frac{v_g}{L}G|\alpha|^2 + (G-1)\frac{\Delta\omega}{2\pi} .$$

The output photon number flow is G times the input photon number flow plus the contribution of amplified spontaneous emission, $(G-1)/(\Delta\omega/2\pi)$. Whereas the input fluctuations are the same as those for the lossy waveguide under the same input conditions, the output fluctuations of the in-phase field component are

$$\langle\hat{E}_2^{(1)2}\rangle - \langle\hat{E}_2^{(1)}\rangle^2 = \frac{1}{4}[\langle(\hat{b}_2^\dagger + \hat{b}_2)^2\rangle - \langle(\hat{b}_2^\dagger + \hat{b}_2)\rangle^2]$$

$$= \frac{1}{4}\{\langle[(\sqrt{G}\hat{a}_1^\dagger + \hat{s}_1^\dagger) + (\sqrt{G}\hat{a}_1 + \hat{s}_1)]^2\rangle\}$$

$$(7.118)$$

$$- \frac{1}{4}\frac{v_g}{L}(\sqrt{G}\alpha^* + \sqrt{G}\alpha)^2 = \frac{1}{4}(2G-1)\frac{\Delta\omega}{2\pi}$$

$$= \langle E_1^{(2)2}\rangle - \langle E_1^{(2)}\rangle^2 .$$

The fluctuations of the quadrature component are found to be equal to those of the in-phase component, as is easily confirmed by a detailed evaluation. The gain increases the fluctuations. In the limit of high gain the fluctuations are twice the value of the amplified zero-point fluctuations at the input. The fact that the mean square fluctuations of the field have twice the value of the amplified zero-point fluctuations has a profound significance related to the quantum theory of simultaneous measurement of two noncommuting observables. A precise measurement of a quantum observable $\hat{A}$ implies total uncertainty in the conjugate observable $\hat{B}$, whose operator does not commute with $\hat{A}$. This statement still allows a less than perfect measurement of the observable $\hat{A}$ that will not totally destroy knowledge of the observable $\hat{B}$.

The problem of a simultaneous measurement of two observables $\hat{A}$ and $\hat{B}$ with $[\hat{A}, \hat{B}] \neq 0$ has been analyzed in a seminal paper by Arthurs and Kelly [15]. They coupled the system containing the observables to a measurement apparatus and showed that an optimal measurement arrangement will arrive at measured values of $\hat{A}$ and $\hat{B}$ with an uncertainty twice that imposed by the uncertainty principle. This is the penalty attached to a "simultaneous measurement". Linear amplifiers of large gain provide an output signal that is classical, one that could be viewed on a scope. The signal can be observed without the disturbance implied by the uncertainty principle. We have shown that the noise accompanying a signal passing through a linear amplifier of large gain is doubled. In the process of amplification, noise has been added. However, the amplifier now permits a simultaneous measurement of the conjugate quadrature component. The fact that the signal-to-noise ratio has been halved and the noise has been doubled is a manifestation of the proof presented by Arthurs and Kelly.

7.10 Optimum Noise Measure of a Quantum Network

The characteristic noise matrix defined in the context of classical networks in Chap. 5 dealt with the available or exchangeable power of a network. The excess noise figure $F - 1$ of a two-port in the classical domain is the available or exchangeable noise power within the bandwidth B at the output of the amplifier divided by the amplifier gain and normalized to $k\theta_o B$.

Turning to quantized linear multiports, we note that (at least some) of the internal noise source operators of active networks described by the scattering-matrix relation (7.102) are creation operators. A consequence of this fact is that active networks emit photons even if no photons are fed into the input. The output of the network contains so-called amplified spontaneous emission. The concept of the power *available* from a port of a network defined in the classical regime is easily generalizable to the quantum case since it involves a thought experiment in which a passive load connected to the port is varied until the power into the load is maximized. The *exchangeable* power from a port with a negative internal resistance involves loading of the port with a source-free negative resistance. Quantum mechanically, a negative resistance cannot be source-free; it has to emit its own amplified spontaneous emission. For this reason we shall limit ourselves in the following discussion to amplifiers with terminal impedances with positive real part. This is not a serious restriction, since all important cases are of this type, e.g. a fiber laser amplifier or a semiconductor laser amplifier. If an amplifier did not meet this condition, it would be embedded into a circulator to ensure a match at input and output for stability, thus ensuring terminal impedances with positive real parts.

The definition of the noise figure of an optical amplifier is still a controversial issue taken up in Chap. 9. Here we take the point of view, justified in Chap. 9, that an excess noise figure can be defined for an optical amplifier as the available power at the amplifier output within the bandwidth B due to the internal noise sources, divided by the gain and normalized to $k\theta_o B$. The normalization itself is not an important issue at this point and will be reconsidered in Chap. 9. From this definition of excess noise figure a noise measure can be defined by division by $1 - 1/G$. The question then arises as to the optimum noise measure of a linear quantum amplifier. This question is answered in this section. Before we do this we introduce the characteristic commutator matrix and determine its connection with the characteristic noise matrix of Chap. 5.

We have found commutators for the noise source operators were we interpret $\hat{s}_i$ as either an annihilation operator or a creation operator. The commutators are c numbers. The commutators are, according to (7.105),

$$[\hat{s}_i, \hat{s}_j^\dagger] = \frac{\Delta\omega}{2\pi}(\delta_{ij} - S_{i\ell}S_{j\ell}^*) \, . \tag{7.119}$$

Equation (7.119) suggests the definition of a characteristic commutator matrix

$$C = (SS^\dagger - 1)^{-1}[\hat{s}, \hat{s}^\dagger] = -\frac{\Delta\omega}{2\pi}1 \, . \tag{7.120}$$

This matrix is proportional to the identity matrix. It reminds one of the characteristic noise matrix defined in Chap. 5, in particular of the characteristic noise matrix applied to a passive network at thermal equilibrium, in which case the characteristic noise matrix is also proportional to the identity matrix. In the present case, the characteristic commutator matrix applies to both passive and active networks. It is easy to see that lossless embeddings as defined in Chap. 5 leave this matrix invariant. It should be emphasized, however, that lossless embeddings imply subtle source transformations that deserve further scrutiny.

First of all, let us suppose that the network is a passive one. Then the matrix $(1 - SS^\dagger)$ is positive definite. All eigenvalues of the characteristic commutator matrix are negative and of equal magnitude. Since the commutator determines the mean square fluctuations, with the noise sources in their ground states, we see immediately that the network emits zero-point fluctuations from every one of its ports. Lossless embeddings transform both $(SS^\dagger - 1)^{-1}$ and the noise sources s_i. The transformations result in linear combinations of the s_i. Thus, the new sources are still formed from annihilation operators.

We may construct a characteristic noise matrix analogous to (5.65) of Chap. 5. The mean square field fluctuations are equal to $(1/4)\langle\hat{s}_i\hat{s}_i^\dagger\rangle$ for each of the two (in-phase and quadrature) field components and thus the proper definition of the characteristic noise matrix as a predictor of the expectation value of the sum of the field amplitudes squared indicated by the subscript "fas" is

$$N_{S,\text{fas}} = \frac{1}{2}(SS^\dagger - 1)^{-1}\langle \hat{s}\hat{s}^\dagger \rangle \, , \tag{7.121}$$

and, according to (7.120), this can be evaluated as

$$N_{S,\text{fas}} = -\frac{\Delta\omega}{4\pi}\mathbf{1} \, . \tag{7.122}$$

This is negative definite and proportional to the identity matrix. Its eigenvalues are all the same. In the classical interpretation of the characteristic noise matrix, its eigenvalues yield the extrema of the noise power emitted into loads under arbitrary variation of the loads. The loads are all passive. In the quantum interpretation, the eigenvalues give the the mean square field fluctuations under arbitrary variation of the passive loads. The mean square field fluctuations of the outgoing waves are all the same and equal to $\Delta\omega/4\pi$. Hence this finding simply confirms that all outgoing waves of a passive network experience standard zero-point fluctuations.

Next, consider a network with a negative definite matrix $(1 - SS^\dagger)$. This is a fully active network. In the ideal case of a perfectly inverted gain medium, all noise sources are creation operators. The ideal minimum available photon flux (the available power divided by $\hbar\omega$) of the network is the same at all ports and equal to $\Delta\omega/2\pi$. A lossless embedding again results in new sources that are linear combinations of the s_i, which are now creation operators. The minimum available photon flux remains unchanged. Of course, a superposition of a creation operator and an annihilation operator could be responsible for a net commutator bracket, as we have seen in the case of an incompletely inverted gain medium. In this case the available photon flux is larger than in the case when all operators $\hat{s}_j$ are pure creation operators. In order to predict the available photon flux, the composition of the operators $\hat{s}_j$ must be known. One may take as a simple example a two-port fiber amplifier. Its scattering matrix is

$$S = \begin{bmatrix} \sqrt{G} & 0 \\ 0 & \sqrt{G} \end{bmatrix} \, . \tag{7.123}$$

Its commutator matrix is, according to (7.119),

$$\langle s, s^\dagger \rangle = -\frac{\Delta\omega}{2\pi} \begin{bmatrix} G-1 & 0 \\ 0 & G-1 \end{bmatrix} \, . \tag{7.124}$$

If the gain medium is perfectly inverted, both noise operators are creation operators and the available photon flux from either of the two ports is

$$\text{available photon flux} = \frac{\Delta\omega}{2\pi}(G-1) \, . \tag{7.125}$$

One may again construct a characteristic noise matrix. Now, however, we cannot appeal to the classical interpretation in terms of a thought experiment in which the loads are varied arbitrarily, and the eigenvalues of the characteristic noise matrix give the extrema of the exchangeable power. These extrema require loading with active terminations, which cannot be noise-free in the quantum limit. We must interpret the eigenvalues in terms of the stationary values of the noise measure. These stationary values may be reached with passive loading if the amplifier is constructed so that the input impedances are all passive (achieved, if necessary, with embeddings using a circulator). In the classical interpretation, the noise measure involves the excess noise figure, which in turn is determined by the noise output power referred to the input by division by the gain. Hence, in the case of an active network with a negative definite matrix $(1 - SS^\dagger)$, the quantum interpretation of the characteristic noise matrix can be in terms of the power (or photon flux), which involves $\langle \hat{s}_i^\dagger \hat{s}_i \rangle$. The characteristic noise matrix, as a predictor of the amplified spontaneous emission (ASE) noise, is

$$N_{S,\text{ASE}} = (SS^\dagger - 1)^{-1}\langle \hat{s}^\dagger \hat{s} \rangle = \frac{\Delta\omega}{2\pi}1 . \tag{7.126}$$

The extrema of the noise measure are given by the eigenvalues of this matrix, which are all identical.

The situation of an indefinite network is more complicated. According to (7.119), the commutators of the noise sources are both positive and negative, i.e. the column matrix consisting of the operators $\hat{s}_j$ contains both creation operators and annihilation operators. This information can be used to evaluate either the zero-point fluctuations of the field or the photon flow from the network. To use it one needs to be specific as to whether one is looking for mean square fluctuations or photon flow. An example may be helpful. Consider the equivalent circuit of an FET in the scattering-matrix formulation. This could also be the equivalent circuit of an optical amplifier (as described by (7.123) and (7.124)), followed by a circulator with a matched termination, as in Fig. 5.8. The scattering matrix is

$$S = \begin{bmatrix} 0 & 0 \\ S_{21} & 0 \end{bmatrix} = \begin{bmatrix} 0 & 0 \\ \sqrt{G} & 0 \end{bmatrix} . \tag{7.127}$$

We find, from the commutator matrix (7.120),

$$\left[\hat{s}, \hat{s}^\dagger\right] = \begin{bmatrix} 1 & 0 \\ 0 & 1-G \end{bmatrix} \frac{\Delta\omega}{2\pi} . \tag{7.128}$$

For $G > 1$, there is one negative and one positive commutator. Hence the two noise sources are represented by a creation operator and an annihilation operator, respectively. From the positive commutator one may evaluate the zero-point fluctuations at the input. From the negative commutator one may obtain the optimum noise measure under conditions of complete inversion,

$$M_{\text{opt}} = \frac{\hbar\omega_o}{k\theta_o} , \tag{7.129}$$

giving a number of the order of 40 for visible or near-infrared light. A microwave traveling-wave tube operating at 10 GHz can have an excess noise figure as low as 1 dB. Thus, optical amplifiers, even under the most ideal conditions, are terribly noisy in comparison with microwave amplifiers. Further, their noise performance is determined by fundamental physical laws. (The quantum limit for traveling-wave tubes is negligible owing to the low energy of a microwave photon.) Yet, long-distance fiber communication is now the major technology for long-distance communications. Why did this happen, when microwave amplifiers have so much better noise performance? The answer lies in the exceedingly low loss and excellent broadband propagation properties of optical fibers. Optical-fiber communication has won out because of the exceptional properties of optical fibers and because it is relatively easy to generate optical signals of sufficiently high power level that a large signal-to-noise ratio can be maintained.

It should be mentioned that it is customary to define noise figures for optical amplifiers normalized to $\hbar\omega_o$, and not to $k\theta_o$. Then, of course, their excess noise figure does not seem so high. Ideal amplification with high gain leads to an excess noise figure of unity, or a noise figure of 2 (3 dB).

7.11 Summary

In this chapter we introduced one of several renormalizations of the creation and annihilation operators. This renormalization was designed to emphasize the correspondence between classical and quantum mechanical linear, noisy networks. The noise, expressed classically as the power in the bandwidth $\Delta\omega/2\pi = B$, was expressed as the photon flux in the same bandwidth.

A linear, lossless, phase-insensitive network has a Hamiltonian that contains sums of the photon number operators, i.e. products of creation and annihilation operators. This Hamiltonian leads to linear equations of motion for the annihilation operators. Integration of Heisenberg's equation of motion yields a unitary scattering matrix that is in one-to-one correspondence with its classical counterpart. Since operators evolve via pre- and postmultiplication by unitary matrices it was of interest to explore how this evolution corresponded to the evolution described by a scattering matrix. We showed this to be the case using some simple functional relations among operator expressions.

While we prefer the Heisenberg formalism to that of Schrödinger, we looked briefly at so-called entangled states, which emerge explicitly only in the Schrödinger formalism. These are nonclassical states that will find application in the analysis of Chap. 14. We studied the characteristic function of a quantum observable which contains the information on the moments of the observable. In the classical regime, the Fourier transform of the characteristic function of two random variables is the joint probability function. In the quantum regime, the Fourier transform may not be positive definite, and thus cannot be interpreted as a probability. We showed with an interferometer example that the interpretation of the Fourier transform as a probability, even when positive definite, can lead to erroneous conclusions. In the case when the two noncommuting observables are position and momentum, the Fourier transform of the characteristic function is the so-called Wigner function. The in-phase and quadrature components of a quantized electric field are equivalent stand-ins for position and momentum.

Coherent states are "classical" states in that they do not exhibit peculiar quantum behavior. Hence, it was of interest to determine how a linear, lossless network transforms an input consisting of coherent states. We found the expected: coherent states remain coherent as they are transformed by a linear, lossless network.

A linear network with loss or gain is not describable by a Hamiltonian. The equations of motion of the annihilation operator are still linear, but photons are not conserved. Conservation of the commutator brackets is provided by operator noise sources. From the commutator relations of the noise sources it was possible to construct a characteristic noise matrix for the network that sets a lower limit on the optimum noise measure achievable with a multiport.

Linear phase-insensitive amplification to a classical level permits the determination of both the in-phase and the quadrature components of the electric field. The operators representing these fields do not commute and thus "are not measurable simultaneously". However, as originally pointed out by Arthurs and Kelly, the measurement is possible at the expense of an uncertainty twice that set by the Heisenberg uncertainty principle. An ideal amplifier permits such a measurement and so does a heterodyne receiver, as shown in the next chapter. It is thus no coincidence that the signal-to-noise ratios of an ideal amplifier of large gain and of a heterodyne detector are the same.

Problems

7.1* Show that the state $(1/\sqrt{2})(|2\rangle|0\rangle - |0\rangle|2\rangle)$ passes through a beam splitter unchanged.

7.2* The photon state $|2\rangle|0\rangle$ enters the input ports of a beam splitter with $|M|T = \phi$. What is the state at the output?

7.3 The coherent-state wave function $|\alpha\rangle|\beta\rangle$ enters the input ports of a 50/50 beam splitter. What is the output wave function?

7.4* Consider the scattering-matrix equivalent circuit of the FET, Fig. 5.13, and use it as a model for a nonreciprocal optical amplifier. Find the commutators of the noise wave generators so that they conserve commutator brackets. What is the optimum noise performance?

It may be worth pointing out that this equivalent circuit applies to a fiber amplifier with a Faraday circulator.

7.5 Evaluate the expectation value of the cosine operator $\hat{C}$ for a coherent state $|\alpha\rangle$. Evaluate the projection $\langle\beta|\hat{C}|\alpha\rangle$. See Appendix A.7.

7.6 Evaluate the probability distribution of the in-phase and quadrature components of the field, $B^{(1)}$ and $B^{(2)}$, at the output of the amplifier described in Sect. 7.10 for a coherent input state $|\alpha\rangle$.

7.7 Evaluate the characteristic function for the in-phase and quadrature components of the output field of an attenuator of loss $\mathcal{L}$ with a single-photon input. Note: you can use the Baker–Hausdorff theorem on the output field since commutators are preserved. Plot the characteristic function as a function of ξ and $\mathcal{L}$. Plot the probability distributions as functions of $B^{(i)}$ and $\mathcal{L}$, $i = 1, 2$.

7.8 Determine the Wigner function of the in-phase and quadrature components of the number state $|1\rangle$ and plot it.

7.9 Find the characteristic function for the number state $|2\rangle$ and plot it.

Solutions

7.1 From (7.49) we find

$$|\psi(T)\rangle = \sum_{n=0}^{\infty} \frac{(-i\phi)^n}{n!} (\hat{A}^\dagger\hat{B} + \hat{B}^\dagger\hat{A})^n \frac{1}{\sqrt{2}}(|2\rangle|0\rangle - |0\rangle|2\rangle) \,.$$

The operator $\hat{A}$ operates on the first wave function in the product, the operator $\hat{B}$ on the second.

$$(\hat{A}^\dagger\hat{B} + \hat{B}^\dagger\hat{A})\frac{1}{\sqrt{2}}(|2\rangle|0\rangle - |0\rangle|2\rangle) = (|1\rangle|1\rangle - |1\rangle|1\rangle) = 0 \,.$$

Thus the series stops at the first term and the state is indeed unchanged.

7.2 The input state can be expressed as the sum of a symmetric and an antisymmetric state

$$|2\rangle|0\rangle = \frac{1}{2}(|2\rangle|0\rangle + |0\rangle|2\rangle) + \frac{1}{2}(|2\rangle|0\rangle - |0\rangle|2\rangle) \,.$$

From (7.49) we find

$|\psi(T)\rangle$

$$= \sum_{n=0}^{\infty} \frac{(-i\phi)^n}{n!} (\hat{A}^\dagger \hat{B} + \hat{B}^\dagger \hat{A})^n \left[\frac{1}{2}(|2\rangle|0\rangle + |0\rangle|2\rangle) + \frac{1}{2}(|2\rangle|0\rangle - |0\rangle|2\rangle) \right] .$$

We follow the evolutions of the symmetric state and the antisymmetric state through the system separately. From Prob. 7.1 we know that the antisymmetric state remains unchanged. The operator $\hat{A}$ operates on the first wave function in the product, the operator $\hat{B}$ on the second:

$$(\hat{A}^\dagger \hat{B} + \hat{B}^\dagger \hat{A})\frac{1}{2}(|2\rangle|0\rangle + |0\rangle|2\rangle) = \sqrt{2}|1\rangle|1\rangle .$$

Operation with the second power gives

$$(\hat{A}^\dagger \hat{B} + \hat{B}^\dagger \hat{A})^2 \frac{1}{2}(|2\rangle|0\rangle + |0\rangle|2\rangle)$$

$$= (\hat{A}^\dagger \hat{B} + \hat{B}^\dagger \hat{A})\sqrt{2}|1\rangle|1\rangle$$

$$= 2(|2\rangle|0\rangle + |0\rangle|2\rangle) .$$

Operation with the third power gives

$$(\hat{A}^\dagger \hat{B} + \hat{B}^\dagger \hat{A})^3 \frac{1}{2}(|2\rangle|0\rangle + |0\rangle|2\rangle)$$

$$= (\hat{A}^\dagger \hat{B} + \hat{B}^\dagger \hat{A})2(|2\rangle|0\rangle + |0\rangle|2\rangle)$$

$$= 4\sqrt{2}|1\rangle|1\rangle .$$

Operation with the fourth power gives

$$(\hat{A}^\dagger \hat{B} + \hat{B}^\dagger \hat{A}]^4 \frac{1}{2}(|2\rangle|0\rangle + |0\rangle|2\rangle)$$

$$= (\hat{A}^\dagger \hat{B} + \hat{B}^\dagger \hat{A})4\sqrt{2}|1\rangle|1\rangle$$

$$= 16\frac{1}{2}(|2\rangle|0\rangle + |0\rangle|2\rangle) .$$

We can now discern the structure of the wave function. The nth odd power gives $2^n(1/\sqrt{2})|1\rangle|1\rangle$. The nth even power gives $2^n(1/2)(|2\rangle|0\rangle + |0\rangle|2\rangle)$. The antisymmetric wave function remains unchanged. We find for the entire series

$|\psi(T)\rangle$

$$= \frac{1}{2}(|2\rangle|0\rangle + |0\rangle|2\rangle) \cos 2\phi - \frac{i}{\sqrt{2}}|1\rangle|1\rangle \sin 2\phi + \frac{1}{2}(|2\rangle|0\rangle - |0\rangle|2\rangle) .$$

It is easily checked that $\langle \psi(T)|\psi(T)\rangle = 1$.

7.4 We use the wave formalism in the transfer matrix formulation, (5.129). We have

$$T = \begin{bmatrix} -2/\mu & 0 \\ 0 & 0 \end{bmatrix}, \quad P_T = \begin{bmatrix} 1 & 0 \\ 0 & -1 \end{bmatrix}.$$

The noise source commutator matrix is

$$\left[\begin{bmatrix} \hat{\gamma}_a \\ \hat{\gamma}_b \end{bmatrix}, \begin{bmatrix} \hat{\gamma}_a^\dagger & \hat{\gamma}_b^\dagger \end{bmatrix} \right] = \begin{bmatrix} [\hat{\gamma}_a, \hat{\gamma}_a^\dagger] & [\hat{\gamma}_a, \hat{\gamma}_b^\dagger] \\ [\hat{\gamma}_b, \hat{\gamma}_a^\dagger] & [\hat{\gamma}_b, \hat{\gamma}_b^\dagger] \end{bmatrix}.$$

The characteristic commutator matrix is

$$C_T = (P_T^{-1} - TP_T^{-1}T^\dagger)^{-1}[\hat{\gamma}, \hat{\gamma}^\dagger] = \begin{bmatrix} \dfrac{[\hat{\gamma}_a, \hat{\gamma}_a^\dagger]}{1 - 4/|\mu|^2} & \dfrac{[\hat{\gamma}_a, \hat{\gamma}_b^\dagger]}{1 - 4/|\mu|^2} \\ -[\hat{\gamma}_b, \hat{\gamma}_a^\dagger] & -[\hat{\gamma}_b, \hat{\gamma}_b^\dagger] \end{bmatrix}.$$

This commutator matrix must be equal to $-\Delta\omega/2\pi$ times the identity matrix. From this requirement we find

$$[\hat{\gamma}_a, \hat{\gamma}_a^\dagger] = -\left(1 - \frac{4}{|\mu|^2}\right)\Delta\omega/2\pi, \quad [\hat{\gamma}_a, \hat{\gamma}_b^\dagger] = [\hat{\gamma}_b, \hat{\gamma}_a^\dagger] = 0,$$

$$[\hat{\gamma}_b, \hat{\gamma}_b^\dagger] = \Delta\omega/2\pi.$$

The operator $\hat{\gamma}_a$ is a creation operator when $|\mu|^2/4 > 1$, when there is gain. The characteristic noise matrix is

$$N_T = (P_T^{-1} - TP_T^{-1}T^\dagger)^{-1} \begin{bmatrix} \langle \hat{\gamma}_a^\dagger \hat{\gamma}_a \rangle & \langle \hat{\gamma}_a^\dagger \hat{\gamma}_b \rangle \\ \langle \hat{\gamma}_b^\dagger \hat{\gamma}_a \rangle & \langle \hat{\gamma}_b^\dagger \hat{\gamma}_b \rangle \end{bmatrix}$$

$$= \frac{\Delta\omega}{2\pi} \begin{bmatrix} 1 & 0 \\ 0 & 0 \end{bmatrix}.$$

The zero eigenvalue is associated with loss. The eigenvalue associated with gain is the standard eigenvalue of an ideal optical amplifier.

8. Detection

In Chap. 6, we studied a measurement of the spontaneous emission of an amplifier. A bolometer detects power directly by measuring the amount of heat generated by the power absorbed. Microwave radiation impinging upon a diode terminating a waveguide induces curents in the diode. The nonlinearity of the diode leads to current or voltage rectification and the d.c. voltage across the diode is a measure of the electric field across the diode. From the electric field, the incident power can be inferred, if there is no reflection or if proper account is taken of the reflection. The power can be calibrated versus the d.c. voltage.

The photons of optical waves impinging upon a photocathode can propel electrons across the potential barrier between the cathode material and the vacuum. The emitted electrons are collected on the anode and their flow is a measure of the incident flow of photons. A p–n junction can act like a vacuum diode. If the photons are absorbed in the depletion region of a p–n junction generating electron–hole pairs, the holes travel to the n side and the electrons to the p side, constituting a photocurrent that is a measure of the absorbed photon flow. The ratio of the number of carriers collected to the number of photons impinging on the photodetector is the so-called quantum efficiency. The quantum efficiency of photodetectors of near-infrared light can approach unity.

In this chapter, we study the noise in detectors in general and photo-detectors in particular. We start with the classical analysis of a square-law detector. Then we look at a photodetector whose current is a measure of the incident photon flux. We determine the signal-to-noise ratio of photodetection. Direct photodetection loses the phase information about the incident optical wave. The phase can be detected by heterodyne detection, which is equivalent to amplification and detection of an incident optical wave. We determine the signal-to-noise ratio of balanced heterodyne detection both classically and quantum mechanically. We also look at homodyne detection, in which the local-oscillator frequency coincides with the signal frequency and in which only one of the two components of the field is detected. The signal-to-noise ratio turns out to be double that of heterodyne detection for reasons that can be traced to the theory of a simultaneous measurement of two noncommuting quantum observables.

8.1 Classical Description of Shot Noise and Heterodyne Detection

Microwave p–n junctions can be used as square-law detectors. The current through the detector is proportional to the square of the electric field. If the detector is not fast enough to follow a microwave cycle, the current can be written

$$i_s(t) = q\gamma|E_s(t)|^2 \,, \tag{8.1}$$

where $E_s(t)$ is the complex electric-field amplitude of the signal, q is the electron charge, and γ is a proportionality constant. The d.c. current is given by the time average. All detectors that produce a time-averaged current flow exhibit shot noise (or higher levels of noise if there is avalanche multiplication) corresponding to the average current in the detector. The mean square current fluctuations of a d.c. current I_o in a bandwidth B are those of shot noise (compare (4.15)):

$$\langle i_n^2 \rangle = 2qI_oB \,. \tag{8.2}$$

If the detector has a resistance R, there may be thermal noise associated with the resistance according to the Nyquist formula (4.76). The signal-to-noise ratio is computed from the ratio of the mean square signal current to the mean square noise current. Suppose that the signal is a steady-state sinusoid $E_s(t) = A_s \exp(-i\omega t)$. The d.c. current is then $I_o = q\gamma|E_s|^2 = q\gamma|A_s|^2$, and the signal-to-noise ratio is

$$\frac{S}{N} = \frac{\langle i_s^2(t) \rangle}{\langle i_n^2 \rangle} = \frac{\gamma|A_s|^2}{2B} \,, \tag{8.3}$$

if thermal noise can be neglected.

An optical detector of unity quantum efficiency can detect, in principle, single photons. The photon flow rate must be low enough that the detector can resolve the incident photons and the thermal noise must be negligible. Such an ideal detector may be considered to be noise-free; it reproduces faithfully the photonic signal. Noise-free detection is consistent with quantum mechanics, since there is no fundamental limit imposed on the accuracy of measurement of an observable. As we have seen earlier, in the example of an optical amplifier, only a simultaneous measurement of two noncommuting observables is accompanied by unavoidable noise.

When a signal is passed through a narrow-band optical preamplifier, as studied in detail in Chap. 9, the ASE photon flow imposes a background noise that is not Poissonian but, rather, has Bose–Einstein statistics; it is not simple shot noise. On the other hand, if the detector is illuminated by attenuated laser light, with photons that are Poisson-distributed, as will be shown in Chap. 9, then the charge current is also Poisson-distributed with a shot noise spectrum. The discussion in this chapter will be limited to laser light with photons and charge carriers that have a Poisson distribution.

An optical power P at an optical frequency ω_o incident upon a photodetector of quantum efficiency η produces a current $i(t)$ according to the formula

$$i(t) = \eta q \frac{P(t)}{\hbar \omega_o} , \tag{8.4}$$

where $P(t)/\hbar \omega_o$ represents the instantaneous photon flow, an identification possible when the optical radiation is sufficiently narrow-band that the assignment of the fixed energy $\hbar \omega_o$ to all photons is legitimate. The physical picture associated with (8.4) is carrier generation in one-to-one correspondence with the incident photon flux. Just how this photon flow is to be defined will be the topic of this chapter. As a simple semiclassical expedient one may write the power in terms of the complex field amplitude $E(t)$, with $E(t)$ so normalized that

$$E^*(t)E(t) = P(t) . \tag{8.5}$$

Note that in (8.4) the absolute magnitude of the complex field amplitude squared is used, not the instantaneous E field squared. This is not an approximation, as it was in the case of a microwave square-law detector, but a consequence of the fact that the process of photodetection responds to the incident photon flux. The photocurrent is thus

$$i(t) = \eta q \frac{E^*(t)E(t)}{\hbar \omega_o} . \tag{8.6}$$

The current fluctuations are those of shot noise accompanying the d.c. current $\eta q P(t)/\hbar \omega_o$. The spectrum of shot noise is white. If the optical power varies with time, the spectrum of the current is composed of a white shot noise background and the spectrum of $P(t)$, as shown in Chap. 4.

All phase information of an optical signal is lost in direct detection. Phase information can be recovered in heterodyne detection. An experimental arrangement for microwave heterodyne detection is shown in Fig. 8.1. The mode incident upon the detector is made up of a local-oscillator mode amplitude $E_o(t)$ and a signal mode amplitude $E_s(t)$, superimposed via a waveguide junction as shown. If the junction is highly transmissive for the signal, no appreciable sacrifice in signal power incident upon the detector need be made. There is, of course, a sacrifice of local-oscillator power, which can be avoided in a balanced detector arrangement as shown later on. The current in the photodetector to first order in the signal field is

$$i(t) \cong q\gamma[E_o^*(t)E_o(t) + E_o^*(t)E_s(t) + E_s^*(t)E_o(t)] = i_o + i_s(t) , \tag{8.7}$$

where we neglect the square of the signal field as very much smaller than the local-oscillator power. The detector current is made up of two parts: a d.c.

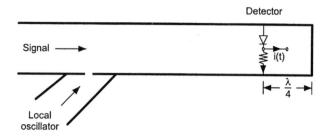

Fig. 8.1. Experimental arrangement for microwave heterodyne detection

current i_o due to the local oscillator and a part due to the beat between the local oscillator and the signal, $i_s(t)$. If the local oscillator produces a simple sinusoid at frequency ω_o and the signal has frequency ω_s, then the fields can be written

$$E_o(t) = A_o \exp(-i\omega_o t) \quad \text{and} \quad E_s(t) = A_s \exp(-i\omega_s t) . \qquad (8.8)$$

The signal current is

$$i_s(t) = q\gamma\{A_o^* A_s \exp[i(\omega_o - \omega_s)t] + A_o A_s^* \exp[-i(\omega_o - \omega_s t]\}$$
$$= 2q\gamma|A_o A_s| \cos[(\omega_o - \omega_s)t + \phi] , \qquad (8.9)$$

where $\phi = \arg(A_o^* A_s)$. The detector current carries both phase and amplitude information. The noise in the detector is the shot noise due to the local-oscillator bias current, which is time-independent, since the small amplitude of the beat term can be ignored (signal-dependent noise is ignored):

$$\langle i_n^2 \rangle = 2q\gamma|A_o|^2 B. \qquad (8.10)$$

The signal-to-noise ratio of heterodyne detection is thus

$$\frac{S}{N} = \frac{\overline{\langle i_s(t)\rangle^2}}{\langle i_n^2 \rangle} = \gamma \frac{|A_s|^2}{B} . \qquad (8.11)$$

The time average of the square of the photodetector current introduces a factor of $1/2$, the average of a cosine-squared function. Note that $\gamma|A_s|^2$ is the number of carriers produced by the signal impinging upon the detector per unit time. The signal-to-noise ratio is equal to the number of carriers produced by the signal in the time interval $1/B$.

The same analysis can be repeated for a photodetector, as shown in Fig. 8.2. Instead of the waveguide junction, a beam splitter is used. The splitting ratio is such that most of the signal is transmitted, but local-oscillator power is sacrificed. We assume that the signal wave and the local-oscillator wave have the same polarization and are phase-coherent across the detector surface.

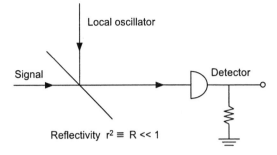

Fig. 8.2. Experimental arrangement for optical heterodyne detector

If not, a mode decomposition is required, and only pairs of modes of the same order and same polarization give a current response. We shall not be concerned with this more complicated situation, since it must be avoided in practice. The analysis is carried through completely analogously. Instead of the coefficient γ, the coefficient $\eta/\hbar\omega_o$ is used, and the signal-to-noise ratio is

$$\frac{S}{N} = \frac{\overline{\langle i_s(t)\rangle^2}}{\langle i_n^2\rangle} = \eta\frac{|A_s|^2}{\hbar\omega_o B} \ . \tag{8.12}$$

Just as in the microwave case, the signal to noise ratio is the the number of charge carriers produced by the signal in the time interval $1/B$.

8.2 Balanced Detection

Heterodyne detection by a local oscillator, coherent with the signal, is a very important processing method for a returning radar signal. However, the detection of a radar signal encounters a serious problem. The signal is the return echo from a powerful pulse, but attenuated by 60 dB or more. In the analysis of detection we have assumed that the noise accompanying the bias current produced by the local oscillator is shot noise. This may not be true, since the local oscillator undergoes disturbances that cause fluctuations of the local oscillator power. Even if they are 60 dB below the local oscillator power level, they become comparable to the level of the returning signal. Balanced heterodyne detection, invented in radar technology, overcomes the problem of oscillator noise. Figure 8.3 shows both the radar implementation and the optical implementation of balanced heterodyne detection. The local oscillator is fed through one waveguide port of a magic T, the signal through the other port. The fields in the outgoing waves in the two waveguides are superpositions of the incident fields, but with sign changes due to the symmetry of the magic T. The local oscillator excites outgoing waves with symmetric electric

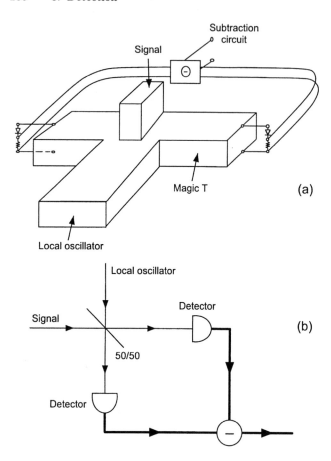

Fig. 8.3. (a) Microwave and (b) optical implementations of balanced heterodyne detection. The magic T is a four-port matched at all ports

fields, the signal excites them antisymmetrically, as can be seen easily by just sketching the field distribution within the magic T. Thus the complex amplitudes of the electric fields impinging upon the detectors are

$$E_1 = \frac{1}{\sqrt{2}}(E_o - E_s) \quad \text{and} \quad E_2 = \frac{1}{\sqrt{2}}(E_o + E_s) \,. \tag{8.13}$$

The currents of the two detectors are subtracted. If the square of the signal is neglected the net output current is

$$i(t) = q\gamma(|E_2|^2 - |E_1|^2) = q\gamma(E_o E_s^* + E_o^* E_s) \,. \tag{8.14}$$

The rectified local-oscillator current cancels. If this current fluctuates, the fluctuations do not appear in the current output. With $E_o = A_o \exp(-i\omega_o t)$ and $E_s = A_s \exp(-i\omega_s t)$, the square of the signal current averaged over time is

$$\langle i_s^2(t) \rangle = q^2 \gamma^2 \langle |A_o A_s^* + A_o^* A_s|^2 \rangle$$

$$= q^2 \gamma^2 4 |A_o|^2 |A_s|^2 \langle \cos^2[(\omega_o - \omega_s)t + \phi] \rangle \tag{8.15}$$

$$= q^2 \gamma^2 2 |A_o|^2 |A_s|^2 \,,$$

where $\phi = \arg(E_o^* E_s)$. The noise is due to the shot noise current in each of the detectors, with a current $\frac{1}{2} q \gamma |E_o|^2$ each, adding to the net shot noise value

$$\langle i_n^2 \rangle = 2 q^2 \gamma |A_o|^2 B \,. \tag{8.16}$$

Thus, the signal-to-noise ratio is

$$\frac{S}{N} = \frac{\langle i_s \rangle^2}{\langle i_n^2 \rangle} = \frac{\gamma |A_s|^2}{B} \,. \tag{8.17}$$

In the numerator is the rate at which charge carriers would be produced by the signal alone impinging upon the detector. Division by B gives the number of carriers that would be produced by the signal alone in a time interval equal to the inverse bandwidth. The signal-to-noise ratio is the same as (8.12) for heterodyne detection with a single detector. The arrangement of the balanced detector has the advantage that it cancels fluctuations of the local-oscillator power to first order and that it uses the total local-oscillator power.

The optical version of the balanced detector, Fig. 8.3b, is entirely analogous. Instead of the magic T, a 50/50 beam splitter is used. The balanced detector utilizes the full local-oscillator power incident upon the 50/50 beam splitter. The factor γ in (8.17) is replaced by $\eta/\hbar\omega_o$. Note that the photon-energy-normalizing factor $\hbar\omega_o$ has not been changed, since detectors can respond only to low beat frequencies. Thus, the energies of the signal photons and local-oscillator photons differ by a negligible amount.

The semiclassical analysis of balanced optical detection is simple. The photons passing through the beam splitter are randomly sent to either one detector or the other. Each of the two detectors experiences the full shot noise associated with the current through it. Fluctuations of the local-oscillator power are coherent at the two detectors and cancel in the subtraction circuit. The shot noise in the two detectors is uncorrelated and the fluctuations add in the subtraction circuit. Hence the difference current has shot noise fluctuations of magnitude equal to the sum of the fluctuations in each detector. In the linearized theory, in which only the local-oscillator current is responsible for the noise, the signal-to-noise ratio of the balanced detector is the same as of the simple heterodyne detector of Fig. 8.2 for equal local-oscillator power as given in (8.12). *Thus the balanced optical heterodyne detector gives the same signal-to-noise ratio as a single heterodyne detector, but with fluctuations of local-oscillator power suppressed.*

Homodyne detection is the degenerate heterodyne detection that occurs when the local-oscillator and signal frequencies are equal. Then, the current is a measure of the electric field that is in phase with the local oscillator. Not all of the shot noise is detected. Shot noise consists of randomly excited amplitudes of cosine and sine waves referred to the phase of the local oscillator. Of these, only the cosine waves are detected. The noise is halved. This fact was pointed out by B. Oliver in 1962 [71]. In a follow-up note by C. H. Townes and the author it was pointed out that the increase in signal-to-noise ratio is accompanied by a loss of information contained in the detected signal [72]. It took some years before the full implication of the difference was grasped in the context of detection of squeezed light, as discussed in Chap. 11.

8.3 Quantum Description of Direct Detection

In the quantum analysis of photocurrent generation, the incident photon flux is responsible for the current, photons are annihilated and carriers are generated. In the quantum analysis of modes, the modes were set up as functions of the propagation constant β_m. We shall now derive the quantum description of the photon current for radiation that consists of a succession of coherent states. The photon number in a quantization interval of length L is $\hat{A}_m^\dagger \hat{A}_m$. A photodetection measurement that converts photons into photoelectrons is, essentially, a measurement describable by the operator $\hat{A}_m^\dagger \hat{A}_m$. In its definition, the quantization interval L plays an essential role. Its choice fixes the increment $\Delta\beta$ of the Fourier decomposition of the modes. This increment also fixes the frequency increment $\Delta\omega = (d\omega/d\beta)\Delta\beta$. Changes in the choice of the interval change the interpretation of the "photons" contained in the optical field. This appears surprising, at first. However, we shall emphasize later on, and in detail in Chap. 14, that the interpretation of the "physical meaning" of a quantum concept requires the specification of the measurement apparatus. The measurement is performed with an apparatus of a certain bandwidth (temporal resolution). It is the bandwidth of the apparatus that dictates the choice of L.

The charge registered by the detector over the time interval L/v_g is

$$\hat{Q} = q\hat{A}_m^\dagger \hat{A}_m , \tag{8.18}$$

a Hermitian operator. For a coherent state, the expectation value of the charge is

$$\langle \alpha|\hat{Q}|\alpha \rangle = q\langle \alpha|\hat{A}^\dagger \hat{A}|\alpha \rangle = q|\alpha|^2 = q\langle n \rangle , \tag{8.19}$$

where $\langle n \rangle$ is the expectation value of the photon number. We shall omit the subscript m henceforth, since we are dealing with one mode only. Let us now determine the fluctuation properties of the charge. We have, from the defining equation and the commutation relations of the creation and annihilation operators of the electromagnetic field,

$$\langle\alpha|\hat{Q}^2|\alpha\rangle = q^2\langle\alpha|\hat{A}^\dagger\hat{A}\hat{A}^\dagger\hat{A}|\alpha\rangle = q^2\langle\alpha|\hat{A}^\dagger\hat{A}^\dagger\hat{A}\hat{A} + \hat{A}^\dagger\hat{A}|\alpha\rangle$$

$$= q^2|\alpha|^4 + q^2|\alpha|^2 = q^2\langle n\rangle^2 + q^2\langle n\rangle \ . \tag{8.20}$$

Here, we have put the operators into normal order: creation operators precede annihilation operators. In the process, the commutation relation is used, which accounts for the added term in the last expression. The terms in normal order are easily evaluated for a coherent state, since the annihilation operators operating on a ket on the right produce an α, and creation operators operating on a bra on the left produce an α^*.

The mean square fluctuations of the charge are

$$\langle\alpha|\hat{Q}^2|\alpha\rangle - \langle\alpha|\hat{Q}|\alpha\rangle^2 = q^2\langle n\rangle \ . \tag{8.21}$$

These are the fluctuations of a Poisson process, consistent with the derivation in Chap. 4 for a completely random flow of charge. One may model the generation of a photocurrent by an optical field as a random generation of photocarriers with the rate of generation determined by the power level of the incident light. Conversely, according to the analysis of Chap. 6, one may view the process as the generation of carriers in one-to-one correspondence with an incident photon flux with a Poisson distribution of photons. Both interpretations are possible at this level of the analysis.

Equation (8.21) shows that the mean square fluctuations of the charge carriers are equal to the photon number. The photon number is evaluated for a length interval L. Photons assigned to a length interval L enter the photon detector within a time interval $T = L/v_g$. Hence the choice of the length interval fixes the time interval of the observation. This time interval, in turn, is related to the time resolution of the detector, the measurement instrument. If the electronic bandwidth of the detector is B, then the detector can resolve changes of the photon flow within a time $T = 1/B$. This implies that the quantization of the incoming photons must choose a length L such that $L = v_g T = v_g/B$. How this assignment is to be interpreted when the optical spectrum has a bandwidth much larger than $\Delta\omega = 2\pi B$ will be discussed in greater detail further on.

It is worth pointing out that the fluctuations have arisen from the commutator of the field operators. Since the commutator is responsible for the zero-point fluctuations, one is justified in interpreting the shot noise as originating from carrier emission fluctuations induced by the zero-point fluctuations of the field. This interpretation is analogous to, yet different from, the interpretation of spontaneous emission in an amplifier as being the emission induced by the zero-point fluctuations of the field. In the amplifier, the zero-point fluctuations induce emission of photons. In the photodetector case, they only contribute to the fluctuations. Zero-point fluctuations by themselves produce no photocurrent.

8.4 Quantum Theory of Balanced Heterodyne Detection

We have presented a classical analysis of heterodyne detection and shown that it detects both the phase and the amplitude of the signal. Now we look at the quantum analysis of heterodyne detection. Consider Fig. 8.3b. The output of the local oscillator impinges upon one port of a beam splitter, the signal on the other port. The beam splitter was analyzed in Sect. 7.2. If only one pair of incident waves is involved, one need not use the full four-by-four scattering matrix; one may use the reduced two-by-two portion of the scattering matrix that is analogous to that of a mirror, (7.10). The waves $\hat{B}_1$ and $\hat{B}_2$ incident upon the two photodetectors are

$$\hat{B}_1 = \frac{1}{\sqrt{2}}(\hat{A}_L - i\hat{A}_s) ,$$

$$\hat{B}_2 = \frac{1}{\sqrt{2}}(-i\hat{A}_L + \hat{A}_s) .$$

(8.22)

The difference between the charges collected by the two detectors is

$$\hat{Q} = q(\hat{B}_1^\dagger \hat{B}_1 - \hat{B}_2^\dagger \hat{B}_2)$$

$$= \frac{q}{2}[(\hat{A}_L^\dagger + i\hat{A}_s^\dagger)(\hat{A}_L - i\hat{A}_s) - (i\hat{A}_L^\dagger + \hat{A}_s^\dagger)(-i\hat{A}_L + \hat{A}_s)]$$

(8.23)

$$= -iq(\hat{A}_L^\dagger \hat{A}_s - \hat{A}_s^\dagger \hat{A}_L) .$$

In the Heisenberg representation, the operators $\hat{A}_L$ and $\hat{A}_s$ are time-dependent, with the time dependences $\exp(-i\omega_L t)$ and $\exp(-i\omega_s t)$, respectively. The expectation value obtained by projection via the coherent states, product states of the local oscillator and signal states, gives

$$\langle \hat{Q} \rangle = -iq\langle \alpha_s|\langle \alpha_L|(\hat{A}_L^\dagger \hat{A}_s - \hat{A}_s^\dagger \hat{A}_L)|\alpha_L\rangle|\alpha_s\rangle$$

$$= 2q|\alpha_L \alpha_s| \sin[(\omega_L - \omega_s)t + \phi] ,$$

(8.24)

where $\phi = \arg(\alpha_s \alpha_L^*)$. The mean square fluctuations of the charge are

$$\langle \hat{Q}^2 \rangle - \langle \hat{Q} \rangle^2 = -\langle \alpha_s|\langle \alpha_L|q^2(\hat{A}_L^\dagger \hat{A}_s - \hat{A}_s^\dagger \hat{A}_L)(\hat{A}_L^\dagger \hat{A}_s$$

$$-\hat{A}_s^\dagger \hat{A}_L)|\alpha_L\rangle|\alpha_s\rangle$$

(8.25a)

$$+q^2\langle \alpha_s|\langle \alpha_L|q^2(\hat{A}_L^\dagger \hat{A}_s - \hat{A}_s^\dagger \hat{A}_L)|\alpha_L\rangle|\alpha_s\rangle^2 .$$

This expression is evaluated by casting the operators into normal order:

$$\langle \hat{Q}^2 \rangle - \langle \hat{Q} \rangle^2 = -\langle \alpha_s | \langle \alpha_L | q^2 (\hat{A}_L^\dagger \hat{A}_L^\dagger \hat{A}_s \hat{A}_s$$

$$- \hat{A}_L^\dagger \hat{A}_L \hat{A}_s^\dagger \hat{A}_s - \hat{A}_s^\dagger \hat{A}_s \hat{A}_L^\dagger \hat{A}_L$$

$$+ \hat{A}_s^\dagger \hat{A}_s^\dagger \hat{A}_L \hat{A}_L) | \alpha_L \rangle | \alpha_s \rangle \tag{8.25b}$$

$$+ q^2 \langle \alpha_s | \langle \alpha_L | q^2 (\hat{A}_s^\dagger \hat{A}_s + \hat{A}_L^\dagger \hat{A}_L) | \alpha_L \rangle | \alpha_s \rangle$$

$$+ q^2 \langle \alpha_s | \langle \alpha_L | (\hat{A}_L^\dagger \hat{A}_s - \hat{A}_s^\dagger \hat{A}_L) | \alpha_L \rangle | \alpha_s \rangle^2$$

$$= q^2 (|\alpha_L|^2 + |\alpha_s|^2) = q^2 (\langle n_L \rangle + \langle n_s \rangle) .$$

The first of the above expressions contains the expectation value of the normally ordered operator $\hat{Q}^2$ and that of the "remainder operator" of the normal-ordering process, minus the expectation value squared of the operator $\hat{Q}$. The expectation value of the normally ordered operator $\hat{Q}^2$ cancels the expectation value squared of the operator $\hat{Q}$. The fluctuations are due entirely to the expectation value of the "remainder operator". The fluctuations are proportional to the sum of the signal and local-oscillator photon numbers. They originate from the commutators. In the classical discussion of heterodyne detection in Sect. 8.2 we attributed all the noise to the local oscillator and ignored the signal-induced noise, which is legitimate if the signal power is much smaller than the local-oscillator power. In the present, more accurate, analysis of the heterodyne detector we find that the fluctuations $q^2 |\alpha_L|^2 = q^2 \langle n_L \rangle$ arise from the commutator of the signal field. One may interpret this term as fluctuations induced by the signal zero-point fluctuations in the charge generated by the local-oscillator photons, and the term $q^2 \langle n_s \rangle$ as the fluctuations produced by the zero-point fluctuations of the local oscillator field in the charge generated by the signal photons.

It should be pointed out that the analysis which led to (8.25) is not complete. We assumed that the input to the beam splitter consisted solely of the local-oscillator output and the signal. A detector tuned to the difference frequency $\Omega = |\omega_s - \omega_L|$ will pick up "signals" at both frequencies $\omega_L \pm \Omega$, i.e. the signal and its "image". Thus, we should have used for the "signal" operator $\hat{A}_s$ in (8.22) the sum of the signal operator $\hat{A}_s$ at frequency ω_s and the image operator $\hat{A}_i$ at frequency $\omega_i = |2\omega_L - \omega_s|$. If the image band is unexcited, then $\langle \hat{A}_i^\dagger \hat{A}_i \rangle = 0$. Yet the presence of $\hat{A}_i$ in (8.22) contributes to the fluctuations. Without writing down explicitly the extended equations (8.23)–(8.25), it is easy to see that the commutator $[\hat{A}_i, \hat{A}_i^\dagger] = 1$ doubles the contribution to the fluctuations of the local oscillator:

$$\langle Q^2 \rangle - \langle Q \rangle^2 = q^2 (2|\alpha_L|^2 + |\alpha_s|^2) . \tag{8.26}$$

8.5 Linearized Analysis of Heterodyne Detection

In the linearized approximation, the local oscillator operator in (8.22) is written as a c number,

$$\hat{A}_L \rightarrow \alpha_L \exp(-i\omega_L t) \, . \tag{8.27}$$

When the replacement (8.27) is entered into (8.23), and we take note that the image band must also be included in the analysis, we find

$$\hat{Q} = -iq[\alpha_L^* \exp(i\omega_L t)(\hat{A}_s + \hat{A}_i) - \alpha_L \exp(-i\omega_L t)(A_s^\dagger + A_i^\dagger)] \, . \tag{8.28}$$

The mean square fluctuations of charge can be obtained in the usual way by putting the creation and annihilation operators in the expression for $\hat{Q}^2$ into normal order and noting that the expectation value of the normally ordered expression cancels against $\langle \hat{Q} \rangle^2$

$$\langle \hat{Q}^2 \rangle - \langle \hat{Q} \rangle^2 = -q^2 \langle \alpha_s | \langle \alpha_i |$$

$$\{\alpha_L^2(\hat{A}_s^\dagger A_s^\dagger + \hat{A}_i^\dagger A_i^\dagger + 2\hat{A}_s^\dagger \hat{A}_i^\dagger)$$

$$+\alpha_L^{*2}(\hat{A}_s A_s + \hat{A}_i \hat{A}_i + 2\hat{A}_s \hat{A}_i)$$

$$-|\alpha_L|^2[2(\hat{A}_s^\dagger + \hat{A}_i^\dagger)(\hat{A}_s + \hat{A}_i) + 2]\}|\alpha_i\rangle|\alpha_s\rangle$$

$$+q^2 \langle \alpha_s | \langle \alpha_i | [\alpha_L^*(\hat{A}_s + \hat{A}_i) - \alpha_L(\hat{A}_s^\dagger + \hat{A}_i^\dagger)]|\alpha_s\rangle|\alpha_i\rangle^2$$

$$= 2q^2|\alpha_L|^2 = 2q^2 \langle n_L \rangle \, . \tag{8.29}$$

We have found a result like (8.26), except for the fact that the contribution to the fluctuations of the signal is missing. If the signal photon number is much smaller than that of the local oscillator, the approximation is legitimate.

In the classical analysis we evaluated the mean square fluctuations of the detector current, rather than the charge. We may convert (8.28) into a current operator by noting that the waves propagate at the group velocity v_g, that the wave packets occupy a length L, and thus that the charge per unit time, namely the current, is v_g/L times the operator (8.28):

$$\hat{i} = q\frac{v_g}{L}\hat{Q} = -iq\frac{v_g}{L}[\alpha_L^*(\hat{A}_s + \hat{A}_i) - \alpha_L(A_s^\dagger + A_i^\dagger)] \, . \tag{8.30}$$

A coherent state has the time dependence $\exp(-i\omega t) = \exp(-i\beta v_g x)$. The expectation value of the current is thus

$$\langle |\hat{i}| \rangle = \frac{2qv_g}{L}\{|\alpha_L \alpha_s| \sin[(\omega_L - \omega_s)t + \phi]\} \, , \tag{8.31}$$

where $\phi = \arg(\alpha_s \alpha_L^*)$. A display of the current on an oscilloscope would show a sinusoidal function of time, lasting a time L/v_g. Since the current operator differs from the charge operator only by a c number factor, the mean square fluctuations of the current may be evaluated in the same way as those of the charge. Accordingly,

$$\langle \hat{i}^2 \rangle - \langle \hat{i} \rangle^2 = q^2 \left(\frac{v_g}{L}\right)^2 (\langle \hat{Q}^2 \rangle - \langle \hat{Q} \rangle^2) = 2q^2 \left(\frac{v_g}{L}\right)^2 \langle n_L \rangle . \tag{8.32}$$

The quantization interval L is determined by the bandwidth of the detector $\Delta \omega = 2\pi B$. It is chosen so that $\Delta \beta = 2\pi/L = (d\beta/d\omega)\Delta \omega = \Delta \omega/v_g$, and thus

$$L = \frac{2\pi v_g}{\Delta \omega} = \frac{v_g}{B} . \tag{8.33}$$

When we introduce (8.33) into (8.32), we obtain the noise current fluctuations:

$$\langle \hat{i}^2 \rangle - \langle \hat{i} \rangle^2 = \langle i_n^2 \rangle = 2qI_L B , \tag{8.34}$$

with

$$I_L = qv_g \frac{\langle n_L \rangle}{L} , \tag{8.35}$$

the d.c. current induced by the local oscillator. Note that the quantum origin of the noise is from the commutators of the signal *and* image. We have mentioned before that this noise can be viewed as detector current fluctuations induced by the zero-point fluctuations of the signal and image.

The signal-to-noise ratio follows from the evaluation of the mean square signal current divided by the mean square noise fluctuations. The signal current is

$$\langle \hat{i}_s \rangle = -\langle \alpha_s | \langle 0_i | iq \frac{v_g}{L}$$

$$\times [\alpha_L^* \exp(i\omega_L t)(\hat{A}_s + \hat{A}_i) - \alpha_L \exp(-i\omega_L t)(\hat{A}_s^\dagger + \hat{A}_i^\dagger)] |0_i\rangle |\alpha_s\rangle$$

$$= 2q \frac{v_g}{L} |\alpha_L \alpha_s| \sin[(\omega_L - \omega_s)t + \phi] , \tag{8.36}$$

where $\phi = \arg(\alpha_s \alpha_L^*)$. The time average of the mean square current is

$$\frac{1}{T} \int_{-T/2}^{T/2} dt \, \langle \hat{i}_s \rangle^2 = 2q^2 \left(\frac{v_g}{L}\right)^2 |\alpha_s|^2 |\alpha_L|^2 , \tag{8.37}$$

and the signal-to-noise ratio is

$$\frac{S}{N} = \frac{(1/T) \int_{-T/2}^{T/2} dt \, \langle \hat{i}_s \rangle^2}{\langle i_n^2 \rangle} = \frac{2q^2 (v_g/L)^2 |\alpha_s \alpha_L|^2}{2q^2 (v_g/L)^2 |\alpha_L|^2} = |\alpha_s|^2 = \langle n_s \rangle . \quad (8.38)$$

The signal-to-noise ratio is equal to the average photon number in one observation time (inverse bandwidth).

In homodyne detection of a signal the noise decreases by a factor of two, since the idler channel merges with the signal channel and thus does not contribute zero-point fluctuations of its own. This is the quantum interpretation of homodyne detection.

Offhand, one might expect that the time average of the signal of a homodyne detector does not incur a reduction by a factor of $1/2$ as in heterodyne detection. However, one must note that the signal is independent of the local oscillator; its phase is not locked with it. From observation time to observation time its phase relative to the local oscillator changes and thus a statistical average of these phase variations will also introduce a factor of $1/2$.

We have shown in Chap. 6 that the signal-to-noise ratio after amplification with a linear amplifier of large gain is equal to the photon number received in a time interval corresponding to the inverse bandwidth. We have found the same result for heterodyne detection. Homodyne detection has twice the signal-to-noise ratio. Now, in the case of a linear amplifier we mentioned that amplification by a phase-insensitive amplifier enables an observer to measure both the in-phase and the quadrature components of the field, two noncommuting observables. The spontaneous noise added in the amplification was the penalty incurred by a simultaneous measurement of two noncommuting observables. Homodyne detection gives information only on the component of the electric field that is in phase with the local oscillator. Thus, a homodyne measurement need not incur the same penalty. Indeed, we found that the fluctuations in the homodyne measurement are just those associated with the zero-point fluctuations of the field being measured, the field having been assumed to be in a coherent state. A homodyne measurement is a noise-free measurement of the input field; no additional noise is added in the process of measurement. *It is a phase-sensitive measurement.* In Chap. 11 we shall study degenerate parametric amplification, which accomplishes noise-free measurement of one component of the input field, and find the same signal-to-noise ratio as for homodyne detection.

It is of interest to ask about the current and its mean square fluctuations in the case when the signal is in a photon number state $|n_s\rangle$. Then we find from (8.30) that $\langle |\hat{i}| \rangle = 0$. Does this mean that a display on the scope of the detector current would show no deflection other than noise? To determine this, let us ask for the mean square fluctuations. We find, in analogy with (8.32),

$$\langle \hat{i}^2 \rangle - \langle \hat{i} \rangle^2 = \langle \hat{i}^2 \rangle = q^2 \frac{v_g^2}{L^2} [n_L(n_s + 2)] . \quad (8.39)$$

These are large fluctuations, proportional to $n_s + 2$. The interpretation is simple. A photon state has a sinusoidal time dependence of the field, with an arbitrary phase. A scope display would show such a sinusoid from sample to sample, but with arbitrary shifts of phase. Hence the average of the current at any instant of time (any value of x) is zero. But the current does vary sinusoidally within each sample, and thus the mean square fluctuations are proportional to $n_s + 2$, roughly proportional to the mean square amplitude of the sinusoids.

8.6 Heterodyne Detection of a Multimodal Signal

In the preceding section we considered heterodyne detection of a sinusoidal signal. If the signal is not sinusoidal (an example is an optical pulse), then the analysis has to be generalized to include a superposition of modes, the sum of which may represent a pulse, in the same way as a Fourier superposition represents a time-dependent signal. We write for the current operator

$$
\hat{i} = -iq\frac{v_g}{L}\left[A_L^*(t)\sum_k(\hat{A}_s + \hat{A}_i)_k - A_L(t)\sum_k(\hat{A}^\dagger + \hat{A}_i^\dagger)_k \right] , \tag{8.40}
$$

where we have replaced the c number α_L of (8.27) by its time-dependent generalization

$$
A_L(t) = \sum_k \alpha_{L,k}\exp(-i\omega_{L,k}t) . \tag{8.41}
$$

We have included the same number of image modes as signal modes, since the detected image band is equal to the signal band. We shall assume that the image band is unexcited, except, of course, for its zero-point fluctuations. The expectation value of the current involves the product of the time-dependent local-oscillator and signal fields:

$$
\langle\hat{i}(t)\rangle = -iq\frac{v_g}{L}[A_L^*(t)A_s(t) - A_L(t)A_s^*(t)] , \tag{8.42}
$$

where

$$
A_s(t) = \left\langle \sum_k(\hat{A}_s)_k \right\rangle . \tag{8.43}
$$

Since the kth component of the signal has a time dependence $\exp(-i\omega_{s,k}t)$, the sum can give an arbitrary waveform.

The fluctuations are obtained by constructing $\langle\hat{i}(t)\hat{i}(t)\rangle - \langle\hat{i}(t)\rangle^2$. The operators are put into normal order using the commutators. When this is done, only the contribution of the commutators remains:

$$\langle\hat{i}(t)\hat{i}(t)\rangle - \langle\hat{i}(t)\rangle^2 = q^2\left(\frac{v_g}{L}\right)^2 |A_L(t)|^2 \sum_k ([\hat{A}_s, \hat{A}_s^\dagger] + [\hat{A}_i, \hat{A}_i^\dagger])_k$$

(8.44)

$$= 2Nq^2\left(\frac{v_g}{L}\right)^2 |A_L(t)|^2 ,$$

where N is the number of modes in the expansion of the signal. If we introduce the expression for the time-dependent current, we find

$$\langle\hat{i}(t)\hat{i}(t)\rangle - \langle\hat{i}(t)\rangle^2 = 2qI_L(t)NB ,$$

(8.45)

where $B = v_g/L$. How are we to interpret this expression? It is the shot noise formula for a time-dependent current and a bandwidth NB. Now, the definition of the bandwidth came from the quantization interval L, chosen large enough to accomodate the modes used for the quantization. If N modes participate, the waveform varies within the time interval $\Delta t = 1/NB$; the net bandwidth is increased by the factor N.

In order to find the fluctuation spectrum, we need to construct the auto-correlation function involving the average of the currents at different times, i.e. the expression $\frac{1}{2}\langle\hat{i}(t)\hat{i}(t') + \hat{i}(t')\hat{i}(t)\rangle$. The current operator was defined within the time interval $\Delta t = 1/B = L/v_g$. When the current waveforms are shifted apart by the time Δt, the fluctuations are uncorrelated and average to zero. Hence one may write

$$\frac{1}{2}\langle\hat{i}(t)\hat{i}(t') + \hat{i}(t')\hat{i}(t)\rangle - \langle\hat{i}(t)\rangle^2 = 2qI_L(t)\delta(t - t') ,$$

(8.46)

where the delta function is of magnitude $1/\Delta t$ in the time interval $|t - t'| \leq \Delta t/2$ and zero outside this time interval. The Fourier transform of this expression gives us the proper shot noise formula. If the bandwidth is increased by a factor N, the fluctuations increase by the same factor.

8.7 Heterodyne Detection with Finite Response Time of Detector

Thus far we have derived relations for the current operator of heterodyne detection without considering the finite response time of the detector. The current produced by the beat between the local oscillator and signal is unaffected by the finite response time if the beat frequency is much smaller than the inverse response time. When this is not true, then the output signal of the detector is reduced. The response times of the fastest detectors are of the order of 10 ps. Within a time interval τ of 10 ps, the optical radiation contains many cycles (of the order of 10,000). Hence the quantization interval $L < v_g\tau$ can be picked long enough that the photon concept can be applied.

If the response of the detector is limited by a simple R–C time constant, the output current operator obeys a simple linear differential equation. The solutions of linear operator equations are the same as the solutions for c-number time functions. Hence, we may describe the output current operator $\hat{I}(t)$ by the convolution integral

$$\hat{I}(t) = \int_{-\infty}^{\infty} dt'\, h(t - t')\hat{i}(t') , \tag{8.47}$$

where $h(t)$ is the detector impulse response and charge conservation dictates that $\int_{-\infty}^{\infty} dt\, h(t) = 1$. The expectation value of the current is

$$\langle \hat{I}(t) \rangle = \int_{-\infty}^{\infty} dt'\, h(t - t')\langle \hat{i}(t') \rangle . \tag{8.48}$$

The expectation value of the current was computed in (8.36). The convolution in the time domain becomes multiplication in the frequency domain. Thus, with

$$H(\omega) = \int dt \exp(i\omega t) h(t) \quad \text{and} \quad \langle \hat{i}(\omega) \rangle = \frac{1}{2\pi} \int dt \exp(i\omega t)\langle \hat{i}(t) \rangle , \tag{8.49}$$

we have from (8.48)

$$\langle \hat{I}(\omega) \rangle = H(\omega)\langle \hat{i}(\omega) \rangle . \tag{8.50}$$

In the case of two sinusoidal signals beating in the balanced heterodyne detector, the current is a sinusoid. The signal is reduced by the factor $|H(\omega_s - \omega_L)|$.

The autocorrelation function is computed analogously:

$$\frac{1}{2}\langle \hat{I}(t)\hat{I}(t') + \hat{I}(t')\hat{I}(t) \rangle$$

$$= \int_{-\infty}^{\infty} dt'' \int_{-\infty}^{\infty} dt''' \, h(t - t'')h(t' - t''')\frac{1}{2}\langle \hat{i}(t'')\hat{i}(t''') + \hat{i}(t''')\hat{i}(t'') \rangle . \tag{8.51}$$

The operators in the autocorrelation function can be put into normal order. If the excitation is by coherent states, the normally ordered part of the expression can be written as a product of expectation values. The term resulting from the commutators is derived as in (8.46). We obtain for the autocorrelation function of the current

$$\frac{1}{2}\langle \hat{I}(t)\hat{I}(t') + \hat{I}(t')\hat{I}(t)\rangle$$

$$= \int_{-\infty}^{\infty} dt'' \int_{-\infty}^{\infty} dt''' \, h(t - t'')h(t' - t''')\langle \hat{i}(t'')\rangle\langle \hat{i}(t''')\rangle \qquad (8.52)$$

$$+ 2qI_L \int_{-\infty}^{\infty} dt'' \, h(t - t'')h(t' - t'') \, .$$

The first part is the signal part; the second part gives the fluctuations. The evaluation of the spectrum of the current is left as a problem at the end of the chapter.

8.8 The Noise Penalty of a Simultaneous Measurement of Two Noncommuting Observables

The theory of quantum measurements has been extensively discussed in the literature and is still a subject of controversy. In later chapters we shall discuss the issues in greater detail and argue that there exists a self-consistent point of view on the meaning of quantum measurements and the concept of "physical reality" as raised by Einstein, Podolsky, and Rosen [73]. At this point, we have investigated a special case of a measurement apparatus, on the basis of which one may gain some insight into the the meaning of a quantum measurement.

A quantum measurement need not introduce noise, or uncertainty. An ideal photodector detects the incoming photon flux and emits carriers that can be counted. In principle, the number of incoming photons can be determined with no uncertainty. The uncertainty underlying quantum theory and stated by Heisenberg's uncertainty principle refers to the properties of the state and not directly to the measurement of the state. The ideal photodetector may be considered noise-free if applied to the measurement of photon states.

Heterodyne detection has been found not to be noise-free. Heterodyne detection permits the simultaneous measurement of the in-phase and quadrature components of an electric field and it is this property of the detector that calls for the addition of noise to the signal by the detector. A similar situation exists with linear amplifiers, which also permit such a simultaneous measurement if they possess large gain.

Arthurs and Kelly [15] addressed the issue of a simultaneous measurement of two noncommuting variables in a classic paper in 1965. They went through a detailed analysis of the coupling of a system containing the observables to a measurement apparatus, and of the measurement carried out with the apparatus. They showed that the estimation of the values of two noncommuting observables from the measurement incurred an uncertainty penalty that at least doubled the uncertainty predicted from Heisenberg's

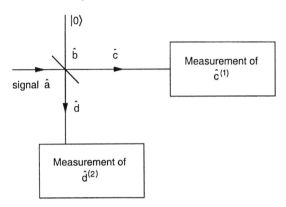

zero-point fluctuations

$|0\rangle$

$\hat{b}$ $\hat{c}$

signal $\hat{a}$

Measurement of $\hat{c}^{(1)}$

$\hat{d}$

Measurement of $\hat{d}^{(2)}$

Fig. 8.4. A beam splitter for simultaneous measurement of amplitude and phase

uncertainty principle. The optical measurements discussed thus far afford a very simple illustration of this general proof.

The homodyne detector does not add noise of its own. Its noise at the output is produced by the fluctuations of the signal. It measures a single observable, the component of the electric field in phase with the local oscillator. Both components of the electric field can be measured in a setup such as shown in Fig. 8.4. A beam splitter splits the incoming signal into two components. Two homodyne detectors measure the two components separately by adjustment of the phases of the respective local oscillators. This is an example of a simultaneous measurement.

In the following we shall show that the operators representing the in-phase and quadrature components of the signal after the beam splitter commute. Thus a simultaneous measurement of these operators is possible without the measurement of one observable affecting the measurement of the other. Physically, this is of course obvious since a measurement apparatus can intercept each of the outgoing beams c and d independently. Mathematically, the finding is of interest since it shows that the commutation has been brought about by the introduction of the vacuum fluctuations of the unused port of the beam splitter.

The beam splitter is represented by a unitary scattering matrix that is power-conserving. In the balanced heterodyne measurement apparatus, we represented the beam splitter by the unitary scattering matrix

$$\frac{1}{\sqrt{2}} \begin{bmatrix} 1 & -i \\ -i & 1 \end{bmatrix} .$$

The response of the beam splitter is thus

$$\hat{c} = \frac{1}{\sqrt{2}}(\hat{a} - i\hat{b}) , \tag{8.53}$$

$$\hat{d} = \frac{1}{\sqrt{2}}(-i\hat{a} + \hat{b}) . \tag{8.54}$$

The in-phase and quadrature operators of the beams c and d, respectively, are

$$\hat{c}^{(1)} = \frac{1}{2}(\hat{c} + \hat{c}^\dagger) = \frac{1}{2\sqrt{2}}(\hat{a} - i\hat{b} + \hat{a}^\dagger + i\hat{b}^\dagger) , \tag{8.55}$$

$$\hat{d}^{(2)} = \frac{1}{2i}(\hat{d} - \hat{d}^\dagger) = \frac{1}{2i\sqrt{2}}(-i\hat{a} + \hat{b} - ia^\dagger - \hat{b}^\dagger) . \tag{8.56}$$

The commutator of $\hat{c}^{(1)}$ and $\hat{d}^{(2)}$ is

$$[\hat{c}^{(1)}, \hat{d}^{(2)}] = \frac{1}{8}\left(-[\hat{a}, \hat{a}^\dagger] + [\hat{b}, \hat{b}^\dagger] + [\hat{a}, \hat{a}^\dagger] - [\hat{b}, \hat{b}^\dagger]\right) = 0 . \tag{8.57}$$

Indeed, the observables commute and there is now no impediment to measuring them simultaneously. What has happened is that the vacuum port of the beam splitter has introduced fluctuations (or commutators) that change the in-phase and quadrature components of the incoming signal into commuting operators. The measurement can now be carried out with two homodyne detectors independently and in a noise-free manner in each beam after the beam splitter. It is clear, however, that a noise penalty has been incurred. Only half of the original signal intensity impinges upon each of the two detectors. The signal-to-noise ratio inferred from the measurement is half of that which would have been attained if the signal impinged directly on one of the detectors.

8.9 Summary

We have studied the current induced in square-law detectors as well as photodetectors. The current is accompanied by shot noise. If shot noise is the most important noise in the detector circuit, then the signal-to-noise ratio of a heterodyne photodection circuit of unity quantum efficiency is equal to the number of photons received in a time corresponding to the inverse bandwidth of the detection. The quantum analysis used the concept of the current operator. The shot noise was interpreted in the quantum analysis as current fluctuations induced by the zero-point fluctuations of the field. The results of the quantum analysis agreed with those of the classical approach.

Balanced detection was discussed as a means of suppressing fluctuations of the local oscillator. The quantum analysis of the balanced detector led us to a surprising interpretation of the source of the shot noise: we found that the cause of the shot noise is the zero-point fluctuations of the signal and image. In this respect, the quantum picture mimics the original purpose of the balanced detector, namely cancellation of the oscillator noise. In the quantum picture we found that the zero-point fluctuations of the oscillator field can be ignored and that the entire noise excitation is attributable to the fluctuations entering the signal port of the beam splitter in the signal and image bands. In homodyne detection the signal and image bands merge and thus the noise is due solely to the zero-point fluctuations of the signal itself. One may consider the homodyne measurement to be a noise-free measurement of the incoming signal field.

Finally, we looked at the noise penalty incurred in a simultaneous measurement of the in-phase and quadrature components of the electric field, two observables with noncommuting operators. We showed that the uncertainty in the values of these two observables inferred from the measurement is double that of the Heisenberg uncertainty.

Problems

8.1 What is the value of the d.c. current producing the same mean square voltage fluctuations due to shot noise across a 50 Ω load resistor as the thermal noise at room temperature? Note that for a detector current greater than this value, the thermal noise can be neglected.

8.2* In a heterodyne receiver, the local oscillator mode at the detector has the profile $\exp(-r^2/w_L^2)\exp(i\phi r^2/w_L^2)/w_L$. The signal has the profile $\exp(-r^2/w_s^2)/w_s$. Show how the signal decreases with deviation from a perfect mode match.

8.3 A non-return to zero (NRZ) bit pattern at the optical carrier frequency ω_o is incident on a detector. NRZ implies that two "ones", represented by two rectangles (of current) that follow each other, merge into a single rectangle of twice the width.

(a) If the pattern is random, i.e. the zeros (blanks) and the ones (rectangles of height A and width τ_o) occur randomly, find the spectrum of the waveform. Under the assumption that the numbers of zeros and ones are equal on average, the average rate of carrier generation (i.e. the average rate of photons) is $\langle R(t) \rangle = (1/2)A$.
(b) Find the spectrum of the detector current.

See Appendix A.14.

8.4* A return-to-zero (RZ) bit pattern at the optical carrier frequency w_o is incident on a detector. RZ implies that the "ones" are pulses, the zeros blanks, and there is a clear separation of two ones following each other. The pulses are Gaussians $A_o \exp(-t^2/2\tau_o^2)$ and their width τ_o is 1/8 of the symbol interval.

(a) If the pattern is random, i.e. the zeros (blanks) and the ones occur randomly, find the autocorrelation function of the waveform. Make the assumption that the numbers of zeros and ones are equal on average.
(b) Find the spectrum of the detector current.

8.5* A Mach–Zehnder interferometer with two 50/50 beam splitters has a phase delay difference between the two paths of $\Delta\theta$. The output beam splitter is followed by a balanced detector.

(a) Find the output annihilation operators $\hat{c}$ and $\hat{d}$ of the Mach–Zehnder interferometer in terms of the input operators $\hat{a}$ and $\hat{b}$.
(b) If the Mach–Zehnder interferometer is excited by two coherent states $|\alpha\rangle|\beta\rangle$, find the charge collected by the detector.

8.6 A photon number state "has no phase". Yet a photon state $|n\rangle$ can interfere with itself. To show this determine the charge $\langle \hat{Q} \rangle$ collected by a balanced detector at the output of a Mach–Zehnder interferometer with the same photon state at input (a) as analyzed in the preceding problem.

8.7 Determine the charge in the setup of the preceding problem for an incident photon state $|n_k\rangle|n_\ell\rangle$, where $|n_k\rangle$ is a photon state of propagation constant β_k and $|n_\ell\rangle$ is a photon state of propagation constant β_ℓ.

Solutions

8.2 The detector, illuminated by a signal field $E_s(r)$ and a local-oscillator field $E_L(r)$, produces a current whose magnitude is proportional to

$$\left| \int dS E_s E_L^* \right|$$

$$\propto \frac{1}{w_L w_s} \left| \int_o^{2\pi} d\phi \int_o^{\infty} r\, dr \exp\left(-\frac{r^2}{w_L^2} - \frac{r^2}{w_s^2}\right) \exp\left(i\varphi \frac{r^2}{w_L^2}\right) \right| .$$

The integral gives

$$\frac{1}{w_L w_s} \int_o^{2\pi} d\phi \int_o^{\infty} r\, dr \exp\left(-\frac{r^2}{w_L^2} - \frac{r^2}{w_s^2}\right) \exp\left(i\varphi \frac{r^2}{w_L^2}\right)$$

$$= \frac{\pi}{\sqrt{(w_s/w_L + w_L/w_s)^2 + \varphi^2 (w_s/w_L)^2}} .$$

This expression has a maximum of $\pi/2$ when $\varphi = 0$ and $w_s = w_L$. Hence the deterioration of the signal is

$$\frac{2}{\sqrt{(w_s/w_L + w_L/w_s)^2 + \varphi^2 (w_s/w_L)^2}} .$$

8.4 The autocorrelation function is, classically,

$$\langle A(t)A(t + \tau) \rangle ,$$

where

$$A(t) = \sum_i A_o \exp -\frac{(t - t_i)^2}{2\tau_o^2} .$$

The ensemble average can be replaced by a time average, if the process is ergodic:

$$\langle A(t)A(t + \tau) \rangle = \frac{1}{T} \int dt\, A(t)A(t + \tau)$$

$$= \frac{1}{T} N \int dt\, A_o^2 \exp -\frac{t^2}{2\tau_o^2} \exp -\frac{(t + \tau)^2}{2\tau_o^2} ,$$

where N is the number of pulses in the long time interval T. We have for N

$$N = \frac{1}{2} \frac{T}{8\tau_o} ,$$

where the $1/2$ is due to the fact that the probability of occurrence of a pulse is $1/2$. The pulses are sufficiently well separated that overlaps can be neglected. Thus,

$$\langle A(t)A(t + \tau) \rangle = \frac{1}{16\tau_o} A_o^2 \int dt \exp -\frac{t^2 + (t + \tau)^2}{2\tau_o^2}$$

$$= \frac{1}{16\tau_o} A_o^2 \int dt \exp -\frac{t^2 + t\tau_o}{\tau_o^2} \exp -\frac{\tau^2}{2\tau_o^2}$$

$$= \frac{\sqrt{\pi}}{16} A_o^2 \exp -\frac{\tau^2}{4\tau_o^2} ,$$

$$\Phi_R(\omega) = \frac{1}{2\pi} \int dt\, e^{i\omega\tau} \frac{\sqrt{\pi}}{16} A_o^2 \exp -\frac{\tau^2}{4\tau_o^2} = \frac{\tau_o}{16} |A_o|^2 \exp -\omega^2\tau_o^2 .$$

We have for $\langle R(t) \rangle$

$$\langle R(t) \rangle = \frac{1}{T} N \int dt\, A_o \exp -\frac{t^2}{2\tau_o^2}$$

$$= \frac{1}{T} N\tau_o A_o \sqrt{2\pi} = \frac{\sqrt{2\pi}}{16} A_o .$$

8.5

(a) The operators pass through the Mach–Zehnder interferometer just like classical complex amplitudes. The scattering matrix is

$$
\frac{1}{\sqrt{2}}
\begin{bmatrix} 1 & -i \\ -i & 1 \end{bmatrix}
\begin{bmatrix} e^{i\Delta\theta/2} & 0 \\ 0 & e^{-i\Delta\theta/2} \end{bmatrix}
\frac{1}{\sqrt{2}}
\begin{bmatrix} 1 & -i \\ -i & 1 \end{bmatrix}
$$

$$
= i
\begin{bmatrix}
\sin\dfrac{\Delta\theta}{2} & -\cos\dfrac{\Delta\theta}{2} \\[2ex]
-\cos\dfrac{\Delta\theta}{2} & -\sin\dfrac{\Delta\theta}{2}
\end{bmatrix}.
$$

The output is

$$
\hat{c} = i\left[\sin\frac{\Delta\theta}{2}\hat{a} - \cos\frac{\Delta\theta}{2}\hat{b} \right]
$$

$$
\hat{d} = -i\left[\cos\frac{\Delta\theta}{2}\hat{a} + \sin\frac{\Delta\theta}{2}\hat{b} \right].
$$

(b) The detector charge difference is

$$
\langle \hat{Q} \rangle = q\langle \hat{c}^\dagger \hat{c} - \hat{d}^\dagger \hat{d} \rangle
$$

$$
= q\langle \cos\Delta\theta(\hat{b}^\dagger\hat{b} - \hat{a}^\dagger\hat{a}) - \sin\Delta\theta(\hat{a}^\dagger\hat{b} + \hat{b}^\dagger\hat{a}) \rangle .
$$

If we take the expectation value for the coherent product state we find

$$
\langle \hat{Q} \rangle = q\langle \hat{c}^\dagger\hat{c} - \hat{d}^\dagger\hat{d} \rangle = q(\cos\Delta\theta(|\beta|^2 - |\alpha|^2) - \sin\Delta\theta(\alpha^*\beta + \beta^*\alpha)) .
$$

9. Photon Probability Distributions and Bit-Error Rate of a Channel with Optical Preamplification

Thus far we have developed the equations of the evolution of operators as transformed by linear systems. We pointed out that strictly linear transformations occur only in passive circuits described by the linear Maxwell equations. Active circuits behave linearly only approximately, up to a maximum intensity. The scattering matrix of the operators of a linear system and the commutators of the noise sources completely describe the system. Average values and mean square fluctuations can be computed.

In this chapter we study the photon statistics of an attenuator represented by a beam splitter. In the process, we introduce generating functions that greatly simplify the analysis. The statistics of spontaneous emission and stimulated emission of photons in an amplifier are determined. Using this analysis we can address the practical problem of determining the bit-error rate of a digitally coded optical communication system using optical preamplification, the kind of system which gives the best high-speed performance in terms of minimum bit-error rate at a given signal power.

When the bit-error rate is very small, say less than 10^{-9}, then it is easy to make the transmission error-free through very simple error correction codes with a negligible increase in message length. For all practical purposes, one may consider the transmission to be error-free in the sense of communication theory and compare the power used in transmission with that predicted by the Shannon formula. We derive Shannon's formula for the number of photons required for transmission of a given amount of information in a noisy environment and compare the result with the performance actually achieved with simple one-bit coding and detection after optical preamplification. Current practice uses a definition of the noise figure for the characterization of optical amplifiers that is based on the mean square fluctuations of the photon number. The definition of the noise figure used in Chap. 5 is based on mean square amplitude fluctuations. We conclude the chapter with a discussion of the relation between the two definitions.

9.1 Moment Generating Functions

In our analysis, we deal with creation and annihilation operators, operators of the electric field, and photon number operators. Creation and annihilation operators, while convenient for analysis, are not Hermitian and thus are not directly observable. The field operators and the photon number operators are Hermitian, and thus observable. One may ask for the expectation values of the moments of these observables. For example, one may ask for the rth moment of the field, $\langle|\hat{E}^r|\rangle$, or the rth moment of the photon number, $\langle|(\hat{A}^\dagger\hat{A})^r|\rangle$. Here we use the creation and annihilation operators $\hat{A}^\dagger$ and $\hat{A}$ as originally defined in Chap. 6. Another important set of moments are the falling and rising factorial moments of photon number, represented by the operators $\hat{A}^{\dagger r}\hat{A}^r$ and $\hat{A}^r\hat{A}^{\dagger r}$. Consider first the expectation value of the operator $\hat{A}^{\dagger r}\hat{A}^r$. If the system is in a superposition of number states $|n\rangle$, each of the r operations by $\hat{A}$ on the ket lowers the photon number by one. A total of r such operations multiplies the final number state $|n-r\rangle$ by $\sqrt{n(n-1)\ldots(n-r+1)}$. An analogous development applies to the operation of the creation operator on the bra. The final result is a term with the bra $\langle n-r|$ in front, the ket $|n-r\rangle$ at the end and a multiplier $n(n-1)\ldots(n-r+1)$. This is the falling factorial moment of order r. Consider a state $\psi = \sum_n c_n|n\rangle$. For this state the expectation value of the falling factorial moment F_r is

$$F_r = \sum_{m=0}^{m=\infty} c_m^* \langle m|\hat{A}^{\dagger r}\hat{A}^r \sum_{n=0}^{n=\infty} c_n|n\rangle$$

$$= \sum_{m,n=0}^{m,n=\infty} c_m^\dagger c_n \sqrt{[m(m-1)\ldots(m-r+1)][n(n-1)\ldots(n-r+1)]}$$

$$\times \langle m-r|n-r\rangle$$

$$= \sum_{n=0}^{m=\infty} p(n)[n(n-1)\ldots(n-r+1)] = \langle n(n-1)\ldots(n-r+1)\rangle .$$

$$(9.1)$$

Here $|c_n|^2 = p(n)$ is the probability of n photons in the state. The complete set of moments gives full information about the probability distribution. It is convenient to construct a generating function for the falling factorial moments by the definition

$$F(\xi) = \sum_{r=0}^{r=\infty} \frac{\xi^r}{r!} F_r = \sum_{r=0}^{r=\infty} \frac{\xi^r}{r!} \langle n(n-1)\ldots(n-r+1)\rangle . \qquad (9.2)$$

The generating function $F(\xi)$ contains full information about all falling factorial moments. In order to determine the rth moment, we expand $F(\xi)$ into a Taylor series; $r!$ times the rth-order term gives the desired moment. In a similar spirit we may define the probability generating function

$$P(\xi) = \sum_{n=0}^{n=\infty} \xi^n p(n) , \qquad (9.3)$$

which contains full information about the probability distribution in its Taylor expansion. The falling-factorial-moment generating function is closely related to the probability generating function. Indeed, from its definition

$$F(\xi) = \sum_{n=0}^{n=\infty} \sum_{r=0}^{r=n} \frac{\xi^r}{r!} p(n) n(n-1) \dots (n-r+1)$$

$$= \sum_{n=0}^{n=\infty} p(n) \left(1 + \xi\right)^n = P(1 + \xi) \,.$$

(9.4)

The falling-factorial-moment generating function is equal to the probability generating function with its independent variable ξ shifted by 1. Another generating function is also of interest; this is the rising-factorial-moment generating function, defined by

$$R(\xi) = \sum_{r=0}^{r=\infty} \frac{\xi^r}{r!} \langle | \hat{A}^r \hat{A}^{\dagger r} | \rangle$$

$$= \sum_{r=0}^{r=\infty} \sum_{n=0}^{n=\infty} \frac{\xi^r}{r!} p(n)(n+1) \dots (n+r) \,.$$

(9.5)

By a manipulation similar to that used to relate the falling-factorial-moment generating function to the probability generating function, we can establish the relation between the rising-factorial-moment generating function and the probability generating function:

$$R(\xi) = \sum_{r=0}^{r=\infty} \sum_{n=0}^{n=\infty} \frac{\xi^r}{r!} p(n)(n+1) \dots (n+r)$$

$$= \frac{1}{1-\xi} \sum_{n=0}^{n=\infty} \left(\frac{1}{1-\xi}\right)^n p(n)$$

(9.6)

$$= \frac{1}{1-\xi} P\left(\frac{1}{1-\xi}\right) \,.$$

The characteristic function is the Fourier transform of the probability distribution and is directly related to the probability generating function $P(\xi)$:

$$C(\xi) = \sum_{n=0}^{n=\infty} e^{in\xi} p(n) = P(e^{i\xi}) \,.$$

(9.7)

Since Fourier transform and inverse transform relations are well known, the characteristic function $C(\xi)$ is particularly useful. Note that the characteristic function can also be written

$$C(\xi) = \langle |\exp(i\xi n)| \rangle . \tag{9.7a}$$

Two important probability distributions of photons have been encountered previously: (a) the Poisson distribution and (b) the Bose–Einstein distribution. We shall now determine their probability generating functions.

9.1.1 Poisson Distribution

The Poisson distribution has been shown to be the photon distribution of a coherent state:

$$p_P(n) = e^{-\langle n \rangle} \frac{\langle n \rangle^n}{n!} . \tag{9.8}$$

The probability generating function of the Poisson distribution is

$$P_P(\xi) = \sum_{n=0}^{n=\infty} \xi^n e^{-\langle n \rangle} \frac{\langle n \rangle^n}{n!} = e^{(\xi-1)\langle n \rangle} . \tag{9.9}$$

9.1.2 Bose–Einstein Distribution

The Bose–Einstein distribution pertains to radiation in thermal equilibrium. The photon number in each mode is governed by the probability

$$p_{B-E}(n) = \frac{1}{1 + \langle n \rangle} \left(\frac{\langle n \rangle}{1 + \langle n \rangle} \right)^n . \tag{9.10}$$

Its probability generating function is

$$
\begin{aligned}
P_{B-E}(\xi) &= \sum_{n=0}^{n=\infty} \xi^n \frac{1}{1 + \langle n \rangle} \left(\frac{\langle n \rangle}{1 + \langle n \rangle} \right)^n \\
&= \frac{1}{1 + \langle n \rangle} \sum_{n=0}^{n=\infty} \left(\frac{\xi \langle n \rangle}{1 + \langle n \rangle} \right)^n \\
&= \frac{1}{1 - \langle n \rangle (\xi - 1)} .
\end{aligned} \tag{9.11}
$$

The relationships among the various generating functions are useful and will be employed through the remainder of this chapter. Thus, (9.4) gives the generating function for falling factorial moments in terms of the probability generating function. For a Bose–Einstein distribution, we have from (9.11) and (9.4)

$$F_{B-E}(\xi) = \frac{1}{1 - \langle n \rangle \xi} . \tag{9.12}$$

Referring back to the definition of $F(\xi)$, and carrying out the expansion of (9.12), we find that, for a Bose–Einstein distribution,

$$F_r = \langle n(n-1)\dots(n-r+1) \rangle = r!\langle n \rangle^r . \tag{9.13}$$

In particular, if we take the second-order factorial moment $F_2 = 2\langle n \rangle^2 = \langle n(n-1) \rangle$ and construct from it the mean square fluctuations of n, we obtain the well-known Bose–Einstein formula:

$$\langle n^2 \rangle = \langle n \rangle + 2\langle n \rangle^2 , \tag{9.14}$$

or

$$\langle \Delta n^2 \rangle = \langle n \rangle + \langle n \rangle^2 . \tag{9.15}$$

The second term is called the classical fluctuation term, and the first term is of quantum origin. Indeed, if one identifies $\hbar\omega\langle n \rangle$ with the energy, then the mean square fluctuations of the energy are given by (9.15),

$$(\hbar\omega)^2 \langle \Delta n^2 \rangle = (\hbar\omega)^2 \langle n \rangle + (\hbar\omega)^2 \langle n \rangle^2 . \tag{9.16}$$

The second term is indeed the square of the average energy, as found from the classical field in (4.109). The first term predominates at low photon number and is of quantum origin, related to the zero-point fluctuations of the field that require a quantum treatment.

The characteristic function, the Fourier transform of the probability distribution, is useful when evaluating the "tails" of probability distributions. Indeed, we know that the behavior of a function $f(t)$ of time t at large values of t affects the behavior of its Fourier transform $f(\omega)$ at small values of ω. When evaluating error rates, we look for a proper representation of the probability distribution in the wings of the distributions. Hence, we are interested in the proper representation of the characteristic function for small values of s.

9.1.3 Composite Processes

Some statistical processes are composites of several independent processes. Thus, one may ask for the probability of n photons if two independent Poisson processes contribute to the process. To answer this question, we show first of all that the probability generating function of a composite process is the product of the generating functions of the individual processes.

Suppose process (1) and process (2) contribute n_1 and n_2 photons each, with probabilities $p_1(n_1)$ and $p_2(n_2)$, respectively. The net number of photons is $n = n_1 + n_2$. The probability of finding n photons is

$$p(n) = \sum_{n_1=0}^{n} p_1(n_1)p_2(n-n_1) . \tag{9.17}$$

It is convenient to rewrite this expression in terms of a sum that extends from minus infinity to plus infinity. This is done by defining the probability functions $p_1(n_1)$ and $p_2(n_2)$ to be zero for negative arguments:

$$p(n) = \sum_{n_1=-\infty}^{+\infty} p_1(n_1)p_2(n-n_1) \tag{9.17a}$$

Now, consider the generating function $P(n)$ of $p(n)$:

$$P(\xi) = \sum_{n=-\infty}^{\infty} \xi^n p(n) = \sum_{n=-\infty}^{+\infty} \sum_{n_1=-\infty}^{+\infty} \xi^{n_1} p_1(n_1) \xi^{n-n_1} p_2(n-n_1)$$

$$\tag{9.18}$$

$$= \sum_{n_1=-\infty}^{\infty} \xi^{n_1} p_1(n_1) \sum_{n_2=-\infty}^{+\infty} \xi^{n_2} p_2(n_2) = P_1(\xi)P_2(\xi) \,.$$

Thus, we have shown that the generating function of the composite process is the product of the generating functions of the individual processes. This result can be used effectively. We first show that a process that is a composite process of two Poisson processes is also a Poisson process. For this purpose we apply (9.18) to (9.8):

$$P(\xi) = P_1(\xi)P_2(\xi) = \exp[-(\xi-1)\langle n_1\rangle] \exp[(\xi-1)\langle n_2\rangle]$$

$$\tag{9.19}$$

$$= \exp[(\xi-1)(\langle n_1\rangle + \langle n_2\rangle)] \,.$$

The compound generating function has a Poissonian dependence with the average photon number equal to the sum of the average photon numbers of each of the two processes.

Another interesting example is the generating function of a composite Bose–Einstein distribution. If the composite process is composed of g independent processes, all with the same average photon number $\langle n\rangle$, then it is said that the Bose–Einstein process is g-fold degenerate. Its generating function is, from (9.10) and a simple generalization of (9.18),

$$P(\xi) = \left(\frac{1}{1 - \langle n\rangle(\xi-1)}\right)^g \,. \tag{9.20}$$

The mean square fluctuations are found from the first and second derivatives of the probability generating functions

$$\frac{d}{d\xi}P(\xi)_{|\xi=1} = \sum_{n=0}^{n=\infty} n\xi^{n-1} p(n)_{|\xi=1} = \sum_{n=0}^{n=\infty} n p(n)_{|\xi=1} = g\langle n\rangle \tag{9.21}$$

and

$$\frac{d^2}{d\xi^2}P(\xi)_{|\xi=1} = \sum_{n=0}^{n=\infty} n(n-1)\xi^{n-2}p(n)_{|\xi=1}$$

$$= \sum_{n=0}^{n=\infty} n(n-1)p(n) \tag{9.22}$$

$$= g(g+1)\langle n\rangle^2 .$$

The mean square fluctuations are

$$\langle \Delta n^2 \rangle = \langle n^2 \rangle - \langle n \rangle^2 = g\langle n\rangle(1 + \langle n\rangle) . \tag{9.23}$$

Remember that $\langle n \rangle$ is the average photon number of each individual Bose–Einstein process. Here, $g\langle n \rangle$ is the average photon number, and $\langle n \rangle$ is the average photon number divided by g. Degenerate Bose–Einstein fluctuations are smaller than the fluctuations of the nondegenerate case of the same average photon number. The same result (9.23) could have been obtained by noting that the mean square fluctuations of independent processes add:

$$\langle \Delta n^2 \rangle = \sum_{j=1}^{j=g} \langle \Delta n_j^2 \rangle = \sum_{j=1}^{j=g} \langle n_j \rangle(1 + \langle n_j \rangle) = g\langle n_j \rangle(1 + \langle n_j \rangle) .$$

If $\langle n \rangle(\xi - 1) \ll 1$, then the probability generating function reduces to

$$P(\xi) = [1 - \langle n\rangle(\xi - 1)]^{-[1/\langle n\rangle(\xi-1)]g\langle n\rangle(\xi-1)} = \exp[g\langle n\rangle(\xi - 1)] . \tag{9.24}$$

This is the generating function of a Poisson distribution with average photon number $g\langle n \rangle$. Thus, we find that a high degeneracy transforms the Bose–Einstein statistics of a process with $\langle n \rangle \ll 1$ into Poisson statistics. This is the reason why detection of light from an incandescent lamp with a photodetector leads to Poisson distributed carriers, since the coherence time τ of the light (of the order of femtoseconds) is much shorter than the inverse bandwidth $1/B$ of the photodetector. The degeneracy factor of the radiation is $g = 1/(B\tau)$. If the photon number per mode is not much smaller than unity, then, according to (9.23), $\langle \Delta n^2 \rangle = g\langle n\rangle(1+\langle n\rangle) > g\langle n\rangle$, and the mean square fluctuations remain larger than Poissonian.

9.2 Statistics of Attenuation

In Chap. 7, we evaluated the field fluctuations produced by attenuation and amplification. The fluctuations of the incident photons determine, at least in part, the noise accompanying detection. In preparation for the study of such detection systems we now concentrate on the photon fluctuations as influenced by attenuation and by amplification. For the evaluation of photon statistics, it is convenient to revert to annihilation and creation operators of photons within a certain time interval, which we indicate with capital letters. We consider only forward-propagating waves (compare Sect. 6.10)

$$\hat{B} = \sqrt{\mathcal{L}}\hat{A} + \hat{N} \; . \tag{9.25}$$

The commutator of the noise source is chosen so that commutators are preserved in passage through the attenuator:

$$[\hat{N}, \hat{N}^\dagger] = 1 - \mathcal{L} \; . \tag{9.26}$$

If the states of the noise operator are in the ground state, the falling-factorial-moment of the output photons has no contribution from the loss reservoir [16, 74]. Indeed

$$F_r = \langle |n_b(n_b - 1) \ldots (n_b - r + 1)| \rangle = \langle |\hat{B}^{\dagger r} \hat{B}^r| \rangle$$
$$= \langle |(\sqrt{\mathcal{L}}\hat{A}^\dagger + \hat{N}^\dagger)^r (\sqrt{\mathcal{L}}\hat{A} + \hat{N})^r| \rangle = \mathcal{L}^r \langle |\hat{A}^{\dagger r} \hat{A}^r| \rangle \; . \tag{9.27}$$

Suppose that the input is in a photon eigenstate $|n_a\rangle$. Then the factorial moment of the output photons is

$$F_{n_b}(\xi) = \sum_{r=0}^{n_a} \frac{\xi^r}{r!} \mathcal{L}^r n_a(n_a - 1) \ldots (n_a - r + 1) = (1 + \xi\mathcal{L})^{n_a} \; . \tag{9.28}$$

The generating function of the probability distribution function is, according to (9.4),

$$P_{n_b}(\xi) = F_{n_b}(\xi - 1) = [1 + \mathcal{L}(\xi - 1)]^{n_a} \; . \tag{9.29}$$

The probability of transmission of n_b photons with n_a incident photons is obtained from the coefficient of mth order in ξ in an expansion of the generating function $P_{n_b}(\xi)$ in powers of ξ:

$$p(n_b|n_a) = \binom{n_a}{n_b} \mathcal{L}^{n_b} (1 - \mathcal{L})^{n_a - n_b} \; . \tag{9.30}$$

We wrote the probability (9.30) as a conditional probability of n_b output photons for n_a input photons. This is the probability of a binomial process. The photons are passed to the output port with probability $\mathcal{L}$ and lost with probability $1 - \mathcal{L}$. The first photon is picked in n_a ways, the second in $n_a - 1$ ways, up to the last, which is picked in $n_a - n_b + 1$ ways. Since the photons are indistinguishable, the probability has to be divided by $n_b!$ It is rather remarkable that the statistics of the binomial photon distribution for an input photon state $|n_a\rangle$ are the result of the noise source, even though, when asking for the falling-factorial-moment generating function, no explicit mention was made of the noise source. Had we asked for the rising-factorial-moment generating function, the noise source would have entered into the computation.

Next, we prove that a Bose–Einstein distribution passing through an attenuator at zero temperature remains Bose–Einstein. This makes physical sense, since the Bose–Einstein distribution is the thermal distribution. A thermal excitation passing through an attenuator at zero temperature must emerge as a thermal distribution at a lower temperature. The generating function of the Bose–Einstein distribution is, from (9.11),

$$P_{B-E}(\xi) = \frac{1}{1 - \langle n \rangle (\xi - 1)} \ . \tag{9.31}$$

Consider next the generating function of the probability distribution $p(n_b)$ at the output of an attenuator, when a Bose–Einstein distribution is fed into the input:

$$
\begin{aligned}
P(\xi) &= \sum_{n_a=0}^{\infty} \sum_{n_b=0}^{\infty} \xi^{n_b} p(n_b | n_a) p_{B-E}(n_a) \\
&= \sum_{n_a=0}^{\infty} P_{n_b}(\xi) \frac{1}{1 + \langle n_a \rangle} \left(\frac{\langle n_a \rangle}{1 + \langle n_a \rangle} \right)^{n_a} \\
&= \sum_{n_a=0}^{\infty} [1 + \mathcal{L}(\xi - 1)]^{n_a} \frac{1}{1 + \langle n_a \rangle} \left[\frac{\langle n_a \rangle}{1 + \langle n_a \rangle} \right]^{n_a} \\
&= \frac{1}{1 - \mathcal{L} \langle n_a \rangle (\xi - 1)} \ .
\end{aligned}
\tag{9.32}
$$

This is the generating function for a Bose–Einstein distribution with an average photon number of $\mathcal{L} \langle n_a \rangle$.

Next, we prove that a Poissonian distribution remains Poissonian after passing through an attenuator. By a method analogous to the derivation for the Bose–Einstein distribution, we obtain

$$
\begin{aligned}
P(\xi) &= \sum_{n_a=0}^{\infty} \sum_{n_b=0}^{\infty} \xi^{n_b} p(n_b | n_a) p_P(n_a) \\
&= \sum_{n_a=0}^{\infty} P_{n_b}(\xi) e^{-\langle n_a \rangle} \frac{\langle n_a \rangle^{n_a}}{n_a!} \\
&= \sum_{n_a=0}^{\infty} [1 + \mathcal{L}(\xi - 1)]^{n_a} e^{-\langle n_a \rangle} \frac{\langle n_a \rangle^{n_a}}{n_a!} = e^{(\xi - 1)\mathcal{L}\langle n_a \rangle} \ .
\end{aligned}
\tag{9.33}
$$

This is the generating function of a Poisson process with average photon number $\mathcal{L} \langle n_a \rangle$.

The important conclusion is that attenuated Poisson and Bose–Einstein distributions remain Poisson and Bose–Einstein distributions, respectively. If, for example, the photons in the pulses of a signal source in a communication system are Poisson-distributed, after attenuation they arrive Poisson-distributed at the receiver.

9.3 Statistics of Optical Preamplification with Perfect Inversion

Optical digital communication uses pulses to represent a "one" and empty time invervals to represent a "zero". The simplest means of detection is direct detection; the received photons are converted into charge carriers. In principle, photodetectors could be noise-free, their current could be a perfect replica of the incident photon flow. In practice, the detectors are noisy. For this reason, the most sensitive high-bit-rate optical receivers use optical preamplification.

Preamplification can remove, or reduce, the influence on the signal-to-noise ratio of the components following the preamplification, including the detector. In this section we study the probability distribution of the photons at the output of an optical preamplifier. The analysis is simple if we assume that, through filtering, only noise within the signal bandwidth is passed on to the receiver. In other words, the photons of spontaneous emission and those of the signal have the same bandwidth. The response of the amplifier is

$$\hat{B} = \sqrt{G}\hat{A} + \hat{N} , \tag{9.34}$$

where the commutator of the noise source required for the conservation of the commutator from input to output is

$$[\hat{N}, \hat{N}^\dagger] = 1 - G < 0 . \tag{9.35}$$

Thus, $\hat{N}$ has to be interpreted as a creation operator and $\hat{N}^\dagger$ as an annihilation operator. If the gain medium is not perfectly inverted, as discussed in the next section, $\hat{N}$ is a superposition of a creation operator and an annihilation operator. The amplifier noise is determined by the state of the noise source. The state of the lasing level can be considered to be the ground state, if it is equilibrated by a reservoir at or near room temperature.

With the noise source in the ground state, the rising factorial moments of the output photon number do not contain a noise source contribution. Thus, it is convenient to determine the rising factorial moment for the evaluation of the probability generating function [16,74]. We start with the rising-factorial-moment generating function for a photon state input with photon number n_a:

$$R(\xi) = \sum_{r=0}^{\infty} \frac{\xi^r}{r!} \langle | \hat{B}^r \hat{B}^{\dagger r} | \rangle = \sum_{r=0}^{\infty} \frac{\xi^r}{r!} G^r \langle | \hat{A}^r \hat{A}^{\dagger r} | \rangle$$

$$= \sum_{r=0}^{\infty} \frac{\xi^r}{r!} G^r (n_a + 1) \dots (n_a + r) = \left(\frac{1}{1 - \xi G} \right)^{n_a + 1} . \tag{9.36}$$

The probability generating function of the output photons n_b for an input photon state n_a is obtained from the rising-factorial-moment generating function through the relation (9.6):

$$P_{n_b}(\xi) = \frac{1}{\xi} R\left(1 - \frac{1}{\xi}\right) = \frac{1}{G - \xi(G-1)} \left[\frac{\xi}{G - \xi(G-1)} \right]^{n_a} . \tag{9.37}$$

Figure 9.1a shows a probability distribution for $n_a = 5$ and $G = 10$. The contribution of ASE is to broaden the distribution for the output photons. Figure 9.1b shows the distribution for $n_a = 20$ and $G = 10$. The distribution is more symmetric than that for the lower input photon number.

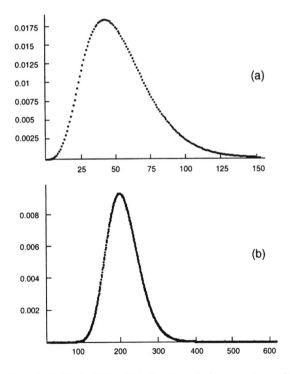

Fig. 9.1. Probability distribution of photons at amplifier output: (**a**) $n_a = 5$, $G = 10$; (**b**) $n_a = 20$, $G = 20$

The probability distribution represented by the generating function (9.37) has a simple interpretation. First of all, note that it is composed of two factors. Since the generating function of the probability distribution of two independent statistical processes is equal to the product of the generating functions of the individual processes, we may disassemble the product, and look at one term at a time and study its statistical properties. Consider first the generating function of photons induced by input photons:

$$P_1(\xi) = \left[\frac{\xi}{G - (G-1)\xi} \right]^{n_a}$$

$$= \xi^{n_a} \sum_{m=0}^{\infty} \left(\frac{n_a(n_a+1)\ldots(n_a+m-1)}{m!} \right) p^m q^{n_a+m} \xi^m , \qquad (9.38)$$

where

$$p = 1 - \frac{1}{G}, \quad q = \frac{1}{G}.$$

The coefficient of the term ξ^{n_a+m} gives the probability of $n_a + m$ output photons. This probability describes a process with its lowest photon number equal to n_a. The photon numbers are obtained by a game of chance with the following rules [74]. The probability of obtaining one more photon with a starting number of n_a photons is the product of the probability $p = 1 - 1/G$ of generating a photon, and the probability $q = 1/G$ of not generating a photon to the power $n_a + 1$, times the number of starting photons. The probability of two additional output photons is, analogously, equal to the probability that one photon has been generated times the probability that the $n_a + 1$ photons will generate one more. Since the two generated photons are indistinguishable, division by 2 is necessary; and so on.

The second statistical process is represented by the generating function

$$P_2(\xi) = \frac{1}{G - (G-1)\xi} = \sum_{m=0}^{\infty} p^m q \xi^m . \qquad (9.39)$$

This process can be described by a game in which a photon is borrowed from the "bank". The game of chance described above is carried out. After the game, the borrowed photon is returned to the "bank".

Consider first the case of zero photon input. Then

$$P(\xi) = \frac{1}{G - \xi(G-1)} . \qquad (9.40)$$

This is the probability generating function of a Bose–Einstein distribution with average photon number $G - 1$. Thus, an amplifier with no input produces a Bose–Einstein distribution, if the observation time is adjusted to fit the inverse filter bandwidth. One may characterize the output by replacing the internally generated noise by a source at the input. If we do this in the present case, then the equivalent source corresponds to one photon per mode. Now, we remember that the power of the equivalent input noise source corresponds to the excess noise temperature. For the erbium amplifier wavelength of 1.54 μm, $\hbar\omega B$ expressed as a noise temperature corresponds to a temperature of 12,000 K.

Consider next the generating function for the output photons of an amplifier with a Poissonian input. We have

$$P_P(\xi) = \sum_{n_b=0}^{\infty} \sum_{n_a=0}^{\infty} \xi^{n_b} p(n_b|n_a) e^{-\langle n_a \rangle} \frac{\langle n_a \rangle^{n_a}}{n_a!}$$

$$= \sum_{n_a=0}^{\infty} P_{n_b}(\xi) e^{-\langle n_a \rangle} \frac{\langle n_a \rangle^{n_a}}{n_a!} \tag{9.41}$$

$$= \frac{1}{G - (G-1)\xi} \exp\left[\frac{\langle n_a \rangle \xi}{G - (G-1)\xi} - \langle n_a \rangle\right] .$$

The generating function can be used to to derive the first and second moments of the output photon distribution. The first-order moment is

$$\langle n_b \rangle = \frac{d}{d\xi} P_P(\xi)|_{\xi=1} = G\langle n_a \rangle + G - 1 . \tag{9.42}$$

The average output photon number consists of the amplified average input photon number and the average spontaneously emitted and amplified photon number. The second falling-factorial-moment is obtained from the second derivative of the probability generating function:

$$\langle n_b(n_b - 1) \rangle = \frac{d^2}{d\xi^2} P_P(\xi)_{\xi=1}$$

$$= 2(G-1)^2 + 4G(G-1)\langle n_a \rangle + G^2 \langle n_a \rangle^2 . \tag{9.43}$$

The mean square fluctuations are

$$\langle \Delta n_b^2 \rangle = G\langle n_a \rangle + 2G(G-1)\langle n_a \rangle + G(G-1) . \tag{9.44}$$

This expression has a simple physical meaning. The mean square fluctuations consist, in part, of the Poisson fluctuations of the signal photon number, $G\langle n_a \rangle$, and in part of the Bose–Einstein fluctuations, $G(G-1)$. Finally, there is an "interference term" of the Bose–Einstein and amplified Poisson fluctuations, $2G(G-1)\langle n_a \rangle$. Note that the noise due to the interference term is much larger than the Poisson value. In the case of large input photon number, $\langle n_a \rangle \gg 1$, and large gain, the fluctuations at the output are entirely due to the interference term: $\langle \Delta n_b^2 \rangle \approx 2G^2 \langle n_a \rangle$.

Finally, consider the generating function in the case when the input is Bose–Einstein distributed. We show that the output is also Bose–Einstein distributed with a new average photon number of $G\langle n_a \rangle + G - 1$:

$$
\begin{aligned}
P_{B-E}(\xi) &= \sum_{n_a=0}^{\infty} P_{n_b}(\xi) \frac{1}{1 + \langle n_a \rangle} \left(\frac{\langle n_a \rangle}{1 + \langle n_a \rangle} \right)^{n_a} \\
&= \sum_{n_a=0}^{\infty} \frac{1}{G - \xi(G-1)} \left[\frac{\xi}{G - \xi(G-1)} \right]^{n_a} \frac{1}{1 + \langle n_a \rangle} \left(\frac{\langle n_a \rangle}{1 + \langle n_a \rangle} \right)^{n_a} \\
&= \frac{1}{G - \xi(G-1)} \frac{1}{1 + \langle n_a \rangle} \sum_{n_a=0}^{\infty} \left[\frac{\xi}{G - \xi(G-1)} \frac{\langle n_a \rangle}{1 + \langle n_a \rangle} \right]^{n_a} \\
&= \frac{1}{G - \xi(G-1)} \frac{1}{1 + \langle n_a \rangle} \frac{1}{1 - \{\xi/[G - \xi(G-1)]\}[\langle n_a \rangle/(1 + \langle n_a \rangle)]} \\
&= \frac{1}{1 - [G\langle n_a \rangle + G - 1](\xi - 1)} \, .
\end{aligned}
\tag{9.45}
$$

Thus, amplification of a Bose–Einstein-distributed photon flow by an amplifier with perfect inversion maintains the Bose–Einstein distribution.

Another important scenario must be included. Thus far we have considered the nondegenerate case, i.e. the signal and the spontaneous emission belong to one single mode. Realized experimentally, this means that a polarizer and an optical filter are used at the amplifier output. The polarizer is aligned with the signal polarization, and the filter passes the signal and the spontaneous-emission photons within the same spectral window. When the filter bandwidth B_f of the optical filter following the amplifier is wider than the signal bandwidth B_s, the noise radiation consists of a set of modes, one each assigned to every spectral slot of width B_s; the total number of modes is $g = B_f/B_s$, with g the so-called degeneracy factor. In this case the spontaneous emission is degenerate. In the absence of a polarizer, $g = 2B_f/B_s$. It is of interest to determine the generating function for the degenerate Bose–Einstein case. More spontaneously emitted photons pass the filter. As long as the detector detects all these photons, they are indistinguishable from the photons within the signal bandwidth and appear as charge carriers in the detector output. We have seen earlier that the probability generating function for n_a incident photons contains a signal part and a spontaneous emission part. When applied to the present case of a broadband filter it becomes

$$
P_{n_a}(\xi) = \left[\frac{1}{G - (G-1)\xi} \right]^g \left[\frac{\xi}{G - (G-1)\xi} \right]^{n_a} .
\tag{9.46}
$$

When the input signal photons are Poisson distributed, the generating function becomes (compare (9.41))

$$P_{P,\text{deg}}(\xi) = \left[\frac{1}{G - (G - 1)\xi} \right]^g \exp \left[\frac{G\langle n_a\rangle\xi}{G - (G - 1)\xi} - \langle n_a\rangle \right]. \tag{9.47}$$

As before, we may evaluate the first- and second-order moments of the output photons, with the result

$$\langle n_b \rangle = G\langle n_a\rangle + g(G - 1) \tag{9.48}$$

and

$$\langle n_b^2 \rangle - \langle n_b \rangle^2 = G\langle n_a\rangle + 2G(G - 1)\langle n_a\rangle + gG(G - 1). \tag{9.49}$$

The amplified signal fluctuations and the coherence term have not changed. Only the spontaneous-emission fluctuations have increased by the factor g. This expression can also be interpreted in the frequency domain. The degeneracy factor represents the bandwidth. The contribution of the spontaneous emission increases linearly with an increase in bandwidth. This is a property of white noise (frequency-independent spectral density).

We have shown that the replacement of the variable ξ in the probability generating function $P(\xi)$ by $\exp(is)$ transforms it into the characteristic function $C(s)$, which is the Fourier transform of the probability distribution. Thus, we may obtain the probability distribution as the inverse Fourier transform of $C(s)$. In this way we find that the probability distribution of a degenerate Bose–Einstein distribution is

$$p(n)_{\text{B-E, deg}} = \frac{\Gamma(n + g)}{\Gamma(n + 1)\Gamma(g)}(1 + \langle n\rangle)^{-g} \left(1 + \frac{1}{\langle n\rangle} \right)^{-n}, \tag{9.50}$$

where Γ is the gamma function. Figure 9.2 shows some of these distributions.

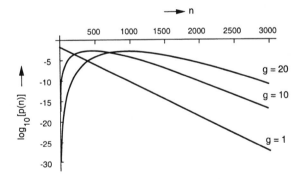

Fig. 9.2. The degenerate Bose–Einstein probability distributions for $G - 1 = 50$, and $g = 1$, 10, and 20

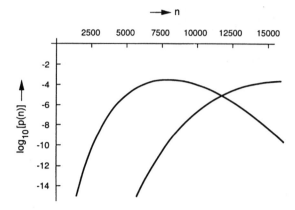

Fig. 9.3. The probability distribution for inputs of $\langle n_a \rangle = 30$ and $\langle n_a \rangle = 160$; $G - 1 = 99$

The probability distribution of a degenerate Bose–Einstein distribution convolved with the Poissonian distribution of an input signal of average photon number $\langle n_a \rangle$ and represented by the generating function (9.47) can also be written down in closed form [76]:

$$p_{\mathrm{P,deg}}(n) = \frac{(G - 1)^n}{G^{n+g}} \exp(-\langle n_a \rangle) L_n^{g-1}\left(-\frac{\langle n_a \rangle}{G - 1}\right) . \tag{9.51}$$

Here L_n^{g-1} is the generalized Laguerre polynomial. This distribution is shown in Fig. 9.3.

9.4 Statistics of Optical Preamplification with Incomplete Inversion

Next we consider the general case of an incompletely inverted medium with a filter of bandwidth wider than the signal bandwidth. Some aspects of the problem are self-evident without a detailed analysis. First of all, an amplifier with incomplete inversion is a system of amplifying layers, representing the upper level, and attenuating layers, representing the absorption by the lower level. The spontaneous-emission noise is still of Bose–Einstein nature, since the amplifying layers produce Bose–Einstein distributed ASE and we have proven that an absorber preserves a Bose–Einstein distribution. If the optical bandwidth is wider than the inverse observation time, the noise has a degenerate Bose–Einstein distribution.

Hence, the analysis of the bit-error rate in the detection of zeros does not change in essence; only the noise levels have to be reevaluated. Of interest is what happens to the interference term, which we determined for the fully inverted medium. More generally, is it possible to derive the full probability distribution of photons for this case? The answer is yes, and in fact the analysis is quite simple if we approach the problem with a little ingenuity.

We have shown in Chap. 6 that the equations for an amplifier with incomplete inversion are (6.137).

$$\hat{B} = \sqrt{G}\hat{A} + \hat{N}_u + \hat{N}_\ell , \tag{9.52}$$

with the commutation relations

$$[\hat{N}_u, \hat{N}_u^\dagger] = \chi(1 - G) \quad \text{and} \quad [\hat{N}_\ell, \hat{N}_\ell^\dagger] = (\chi - 1)(G - 1) . \tag{9.53}$$

The system contains two noise sources, one representing the gain of the upper level, the other the loss caused by absorption by the lower level. The very same equations can be obtained from a system consisting of an absorber of loss $\mathcal{L}_o$, with the equation

$$\hat{B} = \sqrt{\mathcal{L}_o}\hat{A} + \hat{N}_L^o , \tag{9.54}$$

where

$$[\hat{N}_L^o, \hat{N}_L^{o+}] = (1 - \mathcal{L}_o) , \tag{9.55}$$

followed by a gain system of gain G_o, obeying the equations

$$\hat{B} = \sqrt{G_o}\hat{A} + N_G^o , \tag{9.56}$$

where the noise source has the commutation relation

$$[\hat{N}_G^o, \hat{N}_G^{o+}] = (1 - G_o) . \tag{9.57}$$

The combined system has a net gain $G = \mathcal{L}_o G_o$ and obeys the equation

$$\hat{B} = \sqrt{G}\hat{A} + \hat{N}_u + \hat{N}_\ell \quad \text{with} \quad \hat{N}_u = \hat{N}_G^o \quad \text{and} \quad \hat{N}_\ell = \sqrt{G_o}\hat{N}_L^o . \tag{9.58}$$

For any value of the inversion parameter χ and gain G, we may choose a cascade of a loss section and a completely inverted gain section that reproduces the noise source commutators, by choosing

$$G_o = \chi(G - 1) + 1 \quad \text{and} \quad \mathcal{L}_o = \frac{G}{\chi(G - 1) + 1} . \tag{9.59}$$

But this means that we may use the previously derived results for the probability generating function to arrive at the probability generating function for this general case. We note, first of all, that an attenuator section preceding the amplifier still feeds no more than zero-point fluctuations into the amplifier, if the system is unexcited, and Poisson-distributed photons, if it is excited by a Poisson process of average photon number $\langle n_a \rangle$. The average photon number fed into the amplifier is $\mathcal{L}_o \langle n_a \rangle$. Thus, borrowing the probability density generating function (9.46), we find for this case

$$P(\xi) = \left[\frac{1}{G_o - \xi(G_o - 1)}\right]^g \exp\left[\frac{\xi \mathcal{L}_o \langle n_a \rangle}{G_o - \xi(G_o - 1)} - \mathcal{L}_o \langle n_a \rangle\right]$$

$$= \left[\frac{1}{\chi(G-1) + 1 - \xi\chi(G-1)}\right]^g$$

(9.60)

$$\times \exp\left[\frac{\xi G \langle n_a \rangle}{[\chi(G-1) + 1][\chi(G-1) + 1 - \xi\chi(G-1)]}\right.$$

$$\left. - \frac{G}{\chi(G-1) + 1}\langle n_a \rangle\right].$$

From the probability generating function, we may evaluate the first and higher-order moments of the output photon number:

$$\langle n_b \rangle = \frac{d}{d\xi}P(\xi)_{|\xi=1} = g(G_o - 1) + \mathcal{L}_o G_o \langle n_a \rangle = g\chi(G-1) + G\langle n_a \rangle.$$

(9.61)

This is an expected result. The ASE photon number is enhanced by the factor χ, and the input photons are amplified by the factor G. The mean square photon number fluctuations may be evaluated directly from (9.49) by proper identification of parameters:

$$\langle \Delta n_b^2 \rangle = \mathcal{L}_o G_o \langle n_a \rangle + 2\mathcal{L}_o G_o(G_o - 1)\langle n_a \rangle + gG_o(G_o - 1)$$

(9.62)

$$= G\langle n_a \rangle + 2\chi G(G-1)\langle n_a \rangle + g\chi(G-1)[\chi(G-1) + 1].$$

Note that the mean square fluctuations have the expected form. First of all, the last term corresponds to the degenerate Bose–Einstein mean square fluctuations

$$\langle \Delta n^2 \rangle_{B-E} = g\langle n \rangle(1 + \langle n \rangle),$$

(9.63)

with $\langle n \rangle = \chi(G-1)$. Further, the interference term does not contain the degeneracy factor, since interference with the signal occurs only within the signal bandwidth. The interference term is enhanced by the noise enhancement factor χ. The first term represents the amplified fluctuations of the input Poisson process.

The probability distribution for the incompletely inverted medium can be obtained from (9.51) using the same simple substitution method. We know that a cascade of a lossy section with a gain section reproduces completely the governing equations. The output of this structure is produced by a Poissonian input into the gain section with the average photon number $\mathcal{L}_o \langle n_a \rangle$. The gain section with a perfectly inverted medium produces a spontaneous emission of average photon number $g(G_o - 1)$. Thus, we may reuse (9.51) and express the result in terms of the actual gain G and the noise enhancement factor χ:

$$p_{P,deg}(n) = \frac{(G_o - 1)^n}{G_o^{n+g}} \exp(-\mathcal{L}_o\langle n_a\rangle) L_n^{g-1}\left(\frac{\mathcal{L}_o\langle n_a\rangle}{G_o - 1}\right)$$

$$= \frac{[\chi(G-1)]^n}{[\chi(G-1)+1]^{n+g}} \exp\left(-\frac{G\langle n_a\rangle}{\chi(G-1)+1}\right) \qquad (9.64)$$

$$\times L_n^{g-1}\left(\frac{G\langle n_a\rangle}{\chi(G-1)[\chi(G-1)+1]}\right).$$

The derivations thus far are rigorously quantum mechanical. If the signal and noise involve many photons, we expect that the results should approach a classical limit. We now derive the classical limit for the noise and signal intensities. We introduce the classical complex amplitude and the fluctuations around the amplitude δA. In order to define the problem, we have to assign a probability distribution to δA. This is done on the basis of what we have learned from the quantum analysis. We have shown that the photon distribution of amplified spontaneous emission is a Bose–Einstein distribution. The *field* of amplified spontaneous emission is Gaussian-distributed. Thus, amplified spontaneous emission is represented by a δA that is Gaussian-distributed. In order to compare the classical results with the quantum analysis, we still adhere to the normalization in which $|A_o|^2$ represents the photon number. This means, of course, that $|A_o|$ is a very large number. The detector current is equal to the absolute square of the complex amplitude. The detector current is

$$|A_o + \delta A|^2 = |A_o|^2 + A_o\delta A^* + A_o^*\delta A + |\delta A|^2 . \qquad (9.65)$$

It is convenient to fix the phase of the signal so that A_o is real. Then (9.65) becomes

$$|A_o + \delta A|^2 = |A_o|^2 + 2A_o\mathrm{Re}(\delta A) + |\delta A|^2 . \qquad (9.66)$$

The expectation value is

$$\langle |A_o + \delta A|^2\rangle = \langle |A_o|^2\rangle + \langle |\delta A|^2\rangle . \qquad (9.67)$$

Comparison with the quantum result gives

$$\langle |A_o|^2\rangle = G\langle n_a\rangle \quad \text{and} \quad \langle |\delta A|^2\rangle = G - 1 , \qquad (9.68)$$

with $\langle \mathrm{Re}(\delta A)^2\rangle = \frac{1}{2}(G-1)$. If the signal and noise are statistically independent, the mean square power fluctuations are

$$\langle |A_o + \delta A|^4\rangle - \langle |A_o + \delta A|^2\rangle^2$$

$$\qquad (9.69)$$

$$= \langle |A_o^4|\rangle - \langle |A_o^2|\rangle^2 + 4\langle |A_o^2|\rangle\langle |\mathrm{Re}(\delta A)^2|\rangle + \langle |\delta A|^4\rangle - \langle |\delta A|^2\rangle^2 .$$

Now, $\langle|\delta A|^2\rangle = \langle\mathrm{Re}(\delta A)^2\rangle + \langle\mathrm{Im}(\delta A)^2\rangle = G - 1$, and the real and imaginary parts of the fluctuations have equal magnitude. If we ignore the signal fluctuations and fourth-order terms we obtain

$$\text{mean square power fluctuations} = 2G(G - 1)\langle n_a\rangle . \qquad (9.70)$$

This is the interference term in (9.44) obtained from the full quantum analysis. Incidentally, the classical analysis also explains the designation "interference term", since this term arose as the product of the signal and noise. Note that the interference term appears as a beat between the signal and the noise in phase with the signal and lying within the signal bandwidth. This is the reason why the quantum analysis found that the interference term does not change with the degeneracy factor g. A warning is in order, however. If we carried the quasiclassical argument to its logical conclusion, we would represent the photon statistics of the interference term by a Gaussian distribution. This is not a good approximation, as the comparison in Fig. 9.4 shows.

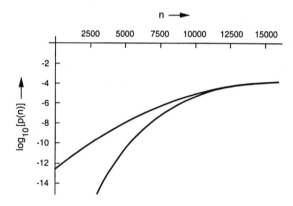

Fig. 9.4. Comparison of degenerate Bose–Einstein distribution for $G - 1 = 99$, $\langle n_a\rangle = 160$, and $\chi = 2$ with a Gaussian distribution that has the same location of the peak and the same mean square deviation

9.5 Bit-Error Rate with Optical Preamplification

9.5.1 Narrow-Band Filter, Polarized Signal, and Noise

We shall assume a receiver that has an optical preamplifier, a polarizer, and a filter, followed by a detector. We make the assumption that the gain of the amplifier is large and that the photon number of the incoming signal is large, $\langle n_a\rangle \gg 1$. We assume that the filter bandwidth is adjusted to be equal to the signal bandwidth B_s. Since the observation time τ is equal to the duration of a pulse, the product of the bandwidth and observation time is unity, $B_s\tau = 1$, and thus the spontaneous emission is nondegenerate, i.e. fully represented by a simple Bose–Einstein distribution. One must note that the assumption of a nondegenerate Bose–Einstein distribution also assumes that the signal and noise are of a single polarization.

The time slots occupied by zeros contain Bose–Einstein-distributed photons. The time slots occupied by the signal have fluctuations given by (9.44). In the case of large gain ($G \gg 1$) and large input photon number ($\langle n_a \rangle \gg 1$), the second term in (9.44) dominates. This interference term has been shown to be the result of the beating of the signal with the noise. The interference term has the mean square deviation

$$\langle \Delta n_b^2 \rangle \simeq 2G(G-1)\langle n_a \rangle \approx 2G^2 \langle n_a \rangle \ . \tag{9.71}$$

First, consider the bit errors in an empty time interval in which only the ASE contributes photons. The probability distribution of the ASE follows a Bose–Einstein distribution of average photon number $G - 1$:

$$p(m) = \frac{1}{1 + \langle n \rangle} \left(\frac{\langle n \rangle}{1 + \langle n \rangle} \right)^m = \frac{1}{G} \left(\frac{G-1}{G} \right)^m \ . \tag{9.72}$$

Suppose that the threshold is set at a photon number $n_{\text{threshold}} = \vartheta G \langle n_a \rangle$, normalized to the amplified input photon number $\langle n_a \rangle$. The threshold parameter ϑ is 0 when the threshold is set to accept all counts greater than 0, and 1 when the threshold is set at the average level of the output signal. The probability of the error of interpreting a zero as a one is

$$p_{\text{B}-\text{E}}(1) = \sum_{\vartheta G \langle n_a \rangle}^{\infty} p(n) = \left(\frac{\langle n \rangle}{1 + \langle n \rangle} \right)^{\vartheta G \langle n_a \rangle} = \left(\frac{G-1}{G} \right)^{\vartheta G \langle n_a \rangle} \ . \tag{9.73}$$

In the limit of large G, this expression approaches

$$\lim_{G \to \infty} p_{\text{B}-\text{E}}(1) = \exp(-\vartheta \langle n_a \rangle) \ . \tag{9.74}$$

Next, consider the probability of error when a pulse is received. We assume at first that the probability of the output in the presence of an input signal can be approximated by a Gaussian with the mean square fluctuations (9.71). Thus we assume the probability distribution of photons in the presence of a signal to be

$$p(n) = \frac{1}{\sqrt{4\pi G^2 \langle n_a \rangle}} \exp \left(- \frac{(n - G\langle n_a \rangle)^2}{4G^2 \langle n_a \rangle} \right) \ . \tag{9.75}$$

If the threshold is set at $n_{\text{threshold}}$, the probability of interpreting a one as a zero is

$$p_{\text{signal}}(0) = \frac{1}{2}\text{erfc}\left(\frac{\sqrt{\langle n_a \rangle}(1-\vartheta)}{2}\right), \tag{9.76}$$

where

$$\text{erfc}(x) \equiv \frac{2}{\sqrt{\pi}}\int_x^\infty e^{-x^2}\,dx\ . \tag{9.77}$$

By proper choice of the threshold ϑ we equate the two error probabilities, $p_{\text{signal}}(0) = p_{\text{B-E}}(1)$, as an optimum detection strategy, if the average rates of transmission of zeros and ones are the same. Figure 9.5 shows the probabilities of error as functions of the threshold. For an input photon number of 110, we find an error probability of 10^{-9} and a value of $\vartheta = 0.188$. The number of photons required for the reception of one bit at a bit-error rate of 10^{-9} is 55 photons, since, on the average, the "ones" occur only half of the time.

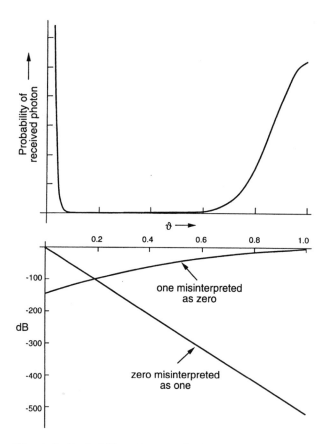

Fig. 9.5. Probability of received photons and probability of error as function of threshold; $\chi = 2$, $\langle n_a \rangle = 290$

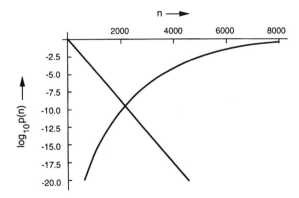

Fig. 9.6. Construction for determination of photon number for a bit-error probability of 10^{-9}. The crossover of the integrals of the probability functions. $G = 100$

We have mentioned that a Gaussian approximation for the photon distribution of a "one" is not a good approximation. If we use the exact probability distribution (9.51) with $g = 1$ we find for the average photon number the value of 40 photons (Fig. 9.6). This is also the value found by Li and Teich [75].

If the amplifier is not perfectly inverted, and instead possesses a noise enhancement factor $\chi = 2$, as is the case for an erbium-doped fiber amplifier pumped at 980 nm wavelength, then the previous theories, both the exact theory and the theory based on a Gaussian assumption, simply multiply the average photon number by the noise enhancement factor. In this way we obtain 110 and 80 photons, respectively, as the required photon number for a bit-error rate of 10^{-9}.

9.5.2 Broadband Filter, Unpolarized Signal

Thus far we have assumed that the detector was preceded by an optical filter with a bandwidth that was equal to the optical bandwidth of the signal. Here we consider the more practical case when an optical filter of bandwidth larger than the signal bandwidth is used. The interference noise of the signal does not change, since the signal beats only with noise that lies within the signal bandwidth. The noise in the zero slot increases by a factor of 2, since it involves two polarizations, and increases further with increasing bandwidth B_f by the total degeneracy factor $g = 2B_f/B_s$.

We shall start with the standard approach that uses the so-called Q factor. This approach is based on the assumption that the noise has a Gaussian distribution for both the "zero" and the "one". In the nondegenerate case it was clear that the distribution of photons in the zero slot was exponential, not Gaussian. Thus we could not use this approach at all. In the case of high

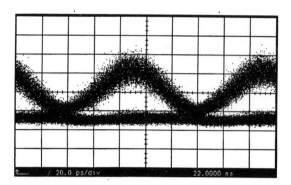

Fig. 9.7. A typical eye diagram of detector voltage with optical preamplification

degeneracy, the distribution in the zero slot is a degenerate Bose–Einstein distribution which starts to resemble a Gaussian. In the slot occupied by a one, the distribution is not Gaussian either, as pointed out earlier, but order-of-magnitude predictions based on such an assumption ought not to be too far from the truth. The widely used Q factor describes the "eye opening" of a so-called "eye diagram". Such eye diagrams are produced by overlapping on an oscilloscope the received detector voltage waveforms for a succession of zeros and ones, as shown in Fig. 9.7. We assume that the distributions are Gaussian, both for the interference noise and for the degenerate spontaneous-emission noise. The two probability distributions are

$$p_o(n) = \frac{1}{\sqrt{2\pi}\sigma_o} \exp\left(-\frac{(n - \langle n_o \rangle)^2}{2\sigma_o^2} \right) , \tag{9.78a}$$

$$p_1(n) = \frac{1}{\sqrt{2\pi}\sigma_1} \exp\left(-\frac{(n - \langle n_1 \rangle)^2}{2\sigma_1^2} \right) , \tag{9.78b}$$

where we treat the photon number as a continuous variable. We assume a signal with equal probabilities of zeros and ones. If the threshold is set at N_ϑ and the probabilities of errors are set equal, we have

$$\frac{1}{\sqrt{2\pi}\sigma_o} \int_{N_\vartheta}^{\infty} dn \exp\left(-\frac{(n - \langle n_o \rangle)^2}{2\sigma_o^2} \right)$$

$$= \frac{1}{\sqrt{2\pi}\sigma_1} \int_{-\infty}^{N_\vartheta} dn \exp\left(-\frac{(n - \langle n_1 \rangle)^2}{2\sigma_1^2} \right) . \tag{9.79}$$

The above relation gives the threshold setting as

$$\frac{N_\vartheta - \langle n_o \rangle}{\sigma_o} = \frac{\langle n_1 \rangle - N_\vartheta}{\sigma_1} = \frac{\langle n_1 \rangle - \langle n_o \rangle}{\sigma_o + \sigma_1} \equiv Q . \tag{9.80}$$

This equation is also the definition of the Q factor. For equal probabilities of zeros and ones, the bit-error rate can be expressed as

$$\text{BER} = \frac{1}{2} \int_{-\infty}^{N_\vartheta} dn\, p_1(n) + \frac{1}{2} \int_{N_\vartheta}^{\infty} dn\, p_2(n)$$

$$= \frac{1}{\sqrt{2\pi}} \int_Q^{\infty} dx \exp\left(-\frac{x^2}{2}\right) = \frac{1}{2}\text{erfc}\left(\frac{Q}{\sqrt{2}}\right) . \tag{9.81}$$

A BER of 10^{-9} requires a Q of 6. We approximate the degenerate spontaneous-emission noise as Gaussian, with the mean square value $\langle \Delta n^2 \rangle = g\langle n_{sp} \rangle (1 + \langle n_{sp} \rangle)$, where $g = B_f/B_s$. We allow for imperfect inversion of a laser, for which $\langle n_{sp} \rangle = \chi(G-1)$, where χ is the inversion factor. The mean square deviation of the zero level is (see 9.62)

$$\sigma_o^2 = g\langle n_{sp} \rangle (1 + \langle n_{sp} \rangle) = g\chi(G-1)[\chi(G-1) + 1] , \tag{9.82}$$

where g is the degeneracy factor. The mean square deviation of the "one" level contains the interference term. In the nondegenerate case we neglected the contribution of the ASE, which is legitimate when $g = 1$. However, for large values of g, this noise cannot be neglected. We thus have

$$\sigma_1^2 = 2\chi G(G-1)\langle n_a \rangle + g\chi(G-1)[\chi(G-1) + 1] . \tag{9.83}$$

The Q factor in the limit of very large gain is thus

$$Q = \frac{\langle n_1 \rangle - \langle n_o \rangle}{\sigma_1 + \sigma_o} = \frac{\langle n_a \rangle}{\sqrt{2\chi\langle n_a \rangle + g\chi^2} + \sqrt{g\chi^2}} . \tag{9.84}$$

The number of input photons required for a BER of 10^{-9} calls for $Q = 6$. We obtain for the required average signal photon number

$$\langle n_a \rangle / 2 = 36 + 6\sqrt{g} . \tag{9.85}$$

A plot of the photon number as a function of the degeneracy for a bit-error rate of 10^{-9}, for $\chi = 1$, appears in Fig. 9.8. The remarkable fact is that this simple expression gives 42 photons for $g = 1$, very close to the exact value of 40. This is in spite of the fact that even the exponential probability distribution of the zero was approximated by a Gaussian. The noise of the zero is underestimated by the approximate analysis, and the noise of the one is overestimated, so that the approximate value for the average photon number is not far from the exact value. For $\chi > 1$, the average photon number is simply multiplied by χ.

A practical receiver has to receive an unpolarized signal since the polarization of the signal cannot be controlled in propagation along a fiber. (Space communication between satellites does not have this difficulty.) This introduces automatically a degeneracy factor of 2. The optical filter bandwidth must, generally, be quite a bit wider than the signal bandwidth to avoid signal distortion on one hand and reduce the effects of environmentally induced

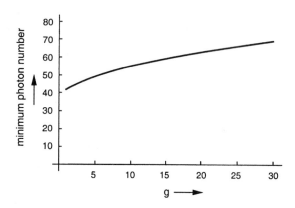

Fig. 9.8. Photon number computed from Q factor for $\chi = 1$

shifts of the filter center frequency on the other hand. A filter bandwidth of 1.0 nm, corresponding to 133 GHz at 1.5 μm wavelength, is quoted in [75], for a bit rate of 10 Gb/s. This gives a degeneracy factor of $2 \times 13 = 26$. From Fig. 4.3 we read off 67 photons. With an an incompletely inverted gain medium with a χ factor of 2, we find 134 photons. In [77] the measured power was -38 dBm. This corresponds to

$$\frac{0.16 \ \mu W}{h\nu \times 10^{10}} = \frac{0.16 \times 10^{-6}}{6.626 \times 10^{-34} \times 2 \times 10^{14} \times 10^{10}} = 121 \ , \tag{9.86}$$

which is quite close to the theoretical value. Reference [78] quotes 137 photons per bit, [79] 155 photons per bit. The lowest number has been quoted in [80], which reported 78 photons per bit. The degeneracy factor in this case, with a filter bandwidth of 70 GHz and a single polarization, was 7. Theory predicts 102 photons for $\chi = 2$. The inversion factor in this experiment may have been close to unity.

Theoretical results also appear in the literature that predict lower photon numbers for a 10^{-9} bit-error rate [81]. These are based on the assumption of shot noise (Poisson-distributed carriers) for the detected ASE noise. This assumption underestimates the actual level of detector noise. An exact computation that takes into account the actual probability distribution of the "zero" and "one" levels as predicted from (3.32) is shown in Fig. 9.9.

9.6 Negentropy and Information

Thus far we have studied the number of photons required for the reception of one bit of information via digital transmission of a pulse (a "one") and a blank (a "zero") at a bit-error rate of 10^{-9}. With a negligible sacrifice of additional bits, the transmission can be made error-free. Thus we may

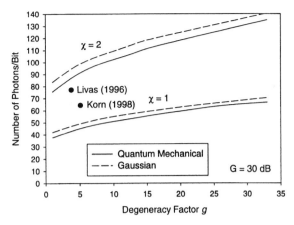

Fig. 9.9. Photon number computed from exact analysis and with Gaussian assumption [80, 86] (computation by W. Wong)

compare the number of photons required to transmit one bit of information with the Shannon formula that predicts the minimum number of photons required for error-free transmission. We follow a derivation first published by J. P. Gordon [82].

The maximum amount of information that can be carried by n photons follows from the negentropy principle of Shannon [83, 84]. The entropy normalized to k (Boltzmann's constant) for a system containing on the average $\langle n \rangle$ photons is

$$\frac{S}{k} \equiv H = -\sum_n p(n) \ln[p(n)] , \tag{9.87}$$

with the probability distribution $p(n)$ so chosen that H is maximized under the constraints

$$\sum_n p(n) = 1 \tag{9.88}$$

and

$$\sum_n np(n) = \langle n \rangle . \tag{9.89}$$

When the maximization is carried out, we find that the probability distribution is Bose–Einstein (compare Sect. 4.8):

$$p(n) = \frac{1}{1 + \langle n \rangle} \left(\frac{\langle n \rangle}{1 + \langle n \rangle} \right)^n . \tag{9.90}$$

This result is not surprising, since the Bose–Einstein distribution is the thermal distribution, the maximally random distribution.

The maximized value of H is

$$H_{\max} = \langle n \rangle \ln\left(1 + \frac{1}{\langle n \rangle}\right) + \ln(1 + \langle n \rangle) . \tag{9.91}$$

The negentropy principle states that the amount of information that can be transmitted with proper encoding in an error-free manner is equal to $H_{\max}$. It is implied that the sender of the information utilizes an encoding in which the probability of transmission is chosen so as to be maximally random, and hence the sender must choose the Bose–Einstein distribution.

If communication takes place in the presence of noise, the entire information content (9.91) cannot be transmitted. According to the negentropy principle of information [83], the maximum amount of information that can be extracted in an error-free manner from signal states in a noise background is equal to the difference of the total entropy (9.91) and the entropy of the noise. The entropy of a thermal background at temperature θ is

$$H_\theta = \langle n_\theta \rangle \ln\left(1 + \frac{1}{\langle n_\theta \rangle}\right) + \ln(1 + \langle n_\theta \rangle) . \tag{9.92}$$

The information content I of the message is thus

$$I = H_{\max} - H_\theta = \langle n \rangle \ln\left(1 + \frac{1}{\langle n \rangle}\right) + \ln(1 + \langle n \rangle)$$

$$- \langle n_\theta \rangle \ln\left(1 + \frac{1}{\langle n_\theta \rangle}\right) - \ln(1 + \langle n_\theta \rangle) . \tag{9.93}$$

It should be emphasized that (9.93) is written in terms of the photons received by an ideal receiver that can distinguish photons. If there is attenuation between transmitter and receiver, $\langle n \rangle$ is the average number of received photons. It is customary to express the quantity of information in terms of the logarithm to base 2, rather than the natural logarithm. The information $I = -\sum_n p(n) \log_2 p(n)$ transmitted by zeros and ones with a probability $1/2$ each is then unity, i.e. one bit.

For a given average photon number, the negentropy principle predicts the amount of information that can be sent per photon, or per binary symbol, without error in a noise background. We have found that for a completely inverted gain medium, 40 photons on average provide a BER of 10^{-9}. If the medium is not completely inverted, e.g., $\chi = 2$, then about 80 photons on average are required. This gives a noise background of $\langle n_\theta \rangle = 2$. It is of interest to ask how much information could be transmitted with photons in a noise background of $\langle n_\theta \rangle = 2$, the equivalent noise background of an erbium-doped preamplifier. In the limit of large average photon number, the information is

$$I \approx \log_2 \left[\frac{1 + \langle n_a \rangle}{(1 + 1/\langle n_\theta \rangle)^{\langle n_\theta \rangle} (1 + \langle n_\theta \rangle)} \right] . \tag{9.94}$$

This formula predicts 2.11 bits for an average photon number $\langle n_a \rangle = 80$ photons, as opposed to the previously predicted 1 bit at an error rate of 10^{-9}. Since it is easy to correct for such a low bit-error rate with an economical code, the comparison is a fair one. It is interesting to note that only a factor of 2.11 is sacrificed by not using a more sophisticated encoding scheme.

9.7 The Noise Figure of Optical Amplifiers

In this chapter, we derived the mean square fluctuations of the charge of a detector illuminated by optical radiation. This very expression is used in a widely accepted measure of the noise performance of optical amplifiers [81]. A signal-to-noise ratio is constructed at the input and at the output of the amplifier, and a noise figure is defined in terms of the ratio of signal-to-noise ratios:

$$F = \frac{\text{input signal-to-noise ratio}}{\text{output signal-to-noise ratio}} . \tag{9.95}$$

The signal-to-noise ratio is constructed from the square of the average signal photon number, divided by the mean square photon number fluctuations. In this section we explore the consequences of this definition of noise figure and show that it leads to a noise figure that is a function of the signal level. Finding this property to be unacceptable, we proceed to define an excess noise figure and noise measure in a way consistent with their use in Chap. 7.

First we present a brief review of the definition of noise figure as standardized by the IRE and accepted by the IEEE [17] for the characterization of electronic amplifiers. The original formulation defined it in concordance with (9.95), using as a measure of the signal-to-noise ratio the signal power divided by the noise power (in photonic terms, the measure of the signal is the average photon number, not its square, and the mean square amplitude fluctuations are the measure of the noise). This fact is of great importance. Indeed, in a linear amplifier, the amplifier noise is additive to the signal and the signal drops out from the definition (9.95):

$$F = \frac{\text{noise power at output}}{G(\text{noise power at input})} \tag{9.96}$$

The noise powers are defined as "available noise powers". In the rare case when the real part of any one of the impedances is negative, exchangeable power is used, as already pointed out in Chap. 5 and [17]. G is the available (or exchangeable) gain. For convenience, the input noise power is standardized to be the thermal noise at room temperature. (We shall concentrate here especially on the spot noise figure, namely the noise figure within a signal bandwidth B_s sufficiently narrow that the properties of the amplifier are frequency-independent within this bandwidth.) The available noise power at the input is $k\theta_o B_s$, where k is Boltzmann's constant and θ_o is the standard "room" temperature of 290 K. Using this fact, (9.95) can be rewritten

$$F = \frac{Gk\theta_o B_s + \text{noise power at output added by amplifier}}{Gk\theta_o B_s}$$

$$= 1 + \frac{\text{noise power at output added by amplifier}}{Gk\theta_o B_s} . \tag{9.97}$$

This definition of noise figure is independent of the signal. Standard noise measurement equipment takes advantage of this fact: it measures the output noise power within a bandwidth B_s with no signal applied to the input, the input being terminated in the source impedance at temperature θ_o. Another convenient definition is the excess noise figure $F - 1$:

$$F - 1 = \frac{\text{noise power at output added by amplifier}}{Gk\theta_o B_s} . \tag{9.98}$$

The excess noise figure (as well as the noise figure) gives full information on the noise power added by the amplifier. Since the amplitude fluctuations of the amplifier noise are (usually) Gaussian, and Gaussian distributions are fully described by the second moment (i.e. power), the noise figure gives the statistics of the amplifier noise.

If the gain of an amplifier is small, cascading with another amplifier may be necessary. Denoting the noise figure of the first amplifier by F_1 and the noise figure of the second amplifier by F_2, the noise figure F of the cascade is [17]

$$F = F_1 + \frac{F_2 - 1}{G_1} . \tag{9.99}$$

These definitions have served the engineering community well in all applications within the "low-frequency" regime, from d.c. to millimeter waves. Electronic preamplifiers that process low signals do not experience saturation. The (unsaturated) gain and the noise figure are reliable attributes of low-noise amplifiers.

The question then arises of how the definition of noise figure should be generalized into the domain of optical laser amplifiers. Laser amplifiers exhibit some features not shared with electronic amplifiers. Current practice [81] is to use the signal-to-noise ratio definition (9.95) of the noise figure and to define the noise at the input as the mean square photon number fluctuations of a Poissonian process of average photon number $\langle n_a \rangle$. The reason for using a Poissonian distribution is based on the fact that a highly attenuated signal will, in general, acquire Poisson statistics. This definition is

$$\text{input signal-to-noise ratio} = \frac{\langle n_a \rangle^2}{\langle \Delta n_a^2 \rangle} = \langle n_a \rangle \, . \tag{9.100}$$

Since the square of the photon number is used as the definition of the signal, the signal at the output is defined as $G^2 \langle n_a \rangle^2$. The noise at the output is given by the mean square fluctuations $\langle \Delta n_b^2 \rangle$ of the photon number at the output as given by (9.62):

$$\langle \Delta n_b^2 \rangle = G \langle n_a \rangle + 2\chi G(G-1)\langle n_a \rangle + g\chi(G-1)[\chi(G-1)+1] \, . \tag{9.101}$$

Thus, the signal-to-noise ratio at the output is

output signal-to-noise ratio

$$\tag{9.102}$$

$$= \frac{G^2 \langle n_a \rangle^2}{G \langle n_a \rangle + 2\chi G(G-1)\langle n_a \rangle + g\chi(G-1)[\chi(G-1)+1)]} \, .$$

The noise figure defined on the basis of photon number fluctuations, F_{pnf}, becomes

$$F_{\text{pnf}} = \frac{1}{G} + 2\chi(1 - 1/G) + g\chi \left[\chi \left(1 - \frac{1}{G} \right) + \left(\frac{1}{G} \right) \right] \left(1 - \frac{1}{G} \right) \frac{1}{\langle n_a \rangle} \, . \tag{9.103}$$

One feature is immediately apparent: the noise figure is not signal-independent. This is in contradistinction to the conventional definition of noise figure for a linear amplifier, which is signal-independent. In common applications one can usually make the approximation that the signal photon number is large enough that the last term in (9.103) can be neglected. It is in this form that the usage has been established and the noise figure becomes signal independent:

$$F_{\text{pnf}} \approx \frac{1}{G} + 2\chi(1 - 1/G) \, . \tag{9.104}$$

However, this definition has problems. The signal photon number is not always that large. An ideal, fully inverted preamplifier requires of the order of 40 photons per bit for a bit-error rate of 10^{-9}. One may envisage situations in which a higher bit-error rate is permitted. Then the signal photon number can become smaller. One may also face the common situation in which the filter following the optical preamplifier has a bandwidth B_f much wider than the signal bandwidth. This is usually the case in practice, since the problem of control of the filter center frequency is alleviated by the choice of a wide bandwidth. Further, this definition does not obey the cascading formula (9.99).

The definition (9.103) does not obey the cascading formula, because it uses as the measure of signal-to-noise ratio a ratio of squares of powers (or energies, or photon numbers). An optical amplifier *is not a linear amplifier of photon number*. Even though the output photon number is G times the input photon number, the noise is not additive. However, the optical laser amplifier is a linear amplifier of the electric and magnetic field of the incident wave, not unlike a microwave traveling-wave tube amplifier. With two optical amplifiers in cascade, the second amplifier amplifies the noise power of the first one and *adds* its own noise. In defining a noise figure of an optical amplifier in the spirit of the IEEE definitions, one can start with a signal-to-noise ratio identified as the ratio of the time-averaged square of the signal amplitude to the mean square fluctuations of the signal amplitude.

In the nondegenerate case, when the signal bandwidth is equal to the noise bandwidth, the amplifier is described by (6.128)

$$\hat{B} = \sqrt{G}\hat{A} + \hat{N}_u + \hat{N}_\ell . \tag{9.105}$$

If the amplifier is excited by a coherent state, the mean square amplitude of the in-phase component of the signal at the input is obtained from

$$\langle \alpha | \frac{1}{2}(\hat{A} + \hat{A}^\dagger)|\alpha\rangle^2 = \frac{1}{4}(\alpha^2 + \alpha^{*2} + 2|\alpha|^2) . \tag{9.106}$$

The first two terms give the time-dependent part of the sinusoidally time-varying signal and have zero average. The quadrature component gives the same average. Thus the signal averaged over time is

$$\text{input signal } = |\alpha|^2 = \langle n_s \rangle . \tag{9.107}$$

The signal at the output is computed from

$$\langle \alpha | \frac{1}{2}(\hat{B} + \hat{B}^\dagger)|\alpha\rangle^2 = G\frac{1}{4}(\alpha^2 + \alpha^{*2} + 2|\alpha|^2) \tag{9.108}$$

and from the amplified quadrature component. Thus

$$\text{output signal } = G|\alpha|^2 = G\langle n_s \rangle . \tag{9.109}$$

The noise at the input is due to zero-point fluctuations in the in-phase and quadrature components, which add up to $1/2$. The noise at the output is composed of the amplified zero-point fluctuations accompanying the signal, $1/2G$, and the noise due to the two noise sources, giving for the in-phase component

$$\frac{1}{4}\langle(\hat{B} + \hat{B}^\dagger)^2\rangle - \frac{1}{4}\langle\hat{B} + \hat{B}^\dagger\rangle^2$$

$$= \frac{1}{4}\langle(\hat{N}_u + \hat{N}_u^\dagger)^2\rangle + \frac{1}{4}\langle(\hat{N}_\ell + \hat{N}_\ell^\dagger)^2\rangle$$

$$= \frac{1}{4}\chi(G-1) + \frac{1}{4}(\chi-1)(G-1) \tag{9.110}$$

$$= \frac{1}{4}(2\chi - 1)(G-1) \,,$$

and the same amount of fluctuations in the quadrature component. Thus the noise figure defined as the ratio of signal-to-noise in terms of squared field amplitudes, F_{fas}, is

$$F_{\text{fas}} = 1 + (2\chi - 1)(1 - 1/G) \,. \tag{9.111}$$

This noise figure is signal independent, as it should be. However, it is not directly measurable. The input signal-to-noise ratio can only be inferred from measurements of the amplifier output.

A viable definition of excess noise figure, one that is directly measurable, is suggested by (9.98). In the classical domain, the excess noise figure gives the amplifier noise power divided by the gain, normalized to thermal noise. The normalization to thermal noise does not make sense for an optical amplifier, since the fluctuations accompanying the signal are zero-point fluctuations that are much larger than thermal noise. Normalization to $\hbar\omega_o B_s$ is suggested by the definition (9.111). A measurement of ASE power gives full information about the photon statistics of the amplifier noise, since they have a Bose–Einstein distribution within the time $1/B_s$. Thus, we may define the excess noise figure for an optical amplifier in terms of the noise power added by the amplifier in one single polarization (the polarization of the signal)

$$F_{\text{ASE}} - 1 \equiv \frac{\text{ASE noise power in bandwidth } B_s}{G\hbar\omega B_s}$$

$$= \frac{\chi(G-1)}{G} = \chi(1 - 1/G) \,. \tag{9.112}$$

With a filter of bandwidth B_s following the amplifier, the optical power measured is $G\hbar\omega_o B_s(F_{\text{ASE}} - 1)$. Thus, with the definition (9.112), we deal with a quantity that is

(a) measurable,
(b) gives full information on the noise statistics that can be used to evaluate system performance, and
(c) obeys the cascading formula.

If no polarizer is used and the bandwidth of the filter is larger than the signal bandwidth, then a measurement of the amplified spontaneous emission still gives full information. We need to note only that the ASE photon statistics are now degenerate Bose–Einstein statistics with a degeneracy factor of $g = 2B_f/B_s$. The knowledge of this photon flow, along with the gain G and degeneracy factor g, is all the information needed to characterize the noise and to predict bit-error rates of detected signals. For example, the output photon fluctuations (9.101) can be determined from this knowledge for any input signal photon number $\langle n_a \rangle$. If the preamplifier is followed by a photodetector of known quantum efficiency and noise of its own, the bit-error rate of the system can be predicted.

If the bit rate of the signal is much higher than the relaxation rate of the gain medium, the amplifier may saturate even for low-level signals. This is a distinct advantage of the erbium fiber amplifier since it prevents inter-symbol interference and provides gain stabilization in long-distance amplifier cascades. With the gain fixed, the amplified spontaneous emission is thus fixed. Thus, in the case of fiber amplifiers it may be necessary to saturate the amplifier to the nominal gain level to arrive at the proper value of the ASE. This can be done by a chopped signal and a measurement of the noise in the time intervals containing no signal.

The proper definition of noise figure for an optical-fiber amplifier can be used to advantage in the choice of pumping schemes for optimum noise performance. The amplifier can be pumped by injection of the pump radiation from either of the two ends, or from both ends. One may also consider the use of more injection points along the amplifier. The purpose is to minimize the noise measure of each segment and, if possible, excite the segments in such a way that the segments of lowest noise measure occur in the front end of the amplifier. The noise measure of a short segment of length Δz of an optical fiber amplifier is $(F_{\mathrm{ASE}} - 1)/(1 - 1/G) = \chi$. The pumping controls the gain and the noise enhancement factor. The aim is thus to minimize the inversion parameter and place the segments with the lowest noise enhancement factor as close to the input as possible.

There is another consideration that favors the definitions of noise figure F_{ASE} and F_{fas}, i.e. the need to connect the definitions of noise figure for linear microwave amplifiers with those for optical amplifiers. In particular, in the far-infrared regime, both quantum effects and classical thermal noise sources contribute to the noise performance. In this frequency regime one needs a definition that covers both quantum-noise and thermal-noise effects.

The noise figure F_{fas} was derived under the assumption that the noise at the input was quantum noise with no thermal contributions. Equation (7.75) gives the mean square in-phase fluctuations when thermal noise of average photon number $\langle n_\theta \rangle$ contributes to the fluctuations:

$$\sigma^2 = \frac{1}{4} + \frac{\langle n_\theta \rangle}{2} .$$

The quadrature fluctuations are of equal magnitude. Thus, the noise at the input is, instead of $1/2$, equal to $1/2 + \langle n_\theta \rangle$. The background of thermal photons is taken to be that of a standard (room) temperature. We find for the generalized noise figure F_{fas}

$$F_{\text{fas}} - 1 = \frac{\chi - 1/2}{1/2 + \langle n_\theta \rangle} \left(1 - \frac{1}{G} \right). \tag{9.113}$$

It is easy to show that this definition approaches the proper classical limit. A classical amplifier adds noise in amounts much larger than those dictated by the penalty for a simultaneous measurement of in-phase and quadrature field components, i.e. a photon number much larger than $G - 1$. Thus, one may describe the operation of a classical amplifier by quantum amplification with a very large χ, i.e. $\chi \gg 1$. Then

$$\lim_{\chi \gg 1} (F_{\text{fas}} - 1) = \frac{\chi(G - 1)}{G(\langle n_\theta \rangle + 1/2)}.$$

Multiplication of the top and bottom by $\hbar \omega_o B$ changes photon numbers into rates of energy, or power flow. Further, in the classical imit, $\langle n_\theta \rangle \gg 1/2$, and

$$\lim_{\text{class}} (F_{\text{fas}} - 1) = \frac{\text{available noise power added by amplifier}}{G k \theta_o B},$$

which is in agreement with (5.85). A similar modification is possible for F_{ASE}. We include the thermal noise by writing

$$F_{\text{ASE}} - 1 = \frac{\chi(G - 1)}{G(\langle n_\theta \rangle + 1)}.$$

Again, this noise figure approaches the proper classical limit and in doing so merges with the definition $F_{\text{fas}} - 1$ in the classical limit.

9.8 Summary

The conservation of commutator brackets in attenuation and amplification calls for the introduction of noise sources into the scattering formalism. If we assume that the reservoirs of the noise sources are in their ground state, we find the minimum noise that is added in amplification or attenuation. The commutators of the noise sources also permit the evaluation of the complete photon statistics of the output photons for a given probability distribution of photons at the input. Attenuation leads to a binomial process: photons are passed with probability $\mathcal{L}$ and lost with probability $1 - \mathcal{L}$, where $\mathcal{L}$ is the loss factor. Both Poisson and Bose–Einstein distributions are preserved in attenuation. Since the Bose–Einstein distribution is the thermal distribution one may interpret attenuation as a form of cooling. Amplification with no input produces a Bose–Einstein distribution of output photons. The noise source does not contribute to the falling factorial moments of the photon number at the output of an attenuator. Similarly, there is no contribution of the noise source to the rising factorial moments of a perfectly inverted amplifier. This finding underscores the fact that the discretization of energy via the photon concept can fully account for quantum effects without the need to invoke zero-point fluctuations.

We were able to generalize the analysis to cover the case of an amplifier with incomplete inversion. The problem can be reduced to a cascade of an attenuator and an amplifier with complete inversion. Since the attenuator preserves a Poisson distribution, the presence of the attenuator is easily taken into account.

The formalism found practical application in the evaluation of the minimum number of photons per pulse required for a bit-error rate of 10^{-9} using a detector with an optical preamplifier. The ideal case is when the signal is polarized, a filter is used with a bandwidth equal to the signal bandwidth, and the amplifier is perfectly inverted. We found that 40 photons are required. When the signal is not polarized, the degeneracy is equal to 2, and when the optical amplifier has a bandwidth wider than the signal, the degeneracy increases further.

One would expect that, in the limit of large signal photon number, the bit-error rate of an optical preamplifier followed by a detector could be predicted by a classical analysis. This is indeed the case. The analysis in [85] gives results that are numerically in good agreement with the quantum analysis.

We looked into the prediction of the Shannon theory for the bit-rate increase if an ideal code were used to overcome the noise and found that an increase of only a factor of 2.11 would be achievable. Finally, we addressed the problem of the definition of the noise figure. Even though we had used the concept and definition of noise figure in Chap. 7 to derive the optimum noise performance of an amplifier, the discussion of the current usage had to wait until we discussed photon number fluctuations. The current use of "noise figure" defines the signal in terms of photon flow squared, rather than field amplitude squared. This definition is designed to determine the signal-to-noise ratio at the detector output. It is not suited to answer the simple question of how to cascade two amplifiers, given their gain and noise figure. In order to answer such questions we need to revert to a generalization of the concept of noise measure as employed in Chap. 7.

Problems

9.1* Show that the characteristic functions of a Bose–Einstein and a Poisson process become approximately equal for a small average photon number, $\langle n \rangle \ll 1$.

9.2 Show that the characteristic function of the photon number of a Poisson distribution approaches the characteristic function of a Gaussian distribution in the limit of large average photon number.

9.3 Show that the probability of a Bose–Einstein distribution approaches that of the energy of a Gaussian distribution in the limit of large average photon number.

9.4* A group velocity dispersion compensator consists of an optical preamplifier of gain G_1 and noise enhancement factor χ_1, followed by a circulator and grating reflector of net loss $\mathcal{L}_2$, and a postamplifier of gain G_3 and noise enhancement factor χ_3 (see Fig. P9.4.1). Determine the noise figures F_{ASE} and F_{fas} of the system.

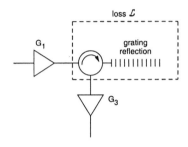

Fig. P9.4.1. Group velocity dispersion compensation

9.5 Consider the circuit of Fig. P9.5.1. It represents a receiving detector in which the current source provides the detector current.

(a) Determine the filter function $H(\omega)$. Remember that $H(\omega)$ represents the frequency dependence of the charge Q_d when the charge Q_s is supplied by the detection process. A circuit with an instantaneous response has a frequency-independent $H(\omega)$, a delta function in the time domain. Thus $H(\omega) = Q_d(\omega)/Q_s(\omega)$.

(b) A non-return-to-zero (NRZ) bit pattern at an optical carrier frequency ω_o is incident on a detector. If the pattern is random, i.e. the zeros (blanks) and the ones (rectangles of height A and width τ_o) occur randomly, find the spectrum of the detector current.

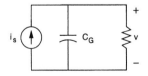

Fig. P9.5.1. Schematic of receiving detector

9.6 In the text we considered generating functions for one, single discrete random process. You are asked to generalize the formalism to a pair of discrete random processes, such as the photon counts of two detectors.

Define the falling-factorial generating function by

$$F(\xi, \eta) = \sum_{m,p} \sum_{n,q} \frac{\xi^p \eta^q}{p!q!} p(m,n) m(m-1)\dots$$

$$\times \dots (m-p+1)n(n-1)\dots(n-q+1) \, .$$

(a) Derive a relation between the probability generating function

$$P(\xi, \eta) = \sum_{m,n} \xi^m \eta^n p(m,n)$$

and the falling-factorial-moment generating function.

(b) Derive the relation between the probability generating function and the rising-factorial-moment generating function

$$R(\xi, \eta) = \sum_{m,p} \sum_{n,q} \frac{\xi^p \eta^q}{p!q!} p(m,n)(m+1)(m+2)$$

$$\dots (m+p)(n+1)(n+2)\dots(n+q) \, .$$

9.7* Show that optical preamplification of high gain G and large signal photon number followed by detection gives higher signal-to-noise ratios than direct detection followed by a microwave amplifier with a noise figure of 3 dB. Assume a bandwidth of 10 GHz and a signal photon number $\langle n_s \rangle = 100$.

9.8 Relate the falling-factorial-moment generating function to the rising-factorial moment generating function.

9.9 Determine the signal photon number $\langle n_s \rangle$ of a "one" required to achieve a Q factor of 6 in heterodyne detection.

Solutions

9.1 The characteristic function for a Poisson process is

$$C_P(\xi) = e^{-\langle n \rangle} \sum_n \frac{\langle n \rangle^n}{n!} e^{i\xi n} = e^{\langle n \rangle [\exp(i\xi) - 1]} \, .$$

The characteristic function for a Bose–Einstein process is:

$$C_{\mathrm{BE}}(\xi) = \frac{1}{1 + \langle n \rangle} \sum_n \left(\frac{\langle n \rangle}{1 + \langle n \rangle} \right)^n e^{i\xi n} = \frac{1}{1 - \langle n \rangle (e^{i\xi} - 1)} \, .$$

For a small expectation value of the photon number we have

$$C_P(\xi) = e^{\langle n \rangle [\exp(i\xi) - 1]} \approx 1 + \langle n \rangle [\exp(i\xi) - 1] \ .$$

and

$$C_{\mathrm{BE}}(\xi) = \frac{1}{1 - \langle n \rangle (e^{i\xi} - 1)} \approx 1 + \langle n \rangle [\exp(i\xi) - 1]$$

$$QED \quad .$$

9.4 We have for the two excess noise figures of an amplifier

$$F_{\mathrm{ASE}} - 1 = \chi(1 - 1/G) \quad \text{and} \quad F_{\mathrm{fas}} - 1 = (2\chi - 1)(1 - 1/G) \ .$$

The attenuator does not add noise power and thus its excess noise figure $F_{\mathrm{ASE}} - 1$ is zero. On the other hand the noise figure of the attenuator defined on the basis of mean square field fluctuations is

$$F_{\mathrm{fas}} = \frac{S_i/N_i}{S_o/N_o} = 1/\mathcal{L} \ ,$$

since the zero-point fluctuations are the same at the input and output, i.e. $N_i = N_o$. The cascading formula works for both definitions of noise figure. We have

$$F - 1 = \left(F_1 - 1 + \frac{F_2 - 1}{G_1} \right) + \frac{F_3 - 1}{G_1 G_2} \ .$$

Thus

$$F_{\mathrm{ASE}} - 1 = \chi_1(1 - 1/G_1) + \frac{\chi_3(1 - 1/G_3)}{G_1 \mathcal{L}_2}$$

and

$$F_{\mathrm{fas}} - 1 = (2\chi_1 - 1)(1 - 1/G_1) + \frac{1}{G_1}\left(\frac{1}{\mathcal{L}_2} - 1 \right)$$

$$+ \frac{(2\chi_3 - 1)(1 - 1/G_3)}{G_1 \mathcal{L}_2} \ .$$

9.7 We determine first the signal-to-noise ratio of a system consisting of an optical preamplifier of gain G followed by direct detection, case (I). If the expectation value of the signal photon number entering the amplifier is $\langle n_s \rangle$, then the signal-to-noise ratio, defined as the ratio of the mean square signal power to the mean square number fluctuations, is

$$\frac{S}{N} = \frac{G^2 \langle n_s \rangle^2}{\langle \Delta n^2 \rangle} = \frac{G^2 \langle n_s \rangle^2}{2\chi G(G - 1)\langle n_s \rangle} = \frac{\langle n_s \rangle}{2\chi(1 - 1/G)} \ , \tag{1}$$

where we have ignored the contribution of ASE to $\langle \Delta n^2 \rangle$, as applicable to large $\langle n_s \rangle$. Next consider direct detection followed by microwave amplification, case (II). For a photon rate r, the current of the photodetector is

$$i = qr \ ,$$

and the signal power P_s emitted by the photodetector into the input resistance R of the amplifier is

$$P_s = i^2 R = q^2 r^2 R .$$

The noise power referred to the microwave amplifier input within a bandwidth B, for an amplifier noise figure F, is

$$P_n = F k \theta_o B .$$

The signal-to-noise ratio is

$$\frac{S}{N} = \frac{P_s}{P_n} = \frac{q^2 r^2 R}{F k \theta_o B} = \frac{1}{F} \frac{q}{k \theta_o} q B R \langle n_s \rangle^2 ,$$

where we have set $r/B = \langle n_s \rangle$. The signal-to-noise ratio is proportional to $\langle n_s \rangle^2$, contrary to case (I). Bit-error rates of better than 10^{-10} are achieved in case (I) with $\langle n_s \rangle \approx 100$, for a bandwidth of $B = 10$ GHz. With an amplifier input resistance of $50 \, \Omega$ we find

$$\frac{S}{N} = \frac{1}{F} \frac{q}{k \theta_o} q B R \langle n_s \rangle^2$$

$$= \frac{1}{F} 40 \times 1.6 \times 10^{-19} \times 10^{10} \times 50 \times 100^2$$

$$= 1.6 \times 10^{-2} .$$

This is a very small number, whereas case (I) achieves a signal-to-noise ratio much greater than unity with $\langle n_s \rangle = 100$. We see that the performance in the case of direct detection followed by microwave amplification is handicapped by the assumption of a $50 \, \Omega$ input impedance and the low photon number. If the input impedance could be set at $5 \, k\Omega$, the detection sensitivities in both cases would be comparable. Also, a higher photon number per bit would help in case (II). However, if the aim is to operate with as small a photon number rate as possible, and microwave amplification at the $50 \, \Omega$ impedance level is required, the disadvantage of case (II) is overwhelming.

10. Solitons
and Long-Distance Fiber Communications

An optical fiber made of silica with a germanium core can support pulses, "solitons", that propagate undistorted if the fiber has negative dispersion at the carrier frequency of the pulses. The self-phase modulation of the pulse by the Kerr effect balances the dispersion [4]. Solitons have been proposed for repeaterless digital communications over transoceanic distances [5]. Instead of a signal composed of pulses and blanks being detected and regenerated every so often, as is done in conventional transoceanic communications, a signal consisting of solitons and empty time intervals would only be amplified every 25 km or so, without regeneration. This revolutionary proposal has been explored in extensive laboratory experiments [7, 87], but the first repeaterless transoceanic fiber communication cable, laid in 1995, uses fiber with close to zero dispersion and functions in what may be called the "linear" regime (self-phase modulation is not utilized, but since it is not entirely avoided, it is combatted by proper dispersion management along the fiber). The operation is called non-return-to-zero, or NRZ for short, as a description of the signal format: if two pulses are adjacent to each other, the pulse amplitude does not return to zero; two rectangular pulses merge into one pulse of double length (see Fig. 10.1). Whereas excellent performance has been achieved with operation in the linear NRZ format, and the bare essentials of the scheme are easily understood, no analytic theory exists for the description of the subtle nonlinear effects and dispersion effects that affect this mode of communications. Much better understanding has been developed for soliton operation, largely owing to the new mathematical methods developed for the analysis of a class of integrable nonlinear partial differential equations derivable from a Hamiltonian [88, 89]. In this chapter we shall study only repeaterless soliton fiber communications, in part since the theory is elegant and well developed and in part because soliton communications may still find implementation, but mainly as a preparation for the analysis of the generation of squeezed radiation using solitons. Further, solitons passing through an amplifier are an interesting example of nonadditive noise, a case that transcends the linear additive noise analyses of the preceding chapters.

In Sects. 10.1 and 10.2 we derive the nonlinear Schrödinger equation that controls soliton propagation and determine the solution for the fundamental soliton. Section 10.3 studies properties of solitons. Soliton perturbations and

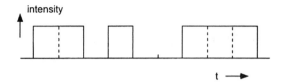

Fig. 10.1. Non-return-to-zero format: a 110100111 sequence

methods for the analysis of solitons perturbed by noise are treated in Sects. 10.4 and 10.5. Long-distance propagation of solitons is analyzed in fiber systems in which the loss is compensated by gain and ASE is generated along the way. The soliton experiences frequency and timing jitter. The frequency jitter poses a particularly serious threat, since it is transformed into a timing jitter by propagation along a dispersive fiber. This effect has become known as the Gordon–Haus effect. Next, we show in Sect. 10.6 how filtering can reduce the effect, and describe the sliding-guiding-filter concept introduced by Mollenauer et al. [90], which greatly extends the error-free propagation distances of solitons. Polarization effects in soliton propagation are considered after that. Finally, we study the continuum generated by a perturbation of the soliton.

10.1 The Nonlinear Schrödinger Equation

The propagation equation of a mode on a dispersive fiber was derived in Sect. 3.5. It was obtained by an expansion of the propagation constant to the second order in frequency deviation $\Delta\omega$ from a carrier frequency. The differential equation for the pulse envelope $a(z,t)$ expressed in terms of the time variable τ contains only the second derivative of the propagation constant with respect to frequency, β'' (see (3.58)):

$$\frac{\partial}{\partial z}a = -\mathrm{i}\frac{\beta''}{2}\frac{\partial^2}{\partial t^2}a \ . \tag{10.1}$$

If the propagation constant β is perturbed by $\delta\beta$ via some other mechanism, then (10.1) becomes

$$\frac{\partial}{\partial z}a = -\mathrm{i}\frac{\beta''}{2}\frac{\partial^2}{\partial t^2}a + \mathrm{i}\delta\beta a \ . \tag{10.1a}$$

The Kerr effect produces a perturbation of the propagation constant by changing the index of the medium in which the wave propagates. The Kerr effect is defined to be positive when the index increases with increasing intensity. The index of the fiber is written (see (3.72)) as

$$n = n_o + n_2 I \ , \tag{10.2}$$

where I is the intensity. A mode of amplitude a, with $|a|^2$ normalized to power, has a nonuniform intensity profile. Since the phase shift in one wavelength is exceedingly small (10^{-7} or smaller), one may use perturbation theory to evaluate the change of index due to the Kerr effect. The field in the fiber is mainly transverse. We denote the normalized field profile by

$$\boldsymbol{E} = a(z)e(x, y) , \tag{10.3}$$

with $|\boldsymbol{E}|^2$ so normalized that its square is equal to the intensity,

$$|\boldsymbol{E}|^2 = I , \tag{10.4}$$

and

$$\int dx \int dy \, |e(x, y)|^2 = 1 . \tag{10.5}$$

From perturbation theory (3.47), we find for the change of propagation constant

$$\begin{aligned}
\frac{\delta\beta}{\beta} &= \frac{\int dx \int dy \, \delta n |\boldsymbol{E}(x, y)|^2}{n_o^2 \int dx \int dy \, |e(x, y)|^2} \\
&= \frac{|a|^2 n_2 \int dx \int dy \, |e(x, y)|^4}{n_o \int dx \int dy \, |e(x, y)|^2} \tag{10.6} \\
&\approx \frac{|a|^2 n_2}{n_o \mathcal{A}_{\text{eff}}} ,
\end{aligned}$$

where we have set $\beta = \omega^2 \mu_o \epsilon = \omega^2 \mu_o \epsilon_o n_o^2$, and the ratio of the integrals of the fourth power and the square of the field patterns defines the inverse of an effective area. Further, we have taken advantage of the fact that the index profile is almost constant at the value n_o. Since $\beta \approx (2\pi/\lambda)n_o$, we find for $\delta\beta$

$$\delta\beta = \frac{2\pi}{\lambda} n_2 \frac{1}{\mathcal{A}_{\text{eff}}} |a|^2 . \tag{10.7}$$

The derivation of the propagation constant change did not address specifically the time dependence of the intensity. In fact, (10.7) is the correct expression when $|a|^2$ is interpreted as the instantaneous power. In order to see this, one must view the self-phase modulation due to the Kerr effect as a degenerate four-wave mixing process in which three waves with frequencies ω, ω', and ω'' combine to give a fourth wave with frequency ω'''. The fourth wave is the result of a product of three waves. If the frequency of the fourth wave is to be close to the frequency of the three waves generating it, of frequencies ω, ω', and ω'', then the product must contain one of the three waves complex conjugated, e.g. the source of the fourth wave, of frequency ω''', is of the form

$$\int d\omega \int d\omega'\, a^*(\omega)a(\omega')a(\omega'') ,$$ (10.8)

where the frequency ω''' is

$$\omega''' = \omega' + \omega'' - \omega .$$ (10.9)

The inverse Fourier transform of the source term in the time domain has a much simpler appearance, since a convolution transforms into a product:

$$\int d\omega''' e^{-i\omega''' t} \int d\omega \int d\omega'\, a^*(\omega)a(\omega')a(\omega'')$$

$$= \int d\omega''' e^{-i(\omega'''+\omega-\omega')t} a(\omega''' + \omega - \omega') \int d\omega\, e^{i\omega t} a^*(\omega) \int d\omega'\, e^{-i\omega' t} a(\omega')$$

$$= a^*(t)a(t)a(t) = |a(t)|^2 a(t) .$$
(10.10)

We can now incorporate the Kerr effect into the mode equation (10.1). In the time domain,

$$i\delta\beta a(z,t) = i\frac{2\pi}{\lambda} n_2 \frac{1}{\mathcal{A}_{\text{eff}}} |a(z,t)|^2 a(z,t) ,$$

and thus (10.1) becomes

$$\frac{\partial}{\partial z} a(z,t) = -i\frac{\beta''}{2}\frac{\partial^2}{\partial t^2} a(z,t) + i\kappa |a(z,t)|^2 a(z,t) ,$$ (10.11)

with

$$\kappa \equiv \frac{2\pi}{\lambda}\frac{n_2}{\mathcal{A}_{\text{eff}}} .$$

This is the nonlinear Schrödinger equation.

Before we conclude this section, a few words of caution are in order. The mode patterns of modes on a fiber are ω-dependent. The present formalism ignores this dependence. This is an approximation, but a good one, since pulses as short as a picosecond contain thousands of wavelengths at an optical (infrared) wavelength of one micron or so. This means that pulses of one picosecond are very narrow-band and the assumption of ω independence of mode profiles is an excellent one over the range of frequency components involved.

10.2 The First-Order Soliton

In the regime of negative group velocity dispersion, $\beta'' < 0$, the nonlinear Schrödinger equation (NLSE) has a "solitary wave" solution of the form (see Fig. 10.2)

$$a_s(z,t) = A_o \, \text{sech}\left(\frac{t}{\tau_o}\right) \exp(i\kappa|A_o|^2 z/2) \, , \tag{10.12}$$

with the constraint

$$|A_o|\tau_o = \sqrt{|\beta''|/\kappa} \, . \tag{10.13}$$

This is a special case of the so-called area theorem [88]. Hence, we have an infinite number of pulse solutions of varying height and width. The energy of the pulse is inversely proportional to its width. Indeed,

$$\int dt \, |a_s(t)|^2 = |A_o|^2 \tau_o \int d\left(\frac{t}{\tau_o}\right) \text{sech}^2\left(\frac{t}{\tau_o}\right) = \left(\frac{|\beta''|}{\kappa}\right)\frac{2}{\tau_o} \, . \tag{10.14}$$

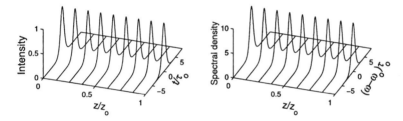

Fig. 10.2. First-order soliton: (a) intensity profile, (b) square of Fourier spectrum

The solution (10.12) has a phase shift due to the Kerr nonlinearity. The phase shift is uniform across the pulse and is given by $\theta(z) = \kappa|A_o|^2 z/2$, as if the average intensity of the pulse were responsible for it. The distance over which the phase shift is $\pi/4$ has become known as the soliton period. This is a rather strange definition since, strictly, the soliton period should be defined for a phase shift of 2π. However, we find that a so-called second-order soliton of initial sech shape, which goes through beats as shown in Fig. 10.3, repeats itself within this period, hence the name.

The soliton period is a measure of the action of the Kerr nonlinearity. It is also a measure of the dispersion effects of the fiber, since the two effects balance each other. If the fiber dispersion varies within distances much smaller than a soliton period, the pulse integrates the effect and only the average dispersion need be considered. Similarly, if loss decreases the pulse amplitude but is compensated by gain within distances much smaller than the soliton period, the loss and gain can be treated as averaged. This recognition has led to great advances in the design of soliton fiber communications. At first it was thought that the loss had to be compensated by distributed gain to obtain proper soliton propagation, namely by Raman gain of the fiber itself [5].

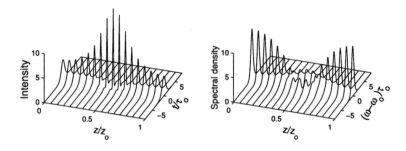

Fig. 10.3. Second-order soliton (**a**) intensity profile, (**b**) square of Fourier spectrum; $|A_o|\tau_o = 2\sqrt{|\beta''|/\kappa}; z_o = 4\pi\kappa|A_o|^2$

Nakazawa [87] recognized that this was not necessary, that lumped erbium-doped fiber amplifiers could be used if spaced by distances much smaller than the soliton period (e.g. 25 km for typical soliton periods of 330 km).

Thus far we have obtained a family of solutions that differ in height and width. The pulses have three more degrees of freedom. First of all, the phase θ_o of the pulse is arbitrary, as already implied by the complex character of A_o. Further, the time of occurrence t_o of the center of the pulse can be arbitrary. Finally, the carrier frequency may deviate by $\Delta\omega$ from the nominal carrier frequency ω_o, with the associated time dependence $\exp(-i\omega_o t)$ that has been removed at the beginning of the analysis. The modified solution can be written down by inspection, and it is left as an exercise for the reader to confirm that it indeed satisfies the nonlinear Schrödinger equation:

$$a_s(z,t) = |A_o| \operatorname{sech}\left(\frac{t - t_o + |\beta''|\Delta\omega\, z}{\tau_o}\right)$$

$$\times \exp\left[i\left(\frac{\kappa|A_o|^2}{2}z - \frac{|\beta''|\Delta\omega^2}{2}z + \phi\right)\right] \exp[-i\,\Delta\omega(t - t_o)].$$

$$(10.15)$$

The effects of the different parameters on the solution are self-evident. A time shift has no consequences. A frequency shift causes a change of the inverse group velocity of $-|\beta''|\Delta\omega$ and a change of the propagation constant of $-|\beta''|\Delta\omega^2/2$, which change the speed of propagation and the phase shift as shown.

If the carrier frequencies of the two solitons differ, they travel at different group velocities and one soliton can pass through the other. Solitons have the remarkable property that they can collide, yet completely recover after a collision. Figure 10.4 shows a computer simulation of a collision of two solitons. Whereas in a linear system two waves of different frequencies do not interact, since the excitation of the two waves is simply the superposition of the excitations of the individual waves, a collision of two wavepackets in a nonlinear

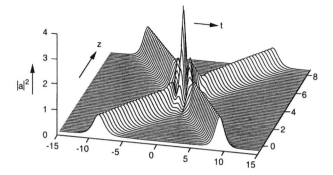

Fig. 10.4. Two colliding first-order solitons

system does not obey the superposition principle. Cross phase modulations occur owing to the nonlinearity. However, the nonlinear Schrödinger equation predicts full recovery of two colliding soliton pulses. The collision does, however, cause phase changes and position changes of the solitons.

The passage of solitons through each other without an effect on their shape has important implications for the use of solitons in wavelength-division-multiplexed (WDM) communications. Pulse streams in two different channels with different carrier frequencies (wavelengths) can pass through each other. While there are position changes due to individual collisions, they are small and average out if the signals in both channels are quasicontinuous. Thus, crosstalk between channels can be avoided. This is not true for multiplexed NRZ. Here the Kerr nonlinearity causes crosstalk which has to be combatted by proper choice of channel wavelengths. For the same reasons, the wavelength spacings of channels in an NRZ system must be wider than for a soliton system, giving the soliton system a bit-rate advantage.

It is customary to normalize the distance variable in (10.1) to a normalizing distance z_n, the time variable to a normalizing time τ_n, and the amplitude $a(z,t)$ to an amplitude A_n. With $|\beta''|z_n/\tau_n^2 = 1$, $\kappa|A_n|^2 z_n = 1$, and $a(z,t)/A_n = u(z,t)$, we obtain

$$-\mathrm{i}\frac{\partial}{\partial z}u(z,t) = \frac{1}{2}\frac{\partial^2}{\partial t^2}u(z,t) + |u(z,t)|^2 u(z,t) \,. \qquad (10.16)$$

We denote the normalized variables as t and z without a subscript so as not to encumber the notation. The normalizing distance is chosen so that the optical Kerr effect produces one radian of phase shift within unit distance. The normalization of the time variable (choice of normalized bandwidth) is chosen so as to produce equal and opposite effects due to GVD and the optical Kerr effect on a standard pulse of unity width. The purpose of the normalization is to arrive at the standard nonlinear Schrödinger equation, with the exception of a factor $1/2$ which has become customary in fiber soliton theory. In Gordon's notation [91], the solution (10.15) in normalized form becomes

$$u_s(z,t) = A \operatorname{sech}(At - q) \exp(iVt + i\phi) ,\qquad(10.17)$$

where

$$\frac{dq}{dz} = AV\qquad(10.18)$$

and

$$\frac{d\phi}{dz} = \frac{1}{2}(A^2 - V^2) .\qquad(10.19)$$

Here we retain z for the (normalized) distance variable and t for the time variable. Gordon uses t for the distance variable and x for the time variable to emphasize the nature of (10.11) as the nonlinear version of the Schrödinger equation. Gordon's notation has mnemonic value. A is the amplitude of the soliton; V is its velocity; q is its position, reminding one of the quantum notation for position; ϕ, of course, is its phase.

10.3 Properties of Solitons

In the preceding section we denoted the solution to the nonlinear Schrödinger equation (NLSE) as a "solitary wave" as well as a "soliton". The term "soliton" is applied, strictly, only to solutions of nonlinear equations that have certain stability properties, e.g. in a collision of two such waves, the two components must emerge unscathed. This is the case with the solitary-wave solutions of the NLSE, and thus the term "soliton" can be rigorously applied. Before we proceed with the study of collisions, we address the remarkable formation process of solitons.

If an input pulse has an area that lies in the range between $\pi/2$ and $3\pi/2$, a soliton forms from the pulse [92]. Figure 10.5 shows the evolution of a soliton from a square pulse of an area obeying this condition. One sees that the soliton "cleans itself out" by shedding continuum. Since the continuum travels away from the soliton in both directions, it has components that are both faster and slower than the soliton. Their frequencies are thus higher and lower, respectively, than the carrier frequency of the soliton. (Note the continuum is of low intensity and thus has linear propagation properties.) These frequency components are, in part, contained in the original excitation, but are also generated in the nonlinear processes partaking in the soliton formation.

The formation process described above has been used to generate solitons at high bit rates [93–100]. The input is a superposition of two continuous waves of equal amplitude and offset by a frequency $\Delta\omega_o$. The result is a sinusoidal beat between the two waves. If the intensities are such that the

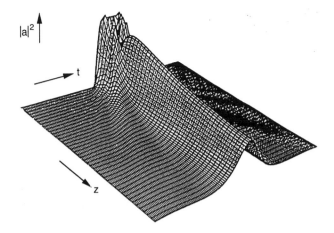

Fig. 10.5. A square pulse evolves into a first order soliton.

area of the excitation between the two nodes of the beat obeys the soliton formation criterion, a bit stream of solitons forms after propagating in a fiber of appropriate length. This pulse formation scheme, followed by an amplitude modulator, has been proposed as a source for soliton communications.

Next, consider soliton collisions. These can be described by a higher-order soliton, a soliton of second order, which is also a closed-form solution of the nonlinear Schrödinger equation, obtainable by the inverse scattering approach of Zakharov and Shabat. It can be written [91] as (note that we have substituted x_j for Gordon's z_j)

$$u = \frac{A_1 e^{i\theta_1}(\rho^* \beta e^{-x_2} + \rho\beta^* e^{x_2}) + A_2 e^{i\theta_2}(\rho^* \beta^* e^{-x_1} + \rho\beta e^{x_1})}{|\rho|^2 \cosh(x_1 + x_2) + |\beta|^2 \cosh(x_1 - x_2) + 4A_1 A_2 \cos(\theta_1 - \theta_2)},$$

$$(10.20)$$

with

$$x_j = A_j t - q_j , \tag{10.21a}$$

$$\theta_j = V_j t + \phi_j , \tag{10.21b}$$

$$\rho = A_1 - A_2 + i(V_1 - V_2) , \tag{10.21c}$$

$$\beta = A_1 + A_2 + i(V_1 - V_2) . \tag{10.21d}$$

The q_j and ϕ_j obey the following equations (compare with (10.18) and (10.19)):

$$\frac{dq_j}{dz} = A_j V_j$$

and

$$\frac{d\phi_j}{dz} = \frac{1}{2}(A_j^2 - V_j^2) .$$

The V_j are normalized velocities or carrier frequencies. If they are picked to be different, (10.20) describes two solitons that are well separated at $t \to -\infty$, collide, and become again well separated as $t \to +\infty$. The solitons experience a timing shift and a phase shift, but otherwise recover fully. The smaller the carrier frequency separation between the two pulses, the larger the shifts are. A collision is shown in Fig. 10.4. When the two pulse envelopes overlap, a beat between the two carrier frequencies is clearly discernible.

The solution in (10.20) can also be used to study the interaction between two solitons when they are well separated. This was done analytically by Gordon [91] and verified experimentally by Mitschke and Mollenauer [101]. The interactions are important in optical communications, since they can also introduce errors in a bit stream of solitons, whose phases may vary randomly, and in which some are randomly omitted. Suffice it to state here that some of these interactions are easily understood. If two solitons of equal phase are placed close to each other, the potential well produced in combination is deeper than when they are widely separated. Thus closeness is energetically favored, which leads to an attractive force. The opposite is true when the solitons are in antiphase; they repel each other.

10.4 Perturbation Theory of Solitons

Next we consider a perturbation of a soliton by a noise source $s(z,t)$. Perturbation theories of solitons can be developed from the inverse scattering transform [89,102–108]. An alternate approach is to start with the linearized form of the NLSE and project out the excitations produced by the perturbation using the adjoint functions [6, 109]. This latter approach provides more direct insight into the physics of the processes involved. The equation for the soliton now reads

$$-i\frac{\partial}{\partial z}u(z,t) = \frac{1}{2}\frac{\partial^2}{\partial t^2}u(z,t) + |u(z,t)|^2 u(z,t) - is(z,t) \ . \tag{10.22}$$

If the source $s(z,t)$ is small, then $u(z,t)$ can be written $u(z,t) = u_s(z,t) + \Delta u(z,t)$, where $\Delta u(z,t)$ is the small deviation of the field from the soliton solution $u_s(z,t)$. The equation obeyed by $\Delta u(z,t)$ to first order is

$$-i\frac{\partial}{\partial z}\Delta u(z,t) = \frac{1}{2}\frac{\partial^2}{\partial t^2}\Delta u(z,t) + 2|u_s(z,t)|^2 \Delta u(z,t)$$

$$+u_s^2(z,t)\Delta u^*(z,t) - is(z,t) \ . \tag{10.23}$$

This is a linear equation with a source. The source can be represented by a sequence of local sources at different positions z', proportional to $\delta(z - z')$. Each excitation can then be expressed in terms of the solutions of the homogenous equation

$$-i\frac{\partial}{\partial z}\Delta u(z,t) = \frac{1}{2}\frac{\partial^2}{\partial t^2}\Delta u(z,t) + 2|u_s(z,t)|^2\Delta u(z,t)$$

$$+u_s^2(z,t)\Delta u^*(z,t) .$$

(10.24)

Linear homogeneous equations are solved by finding their eigenfunctions and writing the solution as a superposition of the eigenfunction excitations. The amplitudes of the excitations are projected out using orthogonality among the eigenfunctions, if the system is self-adjoint. Self-adjointness is a natural consequence of energy conservation: solutions with different time dependences must be orthogonal, since if this were not the case, the energy would be time-varying. Equation (10.22) is not self-adjoint. Even though the NLSE is derivable from a Hamiltonian, the perturbation equation describes excitations in the presence of a pump $u_s^2(z,t)$, and hence the energy of the perturbations need not be conserved. The perturbations may acquire energy from the pump and may lose it to the pump. Orthogonality can be achieved with the solutions of the adjoint equation, whose solutions $\Delta\underline{u}(z,t)$ obey cross-energy conservation:

$$\frac{d}{dz}\text{Re}\left[\int_{-\infty}^{+\infty} dt\,\Delta\underline{u}^*(z,t)\Delta u(z,t)\right] = 0 .$$

(10.25)

It is easily shown that the system adjoint to (10.22) is

$$-i\frac{\partial}{\partial z}\Delta\underline{u}(z,t) = \frac{1}{2}\frac{\partial^2}{\partial t^2}\Delta\underline{u}(z,t) + 2|u_s(z,t)|^2\Delta\underline{u}(z,t)$$

$$-u_s^2(z,t)\Delta\underline{u}^*(z,t) .$$

(10.26)

Note the sign change in the last term. This sign change corresponds to a 90 degree phase shift of the pump. The pump changes the index via the Kerr effect. This phenomenon is an example of a "parametrically" driven system, the parameter that is driven is the index of the fiber. Such systems will be discussed in the next chapter. We shall find that they give rise to solutions that grow or decay exponentially with time, depending upon their phase relative to the pump phase. Or, alternatively, a 90° phase change of the pump transforms a growing solution into a decaying one and vice versa. Whereas growing solutions cannot preserve energy, the cross-energy of growing and decaying solutions can be preserved. From this brief discussion, illustrated more thoroughly in the next chapter, one may surmise that the adjoint of a parametrically driven system is obtained from the original system by a 90° phase change of the pump. This explains the sign change from the original equation (10.24) to its adjoint (10.26).

We write the perturbation as a superposition of changes in the four soliton parameters and of the continuum (we use notation based on Gordon's form of the solution in (10.17)):

$$\Delta u(z,t) = [\Delta A(z)f_A(t) + \Delta\phi(z)f_\phi(t) + \Delta q(z)f_q(t)$$
$$+ \Delta V(z)f_V(t)]e^{iz/2} + \Delta u_c(z,t) , \tag{10.27}$$

where $\Delta u_c(z,t)$ is the continuum. The four perturbation functions $f_P(t), P = A, \phi, q,$ and V, are derivatives of the soliton solution with respect to its four parameters, evaluated at $z = 0$:

$$f_A(t) = \frac{\partial}{\partial A}u_s(0,t) = (1 - t\tanh t)\,\mathrm{sech}\,t , \tag{10.28a}$$

$$f_\phi(t) = \frac{\partial}{\partial \phi}u_s(0,t) = i\,\mathrm{sech}\,t , \tag{10.28b}$$

$$f_q(t) = \frac{\partial}{\partial q}u_s(0,t) = \tanh t\,\mathrm{sech}\,t , \tag{10.28c}$$

$$f_V(t) = \frac{\partial}{\partial V}u_s(0,t) = it\,\mathrm{sech}\,t . \tag{10.28d}$$

With no loss of generality, the unperturbed soliton solution has been assumed to have $A = 1, \phi = 0, q = 0, V = 0$. The adjoint equation has similar solutions. They are orthonormal to the set in (10.28) and are found to be

$$\underline{f}_A(t) = \mathrm{sech}\,t , \tag{10.29a}$$
$$\underline{f}_\phi(t) = i(1 - t\tanh t)\,\mathrm{sech}\,t , \tag{10.29b}$$
$$\underline{f}_q(t) = t\,\mathrm{sech}\,t , \tag{10.29c}$$
$$\underline{f}_V(t) = i\tanh t\,\mathrm{sech}\,t . \tag{10.29d}$$

The adjoint functions must be orthogonal to the continuum. Indeed, at $t \to +\infty$, the continuum is completely dispersed and has no overlap with the functions that occupy the region around the soliton. Because of the conservation law, the orthogonality must hold for all time.

When (10.27) is introduced into the governing equation (10.23), the second derivative of $f_A(t)$ with respect to time produces a term proportional to $f_\phi(t)$. This simply means that a change of amplitude causes a cumulative change of phase, since the contribution from the Kerr effect has changed. Similarly, the second time derivative of $f_V(t)$ produces a term proportional to $f_q(t)$; a change of carrier frequency causes a cumulative change of displacement due to a change in group velocity. The perturbation parameters are projected out by the four adjoint functions. The result is four equations of motion for the soliton parameters:

$$\frac{d}{dz}\Delta A = S_A(z) , \tag{10.30}$$

$$\frac{d}{dz}\Delta\phi = \Delta A + S_\phi(z) , \tag{10.31}$$

$$\frac{d}{dz}\Delta q = \Delta V + S_q(z) \, , \tag{10.32}$$

$$\frac{d}{dz}\Delta V = S_V(z) \, . \tag{10.33}$$

where the sources are given by

$$S_P(z) = \mathrm{Re}\left[\int_{-\infty}^{+\infty} dt\, \underline{f}_P^*(t)e^{-\mathrm{i}z/2}s(z,t)\right] \, . \tag{10.34}$$

These equations can and will be augmented to include filtering. Before we proceed, we take note of the fact that the perturbation analysis permits large changes of ΔA, $\Delta\phi$, Δq, and ΔV, as long as these changes are gradual. Phase shifts, displacements, and frequency shifts leave the soliton envelope unchanged. Even large amplitude changes may be incorporated if the projection functions in (10.26) are generalized to an arbitrary value of the amplitude A, as long as the sources $S_P(z')$ are evaluated at any cross section z' consistent with the state of the soliton at that cross section. Then the parameters ΔA, $\Delta\phi$, Δq, and ΔV are allowed to become large. We emphasize this fact by dropping the prefix Δ henceforth and replacing ΔA by $A - 1$, and ΔV by V:

$$\frac{dA}{dz} = S_A(z) \, , \tag{10.31a}$$

$$\frac{d\phi}{dz} = A - 1 + S_\phi(z) \, , \tag{10.32a}$$

$$\frac{dq}{dz} = V + S_q(z) \, , \tag{10.33a}$$

$$\frac{dV}{dz} = S_V(z) \, . \tag{10.34a}$$

10.5 Amplifier Noise and the Gordon–Haus Effect

The equations of motion of the four soliton perturbation parameters (10.30)–(10.33) contain sources. Linear optical amplifiers have unavoidable noise, as pointed out in Chap. 7. Loss also introduces noise sources that conserve commutator brackets and in doing so conserve zero-point fluctuations. In long-distance soliton transmission, the loss of the fiber is compensated by gain. In practice, the gain is "lumped", i.e. is provided in a fiber amplifier of length negligible compared with the soliton period. The gain can be treated as uniformly distributed if the spacing between amplifiers is much shorter than a soliton period. Since the propagation along the fiber is nonlinear, noise fluctuations added in transit through the amplifier are incorporated into all four soliton perturbations as described by (10.30)–(10.33). The soliton experiences a time displacement and a frequency displacement every time it passes through an amplifier. The time displacements turn out to be too small to worry about, even when accumulated in a fiber of transoceanic length. However, frequency changes are transformed into time displacement via the GVD of the fiber. This frequency-induced timing jitter has become known as the Gordon–Haus effect [6]. This effect imposes a severe limit on long distance soliton propagation unless proper precautions are taken.

We shall start with distributed gain that compensates for the fiber loss. Lumped gain is, of course, the practical case. We shall then consider the effect caused by lumped gain. Suppose the normalized amplitude gain coefficient is α. If the gain compensates perfectly for the loss, (10.17) remains unchanged, assuming that noise can be neglected. However, for long-distance propagation with net gains in the 100 dB range, amplifier noise cannot be neglected. Amplifier noise appears as a source $s(z, t)$ in (10.22). If the amplifier bandwidth is much larger than the signal bandwidth, the source may be considered to be a white noise source.

Equation (10.22) contains only normalized quantities. Hence the correlation function of the noise source

$$\langle s(z, t) s^*(z', t') \rangle = \hbar \omega_o \chi 2 \alpha \delta(z - z') \delta(t - t')$$

must also be normalized. First of all, the amplitude gain coefficient α is normalized by multiplication by z_n, i.e. by replacing αz_n by its normalized counterpart α_n. We do not want to encumber the notation by attaching subscripts "n" to all normalized quantitities. Whether quantities are normalized or not will be obvious from the context. Replacement will be indicated by arrows, e.g. $\alpha z_n \to \alpha$. The photon energy is normalized by division by $|A_o|^2 \tau_n$. The unit impulse functions call for normalizing factors z_n and τ_n. But these will be removed by integration over the normalized z and t, and so they need not be introduced at this stage. Thus, the normalized form of the noise source is

$$\langle s(z, t) s^*(z', t') \rangle = 2 \alpha \mathcal{N} \delta(z - z') \delta(t - t') , \tag{10.35}$$

with $\mathcal{N} = (\hbar \omega_o / |A_o|^2 \tau_n) \chi$. This shows that the normalized noise source is inversely proportional to the photon number in the pulse. The more intense the pulse, the less the influence of the noise source. Since the noise is described in terms of a correlation function, the response must be similarly expressed. The responses take the form of the real part of complex projections, meaning that only the in-phase or the quadrature component of the noise represented by (10.35) contributes to any one of these projections. Since the noise is stationary, the in-phase and quadrature components have equal intensities, each with correlation functions equal to half the value of (10.35). The correlation function of the noise source in (10.34a), driving the frequency parameter V, is

$$\langle S_V^*(z)S_V(z')\rangle = \alpha\mathcal{N}\int_{-\infty}^{+\infty} dt \int_{-\infty}^{+\infty} dt'\,\delta(t-t')\underline{f}_V(t)\underline{f}_V^*(t')\delta(z-z')$$

$$= \alpha\mathcal{N}\int_{-\infty}^{+\infty} dt\,\tanh^2 t\,\text{sech}^2 t\,\delta(z-z')$$

$$= \frac{2}{3}\alpha\mathcal{N}\delta(z-z')\,.$$

$$(10.36)$$

The noise source driving the displacement q (note that $A=1$) is

$$\langle S_q^*(z)S_q(z')\rangle = \alpha\mathcal{N}\int_{-\infty}^{+\infty} dt \int_{-\infty}^{+\infty} dt'\,\delta(t-t')\underline{f}_q(t)\underline{f}_q^*(t')\delta(z-z')$$

$$= \alpha\mathcal{N}\int_{-\infty}^{+\infty} dt\,t^2\,\text{sech}^2 t\,\delta(z-z')$$

$$= \frac{\pi^2}{6}\alpha\mathcal{N}\delta(z-z')\,.$$

$$(10.37)$$

The correlation function of the frequency parameter is

$$\langle V^*(z)V(z')\rangle = \left\langle \int_0^z dz''\,S_V^*(z'') \int_0^{z'} dz'''\,S_V(z''') \right\rangle$$

$$(10.38)$$

$$= \begin{cases} \dfrac{2\alpha\mathcal{N}}{3} \displaystyle\int_0^{z'} dz''' = \dfrac{2\alpha\mathcal{N}}{3}z' & \text{for } z' < z \\[2mm] \dfrac{2\alpha\mathcal{N}}{3} \displaystyle\int_0^{z} dz'' = \dfrac{2\alpha\mathcal{N}}{3}z & \text{for } z' > z \end{cases}\,.$$

The autocorrelation at $z = z'$ grows linearly with z. Thus the frequency fluctuations grow like the displacement in a random walk. The mean square fluctuations of the displacement are produced by frequency fluctuations on one hand, and a noise source driving the displacement directly on the other hand. Since the two noise sources are independent, the mean square fluctuations that they produce are additive. Considering first the mean square fluctuations due to the noise source $S_q(z)$, we have for the fluctuations at a normalized distance L, in analogy with (10.38),

$$\langle q^*(L)q(L)\rangle_q = \left\langle \int_0^L dz'' S_q^*(z'') \int_0^L dz''' S_q(z''') \right\rangle$$

$$= \frac{\pi^2 \alpha \mathcal{N}}{6} \int_0^L dz''' = \frac{\pi^2 \alpha \mathcal{N}}{6} L .$$

(10.39)

These mean square fluctuations grow linearly with L. They correspond to a simple random walk of the displacement variable. The mean square fluctuations caused by the frequency fluctuations are

$$\langle q^*(L)q(L)\rangle_V = \left\langle \int_0^L dz'' V^*(z'') \int_0^L dz''' V(z''') \right\rangle$$

$$= \frac{2\alpha \mathcal{N}}{3} 2 \int_0^L dz'' \int_0^{z''} z''' dz''' = \frac{2\alpha \mathcal{N}}{3} \frac{L^3}{3} .$$

(10.40)

The frequency fluctuations experiencing a random walk (linear growth with L) translate into a growth of the displacement, since pulses with different carrier frequencies travel at different speeds, and this effect becomes severe for large distances of propagation. This is the so-called Gordon–Haus effect [6]. It leads to random displacements of pulses, which may end up in neighboring time slots, causing bit errors.

The analysis thus far has been in normalized units. The standard soliton was sech t, where the normalization time τ_n was equal to the pulse width τ. The mean square displacement fluctuations $\langle q^*(L)q(L)\rangle$ were normalized to the pulse width. The bit-error rate can be computed from these fluctuations directly once the pulse width to bit interval ratio is chosen. The right-hand side of (10.40) is converted to physical dimensions as follows:

$$\langle q^*(L)q(L)\rangle_V = \frac{2\alpha \mathcal{N}}{3} \frac{L^3}{3} \rightarrow \frac{2\alpha z_n}{3} \frac{\hbar \omega_o}{|A_n|^2} \chi \frac{1}{\tau_n} \frac{L^3/z_n^3}{3} .$$

(10.41)

Using the relations $|\beta''| z_n/\tau_n^2 = 1$ and $\kappa |A_n|^2 z_n = 1$, we can write the above in terms of unnormalized parameters:

$$\langle q^*(L)q(L)\rangle_V = \frac{2\alpha}{3} \hbar \omega_o \chi |\beta''| \frac{\kappa L^3}{3\tau^3} .$$

(10.42)

The effect is proportional to the Kerr coefficient κ, indicating that the jitter is Kerr-induced. The jitter increases with the cube of the distance and is proportional to the GVD. Reducing the GVD reduces the effect and this fact is used in the design of the fiber. Ideally, the group velocity dispersion should be made as small as possible. There is a lower limit set by the minimum energy required for a given signal-to-noise ratio. Since the bit rate is proportional to $1/\tau$, we find that the jitter increases with the cube of the bit rate.

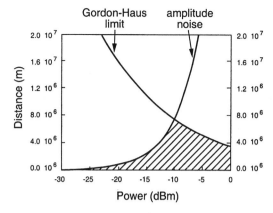

Fig. 10.6. The distance that can be covered with a bit-error rate of 10^{-9}; dispersion 2 ps/(nm km), peak soliton intensity 90 mW; after [110]

Figure 10.6 shows a plot of the distance that can be covered with a bit-error rate equal to or better than 10^{-9} at 5 Gbit/s transmission. The shaded area is the allowed range. There are two boundaries: one, "denoted amplitude noise", is set by requiring a sufficiently large signal-to-noise ratio that "ones" are not mistaken for "zeros" and vice versa. This is due to the additive nature of the ASE noise of the amplifiers. The other boundary is due to the nonlinear aspect of the noise–timing jitter as a result of random frequency shifts. For a trans-Atlantic distance of 4800 km, the allowed range of signal power is quite narrow. This was an aspect of soliton transmission that had to be improved. The improvement came with the introduction of filters.

10.6 Control Filters

In work on noise in fiber ring lasers at MIT, it was found that an effect analogous to the Gordon–Haus effect appeared in such systems as well, but the introduction of filters tended to alleviate it. This work was not published until 1993 [111]. However, the connection was made with long-distance pulse propagation and it was found that the introduction of filters every so often into repeaterless soliton transmission systems [112] could control the Gordon–Haus effect. Independently, Kodama and Hasegawa also arrived at the same conclusion [113]. Let us look at the theory explaining this action of filters.

We consider the simple case of a continuous distribution of filters (lumped filters generate continuum and are thus less ideal; such continuum radiation will be considered later on). The fundamental equation (10.22) is altered to

$$-i\frac{\partial}{\partial z}u(z,t) = \frac{1}{2}\frac{\partial^2}{\partial t^2}u(z,t) + |u(z,t)|^2 u(z,t) - i\frac{1}{\Omega_f^2}\frac{\partial^2}{\partial t^2}u(z,t) - s(z,t) ,$$

$$(10.43)$$

where $1/\Omega_f^2$ expresses the filtering per unit (normalized) distance. The soliton perturbation equation changes accordingly. Since the filtering is assumed to be a small perturbation of the soliton, one may solve the fundamental equation (10.22) without the filtering term, and treat the effect of the filter as a new perturbation term in the linearized perturbed equation of motion (10.24):

$$-i\frac{\partial}{\partial z}\Delta u(z,t) = \frac{1}{2}\frac{\partial^2}{\partial t^2}\Delta u(z,t) + 2|u_s(z,t)|^2\Delta u(z,t)$$

$$+ u_s^2(z,t)\Delta u^*(z,t) - i\frac{1}{\Omega_f^2}\frac{\partial^2}{\partial t^2}u_s(z,t) - s(z,t) .$$

$$(10.44)$$

For the soliton solution we take (compare (10.17))

$$u_s(z,t) = A\operatorname{sech} At\exp[i(V - V_o)t] ,$$ $$(10.45)$$

where $V - V_o$ expresses the frequency deviation of the soliton from the (normalized) center frequency V_o of the filter. We use the ansatz (10.27), except that now we replace ΔV with $V - V_o$. We carry out the projection with the adjoint functions. The filtering term introduces a damping constant into the equation of motion of V:

$$\frac{dV}{dz} = -\gamma(V - V_o) + S_V(z) ,$$ $$(10.46)$$

where

$$\gamma = \frac{4}{3\Omega_f^2} .$$ $$(10.47)$$

As the carrier frequency deviates from the center frequency of the filter, the part of the spectrum farther away from the center experiences greater attenuation than the part nearer the center. The spectrum is pushed towards the center of the filter response. Thus, we have chosen V_o as the steady-state frequency of the soliton.

For the purpose of the analysis in this section, we set $V_o = 0$. The carrier frequency exposed to the driving source of noise does not experience a random walk, since the filter limits the deviation. First, we compute the correlation function of $V(z)$. Since

$$V(z) = e^{-\gamma z}\int_0^z dz'\, e^{\gamma z'} S_V(z') ,$$ $$(10.48)$$

we obtain

$$\langle V^*(z)V(z')\rangle$$

$$= \frac{2}{3}\alpha \mathcal{N} e^{-\gamma(z+z')} \int_0^z dz'' \int_0^{z'} dz''' \, \delta(z'' - z''') e^{\gamma(z''+z''')}$$

$$= \frac{2}{3}\alpha \mathcal{N} e^{-\gamma(z+z')} \int_0^{z'} dz''' \, e^{2\gamma z'''} \tag{10.49}$$

$$= \frac{2}{3}\alpha \mathcal{N} [e^{-\gamma(z-z')} - e^{-\gamma(z+z')}]/2\gamma \quad \text{for } z' < z$$

$$= \frac{2}{3}\alpha \mathcal{N} [e^{-\gamma(z'-z)} - e^{-\gamma(z+z')}]/2\gamma \quad \text{for } z' > z \, .$$

For $\gamma \to 0$, the result agrees with (10.38). When we introduce (10.49) into the equation for the mean square fluctuations of position, we find

$$\langle q^*(L)q(L)\rangle_V = \left\langle \int_0^L dz'' \, V^*(z'') \int_0^L dz''' \, V(z''') \right\rangle$$

$$= \frac{2}{3} 2\alpha \mathcal{N} \int_0^L dz'' \int_0^{z''} dz''' \, e^{-\gamma(z''+z''')} \frac{e^{2\gamma z'''} - 1}{2\gamma} \tag{10.50}$$

$$= \frac{2\alpha \mathcal{N}}{3\gamma^3} \left[\gamma L + \frac{1}{2}(1 - e^{-2\gamma L}) - 2(1 - e^{-\gamma L}) \right] \, .$$

It is clear that as $L \to +\infty$, the position fluctuations grow linearly with L. Introducing physical dimensions, (10.50) becomes, in the limit of large L,

$$\langle q^*(L)q(L)\rangle_V = \frac{2\alpha(\chi)\hbar\omega_o \kappa |\beta''|L}{3\gamma^2 \tau^3} \, , \tag{10.51}$$

where L is the physical length and γ is the filtering constant, $\gamma = 4/3\Omega_f^2 \tau^2$ in units of inverse length.

Figure 10.7 shows a plot of the range of distances reachable with filters in place. The range of permissible powers for trans-Atlantic and trans-Pacific distances is now greatly increased. There is another limit imposed by soliton attraction, as mentioned earlier. This soliton attraction effect can be suppressed by a weak modulation of the soliton energy so that neighboring solitons end up with slightly different peak intensities. As they propagate along the fiber, they experience different Kerr phase shifts. Thus, adjacent solitons are alternately in phase and out of phase; the attractive force gives way to a repulsive force and back again. In this way the soliton attraction effect can be eliminated. It should be noted that the bandwidth of a cascade of filters spanning a transoceanic distance tends to be extremely narrow so

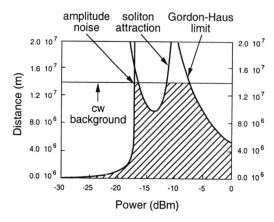

Fig. 10.7. The distance of propagation for a bit-error rate of 10^{-9}; filter bandwidth 0.72 ps^{-1}. The parameters are the same as for Fig. 10.6. After [110]

that "linear" signal transmission at a high bit rate would be impossible. It is the remarkable stability of the solitons that permits them to recover their bandwidth after each filter via the nonlinearity of the fiber, whereas linear signals cannot do so.

Another benefit of the filters is one not associated with noise reduction. Filtering provides stabilization against excessive energy changes of the solitons as they propagate along the fiber cable. An increase of the soliton energy above the design average shortens the soliton and broadens its spectrum. Pulses with a broader spectrum experience excess loss and thus energy increases are reduced by filtering. Energy decreases are similarly combatted. This effect is particularly advantageous when solitons are wavelength-division-multiplexed. The filters for such an application have periodic passbands, one for each channel. Since the gain varies over the erbium bandwidth, different channels experience slightly different gains. The energy stabilization by filtering acts against gain variations.

Filtering, however, is associated with a noise penalty. The solitons require increased gain to compensate for the loss of the filters, which is, per unit length,

$$\frac{\partial}{\partial z} \int_{-\infty}^{+\infty} dt \, |u_s(z,t)|^2$$

$$= \frac{1}{\Omega_f^2} \int_{-\infty}^{+\infty} dt \left[u_s(z,t) \frac{\partial^2}{\partial t^2} u_s^*(z,t) + u_s^*(z,t) \frac{\partial^2}{\partial t^2} u_s(z,t) \right] \qquad (10.52)$$

$$= -\frac{2}{\Omega_f^2} \int_{-\infty}^{+\infty} dt \left| \frac{\partial}{\partial t} u_s(z,t) \right|^2 = -\frac{2}{\Omega_f^2} \, .$$

Noise at the center frequency is not affected by the filters and sees excess gain. This noise eventually limits the propagation distance. Mollenauer and his coworkers [114] arrived at an ingenious way of eliminating this effect by gradually changing the center frequency of the filters along the cable. In its simplest manifestation, the effect of sliding guiding filters is incorporated into (10.46) by noting that V stands for the frequency deviation from the soliton carrier frequency ω_o. If the normalized filter center frequencies $V_o(z)$ are functions of distance along the fiber, then (10.46) becomes

$$\frac{dV}{dz} = -\gamma[V - V_o(z)] + S_V(z) .\tag{10.53}$$

Solitons adapt to the sliding guiding filters of continuously varying $V_o(z)$ by changing their carrier frequency. Because their carrier follows the center frequency of the filters, their loss is less than the loss of the linear noise, which cannot adapt in this way. Off hand, one would expect that up-shifting or down-shifting the filter center frequency along the propagation direction would result in the same amount of noise suppression. If Fabry–Pérot-type filters are used, so as to permit WDM transmission, up-shifting leads to better noise suppression owing to a subtle effect. It is clear that the soliton carrier frequency will deviate from the center frequency of the filter passband: since the solitons are continuously forced to change frequency, their carrier frequency lags behind the shift. As they move off the filter center frequency, higher than second-order GVD is experienced by the solitons. The sign of the third-order GVD is opposite for opposite deviations from the filter center frequency. It so happens that up-shifting the sliding guiding filters introduces a third-order GVD of a sign that is less harmful than that for down-shifting [114].

10.7 Erbium-Doped Fiber Amplifiers and the Effect of Lumped Gain

The amplifying characteristics of an erbium-doped silica fiber amplifier are shown in Fig. 10.8 [115]. Pumping has been demonstrated at 800 nm, 980 nm, and 1480 nm. Note that because of the long lifetime of the metastable state, the pump power required to achieve a 30 dB gain is only of the order of 40 mW, easily supplied by a commercially available diode laser. Three typical amplifier configurations, forward pumping, reverse pumping, and bidirectional pumping, are shown in Fig. 10.9 [116]. The ASE spectrum indicates some gain nonuniformity, which can be made more uniform with additional Al doping [117].

The variation of the pulse energy with distance along the fiber cable for a 25 km spacing of amplifiers is shown in Fig. 10.10. Note that the variation is large, and hence one would assume that the distributed soliton model

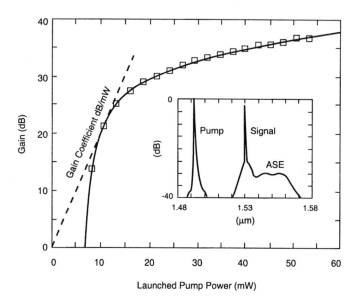

Fig. 10.8. The amplifying characteristics of an erbium-doped fiber amplifier [115]

represented by the NLSE cannot be applied. Fortunately, this is not the case. For typical fiber parameters the soliton period is greater than 200 km. Remember that a soliton experiences a phase shift of 2π over a distance of eight soliton periods. Thus the nonlinear change of the pulse over 25 km is small. Since the nonlinear change balances the linear dispersive change, the latter is small as well. Thus, the 25 km distance may be considered to be a "differential" distance. The propagating pulse is an "average" soliton [118] or a "guiding center" soliton [119]. Its average phase shift is computed from the cumulative small phase shifts in each 25 km segment.

Mollenauer et al. used the experimental setup shown in Fig. 10.11 [120]. A fiber ring with three amplifiers was loaded through a coupler and filled with a pseudorandom sequence of ones and zeros. The excitation was allowed to circulate in the ring and, after a chosen number of transits, coupled out and detected. The microwave spectrum analyzer was a convenient means for measuring the pulse jitter. Figure 10.12 shows the experimental results without the use of filters. The jitter tracks the prediction of the Gordon–Haus effect. Figure 10.13 shows the results of a measurement using the sliding guiding filters. As one can see, the propagation distance for a given bit-error rate has been greatly increased. There are noise contributions other than those attributable to the Gordon–Haus effect. It is believed that these are due to a piezo-optic interaction between solitons [121, 122].

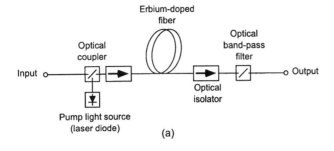

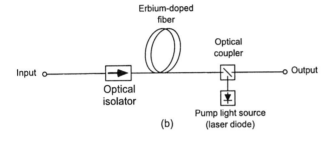

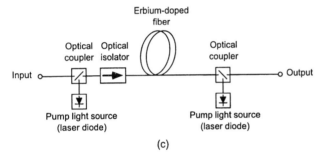

Fig. 10.9. Various amplifier configurations; (**a**) forward pumping, (**b**) backward pumping, and (**c**) bidirectional pumping [116]

10.8 Polarization

Thus far we have not discussed the fact that fibers have a natural linear birefringence (of the order of 10^{-6}–10^{-7}, i.e. a transformation from one linear polarization to the orthogonal polarization occurs in 10^6–10^7 wavelengths). What is the effect of the birefringence on soliton propagation?

If the birefringence were fixed and did not vary randomly along the fiber, the effect would indeed be severe. We have mentioned the remarkable properties of solitons resulting from the integrability of the nonlinear Schrödinger equation. When two polarizations are coupled by birefringence, they are described by two coupled NLSEs, which are not integrable in general. Thus, one might expect that soliton propagation would be possible only

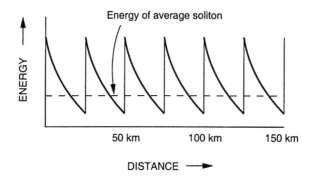

Fig. 10.10. Variation of pulse energy between amplifiers spaced 25 km apart

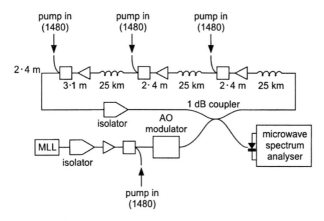

Fig. 10.11. Mollenauer's experimental recirculating loop, after [120]. The acousto-optic modulator is used to reject the signal pulse stream once the loop has been filled

in a polarization-maintaining (PM) fiber, which is more expensive than the regular fiber and also possesses higher losses. Fortunately for soliton communications, a regular fiber will do, for reasons we shall now explain.

The coupled nonlinear Schrödinger equations for the x and y polarizations, represented by the envelopes $v(z,t)$ and $w(z,t)$, are [123,124] (compare 3.71))

$$
\frac{\partial}{\partial z}v(z,t) = -i\frac{1}{2}\beta''\frac{\partial^2}{\partial t^2}v(z,t) + i\frac{\kappa}{3}\{[3|v(z,t)|^2
$$
$$
+2|w(z,t)|^2]v(z,t) + w^2(z,t)v^*(z,t)\}
$$

(10.54)

and

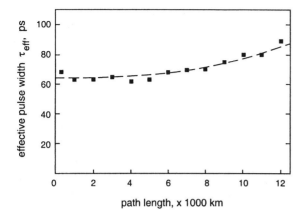

Fig. 10.12. Experimental confirmation of the Gordon–Haus effect, after [120]

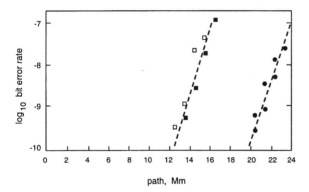

Fig. 10.13. Experimental results with sliding guiding filters, after [114]: 10 Gbit/s (*solid circles*); 2×10 Gbit/s WDM random bit pattern in interfering channel (*squares*); 2×10 Gbit/s WDM regular pattern in interfering channel (*solid squares*). The measured channel always contained a 2^{14} bit pseudorandom word

$$\frac{\partial}{\partial z}w(z,t) = -\mathrm{i}\frac{1}{2}\beta''\frac{\partial^2}{\partial t^2}w(z,t) + \mathrm{i}\frac{\kappa}{3}\{[3|w(z,t)|^2$$

$$+2|v(z,t)|^2]w(z,t) + v^2(z,t)w^*(z,t)\} . \tag{10.55}$$

There are cross-coupling terms, both phase-independent and phase-dependent. The latter are the so-called "coherence terms" $w^2(z,t)v^*(z,t)$ and $v^2(z,t)$ $w^*(z,t)$. In the presence of birefringence, phase coherence is not maintained and the effect of the coherence terms averages out to zero. However, the remaining pair of coupled equations is not integrable. However, if the non-linear effects are much weaker than the birefringence effects, the two polarization states wander all "over the place" within distances short compared with the distance within which soliton effects play a role. Thus, if $\langle v(z,t) \rangle$ and $\langle w(z,t) \rangle$ represent average orthogonal polarization states, rather than linear polarizations, the nonlinear phase shift due to each becomes equal on the average [124]:

$$
\frac{\partial}{\partial z}\langle v(z,t) \rangle = -\mathrm{i}\frac{1}{2}\beta'' \frac{\partial^2}{\partial t^2}\langle v(z,t) \rangle + \mathrm{i}\frac{9\kappa}{8}[|\langle v(z,t) \rangle|^2
$$
$$
+ |\langle w(z,t) \rangle|^2]\langle v(z,t) \rangle ,
$$

(10.56)

$$
\frac{\partial}{\partial z}\langle w(z,t) \rangle = -\mathrm{i}\frac{1}{2}\beta'' \frac{\partial^2}{\partial t^2}\langle w(z,t) \rangle + \mathrm{i}\frac{8\kappa}{9}[|\langle w(z,t) \rangle|^2
$$
$$
+ |\langle v(z,t) \rangle|^2]\langle w(z,t) \rangle .
$$

(10.57)

This is an equation pair that has been shown by Manakov to be integrable [125], and which gives rise to solitons of arbitrary polarization.

Polarization hole burning is another important effect that is related to the saturation properties of erbium-doped amplifiers. Nominally, the amplifier gain is polarization-insensitive. However, if one polarization saturates the amplifier, a slight excess gain is left over in the other polarization. Noise can grow in this polarization and affect the bit-error rate. This effect, first observed by Taylor in the early experiments on open-loop repeaterless systems (namely in experiments in which the pulse stream was propagated over fibers having a length equal to the full transoceanic distance) [126], was later explained by Mazurczyk and Zyskind [127]. The effect is circumvented by varying the input polarization at a rate (> 10 kHz) faster than the relaxation rate of the erbium-doped amplifier [128].

10.9 Continuum Generation by Soliton Perturbation

When a soliton is perturbed by noise or by other causes, such as third-order dispersion, lumped gain, or lumped loss, it sheds continuum. Nicholson and Goldman [129] used conservation laws to compute the soliton radiation due to damping. Kaup found a basis for the continuum states using a perturbation approach [105].

In the ansatz (10.27) the continuum was taken into account by the term $\Delta u_c(z,t)$. The perturbation of the four soliton parameters was evaluated by projections that are orthogonal to the continuum. In this section we evaluate the generation of the continuum. We express the continuum as a quasi-Fourier superposition [130]. The basis functions used were first described by Gordon [131]; they are simple complex exponentials outside the time interval occupied by the soliton and are connected across the time interval occupied by the soliton by solving (10.24).

Let us turn to the derivation of the basis functions. Consider a function $v(t, z)$ that obeys the linear dispersion equation

$$-i\frac{\partial}{\partial z}v(t, z) = \frac{1}{2}\frac{\partial^2}{\partial t^2}v(t, z) .$$
(10.58)

Then the function $f(t, z)$ defined by

$$f(t, z) = -\frac{\partial^2}{\partial t^2}v(t, z) + 2\tanh t\frac{\partial}{\partial t}v(t, z)$$

$$- \tanh^2 t\, v(t, z) + u_s^2(t, z)v^*(t, z)$$
(10.59)

is a solution of the linearized perturbed nonlinear Schrödinger equation (10.24). This can be confirmed by direct substitution of (10.59) in (10.24). One observation can be made immediately about (10.59). Outside the time interval occupied by the soliton, $|t| \geq 1$, $\tanh t = \pm 1$ and $u_s(t, z) = 0$. In this regime the continuum travels unperturbed with no influence of the soliton

$$f(t, z) = -\frac{\partial^2}{\partial t^2}v(t, z) \pm 2\frac{\partial}{\partial t}v(t, z) - v(t, z) .$$
(10.60)

If we take as a special case the exponential solution of (10.58)

$$v(t, z) = c\exp(-i\Omega t)\exp[-i(\Omega^2/2)z]$$
(10.61)

and introduce (10.61) into (10.59), we find

$$f(t, \Omega, z)$$

$$= c(\Omega^2 - 2i\Omega\tanh t - \tanh^2 t)\exp(-i\Omega t)\exp[-i(\Omega^2/2)z]$$
(10.62)

$$+ c^*\mathrm{sech}^2 t\exp(iz)\exp(i\Omega t)\exp[i(\Omega^2/2)z] .$$

The constant c is chosen to be a complex number of magnitude unity. The phase of c fixes the relative phase between the continuum and the soliton. Hence, the t, z dependence of the function changes as the phase of c is changed. It is clear that the function $f(t, z)$ is a simple exponential and of magnitude $1 + \Omega^2$ on both sides of the soliton but of different phase on the two sides. Strictly, two sets of functions $f(t, \Omega, z)$ are required for the expansion; one set is the in-phase set, which we shall denote by the subscript "c" reminiscent of "cosine", for which $c = 1$, and the quadrature set, which we shall denote by the subscript "s" reminiscent of "sine", for which $c = i$.

The functions $f_c(t, \Omega, z)$ and $f_s(t, \Omega, z)$ are used as the basis set of a quasi-Fourier expansion of the continuum. The continuum is constructed from the superposition

$$\Delta u_{\mathrm{cont}}(t, z) = \int_{-\infty}^{\infty}\frac{d\Omega}{2\pi}[F_c(\Omega)f_c(t, \Omega, z) + F_s(\Omega)f_s(t, \Omega, z)] .$$
(10.63)

The coefficients $F_i(\Omega), i = c, s$, are found by projection with the adjoint of (10.63), obeying the adjoint differential equation (10.26). The adjoint functions $\underline{f}_{-i}(t, \Omega, z)$ and the functions $f_i(t, \Omega, z)$ obey the orthogonality relation

$$\text{Re}\left[\int_{-\infty}^{\infty} dt\, \underline{f}_{-i}^*(t, \Omega, z) f_j(t, \Omega', z)\right] = \delta(\Omega - \Omega')\delta_{ij} . \tag{10.64}$$

The orthogonality relation holds not only for $i = c, s$, but also for the entire set of functions $i = c, s, n, \theta, x, p$. Since the adjoint differential equation (10.26) looks like (10.24), except for the sign reversal in front of $u_s^2(t, z)$, the adjoint solutions have the appearance of (10.62) with a sign reversal of $u_s^2(t, z)$:

$$\underline{f}_{-i}(t, \Omega, z) = \underline{c}(\Omega^2 - 2i\Omega \tanh t - \tanh^2 t) \exp(-i\Omega t) \exp[-i(\Omega^2/2)z]$$

$$- \underline{c}^* \text{sech}^2 t \exp(iz) \exp(i\Omega t) \exp[i(\Omega^2/2)z] . \tag{10.65}$$

The functions $f_i(t, z)$ and $\underline{f}_{-i}(t, z)$, $i = c, s$, are defined over a time interval T that is, ideally, infinitely long. In normalizing them over T, one may ignore the short interval occupied by the soliton, over which the functions experience a rapid change. Thus we find for $|\underline{c}|$

$$\lim_{T \to \infty} \text{Re} \int_{-T/2}^{T/2} dt\, \underline{f}_{-i}^*(t, \Omega', z) f_j(t, \Omega, z)$$

$$= \begin{cases} \lim_{T \to \infty} c_j \underline{c}_i^* (\Omega^2 + 1)^2 T & \text{if } \Omega = \Omega' \quad \text{and } i = j \\ 0 & \text{if } \Omega \neq \Omega' \quad \text{and/or } i \neq j \end{cases} \tag{10.66}$$

$$= 2\pi\delta(\Omega - \Omega')\delta_{ij} ,$$

with

$$|\underline{c}_i| = \frac{1}{(\Omega^2 + 1)^2} . \tag{10.67}$$

The delta function has the value $T/2\pi$ over the frequency interval $\Delta\Omega = 2\pi/T$.

Figure 10.14, taken from [130], shows the real part of the basis functions $f_i(t, \Omega, z)$ in two dimensions as functions of t and Ω, for the in-phase and quadrature cases, $c = 1$ and $c = i$.

The coefficients $F_i(\Omega)$ are obtained from (10.63) using the orthogonality condition (10.64):

$$F_i(\Omega) = \text{Re} \int_{-\infty}^{\infty} dt\, \Delta u(t, z) \underline{f}_{-i}^*(t, \Omega, z) . \tag{10.68}$$

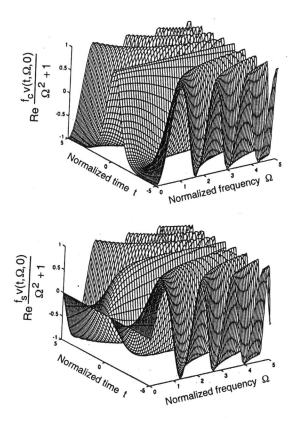

Fig. 10.14. Real part of $f_c(t, \Omega, 0)$ and $f_s(t, \Omega, 0)$ [130]

The quasi-Fourier component amplitudes of the continuum can be evaluated at any cross section. If the change $\Delta u(t)$ is imposed at $z = 0$, then (10.68) can be used to evaluate the continuum generated by this change, setting $z = 0$.

The formalism can be illustrated by a simple and important example: the excitation of continuum in a sudden amplification step. A soliton sech t is changed abruptly into a pulse $(1 + g)$ sech t. The change of the pulse is g sech t. This change is partly incorporated in the new soliton formed from the pulse and partly imparted to the continuum. The continuum portion is evaluated from the projection (10.68) (see Appendix A.15 for the integrals):

$$F_c(\Omega) = \mathrm{Re} \int_{-\infty}^{\infty} dt \, g \, \mathrm{sech} \, t$$

$$\times \frac{1}{(\Omega^2 + 1)^2} \left\{ \begin{array}{l} (\Omega^2 + 2i\Omega \tanh t - \tanh^2 t) \exp(i\Omega t) \\[2mm] -\mathrm{sech}^2 t \exp(-i\Omega t) \end{array} \right. \tag{10.69}$$

$$= -\frac{\pi g}{\Omega^2 + 1} \mathrm{sech} \left(\frac{\pi}{2} \Omega \right)$$

and

$$F_s(\Omega) = 0 \, .$$

This is the continuum generated in one amplifying process. Continuum is generated in the repeated amplifying process of solitons on a transoceanic cable. This process is slightly different from the one just considered, since the continuum excited by the preceding amplifiers accompanies the soliton in its passage through every amplifier. The interested reader is referred to the literature for further details [130, 131].

It should be pointed out that the first-order analysis does not conserve energy. Indeed, the energy in the continuum is of second order and the change of the energy of the soliton is zero in the first-order analysis. However, energy loss from the soliton can be evaluated by noting that most of the energy distribution of the continuum lies outside the interval of the soliton. Thus one may compute the energy in the continuum to second order and conclude that this energy has been extracted from the soliton. In this way, a first-order analysis can yield answers correct to second order.

10.10 Summary

This chapter developed soliton solutions to the nonlinear Schrödinger equation (NLSE). Perturbations of the soliton, such as by noise or by lumped gain, were treated by a perturbation theory based on the linearized NLSE and its adjoint. In this way we derived the Gordon–Haus jitter. This jitter is the consequence of amplifier-noise-induced carrier frequency changes. We showed that the introduction of filters can reduce this jitter. Perturbations of solitons can cause radiation, i.e. excitation of a continuum. The continuum solutions form a Fourier-integral-like basis set, in terms of which the radiation can be expanded. It is of interest to note that the continuum solutions (10.62) are solutions of a linear partial differential equation with scattering wells that are squares of hyperbolic secants. The depths of the wells are in the ratio of 3 to 1 for the in-phase and quadrature components, respectively. What is remarkable about these solutions is that they are traveling waves of equal amplitude on both sides of the well. Thus they represent solutions of a scattering problem in which the wells are "reflection-free": no incident power is reflected. The sole effect of the scattering is to cause a phase change of the incident wave in passage through the well.

As mentioned before, long-distance optical-fiber communication with solitons is competing with existing "linear" transmission schemes using the NRZ format. Solitons are going to prevail only if they offer higher bit rates at no increase of cost per bit. Cost increases with complexity. Thus, soliton systems must use sources and components that are not significantly more complex than those currently employed in NRZ transmission.

The current estimate is that solitons could support twice the bit rate per channel of an NRZ system, and that the channels of a soliton system could be spaced three times more closely [132]. This would give them a sixfold advantage when bit rates much higher than 5 Gbit/s are called for.

With regard to cost, the first issue is the source of the bit stream. Off hand, one would expect that the soliton source would have to emit proper sech-shaped pulses followed by a modulator to represent the ones and zeros. It turns out, however, that a regular NRZ source followed by a phase modulator at an appropriate power level [133] can be used. As the phase-modulated NRZ signal propagates along a fiber equipped with sliding guiding filters, it reshapes itself into the appropriate soliton stream. The continuum generated in the process is eliminated by the filters. Thus no new sources are in fact necessary.

The sliding-guiding-filter concept is an extremely effective way to increase bit rate and/or distance of transmission. It does imply, however, that the amplifier "pods" that are sunk into the ocean are not identical. This is currently a point of contention with the system designers. Nonlinear fiber loops for the suppression of the narrow-band noise have been proposed [134–137]. Incorporation of such loops would make the pods identical, but this additional component makes soliton transmission less attractive compared with NRZ. Other schemes are currently under investigation.

There is also the problem of supervisory control. In a transoceanic cable, one fiber is used for transmission in one direction and another fiber in the opposite direction. At every amplifier stage, a small fraction of the signal propagating in one direction is tapped off and sent in the opposite direction. The bit stream is then amplitude-modulated at a very low rate, enabling one both to obtain information about the state of the amplifiers along the cable and send commands for adjustments in the individual pods [138]. This simple scheme is not acceptable for soliton communications. Since the effect of amplitude modulation is removed by the filtering, no low-level signal can be returned in the fiber carrying the bit stream in the opposite direction. Thus, the supervisory control in a soliton system is an important issue. It has been addressed in a patent [139].

In conclusion, we may state that repeaterless propagation of signals with solitons has made enormous progress in recent years. It is an example of a rapid deployment of a sophisticated physical phenomenon for practical use. The work on soliton transmission has spurred on the development of "linear" NRZ repeaterless optical-fiber transmission, which has already been deployed. The deployment of soliton optical-fiber transmission will have to overcome the stiff competition presented to it by the NRZ systems.

Problems

10.1* Find the energy of a soliton of 20 ps full width at half maximum. Typical fiber parameters are:

$$A_{eff} = 80\,\mu m^2; \lambda = 1.55\,nm; n_2 = 3 \times 10^{-16}\,cm^2/W; |\beta''| = 20\,ps^2/km .$$

10.2* Construct a second-order soliton from (10.20) with $A_1 = 1, A_2 = 3, V_1 = V_2 = 0$, and $q_1(z = 0) = q_2(z = 0) = 0$. Show that it acquires the form of a simple sech. Show also that it repeats its envelope within a distance that is eight times shorter than the distance within which the phase of the fundamental soliton changes by 2π.

10.3 Use the solution (10.20) to study the collision of two solitons with $A_1 = A_2 = 1$, and $V_1 = -V_2 = V$.

10.4 Draw a two-dimensional graph of the amplitude of the second-order soliton of Prob. 10.2.

10.5 Draw a two-dimensional graph of the propagation of the amplitude of the second-order soliton with $A_1 = 1, A_2 = 2, V_1 = V_2 = 0, q_1(z = 0) = 0$, and $q_2(z = 0) = 0$.

10.6 Determine the effect of third-order dispersion β''' on the four soliton parameters, assuming that it acts as a perturbation.

10.7* Determine the effect on a soliton of lumped loss $\mathcal{L}\,(1 - \mathcal{L} \ll 1)$. Which of the four perturbation parameters is affected?

10.8 The loss of a fiber at 1.5 μm is 0.2 dB/km. The loss is compensated by distributed gain. Suppose that the bandwidth of noise accepted is that of the signal bandwidth of 10 GHz.

(a) Find the ASE noise power after 5000 km of propagation.
(b) Suppose that the gain is lumped and the amplifiers are spaced 50 km apart. What is the increase of noise due to the lumping of the amplifiers? Set $\chi = 2$.

Solutions

10.1 The full width at half maximum follows from $\mathrm{sech}^2(\Delta t/\tau_o) = 1/2$, which gives $2\Delta t = 1.76\tau_o$. The balance of the nonlinear propagation equation gives

$$\kappa|A_o|^2 = |\beta''|/\tau_o^2 .$$

This gives an equation for $|A_o|\tau_o$ in terms of the fiber parameters. With $\kappa = (2\pi/\lambda)n_2/A_{\mathrm{eff}}$, we find the energy $2|A_o|^2\tau_o = 2.2$ pJ.

10.2 From (10.20) we find

$$u(x,t) = \frac{4e^{\mathrm{i}t/2}[\cosh(3x) + 3\cosh x\, e^{\mathrm{i}4t}]}{\cosh(4x) + 4\cosh(2x) + 3\cos(4t)} .$$

At $t = 0$,

$$u(x,t = 0) = \frac{4[\cosh(3x) + 3\cosh x]}{\cosh(4x) + 4\cosh(2x) + 3} .$$

Using the relation

$$\cosh(3x) = 4\cosh^3 x - 3\cosh x$$

and

$$\cosh(4x) = 2\cosh^2(2x) - 1 = 8\cosh^4(x) - 8\cosh^2(x) + 1 ,$$

we find

$$u(x,t = 0) = \frac{2}{\cosh x} .$$

From the first equation it is clear that the envelope of this second-order soliton has a period that is eight times shorter than the distance within which the fundamental soliton experiences a phase change of 2π.

10.7 The lumped loss introduces a lumped (delta-function-like) source into the perturbation equation. The perturbation term is symmetric and real. The only parameter affected is the amplitude. We have

$$\Delta A f_A(t) = -(1 - \mathcal{L})\mathrm{Re}\left[\int_{-\infty}^{\infty} dt\, \underline{f}_A^*(t)u_s\right] .$$

The amplitude equation (10.30) acquires the source term

$$S_A(z) = -\delta(z)2(1 - \mathcal{L}) .$$

11. Phase-Sensitive Amplification and Squeezing

Thus far we have discussed phase-insensitive systems, whose response is independent of the phase of the input (initial) excitation. When discussing the example of an amplifier, we assumed that the gain medium was either not saturated or, if saturated, was equilibrated fast enough that the gain was insensitive to the phase of the saturating signal. We also studied a laser resonator below threshold in Chap. 6, in which saturation effects could be neglected.

The response of a nonlinear system is, in general, phase-sensitive. When the gain medium of a laser oscillator saturates, fluctuations of the field in phase with the oscillating field change the power level of the signal and affect the gain saturation. The component of the fluctuation in quadrature to the oscillating field does not change the power to first order and thus does not saturate the gain. The response of the laser to in-phase fluctuations is different from the response to quadrature fluctuations. In this chapter we start with a simple model of a nonlinearity, a medium whose polarization is proportional to the square of the electric field. We consider parametric amplification. Parametric amplification provides gain by variation of a circuit *parameter*, such as capacitance or inductance. This is the origin of the name. If a capacitance is a function of the applied voltage, as it is for example in a reverse-biased junction diode, the capacitance can be varied periodically at the frequency of the applied voltage, the "pump". At optical frequencies, the amplification is achieved by variation of the index of a nonlinear medium by an optical "pump".

A capacitance at thermal equilibrium with its environment stores an energy of $k\theta/2$. It does not generate noise internally; the energy is acquired from its surroundings. Hence, one would expect that parametric amplifiers would function well as low-noise amplifiers, and they do. In fact, we shall show that a degenerate parametric amplifier can provide, in principle, noise-free amplification.

Degenerate parametric amplification can also be used to generate special quantum states, so-called squeezed states. Although these quantum states were discussed in the literature before H. Yuen's work, it was his paper [140] and work by Braginsky [141–145] and Walls [146] that kindled the interest in squeezed states as a means for performing measurements below the so-called "standard quantum limit", taken as shot noise. In this chapter we show how degenerate parametric amplification can generate squeezed states in general, and squeezed vacuum in particular. We then show how squeezed vacuum can be used in an interferometric measurement of phase to lower the noise below the shot noise level. We conclude the chapter with the discussion of a laser above threshold. We obtain the spectrum of the laser output. This analysis leads to the Schawlow–Townes formula for the linewidth of laser radiation. Finally, we show under what conditions the laser can emit squeezed radiation.

11.1 Classical Analysis of Parametric Amplification

A laser with gain requires the introduction of noise sources to conserve commutators, since the presence of gain implies coupling to a reservoir. The consequence is that the equations of motion are not derivable from a Hamiltonian. Parametric amplification can occur without coupling to a reservoir and hence is, ideally, governed by a set of equations that are derivable from a Hamiltonian. We analyze first a traveling-wave parametric amplifier and then develop the multiport scattering formalism for parametric amplification.

A parametric amplifier contains a nonlinear medium. The medium is excited by a pump at the frequency ω_p. The pump modulates the parameters

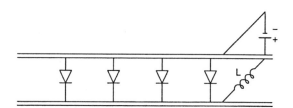

Fig. 11.1. Traveling-wave transmission line parametric amplifier with reverse-biased diodes as variable capacitors

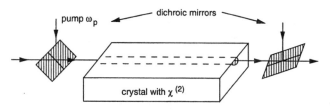

Fig. 11.2. An optical parametric amplifier

of the structure. In a transmission line, this could be the capacitance of the line. In a practical realization of such a traveling-wave transmission line parametric amplifier, the line could be loaded by nonlinear diodes (varactors) whose capacitances then vary at the frequency ω_p (Fig. 11.1). In an optical waveguide, the index may vary with time (Fig. 11.2). In either case, a voltage or electric field of amplitude a_s at the signal frequency ω_s produces new frequency components at $\omega_p + \omega_s$ and $\omega_p - \omega_s$. By proper design of the structure, namely the choice of resonance frequencies in a resonant enclosure or the choice of proper dispersion (frequency dependence of the propagation constant), one may ensure that only the frequency $\omega_p - \omega_s$ is excited. This is the so-called idler frequency. The amount of idler field a_i produced per unit length is given by a simple perturbation of the idler propagation equation:

$$\frac{d}{dz}a_i = \mathrm{i}\beta_i a_i - \mathrm{i}\kappa_{ips}a_p a_s^* . \tag{11.1}$$

The coupling is proportional to the amplitude of the pump. Note that the signal amplitude appears complex-conjugated because the product of the two time dependences $\exp(-\mathrm{i}\omega_p t)$ of the pump amplitude and $\exp(\mathrm{i}\omega_s t)$ of the complex conjugate of the signal amplitude results in the time dependence $\exp\left[-\mathrm{i}(\omega_p - \omega_s)t\right] = \exp(-\mathrm{i}\omega_i t)$ of the idler. Equation (11.1) also reveals the nature of the nonlinearity used to generate the difference frequency. Two fields beat so as to produce a polarization in the medium at the difference frequency, the idler frequency, which then acts as the source of the idler field. The polarization is due to the product of two fields. If the polarization is written as a Taylor expansion in powers of the field,

$$P_i = \chi_{ij}^{(1)} E_j + \chi_{ijk}^{(2)} E_j E_k + \chi_{ijkl}^{(3)} E_j E_k E_l + \ \ldots \tag{11.2}$$

where the coefficients are tensors of progressively higher rank, then the term in this expansion responsible for the coupling term in (11.1) is clearly the second term in the expansion, the second-order nonlinearity. We shall not be concerned about the details of the evaluation of the coupling term χ_{ijk} from the constitutive law (11.2) and the geometry, referring the interested reader to the literature [31, 36] instead.

Analogously, we find the equation for the signal as

$$\frac{d}{dz}a_s = \mathrm{i}\beta_s a_s - \mathrm{i}\kappa_{spi}a_p a_i^* . \tag{11.3}$$

The coefficients κ_{ips} and κ_{spi} are related. This was first proven by Manley and Rowe [147], who derived the so called Manley–Rowe relations, using classical arguments. Later it was shown by Weiss [148] that the Manley–Rowe relations are a consequence of simple quantum mechanical energy conservation arguments. Let us use here the quantum argument of Weiss. Note the energy diagram of Fig. 11.3. The pump photon of energy $\hbar\omega_p$ produces one signal photon and one idler photon, conserving energy in the process, since

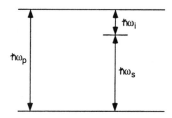

Fig. 11.3. Energy diagram for parametric amplification

$$\hbar\omega_p = \hbar\omega_s + \hbar\omega_i \ . \tag{11.4}$$

Hence, the number of signal photons generated per unit length must be equal to the number of idler photons per unit length:

$$\frac{1}{\hbar\omega_s}\frac{d}{dz}|a_s|^2 = \frac{1}{\hbar\omega_i}\frac{d}{dz}|a_i|^2 \ . \tag{11.5}$$

When (11.1) and (11.3) are used in (11.5), we find

$$\frac{\kappa_{ips}}{\omega_i} = \frac{\kappa_{spi}}{\omega_s} \quad \text{or} \quad \sqrt{\frac{\omega_s}{\omega_i}}\kappa_{ips} = \sqrt{\frac{\omega_i}{\omega_s}}\kappa_{spi} \ . \tag{11.6}$$

Next, consider the spatial dependence of the signal and idler dictated by (11.1) and (11.3). In an optical system, the pump is usually a traveling wave as well, with the spatial dependence

$$a_p(z) = A_p \exp(i\beta_p z) \ . \tag{11.7}$$

If one neglects pump depletion, A_p can be treated as a constant. Since the natural spatial dependences of the signal and idler are $\exp(i\beta_i x)$ and $\exp(i\beta_s x)$, respectively, the coupling gets out of phase rapidly as one proceeds along the waveguide, unless

$$\beta_p = \beta_s + \beta_i \ . \tag{11.8}$$

This is the condition of phase matching. It is accomplished in practice by proper choice of the waveguide dispersion or of the birefringence of the nonlinear crystals employed in a parametric amplifier. Equation (11.8) is also known as the momentum conservation condition, since quantum mechanically the momentum of a photon is known to be $\hbar\beta$. Thus, not surprisingly, in the photon-scattering picture, energy and momentum must be conserved.

Let us now return to the differential equations. Setting $a_s(x) = A_s(x)\exp(i\beta_s x)$ and $a_i(x) = A_i(x)\exp(i\beta_i x)$, taking advantage of (11.6), and noting that the complex conjugate of the idler wave couples to the signal wave, we obtain the two coupled differential equations

$$\frac{d}{dz}\frac{A_s}{\sqrt{\omega_s}} = -i\frac{\sqrt{\omega_i}\kappa_{spi}A_p}{\sqrt{\omega_s}}\frac{A_i^*}{\sqrt{\omega_i}} \ , \tag{11.9}$$

$$\frac{d}{dz}\frac{A_i^*}{\sqrt{\omega_i}} = i\frac{\sqrt{\omega_i}\kappa_{spi}^*A_p}{\sqrt{\omega_s}}\frac{A_s}{\sqrt{\omega_s}} \ . \tag{11.10}$$

Consider a system of length L, excited by a signal of amplitude $A_s(0)$ and an idler of amplitude $A_i(0)$ at the input at $z = 0$. Set $i\sqrt{\omega_i/\omega_s}\kappa_{spi}A_p = -i\sqrt{\omega_s/\omega_i}\kappa_{spi}^*A_p^* = \gamma$ with γ real and positive for simplicity, by choice of the definition of the phase of the pump. Then the output is

$$\begin{bmatrix} \dfrac{A_s(L)}{\sqrt{\omega_s}} \\[2ex] \dfrac{A_i^*(L)}{\sqrt{\omega_i}} \end{bmatrix} = \begin{bmatrix} \cosh(\gamma L) & -\sinh(\gamma L) \\[1ex] -\sinh(\gamma L) & \cosh(\gamma L) \end{bmatrix} \begin{bmatrix} \dfrac{A_s(0)}{\sqrt{\omega_s}} \\[2ex] \dfrac{A_i^*(0)}{\sqrt{\omega_i}} \end{bmatrix} \ . \tag{11.11}$$

The signal grows exponentially along with the idler. The energy required for the growth is provided by the pump.

11.2 Quantum Analysis of Parametric Amplification

The quantum analysis of wavepacket propagation along a parametrically excited system proceeds in the time domain. We start with a Hermitian coupling Hamiltonian of the form

$$\hat{H} = \hbar\left(\chi_{spi}\hat{a}_s^\dagger\hat{a}_p\hat{a}_i^\dagger + \chi_{spi}^*\hat{a}_i\hat{a}_p^\dagger\hat{a}_s\right) \ . \tag{11.12}$$

We use the same normalization for the creation and annihilation operators as in Chap. 6 for linear phase-insensitive amplification. The equations of motion for the signal and idler operators are

$$\frac{d\hat{a}_s}{dt} = -i\chi_{spi}\hat{a}_p\hat{a}_i^\dagger \ , \tag{11.13}$$

$$\frac{d\hat{a}_i^\dagger}{dt} = i\chi_{spi}^*\hat{a}_p\hat{a}_s \ . \tag{11.14}$$

When the pump is very intense and its depletion can be ignored, we may replace the pump operator $\hat{a}_p$ with a c-number amplitude A_p. Then, the quantum form of the equations is in close analogy with the classical equations of motion. Note that no normalizing factors involving the square root of the idler and signal frequencies appear in the quantum version. This is because the expectation values of the operator products represent photon numbers, not power as is the case for products of the classical amplitudes. The quantum form of the equations ensures automatically photon number generation or annihilation in pairs.

Integration of the equations over an interaction time T, cast into a scattering-matrix form by defining the output operators

$$\begin{bmatrix} \hat{b}_s \\ \hat{b}_i^\dagger \end{bmatrix} \equiv \begin{bmatrix} \hat{a}_s(T) \\ \hat{a}_i^\dagger(T) \end{bmatrix}$$

in terms of the input operators

$$\begin{bmatrix} \hat{a}_s \\ \hat{a}_i^\dagger \end{bmatrix} \equiv \begin{bmatrix} \hat{a}_s(0) \\ \hat{a}_i^\dagger(0) \end{bmatrix} ,$$

gives

$$\hat{b} \equiv \begin{bmatrix} \hat{b}_s \\ \hat{b}_i^\dagger \end{bmatrix} = S \begin{bmatrix} \hat{a}_s \\ \hat{a}_i^\dagger \end{bmatrix} \equiv S\hat{a} , \tag{11.15}$$

with the scattering matrix

$$S = \begin{bmatrix} \cosh(\delta T) & -\sinh(\delta T) \\ -\sinh(\delta T) & \cosh(\delta T) \end{bmatrix} , \tag{11.16}$$

where we have set $\delta \equiv i\chi_{spi}A_p$ and assumed that δ is real and positive. The analogy with the classical result is unmistakable. Note that the scattering formalism (11.15) does not involve an internal noise source, because commutators are preserved in a system described by a simple Hamiltonian (as opposed to a system whose Hamiltonian contains coupling to resonator modes that do not appear explicitly in the final scattering-matrix representation of the system).

The scattering formalism of a parametric amplifier does not contain noise sources. Yet the amplifier provides gain. Is the gain then noise-free? The answer is no as we shall now show. Let us ask about the signal photons in the output when there is zero signal input, a situation analogous to a linear amplifier, which emits amplified spontaneous emission at its output when there is no input. The photon number in the signal channel is

$$\langle 0_s | \langle 0_i | \hat{a}_s^\dagger(T)\hat{a}_s(T) | 0_i \rangle | 0_s \rangle$$

$$= \langle 0_s | \langle 0_i |$$

$$\times \{\cosh^2(\delta T)\hat{a}_s^\dagger(0)\hat{a}_s(0) - \cosh(\delta T)\sinh(\delta T)[\hat{a}_s^\dagger(0)\hat{a}_i^\dagger(0) + \hat{a}_s(0)\hat{a}_i(0)]$$

$$+ \sinh^2(\delta T)\hat{a}_i(0)\hat{a}_i^\dagger(0)\} | 0_i \rangle | 0_s \rangle$$

$$= \sinh^2(\delta T)\left(\frac{\Delta\omega}{2\pi}\right) = (G - 1)\frac{\Delta\omega}{2\pi} ,$$

$$\tag{11.17}$$

where $G \equiv \cosh^2(\delta T)$ is the gain of the system. We get the same formula for the photons at the output of a parametric amplifier as we have for a linear amplifier with complete inversion. In the parametric-amplifier case, the output photons are due to the zero-point fluctuations of the idler channel, which stimulate pump photons to break into signal and idler photon pairs.

Note that the scattering matrix S of (11.16) is not unitary. The reason for this is that it does not describe a system which conserves photon number, but rather a system that generates photons in pairs. If one were to assign to the photon number in the idler channel a negative value, then the sum of these "negative" idler photons and positive signal photons would be conserved. Formally, we may express this conservation law by defining the parity matrix P,

$$P \equiv \begin{bmatrix} 1 & 0 \\ 0 & -1 \end{bmatrix} , \tag{11.18}$$

and the associated photon number operator,

$$\text{photon number operator} \equiv \hat{b}^\dagger P \hat{b} . \tag{11.19}$$

Then we may confirm that the sum of the photon numbers thus defined is conserved:

$$\hat{b}^\dagger P b = \hat{a}^\dagger S^\dagger P S a = \hat{a}^\dagger P \hat{a} , \tag{11.20}$$

since the matrix S of (2.5) obeys the condition

$$S^\dagger P S = P . \tag{11.21}$$

Of course, commutators are preserved as well. This is a direct consequence of the derivation of the equations of motion from a Hamiltonian. It is informative to see how commutator conservation follows from commutator matrix manipulation. First of all, the commutator matrix of the signal–idler column matrices must be written:

$$[\hat{b}, \hat{b}^\dagger] = [\hat{a}, \hat{a}^\dagger] = P , \tag{11.22}$$

since the idler operator is entered into the column matrix as a creation operator, and not an annihilation operator. Direct manipulation yields

$$[\hat{b}, \hat{b}^\dagger] = [S\hat{a}, \hat{a}^\dagger \hat{S}^\dagger] = S[\hat{a}, \hat{a}^\dagger]S^\dagger = SPS^\dagger = P , \tag{11.23}$$

where we have used the fact that (11.21) implies as well the relation

$$SPS^\dagger = P . \tag{11.24}$$

The assignment of a negative sign to the idler photons, so as to ensure conservation of photon number, has an analogy in plasma physics and microwave electron beam amplifiers. Waves traveling in a moving plasma or an electron beam may be assigned positive or negative energies depending upon whether their excitation raises or lowers the translational kinetic energy of the plasma or the electron beam. This property is basic to L. J. Chu's small-signal kinetic power theorem [149]. Coupling of a wave with positive energy to one with negative energy leads to exponential growth of both waves. Energy is conserved since the positive-energy wave grows at the same rate as the negative-energy wave, the sum of their energies remaining zero. Even more generally, the analogy extends to the evolution of the universe from an initial vacuum fluctuation via the "big bang". Gravitational energy is negative, radiational and particle energies are positive. The universe is, in the words of Alan Guth, "the ultimate free lunch": it was generated from zero energy and even today has zero net energy [150].

11.3 The Nondegenerate Parametric Amplifier as a Model of a Linear Phase-Insensitive Amplifier

A parametric amplifier is called nondegenerate when the signal and idler frequencies are different. This was the case analyzed in the preceding section. The amplification was provided by pump photons that split into signal photons and idler photons. Signal and idler photons are generated in pairs. We have set up a formal photon conservation law by assigning a negative sign to the idler photon number; then the sum of signal and idler photons is conserved. The signal and idler photon numbers can increase exponentially.

If no attention is paid to the details of the amplification process, and one considers only the signal excitation, one arrives at the equation

$$\hat{b}_s = \sqrt{G}\hat{a}_s + \hat{n}_s , \tag{11.25}$$

with $\sqrt{G} = \cosh(\delta T)$, and $\hat{n}_s = -\sinh(\delta T)\hat{a}_i^\dagger$.

The parametric-amplifier model developed here is indistinguishable in its operation from a linear amplifier. The noise source is caused by the idler, and the commutation relation for the noise source is the one required to maintain the commutator of the signal operator:

$$[\hat{n}_s, \hat{n}_s^\dagger] = \sinh^2(\delta T)[\hat{a}_i^\dagger, \hat{a}_i] = -\sinh^2(\delta T)\frac{\Delta\omega}{2\pi} = (1 - G)\frac{\Delta\omega}{2\pi} . \tag{11.26}$$

The parametric-amplifier equations follow from a Hamiltonian description. On the other hand, the laser amplifier equations can be derived from Hamiltonians for the field and for the gain medium. Many approximations have to be made in the analysis of the medium before the complete description of the gain process is cast into that of a linear amplifier. The noise source arises from the quantum fluctuations in the gain medium. Yet, when all is said and done, and the transfer function of the system is a linear relation between input and output, the commutator of the signal must be preserved, since the commutator is an attribute of the signal wave. Therefore, the commutator of the noise source is fixed by very fundamental physical considerations.

Whereas the commutator of the noise source must have a prescribed value, the nature of the noise operator, as a single creation operator or a superposition of creation and annihilation operators, depends on the physical situation. We modeled the case of a partially inverted laser medium as a combination of gain and loss media. As a result, we obtained a noise source composed of annihilation and creation operators. The case of the parametric amplifier gave us a noise source consisting of a creation operator only. This is the ideal case of a lossless parametric amplifier. When loss is present, additional noise sources have to be introduced. However, the net commutator must still obey the relationship

$$[\hat{n}_s, \hat{n}_s^\dagger] = (1 - G)\frac{\Delta\omega}{2\pi} , \tag{11.27}$$

where G is the net gain.

It is possible to derive the equations of an active N-port and the commutators of the noise sources from a parametric $2N$-port, where N ports are excited at the signal frequency and N ports at the idler frequency. By suppressing all references to the idler ports, one obtains the scattering matrix of an N-port with associated noise sources. The noise sources have the same commutators as those obtained from the requirement of commutator conservation. The details are shown in Appendix A.16.

11.4 Classical Analysis of Degenerate Parametric Amplifier

A parametric amplifier is called degenerate when the signal and idler frequencies coincide. The pump photons split into two photons of equal energy. Then, the signal and idler occupy the same frequency band. They must be distinguished, however, because they still represent two electric fields that are differently phased. We may replace the subscript "i" with the subscript "s", but we must treat A_s and A_s^* as *independent excitations* that couple to each other. In analogy with (11.9) and (11.10), we have

$$\frac{d}{dz}\frac{A_s}{\sqrt{\omega_s}} = -i\kappa_{\mathrm{sps}}A_p\frac{A_s^*}{\sqrt{\omega_s}} , \tag{11.28}$$

$$\frac{d}{dz}\frac{A_s^*}{\sqrt{\omega_s}} = i\kappa_{\mathrm{sps}}^*A_p^*\frac{A_s}{\sqrt{\omega_s}} . \tag{11.29}$$

Set $i\kappa_{\mathrm{sps}}A_p = \gamma e^{i\psi}$ with γ real and positive. Then integration of these equations over a distance of propagation L gives

$$
\begin{bmatrix} \dfrac{A_s(L)}{\sqrt{\omega_s}} \\[2ex] \dfrac{A_s^*(L)}{\sqrt{\omega_s}} \end{bmatrix} = \begin{bmatrix} \cosh(\gamma L) & -\sinh(\gamma L)e^{i\psi} \\ -\sinh(\gamma L)e^{-i\psi} & \cosh(\gamma L) \end{bmatrix} \begin{bmatrix} \dfrac{A_s(0)}{\sqrt{\omega_s}} \\[2ex] \dfrac{A_s^*(0)}{\sqrt{\omega_s}} \end{bmatrix} . \tag{11.30}
$$

These equations look very similar to (11.11) for the nondegenerate amplifier. One must note, however, one very important difference: the excitations $A_s(0)$ and $A_s^*(0)$ lie in the same frequency band and thus determine jointly the input signal excitation.

The physics of the degenerate parametric amplifier is brought out more explicitly if we revert to canonical, decoupled variables in (11.28) and (11.29). We define

$$
A_s^{(1)} = \frac{1}{2}(A_s e^{-i\psi/2} + A_s^* e^{i\psi/2}) \quad \text{and} \quad A_s^{(2)} = \frac{1}{2i}(A_s e^{-i\psi/2} - A_s^* e^{i\psi/2}) . \tag{11.31}
$$

Note that ψ can be made equal to zero by proper choice of the pump phase. Then the definition of the canonical variables is particularly simple. We can write (11.28) and (11.29) in terms of the canonical variables:

$$
\frac{d}{dz}\frac{A_s^{(1)}}{\sqrt{\omega_s}} = -\gamma\frac{A_s^{(1)}}{\sqrt{\omega_s}} , \tag{11.32}
$$

$$
\frac{d}{dz}\frac{A_s^{(2)}}{\sqrt{\omega_s}} = \gamma\frac{A_s^{(2)}}{\sqrt{\omega_s}} . \tag{11.33}
$$

These equations predict exponential spatial growth of $A_s^{(2)}$ and exponential decay of $A_s^{(1)}$. The two excitations are clearly 90° out of phase. The solution of (11.32) and (11.33) is

$$
\begin{bmatrix} \dfrac{A_s^{(1)}(L)}{\sqrt{\omega_s}} \\[2ex] \dfrac{A_s^{(2)}(L)}{\sqrt{\omega_s}} \end{bmatrix} = \begin{bmatrix} \exp(-\gamma L) & 0 \\ 0 & \exp(\gamma L) \end{bmatrix} \begin{bmatrix} \dfrac{A_s^{(1)}(0)}{\sqrt{\omega_s}} \\[2ex] \dfrac{A_s^{(2)}(0)}{\sqrt{\omega}} \end{bmatrix} . \tag{11.34}
$$

Degenerate parametric amplification is, in fact, a well-known physical phenomenon. A child on a swing can amplify the motion of the swing by pumping it at twice the frequency of the resonance of the pendulum formed by the child on the swing. The child can also bring the swing to a stop, without touching the ground, by changing the phase with which it pumps the swing relative to the phase of the motion of the swing.

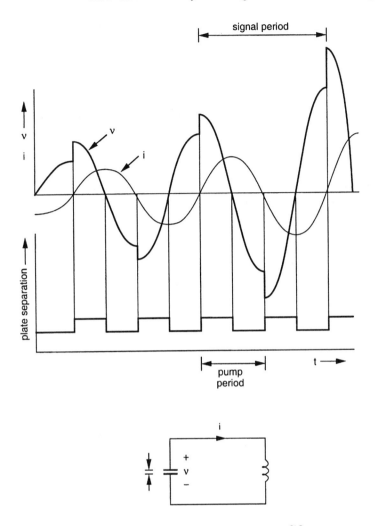

Fig. 11.4. An electrical degenerate parametric amplifier

There also exists a very simple electrical model of the degenerate parametric amplifier, shown in Fig. 11.4. It is an L–C circuit whose capacitance
varies with time. Such a time dependence could be produced mechanically by
varying the plate spacing of a capacitor, or electrically by a time-dependent
voltage applied to a nonlinear capacitor. For simplicity, we assume that the
time dependence of the capacitance is square-wave-like and in synchronism
with the voltage–current excitation in the circuit. Suppose the capacitance is
decreased mechanically; the capacitor plates are pulled apart at the instant
of time when the voltage across the capacitance is a maximum. Work is being
done against the attractive force between the capacitor plates. Suppose the

capacitance varies between C_1 and C_2 as shown. The voltage increases at constant charge (no current can flow through the inductor if the motion is very fast). The energy $Cv^2/2$ of the increased voltage and decreased capacitance is raised by the factor C_1/C_2. This energy is transferred to the inductor energy $Li^2/2$. The current increases in the ratio $\sqrt{C_1/C_2}$. When the voltage across the capacitor is zero, the plates are pushed back, and no work is done. One full cycle results in the growth of both the peak voltage and the peak current amplitudes by the factor $\sqrt{C_1/C_2}$. A pump drive of opposite phase causes decay of the excitation in the L–C circuit. When the ratio $\sqrt{C_1/C_2}$ is only slightly greater than unity, the growth becomes exponential with time. Also, one may disregard the slight change in the period of the L–C circuit and make the time intervals of the two half cycles of the pump equal to each other.

11.5 Quantum Analysis of Degenerate Parametric Amplifier

It is not difficult to develop a quantum description of the degenerate parametric amplifier in the time domain. We start with a Hermitian Hamiltonian of the form

$$\hat{H} = \frac{\hbar}{2} \left(\chi_{\text{sps}} \hat{A}_s^\dagger \hat{A}_p \hat{A}_s^\dagger + \chi_{\text{sps}}^* \hat{A}_s \hat{A}_p^\dagger \hat{A}_s \right) , \qquad (11.35)$$

where we use the normalization of creation and annihilation operators with the standard commutation relation $[\hat{A}, \hat{A}^\dagger] = 1$. The coefficient χ_{sps} has units different from those of the coefficients χ_{spi} of (11.12). The Heisenberg equations of motion lead to the coupled equations

$$\frac{d\hat{A}_s}{dt} = -i\chi_{\text{sps}} \hat{A}_p \hat{A}_s^\dagger , \qquad (11.36)$$

$$\frac{d\hat{A}_s^\dagger}{dt} = i\chi_{\text{sps}}^* \hat{A}_p^\dagger \hat{A}_s . \qquad (11.37)$$

Note that the Heisenberg equations of motion quite naturally dictate the coupled evolution of the operators $\hat{A}_s$ and $\hat{A}_s^\dagger$. When the pump is very intense and its depletion can be ignored, one may replace the pump operator $\hat{A}_p$ with a c-number amplitude A_p. Integration of the equations over an interaction time T, cast into a scattering-matrix form by defining the output operators

$$\begin{bmatrix} \hat{B}_s \\ \hat{B}_s^\dagger \end{bmatrix} \equiv \begin{bmatrix} \hat{A}_s(T) \\ \hat{A}_s^\dagger(T) \end{bmatrix}$$

in terms of the input operators

$$\begin{bmatrix} \hat{A}_s \\ \hat{A}_s^\dagger \end{bmatrix} \equiv \begin{bmatrix} \hat{A}_s(0) \\ \hat{A}_s^\dagger(0) \end{bmatrix} ,$$

gives

$$\begin{bmatrix} \hat{B}_s \\ \hat{B}_s^\dagger \end{bmatrix} = S \begin{bmatrix} \hat{A}_s \\ \hat{A}_s^\dagger \end{bmatrix} , \tag{11.38}$$

with the scattering matrix

$$S = \begin{bmatrix} \mu & \nu \\ \nu^* & \mu \end{bmatrix} , \tag{11.39}$$

where we have set $i\chi_{\text{sps}} A_p = \delta e^{i\psi}$, $\mu = \cosh(\delta T)$, and $\nu = -\sinh(\delta T)e^{i\psi}$. Note that the operator $\hat{B}_s$ is related to the operators $\hat{A}_s$ and $\hat{A}_s^\dagger$ by the transformation

$$\hat{B}_s = \mu \hat{A}_s + \nu \hat{A}_s^\dagger \quad \text{with} \quad |\mu|^2 - |\nu|^2 = 1 . \tag{11.40}$$

This is a so-called Bogolyubov transformation. The in-phase and quadrature field operators have been defined in Chap. 6. Using these definitions, and setting $\psi = 0$ by proper choice of the pump phase, we find

$$\begin{bmatrix} \hat{A}_s^{(1)}(T) \\ \hat{A}_s^{(2)}(T) \end{bmatrix} = \begin{bmatrix} \exp(-\delta T) & 0 \\ 0 & \exp(\delta T) \end{bmatrix} \begin{bmatrix} \hat{A}_s^{(1)}(0) \\ \hat{A}_s^{(2)}(0) \end{bmatrix} . \tag{11.41}$$

The analogy with the classical result is unmistakable. One component of the electric field grows, the other decays. Suppose we have an input in the state $|\alpha\rangle$. The expectation values of the amplitude are

$$\begin{bmatrix} \langle\alpha|\hat{A}_s^{(1)}(T)|\alpha\rangle \\ \langle\alpha|\hat{A}_s^{(2)}(T)|\alpha\rangle \end{bmatrix} = \begin{bmatrix} \exp(-\delta T)\dfrac{1}{2}(\alpha + \alpha^*) \\ \exp(\delta T)\dfrac{1}{2i}(\alpha - \alpha^*) \end{bmatrix} . \tag{11.42}$$

The mean square fluctuations of the output state are

$$\begin{bmatrix} \langle\alpha|\hat{A}_s^{(1)}(T)^2|\alpha\rangle - \langle\alpha|\hat{A}_s^{(1)}(T)|\alpha\rangle^2 \\ \langle\alpha|\hat{A}_s^{(2)}(T)^2|\alpha\rangle - \langle\alpha|\hat{A}_s^{(2)}(T)|\alpha\rangle^2 \end{bmatrix}$$

$$= \begin{bmatrix} \exp(-2\delta T)\left(\langle\alpha|\hat{A}_s^{(1)}(0)^2|\alpha\rangle - \langle\alpha|\hat{A}_s^{(1)}(0)|\alpha\rangle^2 \right) \\ \exp(2\delta T)\left(\langle\alpha|\hat{A}_s^{(2)}(0)^2|\alpha\rangle - \langle\alpha|\hat{A}_s^{(2)}(0)|\alpha\rangle^2 \right) \end{bmatrix} \tag{11.43}$$

$$= \frac{1}{4} \begin{bmatrix} \exp(-2\delta T) \\ \exp(2\delta T) \end{bmatrix} .$$

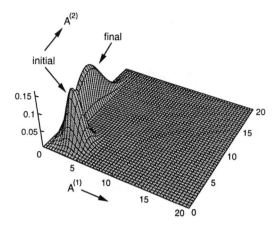

Fig. 11.5. Initial and final Wigner distribution of degenerately amplified signal. Initial phasor at 45°; in-phase component at 90°, quadrature component at 0°

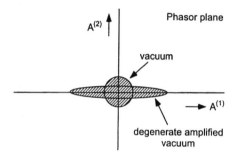

Fig. 11.6. Vacuum and squeezed vacuum

The quadrature component has an exponentally larger fluctuation; the fluctuation of the in-phase component is exponentially smaller. Figure 11.5 illustrates the phasor diagrams at the input and output. The signal has been amplified and attenuated depending upon its relative phase. The diagram illustrates what is known as a squeezed state. It has an amplitude that exhibits fluctuations with a two-dimensional Gaussian distribution, with its root mean square fluctuations lying on an ellipse. The state has four free parameters. One is the amplitude; the second is the phase of the phasor. The third is the angle of the major axis of the fluctuation ellipse with respect to the phasor. The fourth parameter is the ratio $\exp(2\delta T)$ of the major and minor axes. Note that the product of the axes remains $1/4$. Squeezed vacuum has no phasor and is illustrated in Fig. 11.6. It is described by two parameters, $|\mu|$ and the orientation of the ellipse with respect to the real axis. The distribution of the phasor endpoints is a two-dimensional Gaussian. Appendix A.17 goes into further details on two-dimensional Gaussian probability distributions.

11.6 Squeezed Vacuum and Its Homodyne Detection

Degenerate parametric amplification is phase-sensitive; the signal component with the proper phase relative to the pump experiences gain, the component in quadrature experiences loss. If no signal is fed into the amplifier, the ouput is so-called "squeezed vacuum". Even though the word "vacuum" seems to imply only zero-point fluctuation energy and the absence of photons, the fact is that squeezed vacuum contains photons and produces current in a photodetector. In this section we look at the nature of squeezed vacuum in greater detail.

Squeezed vacuum is described by the Bogolyubov transformation

$$\hat{B} = \mu \hat{A} + \nu \hat{A}^\dagger \,, \tag{11.44}$$

where μ and ν obey the constraint

$$|\mu|^2 - |\nu|^2 = 1 \,, \tag{11.45}$$

and where the expectation values of $\hat{A}$ satisfy the vacuum conditions

$$\langle |\hat{A}^2| \rangle = \langle |\hat{A}^{\dagger 2}| \rangle = \langle \hat{A}^\dagger \hat{A} \rangle = 0 \quad \text{and} \quad \langle \hat{A} \hat{A}^\dagger \rangle = 1 \,. \tag{11.46}$$

Condition (11.46) shows that no photons have been fed into the input of the parametric amplifier when squeezed vacuum is generated. However, squeezed vacuum contains photons which are derived from the pump in the parametric amplification process. Indeed, if we ask for the photon number of the output $\langle n_{\text{sq. vac}} \rangle$ using (11.46), we find

$$\langle \hat{B}^\dagger \hat{B} \rangle = \langle n_{\text{sq. vac}} \rangle = \langle |(\mu^* \hat{A}^\dagger + \nu^* \hat{A})(\mu \hat{A} + \nu \hat{A}^\dagger)| \rangle = |\nu|^2 \,. \tag{11.47}$$

Thus squeezed vacuum contains photons; it contains more photons the greater the degree of squeezing and the more elongated the ellipse in the phasor plane.

Next let us study the measurement of squeezed vacuum via homodyne detection. We assume a standard balanced detector arrangement as shown in Fig. 11.7. We linearize the formalism, treating the local oscillator as a classical variable. The current operator is given by (compare (8.30))

$$\hat{i} = -\mathrm{i} q v_g \frac{1}{L} (\alpha_L^* \hat{B} - \alpha_L \hat{B}^\dagger) \,. \tag{11.48}$$

The expectation value of $\hat{i}$ is zero.The expectation value of the square of the current operator is given by

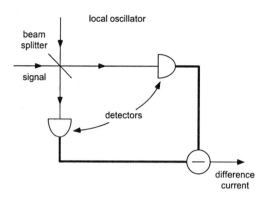

Fig. 11.7. Balanced detector

$$\langle \hat{i}^2 \rangle = q^2 v_g^2 \frac{1}{L^2}[-(\alpha_L^*)^2 \mu \nu - (\alpha_L)^2 \mu^* \nu^* + |\alpha_L|^2 (|\mu|^2 + |\nu|^2)]$$

(11.49)

$$= q^2 v_g^2 \frac{|\alpha_L|^2}{L^2}(-2|\mu\nu|\cos\phi + |\mu|^2 + |\nu|^2) ,$$

where $\phi = \arg(\alpha_L^2) - \arg(\mu\nu)$. The mean square fluctuations vary from

$$\langle \hat{i}^2 \rangle = q^2 v_g^2 \frac{|\alpha_L|^2}{L^2}(|\mu| - |\nu|)^2$$

(11.50a)

to

$$\langle \hat{i}^2 \rangle = q^2 v_g^2 \frac{|\alpha_L|^2}{L^2}(|\mu| + |\nu|)^2 ,$$

(11.50b)

depending upon the phase between the local oscillator and the squeezed vacuum. The shot noise level corresponds to $|\mu| = 1$ and $\nu = 0$. The fluctuations (11.50a) can be written in terms of the number of photons in the squeezed vacuum. Indeed, from (11.45) we have $|\mu| = \sqrt{1 + |\nu|^2}$, and thus (11.50a) can be written

$$\langle \hat{i}^2 \rangle = q^2 v_g^2 \frac{|\alpha_L|^2}{L^2}(1 + 2|\nu|^2 \mp 2|\nu|\sqrt{1 + |\nu|^2}) .$$

(11.51)

It is of interest to cast the last expression into the familiar shot noise form so that we gain a direct comparison. The expectation values have been obtained for a mode in the interval $\Delta\beta = L/2\pi$. Thus, if we ask for the fluctuations in a frequency band $\Delta\omega = (d\omega/d\beta)\Delta\beta = v_g\Delta\beta = v_g 2\pi/L$, we write for the right-hand side of (11.51)

$$\langle \hat{i}^2 \rangle = q^2 v_g \frac{\langle n_L \rangle}{L}(1 + 2|\nu|^2 \mp 2|\nu|\sqrt{1 + |\nu|^2})\Delta\omega/2\pi$$

(11.52)

$$= qI_L(1 + 2|\nu|^2 \mp 2|\nu|\sqrt{1 + |\nu|^2})B ,$$

where $I_L = qv_g|\alpha_L|^2/L$ is the current induced by the local oscillator, and B is the bandwidth.

When the squeezing is strong and $|\nu|$ is very large, we find for the two extrema

$$\langle \hat{\imath}^2 \rangle \approx \frac{1}{4\langle n_{\text{sq. vac}} \rangle} qI_L B \tag{11.53a}$$

and

$$\langle \hat{\imath}^2 \rangle \approx 4\langle n_{\text{sq. vac}} \rangle qI_L B . \tag{11.53b}$$

Several observations are in order. First of all, in the absence of squeezing, i.e. $\nu = 0$, (11.42) gives half the shot noise value. This is the consequence of homodyne detection, detection of a signal of the same frequency as the local oscillator. In this case, the idler merges with the signal and the zero-point fluctuations of the "idler" become part of the signal fluctuation. As the squeezing increases, the square root of the product of the maximum and minimum fluctuations remains at half the shot noise value, as can be seen from (11.50a) with $|\mu|^2 - |\nu|^2 = 1$. The value of the squeezed fluctuations is inversely proportional to the photon number of the squeezed radiation.

11.7 Phase Measurement with Squeezed Vacuum

We now determine the signal-to-noise ratio in a measurement of an interferometric phase with homodyne detection using squeezed vacuum as shown in Fig. 11.8. A Mach–Zehnder interferometer is unbalanced by small phase shifts of $\Delta\Phi/2$ in one arm and $-\Delta\Phi/2$ in the other arm. Into the input port (a) is fed a probe wave in a coherent state. The local oscillator in the balanced detector is also supplied by the same coherent-state source. Squeezed vacuum enters the vacuum port (b). In order to understand how this excitation can be accomplished experimentally, we look at the arrangement of Fig. 11.9, which gives the details of the generation of the different excitations. One starts with a single-frequency source (ideally in a coherent state) and splits off one part to serve as a probe and local oscillator, and another part which is frequency-doubled and serves as the pump for a degenerate parametric amplifier. Only by deriving the fields from one common source can one ensure coherence among the squeezed vacuum, probe, and local oscillator.

Returning to the measurement setup of Fig. 11.8, we now follow the evolution of the operators as they pass through the interferometer and the balanced detector. Owing to the interferometer imbalance the output from port (d) is composed of a signal part and the contributions from the squeezed vacuum port:

$$\hat{D} = i\left[\sin\left(\frac{\Delta\Phi}{2}\right)\hat{A} + \cos\left(\frac{\Delta\Phi}{2}\right)\hat{B}\right] \simeq i\left(\frac{\Delta\Phi}{2}\hat{A} + \hat{B}\right) . \tag{11.54}$$

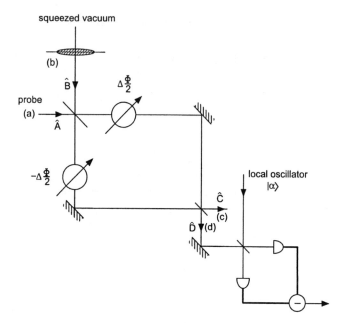

squeezed vacuum

Fig. 11.8. Measurement of phase shift $\Delta\Phi$

The balanced-detector charge operator is

$$\hat{Q} = -iq(\alpha_L^* \hat{D} - \alpha_L \hat{D}^\dagger) \,. \tag{11.55}$$

The probe $\hat{A}$ is in the state $|\alpha_p\rangle$; the input $\hat{B}$ is in a squeezed state. The signal produces an output

$$\langle \hat{Q}_s \rangle = q(\alpha_L^* \alpha_p + \alpha_L \alpha_p^*) \left(\frac{\Delta\Phi}{2}\right) = 2q|\alpha_L||\alpha_p| \cos\psi \left(\frac{\Delta\Phi}{2}\right) , \tag{11.56}$$

where $\psi = \arg(\alpha_L) - \arg(\alpha_p)$. In order to maximize the response one adjusts the probe phase with respect to the local-oscillator phase so that $\psi = 0$ or $\pm\pi$. The noise is due to the squeezed-vacuum fluctuations and the noise of the probe, with an amplitude proportional to $\Delta\Phi$ for small values of $\Delta\Phi$. If we ignore signal-dependent noise, then the contribution of the probe can be neglected and the cosine of $\Delta\Phi$ can be set to one:

$$\langle \hat{Q}^2 \rangle = q^2 \langle |\alpha_L|^2 (\hat{B}^\dagger \hat{B} + \hat{B}\hat{B}^\dagger) - \alpha_L^{*2}\hat{B}^2 - \alpha_L^2 \hat{B}^{\dagger 2} \rangle$$

$$= q^2 |\alpha_L|^2 [|\mu|^2 + |\nu|^2 - 2|\mu\nu| \cos\phi] , \tag{11.57}$$

where $\phi = \arg(\alpha_L^2) - \arg(\mu\nu)$. Clearly, the noise is minimized when the phase of the squeezed vacuum is adjusted so that $\phi = 0$. The optimum signal-to-noise ratio is

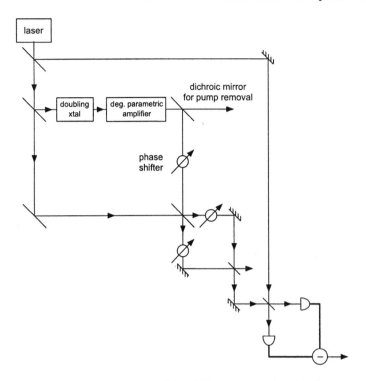

laser

doubling xtal

deg. parametric amplifier

dichroic mirror for pump removal

phase shifter

Fig. 11.9. The generation of the different excitations in the measurement of phase

$$\frac{S}{N} = \Delta\Phi^2 \frac{\langle n_p \rangle}{(|\mu| - |\nu|)^2} \ . \tag{11.58}$$

The signal-to-noise ratio is proportional to the number of probe photons and inversely proportional to $(|\mu| - |\nu|)^2$. If we express the denominator in terms of the photon number of the squeezed vacuum, in the limit of large squeezing we find

$$\frac{S}{N} = 4\Delta\Phi^2 \langle n_p \rangle \langle n_{\text{sq. vac}} \rangle^2 \ . \tag{11.59}$$

The signal-to-noise ratio is proportional to the square of the photon number in the squeezed vacuum.

This simple example shows how squeezed vacuum can be used in an interferometer to obtain signal-to-noise ratios below the shot noise limit. Here we have studied generation of squeezed vacuum with a $\chi^{(2)}$ process (second-order nonlinearity), which uses a pump at twice the signal frequency. In the next chapter we shall show how squeezed vacuum can be generated with a third-order nonlinearity. In such a process the pump can be at the signal wavelength. Further, an optical fiber can serve as a convenient nonlinear propagation medium, providing a guide that confines the radiation to a small cross section over long propagation distances in which the cumulative non-linear effects can be made large.

11.8 The Laser Resonator Above Threshold

In Chap. 6, we analyzed a resonator with loss containing a gain medium, using the requirement of commutator conservation. This requirement gave us the commutators of the noise sources. The assumption that the noise sources were in their ground states gave us the mean square fluctuations of the resonator output. The system was linear and phase-insensitive. As the gain approached full compensation of the loss, the output approached infinity. We pointed out that this limit is not reached, because of gain saturation. Gain saturation occurs when the rate of depletion of the inversion of the laser medium by the radiation approaches the rate of replenishment of the inversion by the pumping mechanism. We start with the analysis of gain saturation.

The laser gain is provided by a medium with two energy levels, the upper of which is more highly occupied than the lower level. The rate of change of the population N_u of the upper level is

$$\frac{d}{dt}N_u = -\sigma_u N_u - \gamma(N_u - N_\ell)S + P \ . \tag{11.60}$$

Here σ_u is the decay rate of the upper level, N_ℓ is the population in the lower level, S is the photon number, γ is the gain cross section, and P is the pumping rate. The term $\gamma(N_u - N_\ell)S$ is the rate of depletion of the upper level, which is proportional to the product of the population difference and the photon number. The gain action lifts the population in the lower level into the upper level and also depletes the upper level. The rate equation for the lower level is

$$\frac{d}{dt}N_\ell = -\sigma_\ell N_\ell + \gamma(N_u - N_\ell)S \ . \tag{11.61}$$

In the steady state, the rates of change are zero, and we find from (11.60) and (11.61)

$$N_u - N_\ell = \frac{P/\sigma_u}{1 + \gamma\left(1/\sigma_u + 1/\sigma_\ell\right)S} \ . \tag{11.62}$$

The population inversion is proportional to the pumping rate and decreases with increasing photon number. The rate at which photons are generated is

$$\text{rate of photon generation} = \gamma(N_u - N_\ell)S = \frac{\gamma P/\sigma_u}{1 + \gamma\left(1/\sigma_u + 1/\sigma_\ell\right)S}S \ . \tag{11.63}$$

The rate of photon generation is proportional to the energy growth rate due to gain, $2/\tau_g$, as defined earlier in the treatment of the resonator. We may write

$$\frac{2}{\tau_g} = \frac{\gamma P}{\sigma_u} \frac{1}{1 + \gamma(1/\sigma_u + 1/\sigma_\ell)S} = \frac{2}{\tau_g^o} \frac{1}{1 + S/S_{\text{sat}}} \,, \tag{11.64}$$

where

$$S_{\text{sat}} = \frac{1}{\gamma(1/\sigma_u + 1/\sigma_\ell)}$$

is the so-called saturation photon number; it is the number of photons that reduces the gain to half its value. The unsaturated value of the gain corresponds to the small-signal growth rate $2/\tau_g^o$, where

$$\frac{2}{\tau_g^o} = \frac{\gamma P}{\sigma_u} \,. \tag{11.65}$$

The rate of growth changes with changes of the photon number density. These changes can cause noise. If the fluctuations of S are small we may write

$$S = S_o + \Delta S \,, \tag{11.66}$$

and the growth rate expanded to first order in ΔS is

$$\frac{2}{\tau_g} = \frac{2}{\tau_g^o} \frac{1}{1 + S_o/S_{\text{sat}}} - \frac{2}{\tau_g^o} \frac{1}{(1 + S_o/S_{\text{sat}})^2} \frac{\Delta S}{S_{\text{sat}}} \,. \tag{11.67}$$

Fluctuations of the photon number can induce fluctuations of the growth rate. The analysis thus far has been classical. Quantum equations are obtained when the complex c-number amplitudes are replaced by annihilation operators of the electric field and noise sources are introduced that ensure conservation of commutators. The equation of the resonator introduced in Chap. 6 reads

$$\frac{d\hat{U}}{dt} = -\left(i\omega_o + \frac{1}{\tau_e} + \frac{1}{\tau_o} - \frac{1}{\tau_g}\right)\hat{U} + \sqrt{\frac{2}{\tau_e}}\hat{a} + \hat{N} \,, \tag{11.68}$$

where we combine in $\hat{N}$ all the noise sources associated with loss and gain. The laser saturation introduces a new aspect into the equation. The rate of growth due to the gain itself experiences fluctuations in response to the fluctuations of the mode amplitude. Further, the responses to the in-phase and quadrature components of the mode field are different. In anticipation of the separation of the equations into in-phase and quadrature components, we remove the natural time dependence by the replacement $\hat{U} \rightarrow \hat{U} \exp(-i\omega_o t)$

(without change of notation!) and do the same with the operators $\hat{a}$ and $\hat{N}$. We express the mode amplitude operator as the sum of a (classical) c-number amplitude and an operator perturbation:

$$\hat{U} = U_o + \Delta\hat{U} . \tag{11.69}$$

The photon number operator is, to first order,

$$\hat{U}^\dagger\hat{U} = |U_o|^2 + \Delta\hat{U} U_o^* + \Delta\hat{U}^\dagger U_o . \tag{11.70}$$

This operator is identified with the c number $S = S_o + \Delta S$. When this is done we obtain equations of motion for the c number U_o:

$$\frac{dU_o}{dt} = -\left(\frac{1}{\tau_e} + \frac{1}{\tau_o} - \frac{1}{\tau_g^o}\frac{1}{1 + |U_o|^2/S_{\text{sat}}}\right) U_o . \tag{11.71}$$

Here we have omitted the noise source, since it drives the perturbation operator, and the incident wave, since it is assumed to be unexcited except, of course, for its zero-point fluctuations, which drive the perturbation operator $\Delta\hat{U}$. In the steady state, the amplitude has to remain constant and the growth and decay rates must balance. This serves to determine the steady-state amplitude:

$$\frac{|U_o|^2}{S_{\text{sat}}} = \frac{1/\tau_g^o}{1/\tau_e + 1/\tau_o} - 1 . \tag{11.72}$$

Threshold is reached when the unsaturated gain, represented by the growth rate $1/\tau_g^o$, becomes equal to the decay rate $1/\tau_e + 1/\tau_o$. The equation of motion for the perturbation operator $\Delta\hat{U}$ becomes

$$\begin{aligned}
\frac{d\Delta\hat{U}}{dt} = &-\left(\frac{1}{\tau_e} + \frac{1}{\tau_o} - \frac{1}{\tau_g}\right)\Delta\hat{U} \\
&- \frac{1}{\tau_g^o}\frac{1}{(1 + |U_o|^2/S_{\text{sat}})^2}\frac{|U_o|^2}{S_{\text{sat}}}(\Delta\hat{U}^\dagger + \Delta U) \\
&+ \sqrt{\frac{2}{\tau_e}}\hat{a} + \hat{N} ,
\end{aligned} \tag{11.73}$$

where we have assumed that the phase of U_o is zero. The operator $\Delta\hat{U}$ couples to its Hermitian conjugate. This is an indication of a phase-sensitive process.

In order to solve (11.73) we introduce in-phase and quadrature components

$$\Delta\hat{U}^{(1)} \equiv \frac{1}{2}(\Delta\hat{U} + \Delta\hat{U}^\dagger), \qquad \Delta\hat{U}^{(2)} \equiv \frac{1}{2i}(\Delta\hat{U} - \Delta\hat{U}^\dagger) . \tag{11.74}$$

In the steady state

$$\frac{1}{\tau_e} + \frac{1}{\tau_o} - \frac{1}{\tau_g} = 0 . \tag{11.75}$$

We then find

$$\frac{d\Delta\hat{U}^{(1)}}{dt} = -\frac{1}{\tau_s}\Delta\hat{U}^{(1)} + \sqrt{\frac{2}{\tau_e}}\hat{a}^{(1)} + \hat{N}^{(1)} , \tag{11.76}$$

and

$$\frac{d\Delta\hat{U}^{(2)}}{dt} = \sqrt{\frac{2}{\tau_e}}\hat{a}^{(2)} + \hat{N}^{(2)} ,$$

where we have defined

$$\frac{2}{\tau_g^o}\frac{1}{(1 + |U_o|^2/S_{\text{sat}})^2}\frac{|U_o|^2}{S_{\text{sat}}} \equiv \frac{1}{\tau_s} . \tag{11.77}$$

Equations (11.75) and (11.76) assign different time dependences to the in-phase and quadrature components. This is a novel situation that requires closer scrutiny. In order to investigate this case we generalize to the situation in which the in-phase and quadrature components have decay rates γ_1 and γ_2 [151, 152]:

$$\frac{d}{dt}\Delta\hat{U}^{(1)} = -\gamma_1\Delta\hat{U}^{(1)} + \hat{N}^{(1)} ,$$

$$\frac{d}{dt}\Delta\hat{U}^{(2)} = -\gamma_2\Delta\hat{U}^{(2)} + \hat{N}^{(2)} . \tag{11.78}$$

The commutator of the noise source and the expectation values of its moments are obtained in a sequence of steps. We first look at the commutator of the observables $\Delta\hat{U}^{(1)}(t)$ and $\Delta\hat{U}^{(2)}(t)$:

$$[\Delta\hat{U}^{(1)}(t), \Delta\hat{U}^{(2)}(t)] = \frac{i}{2} . \tag{11.79}$$

From (11.78) we obtain for the rate of change:

$$\frac{d}{dt}\left[\Delta\hat{U}^{(1)}(t), \Delta\hat{U}^{(2)}(t)\right]$$

$$= -(\gamma_1 + \gamma_2)\left[\Delta\hat{U}^{(1)}(t), \Delta\hat{U}^{(2)}(t)\right] + \left[\Delta\hat{U}^{(1)}(t), \hat{N}^{(2)}(t)\right] \tag{11.80}$$

$$+ \left[\hat{N}^{(1)}(t), \Delta\hat{U}^{(2)}(t)\right] = 0 .$$

Within the time interval Δt, the excitations $\Delta \hat{U}^{(1)}(t)$ and $\Delta \hat{U}^{(2)}(t)$ acquire contributions $(1/2)\Delta t \, \hat{N}^{(1)}(t)$ and $(1/2)\Delta t \, \hat{N}^{(2)}(t)$, respectively, from the noise sources. Thus, one finds from (11.80)

$$-(\gamma_1 + \gamma_2)\left[\Delta \hat{U}^{(1)}(t), \Delta \hat{U}^{(2)}(t)\right] + 2\frac{1}{2}\Delta t \left[\hat{N}^{(1)}(t), \hat{N}^{(2)}(t)\right] = 0 ,$$

(11.81)

from which we infer

$$\left[\hat{N}^{(1)}(t), \hat{N}^{(2)}(t')\right] = \frac{i}{2}(\gamma_1 + \gamma_2)\delta(t - t') .$$

(11.82)

One may construct an annihilation operator from the in-phase and quadrature noise operators

$$\hat{N}(t) = \hat{N}^{(1)}(t) + i\hat{N}^{(2)}(t) .$$

(11.83)

Its commutator with the Hermitian conjugate creation operator is

$$\left[\hat{N}(t), \hat{N}^{\dagger}(t')\right] = (\gamma_1 + \gamma_2)\delta(t - t') .$$

(11.84)

The commutator of a pair of operators assigns a minimum uncertainty to the product of the mean square fluctuations of the observables, but does not put limits on the mean square value of the fluctuations of either of the two observables. However, if the noise source equilibrates rapidly compared with the characteristic times of the system, the noise should be stationary and phase-insensitive, and the mean square fluctuations of the in-phase and quadrature components have to be equal:

$$\langle |(\hat{N}^{(1)})^2| \rangle = \langle |(\hat{N}^{(2)})^2| \rangle .$$

(11.85)

We must remember that the noise source $\hat{N}(t)$ is constructed from a sum of annihilation and creation operators. Whereas the contributions to the commutator of the creation and annihilation operators are of opposite signs, their contributions add in the mean square fluctuations. Specifically, in the case of the laser oscillator,

$$\gamma_1 = \frac{1}{\tau_s} \quad \text{and} \quad \gamma_2 = 0 .$$

(11.86)

The question then arises of how the mean square fluctuations are to be assigned to the individual noise sources. The noise source associated with the loss has the usual autocorrelation function

$$\langle |\hat{N}_o^{(1)}(t)\hat{N}_o^{(1)}(t')| \rangle = \langle |\hat{N}_o^{(2)}(t)\hat{N}_o^{(2)}(t')| = \frac{1}{4}\frac{2}{\tau_o}\delta(t - t') .$$

(11.87)

The noise source assigned to the gain represents the ASE of the inverted medium. When the relaxation rate of the lower level of the gain medium is very fast, $\sigma_\ell/\sigma_u \to \infty$, the lower level remains unoccupied. Then the mean square fluctuations of the noise source are

$$\langle|\hat{N}_g^{(1)}(t)\hat{N}_g^{(1)}(t')|\rangle = \langle|\hat{N}_g^{(2)}(t)\hat{N}_g^{(2)}(t')|\rangle = \frac{1}{4}\frac{2}{\tau_g}\delta(t-t') .\qquad(11.88)$$

As saturation sets in, the commutator of the noise source associated with the gain medium changes. The commutator by itself sets only the minimum value of the product of the mean square fluctuations of the in-phase and quadrature components. Stationarity sets the two fluctuations equal. But the question remains as to how large the actual fluctuations are. We shall use two models in the remainder of this discussion. The simplest model is one in which the noise contributed by the gain medium is assumed to be independent of the degree of saturation and given by (11.88). This is in spite of the fact that the commutator of the gain medium has to accommodate the saturation-induced rate of decay of an amplitude perturbation, $1/\tau_s$. One may imagine that saturation is accompanied by a buildup of the lower level, which contributes noise of its own so as to offset the decrease of the magnitude of the commutator of $\hat{N}_g$. We call this model "model I". We shall also consider the case of minimum noise, "model II", in which a single noise source is assigned to the gain medium, a creation operator with the commutator

$$[\hat{N}_g(t), N_g^\dagger(t')] = -\left(\frac{2}{\tau_g} - \frac{1}{\tau_s}\right)\delta(t-t') .\qquad(11.89)$$

As the decay rate $1/\tau_s$ increases, the commutator of the gain medium decreases in magnitude. In the limit of very strong saturation, the gain medium becomes noise-free. This is an interesting model of the laser, which will be discussed in greater detail in Sect. 11.11.

11.9 The Fluctuations of the Photon Number

Equations (11.78) are particularly suitable for an initial-value problem. The evolution in time can also model an evolution in space of a wavepacket that travels with the group velocity v_g, covering the distance $L = v_g T$ in the time interval T. Linear differential equations are solved in the time domain by superposition of their impulse responses. Denote the impulse responses of (11.78) by $h_1(t)$ and $h_2(t)$. Then

$$\Delta\hat{U}^{(i)}(t) = \int dt'\, h_i(t-t')\hat{N}^{(i)}(t'), \quad \text{where} \quad i = 1, 2 .\qquad(11.90)$$

The commutator evolves according to

$$[\Delta\hat{U}^{(1)}(t), \Delta\hat{U}^{(2)}(t')]$$

$$= \int dt'' \int dt''' \, h_1(t - t'')h_2(t' - t''')[\hat{N}^{(1)}(t''), \hat{N}^{(2)}(t''')]$$

$$= \frac{i}{2}(\gamma_1 + \gamma_2) \int dt'' \int dt''' \, h_1(t - t'')h_2(t' - t''')\delta(t'' - t''')$$

$$= \frac{i}{2}(\gamma_1 + \gamma_2) \int dt'' \, h_1(t - t'')h_2(t' - t'') .$$

(11.91)

When we introduce the specific impulse response functions, we find that the commutator is a function of the time difference $t - t'$, i.e. is stationary, and has the value

$$[\Delta\hat{U}^{(1)}(t), \Delta\hat{U}^{(2)}(t')] = \begin{cases} \dfrac{i}{2}\exp[-\gamma_1(t - t')] & \text{for } t > t' \\[2mm] \dfrac{i}{2}\exp[-\gamma_2(t' - t)] & \text{for } t < t' . \end{cases}$$

(11.92)

The excitation in the resonator "remembers" its drive. The memory dies out at the decay rate of the in-phase component for $t > t'$, and at the decay rate of the quadrature component for $t < t'$.

The commutator at equal times, $t = t'$, is $1/2$ as it should be. Consider next the expectation value of the product of the perturbation operators $\Delta\hat{U}^{(1)}(t)$ and $\Delta\hat{U}^{(2)}(t')$ of the fluctuations in the resonator:

$$\langle\Delta\hat{U}^{(i)}(t)\Delta\hat{U}^{(i)}(t')\rangle$$

$$= \int dt'' \, h_i(t - t'') \int dt''' \, h_i(t' - t''')\langle|\hat{N}^{(i)}(t'')\hat{N}^{(i)}(t''')|\rangle$$

$$= \frac{1}{4}\int dt'' \, h_i(t - t'') \int dt''' \, h_i(t' - t''')\left(\frac{2}{\tau_e} + \frac{2}{\tau_g} + \frac{2}{\tau_o}\right)\delta(t'' - t''')$$

$$= \frac{1}{4}\left\{\frac{2}{\tau_e} + \frac{2}{\tau_g} + \frac{2}{\tau_o}\right\}\int dt'' \, h_i(t - t'')h_i(t' - t'') .$$

(11.93)

With the expression for the in-phase impulse response, we find for the auto-correlation of the in-phase component

$$\frac{1}{2}\left[\langle\Delta\hat{U}^{(1)}(t)\Delta U^{(1)}(t')\rangle + \langle\Delta\hat{U}^{(1)}(t')\Delta U^{(1)}(t)\rangle\right]$$

$$= \frac{1}{4}\left(\frac{2}{\tau_e} + \frac{2}{\tau_g} + \frac{2}{\tau_o}\right)\frac{\tau_s}{2}\exp\left(-\frac{|t - t'|}{\tau_s}\right) .$$

(11.94)

Setting $t = t'$, one obtains the mean square fluctuations. It is of interest to note that these become smaller and smaller the higher the degree of saturation. In the ideal limit of a very high degree of saturation, $1/\tau_s \approx 2/\tau_g$, and for negligible internal loss, $\tau_e/\tau_o \to 0$, the mean square fluctuations approach $1/4$. This is the value for the field fluctuations of a Poissonian photon number distribution.

The relaxation rate of the quadrature component is zero, the relaxation time infinite. Equation (11.93) applied to the case of an infinite relaxation time gives infinity. How does one obtain physically meaningful information in this case? First of all we note that the derivation of (11.94) assumed a stationary steady state; the limit in the integral over the impulse response went from $-\infty$ to $+\infty$. When the relaxation rate goes to zero, an infinite disturbance builds up. Now, our linearization approximation assumed that the perturbations $\Delta \hat{U}^{(i)}$ are small compared with U_o. This assumption does not permit arbitrarily large perturbations. However, the situation changes if one defines the phase operator $\Delta \hat{\theta} \equiv \Delta \hat{U}^{(2)}/U_o$. The linearization is still valid if the changes of phase are small within any given finite time interval. Further, the phase can grow without bound, whereas the quadrature component referred to an unchanging phasor cannot. It is in this spirit that the quadrature fluctuations have to be interpreted, namely as phase fluctuations.

The autocorrelation of the phase is

$$\frac{1}{2}[\langle \Delta\hat{\theta}(t)\Delta\hat{\theta}(t')\rangle + \langle \Delta\hat{\theta}(t')\Delta\hat{\theta}(t)\rangle]$$

$$= \frac{1}{2|U_o|^2}[\langle \Delta\hat{U}^{(2)}(t)\Delta\hat{U}^{(2)}(t')\rangle + \langle \Delta\hat{U}^{(2)}(t')\Delta\hat{U}^{(2)}(t)\rangle] \, . \tag{11.95}$$

For the evaluation of the laser linewidth, we shall be interested in the mean square value of the phase difference at two times, $\langle |\Delta\theta(t) - \Delta\theta(t+\tau)|^2\rangle$. One simple way is to introduce a finite relaxation time, use an expression of the form of (11.94) for the correlation function, and then go to the limit of an infinite relaxation time. The infinities cancel and one finds

$$\langle |\Delta\hat{\theta}(t) - \Delta\hat{\theta}(t+\tau)|^2\rangle$$

$$= \langle \Delta\hat{\theta}^2(t)\rangle + \langle \Delta\hat{\theta}^2(t+\tau)\rangle - \langle \Delta\hat{\theta}(t)\Delta\hat{\theta}(t+\tau)\rangle - \langle \Delta\hat{\theta}(t+\tau)\Delta\hat{\theta}(t)\rangle$$

$$= \frac{1}{2|U_o|^2}\left(1 + \frac{\tau_e}{\tau_g} + \frac{\tau_o}{\tau_o}\right)\frac{|\tau|}{\tau_e} \, . \tag{11.96}$$

The spectrum of the amplitude fluctuations is obtained by Fourier transformation of the autocorrelation function:

$$\frac{1}{2\pi} \int d\tau \exp(i\omega\tau)\frac{1}{2}[\langle \Delta \hat{U}^{(1)}(t)\Delta \hat{U}^{(1)}(t+\tau)\rangle + \langle \Delta \hat{U}^{(1)}(t+\tau)\Delta U^{(1)}(t)\rangle]$$

$$= \frac{1}{2\pi}\frac{1}{4}\left(\frac{2}{\tau_e} + \frac{2}{\tau_g} + \frac{2}{\tau_o}\right)\frac{2\tau_s^2}{\omega^2\tau_s^2 + 1} \ .$$

$$(11.97)$$

The spectrum of the phase fluctuations requires greater care. Since the relaxation time of the phase is infinite, the mean square phase fluctuations increase with time; the process is not stationary. A nonstationary process does not allow for the simple Fourier transform relation between autocorrelation function and spectrum. Yet there is an aspect of stationarity in the phase diffusion process: starting with a particular phase $\theta(t)$ at a time t, the phase at some later time walks away from the initial value in the same way, independent of the starting time. This aspect justifies the step which we now undertake. We use the expression for the autocorrelation function of the amplitude and take the limit of $\tau_s \to \infty$. In this way we obtain for the spectrum of the phase

$$\frac{1}{2\pi} \int d\tau \exp(i\omega\tau)\frac{1}{2}[\langle \Delta\hat{\theta}(t)\Delta\hat{\theta}(t+\tau)\rangle + \langle \Delta\hat{\theta}(t+\tau)\Delta\hat{\theta}(t)\rangle]$$

$$(11.98)$$

$$\approx \frac{1}{2\pi}\frac{1}{2|U_o|^2}\left(\frac{2}{\tau_e} + \frac{2}{\tau_g} + \frac{2}{\tau_o}\right)\frac{1}{\omega^2} \ .$$

This is the spectrum of a random walk. It has a singularity at the origin.

11.10 The Schawlow–Townes Linewidth

The spectrum of the radiation in the laser can be evaluated from the autocorrelation function of the time-dependent field amplitude $U_o + \Delta U$. We follow here the standard classical analysis for the evaluation of the spectrum of an oscillator [153]:

$$U(t) = [U_o + \Delta U(t)]\cos[(\omega_o t + \Delta\theta(t)]$$

$$= \frac{1}{2}[U_o + \Delta U(t)]\{\exp[i(\omega_o t + \Delta\theta(t))] + \exp[-i(\omega_o t + \Delta\theta(t))]\} \ .$$

$$(11.99)$$

Its autocorrelation function is

$$\langle U(t)U(t+\tau)\rangle$$

$$= \frac{1}{4}[U_o^2 + \langle \Delta U(t)\Delta U(t+\tau)\rangle]$$

$$\times \{\langle \exp[i(\Delta\theta(t) - \Delta\theta(t+\tau))]\rangle \exp(i\omega_o\tau)$$

$$+ \langle \exp[-i(\Delta\theta(t) - \Delta\theta(t+\tau))]\rangle \exp(-i\omega_o\tau)\} .$$

(11.100)

Since the in-phase and quadrature fluctuations are independent, the expectation value of the product is equal to the product of the expectation values. The exponentials are related to the characteristic function

$$C_\theta(\xi) = \langle \exp\{i\xi[\Delta\theta(t) - \Delta\theta(t+\tau)]\}\rangle$$

(11.101)

of the phase change $\Delta\hat{\theta}(t) - \Delta\hat{\theta}(t+\tau)$. The phase distribution is Gaussian, and thus we know that $C_\theta(\xi)$, the Fourier transform of the probability distribution, is also Gaussian:

$$C_\theta(\xi) = \exp\left(-\frac{\xi^2\sigma_\theta^2}{2}\right) ,$$

(11.102)

where the mean square deviation has been evaluated in (11.96) Thus, we obtain for the autocorrelation function

$$\langle U(t)U(t+\tau)\rangle$$

$$= \frac{1}{2}[U_o^2 + \langle \Delta U(t)\Delta U(t+\tau)\rangle]\exp\left[\frac{-(1 + \tau_e/\tau_g + \tau_e/\tau_o)|\tau|}{4U_o^2\tau_e}\right]\cos(\omega_o\tau) .$$

(11.103)

The spectrum is the Fourier transform of the autocorrelation function. Since the autocorrelation function is a product of the autocorrelations of the amplitude and carrier, the spectrum is the convolution of the respective spectra. The evaluation of the spectrum of (11.103) is left as an exercise. Here we concentrate on the spectrum of the carrier, which is the dominant contribution to the lineshape:

$$\frac{1}{2\pi} \int d\tau \exp(i\omega\tau) \exp\left[\frac{-(1 + \tau_e/\tau_g + \tau_e/\tau_o)|\tau|}{4U_o^2\tau_e}\right]\cos(\omega_o\tau)$$

$$= \frac{1}{2\pi}\Delta\Omega\left[\frac{1}{(\omega - \omega_o)^2 + \Delta\Omega^2} + \frac{1}{(\omega + \omega_o)^2 + \Delta\Omega^2}\right] ,$$

(11.104)

where

$$\Delta\Omega = \frac{1 + \tau_e/\tau_g + \tau_e/\tau_o}{4U_o^2\tau_e}$$

(11.105)

is the so called Schawlow–Townes linewidth. This linewidth is, generally, very small since it is inversely proportional to the photon number in the cavity and the external Q of the resonator. Jaseja et al. attempted to measure it on a He–Ne laser early after its invention [154], to no avail. The thermal vibrations of the rods supporting the laser cavity broadened the line much beyond the value of (11.105). This gave the impetus to C. Freed and the author to measure spontaneous-emission effects in the amplitude noise near threshold, where these effects emerge above the background of environmentally produced fluctuations [155]. Later, after the invention of the semiconductor laser, in which the Q is much lower and the photon number in the resonator is much smaller, Freed was able to observe the quantum limit set by (11.105) [156, 157].

11.11 Squeezed Radiation from an Ideal Laser

The study of a laser above threshold on the basis of model I, in which the noise generated by the gain medium was assumed independent of the saturation level, showed that the lowest fluctuations internal to the laser were those of a Poisson process. Model II, in which it is assumed that the noise of the gain decreases with increasing degree of saturation, can give sub-Poissonian outputs as we now show. For this purpose it is necessary to study the radiation emitted by the laser as represented by the operator $\hat{b}$. We use the resonator–waveguide coupling equation (6.161)

$$\hat{b} = -\hat{a} + \sqrt{\frac{2}{\tau_e}}\hat{U} \ . \tag{11.106}$$

When the excitation in the resonator is linearized, the incident wave $\hat{a}$ acts as a noise source. The outgoing wave consists of two parts: (a) the outgoing, c-number, steady-state laser signal $\sqrt{2/\tau_e}U_o$, and (b) the fluctuation operator $\Delta\hat{b}$. Thus, the linearized form of (11.106), separated into in-phase and quadrature components, is

$$\Delta\hat{b}^{(i)} = -\hat{a}^{(i)} + \sqrt{\frac{2}{\tau_e}}\Delta\hat{U}^{(i)} \ , \quad i = 1, 2 \ . \tag{11.107}$$

The commutator of the resonator mode has an exponential time dependence. On the other hand, the wave $\Delta\hat{b}$ represents a wave on an open waveguide and thus has a prescribed commutator

$$[\Delta\hat{b}^{(1)}(t), \Delta\hat{b}^{(2)}(t')] = \frac{i}{2}\delta(t - t') \ . \tag{11.108}$$

Let us determine the mean square fluctuations of the output wave. Consider $\langle\Delta\hat{b}^{(i)}(t)\Delta\hat{b}^{(i)}(t')\rangle$. We have

$$\langle \Delta \hat{b}^{(i)}(t) \Delta \hat{b}^{(i)}(t') \rangle$$

$$= \langle \hat{a}^{(i)}(t) \hat{a}^{(i)}(t') \rangle - \sqrt{\frac{2}{\tau_e}} [\langle \hat{a}^{(i)}(t) \Delta \hat{U}^{(i)}(t') \rangle + \langle \Delta \hat{U}^{(i)}(t) \hat{a}^{(i)}(t') \rangle]$$

$$+ \frac{2}{\tau_e} \langle \Delta \hat{U}^{(i)}(t) \Delta \hat{U}^{(i)}(t') \rangle, \quad i = 1, 2 .$$

$$(11.109)$$

There is partial cancellation between the term in brackets and the last term. The response of $\Delta \hat{U}^{(1)}$ to the source $\hat{a}^{(1)}$ is given by (11.90). Thus

$$\sqrt{\frac{2}{\tau_e}} \langle \Delta \hat{U}^{(i)}(t) \hat{a}^{(i)}(t') \rangle$$

$$= \frac{2}{\tau_e} \int dt'' \, h_i(t - t'') \langle \hat{a}^{(i)}(t'') \hat{a}^{(i)}(t') \rangle \frac{1}{4} \frac{2}{\tau_e} \int dt'' \, h_i(t - t'') \delta(t'' - t')$$

$$= \begin{cases} \dfrac{1}{4} \dfrac{2}{\tau_e} h_i(t - t') & \text{for} \quad t > t' \\ \\ 0, & \text{for} \quad t < t' . \end{cases}$$

$$(11.110)$$

Similarly

$$\sqrt{\frac{2}{\tau_e}} \langle \hat{a}^{(i)}(t) \Delta \hat{U}^{(i)}(t') \rangle = \begin{cases} \dfrac{1}{4} \dfrac{2}{\tau_e} h_i(t' - t) \,; & \text{for} \quad t < t' \\ \\ 0 & \text{for} \quad t > t' . \end{cases}$$

$$(11.111)$$

Finally, using (11.93) with the noise of model II, we may write down the expectation value of $\Delta \hat{b}^{(i)}(t) : \Delta \hat{b}^{(i)}(t')$

$$\langle \Delta \hat{b}^{(i)}(t) \Delta \hat{b}^{(i)}(t') \rangle$$

$$= \frac{1}{4} \left[\delta(t - t') - \frac{2}{\tau_e} h_i(|t - t'|) + \frac{2}{\tau_e} \left(\frac{2}{\tau_e} + \frac{2}{\tau_g} - \frac{1}{\tau_s} + \frac{2}{\tau_o} \right) \right. \tag{11.112}$$

$$\left. \times \int dt'' \, h_i(t - t'') h_i(t' - t'') \right] .$$

When this expression is applied to the in-phase and quadrature components, we find along the lines of the analysis in Sect. 11.9, the autocorrelation function of the in-phase component of the outgoing wave as

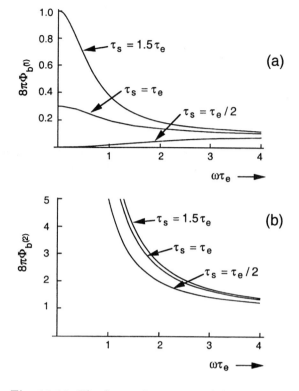

Fig. 11.10. The fluctuation spectra of the in-phase and quadrature components

$$\frac{1}{2}\langle \Delta\hat{b}^{(1)}(t)\Delta\hat{b}^{(1)}(t') + \Delta\hat{b}^{(1)}(t')\Delta\hat{b}^{(1)}(t)\rangle$$

$$= \frac{1}{4}\left[\delta(t-t') + \frac{2}{\tau_e}\left(\frac{2/\tau_e + 2/\tau_g - 1/\tau_s + 2/\tau_o}{2/\tau_s} - 1\right)\exp\left(-\frac{|t-t'|}{\tau_s}\right)\right].$$
$$(11.113)$$

The quadrature component is

$$\frac{1}{2}\langle \Delta\hat{b}^{(2)}(t)\Delta\hat{b}^{(2)}(t') + \Delta\hat{b}^{(2)}(t)\Delta\hat{b}^{(2)}(t')\rangle$$

$$= \begin{cases} \frac{1}{4}\left[\delta(t-t') + \frac{2}{\tau_e}\left(\frac{2}{\tau_e} + \frac{2}{\tau_g} - \frac{1}{\tau_s} + \frac{2}{\tau_o}\right)t\right] & \text{for } t \le t' \\[2ex] \frac{1}{4}\left[\delta(t-t') + \frac{2}{\tau_e}\left(\frac{2}{\tau_e} + \frac{2}{\tau_g} - \frac{1}{\tau_s} + \frac{2}{\tau_o}\right)t'\right] & \text{for } t' \le t \end{cases}$$
$$(11.114)$$

The spectra can be obtained analogously to the analysis in Sect. 11.9:

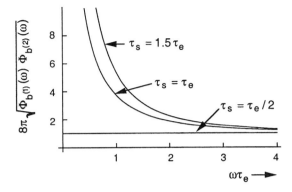

Fig. 11.11. The product of the spectra of the in-phase and quadrature components

$$\Phi_{b^{(1)}}(\omega) = \frac{1}{2\pi}\frac{1}{4}\left[1 + \frac{2}{\tau_e}\left(\frac{2/\tau_e + 2/\tau_g - 1/\tau_s + 2/\tau_o}{2/\tau_s} - 1\right)\frac{2\tau_s}{\omega^2\tau_s^2 + 1}\right],$$
$$(11.115)$$

$$\Phi_{b^{(2)}}(\omega) = \frac{1}{2\pi}\frac{1}{4}\left[1 + 2\left(1 + \frac{\tau_e}{\tau_g} - \frac{\tau_e}{2\tau_s} + \frac{\tau_e}{\tau_o}\right)\frac{2}{\omega^2\tau_e^2}\right]. \qquad (11.116)$$

In the ideal limit of strong saturation, when $2/\tau_g = 2/\tau_e = 1/\tau_s$, and when the loss is negligible, the spectrum of the in-phase component vanishes at zero frequency. This corresponds to perfect amplitude squeezing. The amplitude squeezing is at the expense of the phase, whose spectrum diverges at the origin of the frequency. Figure 11.10 shows the two spectra for different values of saturation, in the ideal limit of zero loss. When $1/\tau_s = 2/\tau_g$, the amplitude fluctuations near zero frequency go to zero, indicating perfect amplitude squeezing. Fig. 11.11 shows the square root of the product of the two spectra. As we may see, the product never dips below the minimum uncertainty value of $1/8\pi$, as is required by the uncertainty principle.

The preceding analysis is an idealized model of the amplitude squeezed radiation observed by Yamamoto and Machida [158]. They employed a semiconductor laser, current-excited through a high series resistance. The high resistance reduces the current fluctuations below the shot noise level and thus the carrier injection into the laser becomes sub-Poissonian. If the probability of induced emission is very high, then the sub-Poissonian character of the injection current manifests itself in a sub-Poissonian photon emission. In the limit of zero current fluctuations, the squeezing would become perfect.

11.12 Summary

Phase-sensitive amplification leads to different fluctuation spectra of the components in phase and in quadrature with a pump. In the case of the parametric amplifier, the pump is at a frequency different from the signal frequency. In the case of the laser above threshold the role of the pump is played by the c.w. signal amplitude of the oscillator.

Parametric amplifiers are special cases of multiports described by a scattering equation that does not contain noise sources. This does not mean that noise-free amplification can be achieved with a parametric amplifier. The output noise is generated from the zero-point fluctuations of the idler input.

We found it convenient to assign negative photon numbers to the idler and positive photon numbers to the signal. In this way a conservation principle of photon number was obtained. This situation is analogous to a widely used formalism in plasma physics in which growing and decaying wave solutions are ascribed to coupling of waves with positive and negative effective energies. In plasma physics, a negative energy is assigned to an excitation of an electron or ion beam when the kinetic energy is reduced by the excitation. Thus, this assignment is a matter of convenience. There is one physical situation in which negative energy has an unequivocal meaning. The theory of the evolution of the universe after the big bang is, in fact, based on the recognition of the negative energy of gravity. As matter evolves from vacuum, the negative energy of the gravitational field balances the positive energy of everything else. Thus the evolution of the universe is, in the words of Alan Guth, "the ultimate free lunch".

We found that degenerate parametric amplification produces squeezed states. We analyzed the fluctuations of the current in a balanced detector illuminated with squeezed vacuum. Finally, we showed how squeezed vacuum can be used to improve the signal-to-noise ratio of an interferometric phase measurement.

It is of interest to note that the equations for the mechanical degenerate parametric amplifier, the pumped capacitor in an L–C circuit, are identical with the equations for the optically pumped degenerate parametric amplifier. Clearly, the mechanical pumping can be replaced by electric pumping of a varactor amplifier, in which the width of the depletion region of a reverse-biased junction is changed by an applied voltage. Conversely, the role of the optical pump in the variation of the dielectric constant could be taken over by a distribution of Maxwell demons moving the atoms of the medium back and forth and producing index variations in this manner. Even though this appears to be an extreme version of a "Gedankenexperiment", it is useful to pursue some of its consequences. Instead of the atoms in the medium being moved, the perfectly reflecting walls of the resonator could be moved back and forth. This is equivalent to the variation of the capacitor in the L–C circuit example. In the absence of an optical excitation, the motion of the atoms or the walls affects only the zero-point fluctuations within the optical resonator.

The analysis shows that photons are generated in a degenerate parametric amplifier without any initial (input) photons. Thus, photons are generated from vacuum when the boundary conditions of the zero-point fluctuations are changed in a way that involves acceleration. (Uniform motion of a mirror Doppler-shifts the frequency of an incident photon flow, but does not generate new photons.)

We developed two models for the laser gain. One led to fluctuations of the laser output in excess of the zero-point fluctuations. The other, ideal model was constructed from the postulate that the noise should be as small as was compatible with the requirement of commutator conservation. This latter model gave a laser that emits amplitude-squeezed radiation.

Problems

11.1* Squeezed vacuum with the parameters μ and ν passes through a 50/50 beam splitter. The second port is excited by regular vacuum.

(a) Find the major and minor axes of the uncertainty ellipses of the electric field in the two output ports.
(b) If the input is perfectly squeezed, i.e. $|\mu| \to \infty$, determine the major and minor axes of the uncertainty ellipses.

11.2* The saturation analysis of Sect. 11.8 permits the evaluation of the noise enhancement factor $\chi = N_u/(N_u - N_\ell)$. Obtain an expression in terms of the medium parameters and the photon number.

11.3 Rederive the Schawlow–Townes linewidth of Sect. 11.10 for the ideal laser of Sect. 11.11.

11.4* Set up the Hamiltonian and the Heisenberg equations of motion for parametric upconversion. In the upconversion process, a pump photon at frequency ω_p combines with a signal photon to yield a photon at the so-called anti-Stokes frequency ω_a: $\omega_a = \omega_p + \omega_s$.

(a) Derive the equation of photon conservation.
(b) Find a general solution to the equation of motion assuming that the signal is small and the pump amplitude can be treated as a time independent c number.

11.5 Consider the classical equations of a resonator with two resonant modes, of frequencies ω_{0s} and ω_{0i} and decay rates $1/\tau_{os}$ and $1/\tau_{oi}$. A pump $U_p \exp(-i\omega_p t)$ is applied and produces sources in the equations for the two modes, $\kappa_{spi} U_p U_i^* \exp[-i(\omega_p - \omega_i)t]$ and $\kappa_{ips} U_p U_s^* \exp[-i(\omega_p - \omega_s)t]$, respectively. Normalize the amplitudes so that their squares are photon numbers.

(a) From photon number conservation, write down the relation between the coefficients κ_{spi} and κ_{ips}.

(b) For the source $\sqrt{2/\tau_e}s_s$ from an input port at frequency $\omega_s \neq \omega_{os}$, derive the excitations of the signal and idler. Assume that the idler is not coupled to the output port.

(c) Plot $\tau_e/2|U_s/s|^2$ versus $\Delta\omega\,\tau_{os} = (\omega_s - \omega_{os})\tau_{os}$ under the assumption that $\omega_{os} + \omega_{oi} = \omega_p$, $\tau_{os} = \tau_{oi}$, and $|\kappa_{spi}U_p|^2\tau_{os} = 0.25$.

11.6 The rate of growth of the mechanical parametric circuit was obtained for abrupt, step pumping. Show that, in the limit of small steps, we may obtain the same rate of growth with sinusoidal pumping, the amplitude of the sinusoidal variation of the capacitor being equal to the first harmonic of the Fourier expansion of the step excitation. Write $C = C_o[1 + M\sin(\omega_p t)]$ and retain only terms of frequency $\omega_s = \omega_p/2$.

11.7 Consider a degenerate parametric amplifier, neglecting pump depletion. Its Hamiltonian is then (compare 11.35)

$$\hat{H} = \hbar(\chi_{\text{sps}}A_p\hat{A}_s^\dagger\hat{A}_s^\dagger + \chi_{\text{sps}}^*A_p^*\hat{A}_s\hat{A}_s) ,$$

with A_p a constant. The interaction time is T. With vacuum as input, show that the photons are emitted in pairs by looking at the wave function $|\psi(T)\rangle$. You need not find the actual probability distribution.

11.8 Evaluate the spectrum of the autocorrelation function (11.103).

Solutions

11.1 Denote the outputs of the beam splitter as $\hat{C}$ and $\hat{D}$. Then

$$\hat{C} = \frac{1}{\sqrt{2}}[(\mu A + \nu\hat{A}^\dagger) - i\hat{B}] , \quad \hat{D} = \frac{1}{\sqrt{2}}[-i(\mu A + \nu\hat{A}^\dagger) + \hat{B}] .$$

We find for the in-phase and quadrature components of $\hat{C}$

$$\hat{C}_1 = \frac{1}{2}(\hat{C} + \hat{C}^\dagger)$$

$$= \frac{1}{2\sqrt{2}}(\mu\hat{A} + \mu^*\hat{A}^\dagger + \nu\hat{A} + \nu^*\hat{A}^\dagger - i\hat{B} + i\hat{B}^\dagger) .$$

The expectation value of $\hat{C}_1^2$ is obtained by putting the resulting expression into normal order and noting that $\langle|\hat{A}^\dagger\hat{A}|\rangle = \langle|\hat{B}^\dagger\hat{B}|\rangle = 0$. In this way, we obtain:

$$\langle|\hat{C}_1^2|\rangle = \frac{1}{8}(|\mu|^2 + |\nu|^2 + \mu\nu + \mu^*\nu^* + 1) .$$

In a similar way we find for the expectation value of the square of the quadrature component

$$\langle|\hat{C}_2^2|\rangle = \frac{1}{8}(|\mu|^2 + |\nu|^2 - \mu\nu - \mu^*\nu^* + 1) .$$

If we phaseshift the components $\hat{C}_1$ and $\hat{C}_2$ so as to maximize one and minimize the other, we find for this new reference phase

$$\frac{1}{8}[(|\mu| \pm |\nu|)^2 + 1] \ .$$

These are the major and minor axes of the squeezing ellipse. If the squeezing is perfect, and $|\mu| - |\nu| \to 0$, we find that the noise reduction in the output is only 3 dB.

A completely analogous derivation finds the same major and minor axes of the squeezing ellipse in port (d).

11.2 The two rate equations are

$$\frac{dN_u}{dt} = -\sigma_u N_u - \gamma(N_u - N_\ell)S + P \ ,$$

$$\frac{dN_\ell}{dt} = -\sigma_\ell N_\ell + \gamma(N_u - N_\ell)S \ .$$

In the steady state $d/dt = 0$, and we find

$$N_u - N_\ell = \frac{P/\sigma_u}{1 + \gamma(1/\sigma_u + 1/\sigma_\ell)S}$$

and

$$N_u = \left(1 + \frac{\gamma}{\sigma_\ell}S\right)\frac{P/\sigma_u}{1 + \gamma(1/\sigma_u + 1/\sigma_\ell)S} \ .$$

Thus we find

$$\chi = \frac{N_u}{N_u - N_\ell} = 1 + \frac{\gamma}{\sigma_\ell}S \ .$$

The noise enhancement factor approaches the ideal value of unity when the relaxation rate of the lower level is much faster than the induced transition rate of the upper level.

11.4 The Hamiltonian is of the time-independent form

$$\hat{H} = \hbar\chi_{psa}\hat{A}_p\hat{A}_s\hat{A}_a^\dagger + \text{H.c.} \ .$$

The Heisenberg equations of motion are

$$\frac{d}{dt}\hat{A}_s = \frac{i}{\hbar}[\hat{H}, \hat{A}_s] = -i\chi_{psa}^*\hat{A}_p^\dagger\hat{A}_a \ ,$$

$$\frac{d}{dt}\hat{A}_a = \frac{i}{\hbar}[\hat{H}, \hat{A}_a] = -i\chi_{psa}\hat{A}_p\hat{A}_s \ .$$

These two equations imply conservation of photon number, i.e. a Stokes photon is exchanged for an anti-Stokes photon and vice versa.

$$\frac{d}{dt}(\hat{A}_s^\dagger\hat{A}_s) + \frac{d}{dt}(\hat{A}_a^\dagger\hat{A}_a) = 0$$

The equations can be integrated, when linearized by replacing the pump amplitude by a time-independent c number. We find

$$\hat{A}_s(t) = \cos(\kappa t)\hat{A}_s(0) + e^{i\psi}\sin(\kappa t)\hat{A}_a(0) ,$$

$$\hat{A}_a(t) = \cos(\kappa t)\hat{A}_a(0) - e^{-i\psi}\sin(\kappa t)\hat{A}_s(0) ,$$

where $\kappa \equiv |\chi_{psa}A_p|$ and $\psi = \arg(-i\chi_{psa}^*A_p^*)$.

12. Squeezing in Fibers

In the preceding chapter we showed how degenerate parametric amplification can lead to squeezing. Further, we gave an example of how squeezed vacuum, in conjunction with a coherent probe, can be used to improve the sensitivity of an interferometric measurement. In order to provide coherence between the probe and the squeezed vacuum, the pump radiation, at twice the frequency of the probe, had to be generated in a doubling crystal, coherently with the probe. The experiment is not a simple one; in particular, since coherence must be maintained across the entire experimental setup, changes in the different optical paths must be kept much smaller than an optical wavelength. It is difficult to maintain this coherence in adverse environmental conditions. For this reason one may look for methods of squeezing that are less subject to environmental effects. This can be done with fibers that possess the Kerr nonlinearity.

The optical version of parametric amplification uses a nonlinear medium with a second-order nonlinearity. This nonlinearity occurs only in crystals with no inversion symmetry. Indeed, a quadratic response to an electric field is produced only when reversal of the field does not result in a reversal of the response, i.e. the polarization. The medium must not be invariant under the symmetry operation of inversion. Media with inversion symmetry have no second-order nonlinearity; their lowest-order nonlinearity is of third order. The material of silica fibers is isotropic and thus its lowest-order nonlinearity is of third order. A third-order nonlinearity produces a polarization density according to the law

$$P_i(\omega) = \epsilon_o \chi_{ijkl}^{(3)} E_j^*(\omega') E_k(\omega'') E_l(\omega''') \quad \text{with} \quad \omega = \omega'' + \omega''' - \omega' . \quad (12.1)$$

The complex conjugates are placed so that the response is at and near the fundamental, rather than the third harmonic. There is, of course, some third-harmonic generation in a fiber that shows the Kerr effect. However, it is not phase-matched and is thus of neglible magnitude compared with the fundamental, which is automatically phase-matched when $\omega'' = \omega''' = \omega' = \omega$, and close to phase-matched for the range of frequencies encompassed by a pulse containing many cycles. The third-order nonlinearity is described by a third-rank tensor. However, usually, a much simpler description is satisfactory. Clearly, the generation of a polarization, in addition to the linear response and at the same frequency, can be represented by a change of the index proportional to the intensity. This is the common way of describing the optical Kerr effect. As in Sect. 10.1, we write for the index

$$n = n_o + n_2 I \, , \tag{12.2}$$

where n_2 is the nonlinear index coefficient. A pump of significant intensity changes the index of a fiber. In the process phase shifts of the field are produced. These phase shifts can generate squeezed radiation. Note that no frequency doubling is involved, the pump can be "recycled" as the local oscillator.

The Kerr effect is one of the phenomena of four-wave mixing. Indeed, as (12.1) shows, the beating of three fields produces a polarization source at a fourth frequency. In the quantum description, the three modes that mix are expressed as three waves with three propagation constants. Either classically or quantum mechanically, a superposition over all frequencies leads to a polarization written as the convolution of three spectra. Classically, an inverse Fourier transformation into the time domain transforms the convolution into a product of time functions. Quantum mechanically, an inverse Fourier transformation puts the convolution into product form, a product of three functions of space, of the x coordinate. We start with a careful study of the Fourier transforms of operators. Then we set up the quantum form of the Kerr effect. The theory is applied to the generation of squeezed vacuum in a nonlinear Mach–Zehnder interferometer under the action of the Kerr effect. The Mach–Zehnder interferometer is replaced by a Sagnac fiber loop reflector, which performs the same function as the Mach–Zehnder interferometer but which is self-stabilized against changes of index in the interferometer due to environmental effects that are slow compared with the transit time through the loop. We present experiments that have demonstrated appreciable amounts of squeezing and shot noise reduction. We conclude with an experiment that demonstrated measurement of the phase of an interferometer at a level below the shot noise level by the injection of squeezed vacuum.

12.1 Quantization of Nonlinear Waveguide

The modes of an optical fiber were derived in Chap. 3. The quantization of the electromagnetic field in a waveguide or a fiber has been treated in Sect. 6.3. In a single-mode fiber, modes of two polarizations have to be distinguished, and to each of the modes of a particular polarization creation and annihilation operators are assigned. The evolution of the complex field amplitude operator in time is described by the Heisenberg equation of motion. A forward wave "packet" is selected, occupying a length L (taken as very long so that the wave can be considered monochromatic), and its propagation in time is followed. This wave travels forward at the group velocity and occupies different spatial regions as it proceeds.

The complex field amplitude is proportional to the photon annihilation operator $\hat{a}(\beta)$. The operator $\hat{a}^\dagger(\beta)$ is a creation operator, the Hermitian conjugate operator to $\hat{a}(\beta)$. The following commutation relation holds:

$$[\hat{a}(\beta), \hat{a}^\dagger(\beta')] = \frac{1}{2\pi}\delta(\beta - \beta') \ . \tag{12.3}$$

The Hamiltonian of the mode is

$$\hat{H} = 2\pi\hbar \int d\beta\, \omega(\beta)\hat{a}^\dagger(\beta)\hat{a}(\beta) \ . \tag{12.4}$$

Using the Heisenberg equation and the commutator relation (12.1), we find the equation of motion for the operator as

$$\frac{d}{dt}\hat{a}(\beta) = -\mathrm{i}\omega(\beta)\hat{a}(\beta) \ . \tag{12.5}$$

In order to treat the Kerr nonlinearity in the simplest possible manner, it is necessary to introduce operators that are functions of the Fourier transform variable x. The Kerr effect was described classically in Chap. 10 as a process of four-wave mixing. Three Fourier components of the field at frequencies ω, ω', and ω'' produce a fourth one at frequency $\omega''' = \omega' + \omega'' - \omega$. The Kerr effect was written as a convolution of these three Fourier spectra.

In the quantum description, the evolution of the operators of given propagation constant(s) is a function of time. The four-wave mixing process is described in terms of a convolution of operator amplitude spectra of the propagation constant, rather than the frequency. If the medium is dispersion-free, as we shall assume to be the case, then the energy conservation relation $\hbar\omega''' = \hbar\omega' + \hbar\omega'' - \hbar\omega$ also implies the momentum conservation condition $\hbar\beta''' = \hbar\beta' + \hbar\beta''$ obeyed by the propagation constants. The Kerr effect can be described by the interaction Hamiltonian $\hat{H}_K$:

$$\hat{H}_K = -\frac{\hbar}{2} 2\pi K \int d\beta \int d\beta' \int d\beta''\hat{a}^\dagger(\beta)\hat{a}^\dagger(\beta')\hat{a}(\beta'')\hat{a}(\beta + \beta' - \beta'') \ . \tag{12.6}$$

Note the minus sign in front of the integral. The energy associated with the Kerr effect is negative. In the dispersion-free case, this Hamiltonian is time-independent, as it should be. Since the commutator of the operator $\hat{a}(\beta''')$ with the interaction Hamiltonian is

$$[\hat{H}_K, \hat{a}(\beta''')] = \hbar K \int d\beta' \int d\beta''\hat{a}^\dagger(\beta')\hat{a}(\beta'')\hat{a}(\beta''' + \beta' - \beta'') \ , \tag{12.7}$$

the Heisenberg equation of motion is now

$$\frac{d}{dt}\hat{a}(\beta) = -\mathrm{i}\omega\hat{a}(\beta) + \mathrm{i}K \int d\beta' \int d\beta''\, \hat{a}^\dagger(\beta')\hat{a}(\beta'')\hat{a}(\beta + \beta' - \beta'') \ . \tag{12.8}$$

The Kerr effect is expressed as a convolution. Just as in the classical case, it is convenient to Fourier transform the operators so as to convert the convolution into a product. The Fourier transform is now with respect to the propagation constant, and not the frequency. The Fourier transform is expressed as a function of position x, rather than time t. We call this the x representation. The next two sections are devoted to the discusion of the x representation of operators.

12.2 The x Representation of Operators

A general spatial dependence can be built up from a superposition of modes. Optical systems transmit radiation that possesses a carrier frequency ω_o and a carrier wavelength β_o, and can be of very short duration, i.e. pulse-like. Even so, the pulse contains many cycles and can be still considered narrow-band, except in the case of ultrashort pulses containing only very few cycles [159–161]. Such ultrashort pulses can only propagate undistorted in free space or through very thin slabs of materials, not in optical fibers. In the quantum analysis to follow, we exclude such extremely short pulses. Pulses containing many optical cycles can be treated within the slowly-varying-envelope approximation (SVEA). It should be noted further that annihilation operators have the time variation $\exp(-i\omega t)$, with $\omega > 0$, and thus only excitations of positive β are represented in the following equations. The derivations will be more transparent if we use the mode amplitude operators $\hat{A}_m$, rather than the operators $\hat{a}(\beta)$. The two are connected by the simple renormalization (6.94). Consider the following superposition of mode operators:

$$\frac{1}{\sqrt{L}} \sum_m \hat{A}_m e^{i\beta_m x} = \frac{1}{\sqrt{L}} \sum_m \hat{A}_m e^{i\,\delta\beta_m x} e^{i\beta_o x}$$

$$= \hat{a}(x) e^{i\beta_o x} \ . \tag{12.9}$$

The inverse transform is

$$\hat{A}_m = \frac{1}{\sqrt{L}} \int_{-L/2}^{L/2} dx\, \hat{a}(x) e^{-i\,\delta\beta_m x} \ , \tag{12.10}$$

where $\beta_o + \delta\beta_m = \beta_m$. The photon operator, expressed in terms of $\hat{a}(x)$ and $\hat{a}^\dagger(x)$, is

$$\sum_m \hat{A}_m^\dagger \hat{A}_m = \frac{1}{L} \sum_m \int_{-L/2}^{L/2} dx \int_{-L/2}^{L/2} dx'\, \hat{a}^\dagger(x)\hat{a}(x') \exp[i\,\delta\beta_m(x - x')]$$

$$\approx \int_{-L/2}^{L/2} dx\, \hat{a}^\dagger(x)\hat{a}(x) \ . \tag{12.11}$$

The last expression is valid if L is chosen large enough compared with the spatial extent of the wavepacket. This expression suggests that the operator $\hat{a}^\dagger(x)\hat{a}(x)$ can be interpreted as a photon number density operator. Normally we think of photons as monochromatic, namely pertaining to wave packets of length so long that the excitation has a well-defined propagation constant and frequency. By extending the photon concept to a length smaller than L, one stretches the interpretation of photons beyond their usual definition, they cannot be considered to possess a sharply defined frequency. Just how the length L is chosen, and the local operator $\hat{a}(x)$ defined, depends on the measurement apparatus. This will emerge in the discussion of specific measurements.

The photon flow rate is

$$\text{photon flow rate} = v_g \hat{a}^\dagger(x)\hat{a}(x) . \tag{12.12}$$

The commutator of the operators $\hat{a}(x)$ and $\hat{a}^\dagger(x')$ is

$$[\hat{a}(x), \hat{a}^\dagger(x')] = \frac{1}{L} \sum_{m,n} \exp[i(\delta\beta_m x - \delta\beta_n x')][\hat{A}_m, \hat{A}_n^\dagger]$$

$$= \frac{1}{L} \sum_{m,n} \exp[i(\delta\beta_m x - \delta\beta_n x')]\delta_{mn}$$

$$= \frac{1}{L} \sum_{m} \exp[i\,\delta\beta_m(x - x')] \tag{12.13}$$

$$= \frac{1}{2\pi} \int_{-\Delta}^{\Delta} d\delta\beta \exp[i\,\delta\beta(x - x')]$$

$$= \frac{\sin[\Delta(x - x')]}{\Delta(x - x')} \frac{\Delta}{\pi} ,$$

where the spectrum of the propagation constant has been assumed to extend from $\beta_o - \Delta$ to $\beta_o + \Delta$. The commutator is a Nyquist function. Thus the photons are localized to a spatial interval determined by the spectral width of the spectrum in β space. If the Nyquist function appears in an expression involving functions of much slower x variation, one may replace the Nyquist function by a delta function:

$$\frac{\sin[\Delta(x - x')]}{\Delta(x - x')} \frac{\Delta}{\pi} \to \delta(x - x') . \tag{12.14}$$

Its Fourier transform gives a flat spectrum up to its cutoff at Δ. In this notation, we have for the commutator

$$[\hat{a}(x), \hat{a}^\dagger(x')] = \delta(x - x') . \tag{12.15}$$

The transformation of the states by a Fourier transform is discussed in Appendix A.18.

The preceding discussion shows the analogy between a Fourier transform and a physical transformation of a column matrix of operators by propagation through a conservative system. One physical process that takes a Fourier transform is the process of diffraction through a lens, for which the field at the second focus is the Fourier transform of the field at the first focus. This same mechanism can be implemented in a fiber with dispersion, with the lens replaced by an appropriate phase-shifting filter.

12.3 The Quantized Equation of Motion of the Kerr Effect in the x Representation

The transformation into the x representation was carried out with the operators $\hat{A}_m$ for clarity. In particular, it was noted that the mode spectrum clusters around the average propagation constant β_o. The dependence $\exp(i\beta_o x)$ was explicitly factored out from the x-dependent envelope. The transformation was written in terms of the deviation $\delta\beta$ of the propagation constant β from β_o. The discrete spectrum of $\hat{A}_m$ called for a discrete Fourier transform. This notation is somewhat cumbersome. To simplify notation, we shall henceforth replace the summations by integrals. We shall use the renormalization (6.94), and use the operator $\hat{a}(\delta\beta)$. In effect we have defined a new operator function $\hat{a}(\beta) \rightarrow \hat{a}(\delta\beta)\exp(i\beta_o x)$. This redefinition changes the equation of motion (12.5). The equation of motion for $\hat{a}(\delta\beta)$ is

$$\frac{d}{dt}\hat{a}(\delta\beta) = -i(\omega_o + v_g\delta\beta)\hat{a}(\delta\beta) , \tag{12.16}$$

where $v_g = d\omega/d\beta$ is the group velocity. This equation simplifies when the time dependence $\exp(-i\omega_o t)$ is removed; $\hat{a}(\delta\beta) \rightarrow \hat{a}(\delta\beta)\exp(-i\omega_o t)$. We shall make this substitution, again without a change of notation. The operator $\hat{a}(\delta\beta)$ then functions as an envelope function, which has to be multiplied by $\exp(-i\omega_o t + i\beta_o x)$ in order to obtain the actual space–time dependence. We shall simplify the notation by dropping the δ from the propagation-constant difference $\delta\beta$ by the replacement $\delta\beta \rightarrow \beta$. The spectrum of $\hat{a}(\beta)$ is now positioned at and around $\beta = 0$. We shall also dispense with the distinction between the Nyquist function (12.14) and the delta function, under the stipulation that we are dealing with pulses that possess a bandwidth Δ (measured in the propagation-constant coordinate) much larger than the quantization interval $2\pi/L$. With this simplification in notation we may treat $\hat{a}(x)$ and $\hat{a}(\beta)$ as Fourier transform pairs

$$\hat{a}(x) = \frac{1}{2\pi}\int d\beta\, \hat{a}(x)e^{-i\beta x} , \tag{12.17}$$

with the inverse Fourier transform

$$\hat{a}(\beta) = \int dx\, \hat{a}(x)e^{i\beta x} \,. \tag{12.18}$$

The operators $\hat{a}(x)$ and $\hat{a}^\dagger(x)$ obey the commutation relation

$$[\hat{a}(x), \hat{a}^\dagger(x')] = \delta(x - x') \,. \tag{12.19}$$

The integral in (12.17) extends over positive and negative values of β since β now represents the deviation of the propagation constant from β_o. In consonance with the replacement of the Nyquist function with the delta function, the limits of the integral are extended to $-\infty$ and $+\infty$. In this way the correspondence with Fourier integral theory is made complete. In this new notation, (12.16) assumes the form

$$\frac{d}{dt}\hat{a}(\beta) = -i\beta v_g \hat{a}(\beta) \tag{12.20}$$

or, if Fourier transformed,

$$\frac{\partial}{\partial t}\hat{a}(x) = -v_g \frac{\partial}{\partial x}\hat{a}(x) \,. \tag{12.21}$$

This equation is the quantum version of the classical propagation equation for the mode envelope. It is the Heisenberg equation of motion of a system with the Hamiltonian

$$\hat{H} = \frac{1}{2}i\hbar v_g \int dx\left[\left(\frac{\partial \hat{a}^\dagger(x)}{\partial x}\right)\hat{a}(x) - \hat{a}^\dagger(x)\left(\frac{\partial \hat{a}(x)}{\partial x}\right)\right] \,. \tag{12.22}$$

Indeed, use of the commutator (12.15) and integration by parts leads to (12.21). Note that the kernel of the Hamiltonian is reminiscent of the current operator in second quantization.

Next we turn to the x representation of the Kerr effect. The convolution (12.6) is transformed into products:

$$\hat{H}_K = -\frac{\hbar}{2}\frac{K}{(2\pi)^3}\int d\beta \int d\beta' \int d\beta'' \int dx\, e^{i\beta x}\hat{a}^\dagger(x)$$

$$\times \int dx'\, e^{i\beta' x'}\hat{a}^\dagger(x') \int dx''\, e^{-i\beta'' x''}\hat{a}(x'') \tag{12.23}$$

$$\times \int dx'''\, e^{-i(\beta+\beta'-\beta'')x'''}\hat{a}(x''') \,.$$

The integrals over the βs can be transformed piecewise into delta functions:

$$\int d\beta\, e^{i\beta x} e^{-i\beta x'''} = 2\pi\delta(x - x''') \, ,$$

$$\int d\beta'\, e^{i\beta' x'} e^{-i\beta' x'''} = 2\pi\delta(x' - x''') \, ,$$

$$\int d\beta''\, e^{-i\beta'' x''} e^{i\beta'' x'''} = 2\pi\delta(x'' - x''') \, .$$

The Hamiltonian simplifies to a single integral and becomes

$$\hat{H}_K = -\hbar\frac{K}{2}\int dx\, \hat{a}^\dagger(x)\hat{a}^\dagger(x)\hat{a}(x)\hat{a}(x) \, . \tag{12.24}$$

This is expressed in terms of a Hamiltonian density integrated over all x. The total Hamiltonian is the combination of (12.22) and (12.24):

$$\hat{H} = \frac{1}{2}i\hbar v_g \int dx\left[\left(\frac{\partial\hat{a}^\dagger(x)}{\partial x}\right)\hat{a}(x) - \hat{a}^\dagger(x)\left(\frac{\partial\hat{a}(x)}{\partial x}\right)\right]$$

$$-\hbar\frac{K}{2}\int dx\, \hat{a}^\dagger(x)\hat{a}^\dagger(x)\hat{a}(x)\hat{a}(x) \, . \tag{12.25}$$

The Heisenberg equation of motion is

$$\frac{\partial}{\partial t}\hat{a}(x) = -v_g\frac{\partial}{\partial x}\hat{a}(x) + iK\hat{a}^\dagger(x)\hat{a}(x)\hat{a}(x) \, . \tag{12.26}$$

Again we may simplify this equation by a change of variables that transforms the coordinates into the frame moving with the group velocity, $t \to t - x/v_g$ and $x \to x$. Without changing notation, we obtain

$$\frac{\partial}{\partial t}\hat{a}(x) = iK\hat{a}^\dagger(x)\hat{a}(x)\hat{a}(x) \, . \tag{12.27}$$

It is apparent that the introduction of the operator $\hat{a}(x)$ has greatly simplified the Heisenberg equation of motion involving the nonlinear Kerr effect. Comparison of (12.27) with the classical counterpart identifies the coefficient K as

$$K = \hbar\omega_o v_g\kappa \, , \tag{12.28}$$

since $\hat{a}^\dagger(x)\hat{a}(x)$ is the photon density operator and, classically, $|a(t)|^2$ stands for the intensity.

12.4 Squeezing

Radiation propagating along a fiber with a nonlinear Kerr coefficient becomes squeezed. The locus of the $e^{-1/2}$ points of the probability distribution of the complex amplitude starts out as a circle, if the input is in a coherent state. The phase of an excitation of a given amplitude shifts proportionally to the amplitude. The initial circular probability distribution can be sliced into segments of constant amplitude. Each segment of a given amplitude is phase-translated proportionally to the amplitude. The locus of the $e^{-1/2}$ points distorts into an ellipse of the same area as the original circle. As we shall show, this is a manifestation of squeezing.

Consider a nonlinear fiber excited initially by a coherent state. The Heisenberg equation of motion (12.27) conserves photon number and leaves the operator $\hat{a}^{\dagger}(x)\hat{a}(x)$ invariant, independent of t. We may integrate (12.27) from $t = 0$ to $t = T$ directly to obtain

$$\hat{a}(T, x) = \exp[iKT\hat{a}^{\dagger}(0, x)\hat{a}(0, x)]\hat{a}(0, x) . \tag{12.29}$$

We may linearize (12.29) by setting

$$\hat{a}(t, x) = a_o(t, x) + \Delta\hat{a}(t, x) , \tag{12.30}$$

and by dropping all terms of order higher than first in $\Delta\hat{a}(t, x)$. The function $a_o(t, x)$ is a c number that follows the classical evolution of the complex field:

$$a_o(T, x) = \exp(i\Phi)a_o(0, x) , \tag{12.31}$$

with $\Phi = K|a_o(0, x)|^2 T$, the classical Kerr phase shift. The operator $\Delta\hat{a}(t, x)$ acquires the commutator of $\hat{a}(t, x)$. We have

$$a_o(T, x) + \Delta\hat{a}(T, x)$$

$$= \exp\{[iKT[a_o^*(0, x) + \Delta\hat{a}^{\dagger}(0, x)][a_o(0, x) + \Delta\hat{a}(0, x)]\}$$

$$\times [a_o(0, x) + \Delta\hat{a}(0, x)]$$

$$\approx \exp\{iKT[a_o^*(0, x)a_o(0, x) + \Delta\hat{a}^{\dagger}(0, x)a_o(0, x) + \Delta\hat{a}(0, x)a_o^*(0, x)]\}$$

$$\times [a_o(0, x) + \Delta\hat{a}(0, x)]$$

$$\approx \exp\{iKT[a_o^*(0, x)a_o(0, x)]\}\{a_o(0, x) + [1 + iKT|a_o(0, x)|^2]\Delta\hat{a}(0, x)$$

$$+ iKTa_o^2(0, x)\Delta\hat{a}^{\dagger}(0, x)\} . \tag{12.32}$$

Equating zeroth-order and first-order terms, we end up with (12.31) as the solution for $a_o(t, x)$, and with the Bogolyubov transformation for $\Delta\hat{a}(0, x)$:

$$\Delta\hat{a}(T, x) \approx \exp(i\Phi)[\mu \, \Delta\hat{a}(0, x) + \nu \, \Delta\hat{a}^{\dagger}(0, x)] , \tag{12.33}$$

with

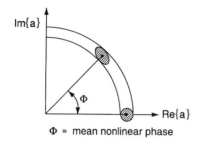

Φ = mean nonlinear phase

Fig. 12.1. The $1/e$ loci of the distribution of endpoints in the phasor plane

$$\mu = 1 + i\Phi, \quad \nu = i\Phi \exp\{2\arg[a_o(0,x)]\} \ . \tag{12.34}$$

Figure 12.1 shows the evolution of the locus of the Gaussian distribution of phasor endpoints in the complex phasor plane for a Kerr phase shift Φ and an initial coherent state of zero phase, $\arg(a_o) = 0$. The endpoint of the phasor has a distribution that starts with a circular $1/e^{1/2}$ locus for $T = 0$ (Fig. 12.2) and distorts into an ellipse. The ellipse remains tangential to the concentric circles drawn from the extrema of the uncertainty circle, since the phase modulation by the Kerr effect leaves amplitudes unaffected. The area of the ellipse remains the same as that of the circle, since the Bogolyubov transformation preserves commutators. In the absence of noise sources, the initial state being one of minimum uncertainty, the final state must remain a minimum-uncertainty state. The Bogolyubov transformation is analogous to that associated with degenerate parametric amplification. It should be noted, however, that squeezing via degenerate parametric amplification is described adequately by the Bogolyubov transformation for all levels of squeezing as long as pump depletion can be neglected. The Kerr process, on the other

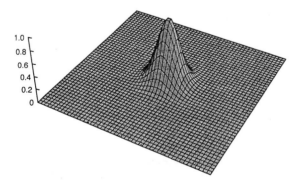

Fig. 12.2. A three-dimensional plot of the initial Gaussian

hand, leaves amplitudes strictly unaffected and hence, when the locus elongates so as to span an appreciable angular segment, the locus distorts into a meniscus and the simple Bogolyubov transformation (12.33) ceases to be an adequate description.

12.5 Generation of Squeezed Vacuum with a Nonlinear Interferometer

Squeezing cannot be utilized for measurements with improved sensitivity unless the noise is separated from the pump, phase shifted, and subsequently interfered with the pump, used as a local oscillator in a homodyne experiment. This can be done with a nonlinear Mach–Zehnder interferometer (Fig. 12.3). The beam splitter is a four-port. However, if one considers only the two incident waves in ports (1) and (4) in Fig. 7.3, and the outgoing waves in ports (2) and (3) of the same figure, the 50/50 beam splitter can be represented by the reduced scattering matrix

$$S = \frac{1}{\sqrt{2}} \begin{bmatrix} 1 & -i \\ -i & 1 \end{bmatrix} . \tag{12.35}$$

If two beam splitters are used in cascade, then the net scattering matrix for the output is

$$S^2 = \frac{1}{2} \begin{bmatrix} 1 & -i \\ -i & 1 \end{bmatrix} \begin{bmatrix} 1 & -i \\ -i & 1 \end{bmatrix} = -i \begin{bmatrix} 0 & 1 \\ 1 & 0 \end{bmatrix} . \tag{12.36}$$

The output port that would be reached by two reflections in Fig. 12.3 suppresses the input. All of the input from port (a) goes to output port (c), which is reached by one reflection and one transmission in each of the two paths of the interferometer. Likewise, the input from port (b) emerges from port (d).

When Kerr media are introduced into the two arms of the interferometer, the fluctuations of the input in port (a) cause an imbalance of the interferometer and some of the input of port (a) appears in the output port (d). Here we present a linearized analysis of a fiber interferometer operating at a carrier wavelength equal to the zero-dispersion wavelength of the fiber. If pulses are used for the excitation, as they have to be if the average power

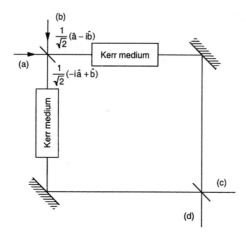

Fig. 12.3. Schematic of Mach–Zehnder interferometer

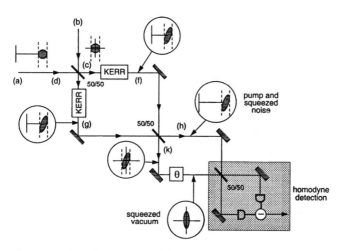

Fig. 12.4. The description of the phasors at different reference planes of the Mach–Zehnder interferometer

levels are to be kept low, we can analyze the action of rectangular segments of the pulse, each of approximately constant intensity. Figure 12.4 shows in the insets the evolution of the phasors in the phasor plane. For simplicity, the classical Kerr phase shift has been dropped, so that all phasors are shown horizontal.

Now, turning to the mathematical analysis of the operation of the nonlinear Mach–Zehnder interferometer of Fig. 12.4, we note that the input from ports (a) and (b) produces the output operators $\hat{c}$ and $\hat{d}$, and where

$$\hat{c} = \frac{1}{\sqrt{2}}(\hat{a} - \mathrm{i}\hat{b}) \quad \text{and} \quad \hat{d} = \frac{1}{\sqrt{2}}(\mathrm{i}\hat{a} + \hat{b}) \,. \tag{12.37}$$

These new operators commute, as is easily checked by evaluating the commutator and by finding that it vanishes; $[\hat{c}, \hat{d}^\dagger] = 0$. This means that the operators $\hat{c}$ and $\hat{d}$ have standard vacuum fluctuations that are uncorrelated. The situation is analogous to the action of a beam splitter on thermal noise. If thermal noise impinges on the two input ports of a beam splitter, the excitations at the output ports are uncorrelated and at the thermal noise level.

A consequence of the independence of the noise excitations in the two arms of the interferometer is that the transformations of $\hat{c}$ and $\hat{d}$ by the Kerr media can be treated independently. We linearize the equations by expressing the operators as sums of c numbers and perturbation operators; $\hat{c} = c_o + \Delta\hat{c}$, $\hat{d} = d_o + \Delta\hat{d}$. The transformation by a Kerr medium is described in (12.32). Using this result, we find the operators $\hat{f}$ and $\hat{g}$ at the output ports of the Kerr media:

$$\hat{f} = \exp(\mathrm{i}\Phi)(c_o + \mu\,\Delta\hat{c} + \nu\,\Delta\hat{c}^\dagger) \,, \tag{12.38}$$

$$\hat{g} = \exp(\mathrm{i}\Phi)(d_o + \mu\,\Delta\hat{d} + \nu\,\Delta\hat{d}^\dagger) \,, \tag{12.39}$$

where

$$\Phi = KT|c_o|^2 = KT|d_o|^2 = KT|a_o|^2/2 \,, \mu = 1 + \mathrm{i}\Phi \,, \text{ and } \nu = \mathrm{i}\Phi \,,$$

and the phase of a_o has been set equal to zero. The parameter Φ is the classical phase shift produced by the pumps (the c-number parts of the excitations) in each of the Kerr media. The outputs are superpositions of c-number amplitudes and perturbation operators. The perturbation operators are uncorrelated and their states are vacuum states.

Finally, consider the outputs $\hat{h}$ and $\hat{k}$ of the interferometer:

$$\hat{h} = \frac{1}{\sqrt{2}}(\hat{g} - \mathrm{i}\hat{f}) \quad \text{and} \quad \hat{k} = \frac{1}{\sqrt{2}}(-\mathrm{i}\hat{g} + \hat{f}) \,. \tag{12.40}$$

The c-number amplitudes add in the output $\hat{h}$ and cancel in the output $\hat{k}$, since

$$f_o = \exp(\mathrm{i}\Phi)a_o/\sqrt{2} \quad \text{and} \quad g_o = -\mathrm{i}\exp(\mathrm{i}\Phi)a_o/\sqrt{2} \,. \tag{12.41}$$

We obtain:

$$\hat{h} = -\mathrm{i}\exp(\mathrm{i}\Phi)a_o + \frac{1}{\sqrt{2}}\exp(\mathrm{i}\Phi)[-\mathrm{i}(\mu\,\Delta\hat{c} + \nu\,\Delta\hat{c}^\dagger) + (\mu\,\Delta\hat{d} + \nu\,\Delta\hat{d}^\dagger)] \,, \tag{12.42}$$

$$\hat{k} = \frac{1}{\sqrt{2}} \exp(\mathrm{i}\Phi)[(\mu\,\Delta\hat{c} + \nu\,\Delta\hat{c}^\dagger) - \mathrm{i}(\mu\,\Delta\hat{d} + \nu\,\Delta\hat{d}^\dagger)] \ . \tag{12.43}$$

One of the interferometer outputs is squeezed vacuum. The other output is the phase-shifted pump amplitude accompanied by squeezed vacuum. This output can be used as the local oscillator in the balanced detector. An adjustable phase shifter imparts a phase delay ψ to the local oscillator so that the squeezed vacuum can be projected out along any phase direction. The (squeezed vacuum) noise of the local oscillator cancels in the balanced detector. The detector current autocorrelation function is

$$\frac{1}{2}\langle \hat{i}(x)\hat{i}(x') + \hat{i}(x')\hat{i}(x)\rangle$$

$$= q^2 v_g^2 \frac{1}{L}|a_o|^2 \delta(x - x')(|\mu|^2 + |\nu|^2 - 2|\mu\nu|\cos\vartheta) \tag{12.44}$$

$$= q^2 v_g^2 \frac{1}{L}|a_o|^2 \delta(x - x')(1 + 2\Phi^2 - 2\Phi\sqrt{1 + \Phi^2}\cos\vartheta) \ ,$$

where $\vartheta = \arg(\mu) + \arg(\nu) - 2\psi$. This expression, normalized to shot noise, is plotted in Fig. 12.5 for $\vartheta = 0$ for optimum adjustment of the local oscillator phase, and for $\vartheta = \pi$. Depending upon the phase adjustments between the local oscillator and the squeezed vacuum, the fluctuations are either below the shot noise level by a factor $(|\mu| - |\nu|)^2$, or above by a factor $(|\mu| + |\nu|)^2$.

The analysis has shown that the action of the interferometer can be inferred rather easily from the operator of the Kerr media alone. The Kerr media generate squeezed states consisting of a classical phasor superimposed on squeezed vacuum. The sole purpose of the interferometer is to provide one output from which the phasor has been removed.

The preceding analysis assumed a unique phase Φ across the entire pulse, i.e. the pulse was treated as rectangular. In practice, one uses either Gaussian

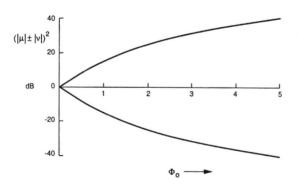

Fig. 12.5. The amount of squeezing as a function of the peak Kerr phase shift

pulses, generated from an actively mode-locked laser, or secant hyperbolic pulses from a passively mode-locked laser. In either case, the squeezing varies across the intensity profile of the pulse. A balanced homodyne detector excited by the pump used as a local oscillator automatically cancels the Kerr phase factor $\exp(i\Phi)$, which determines the location of the pump phasor in the phasor plane and also appears as a phase factor of the squeezed amplitudes. It does not correct for the change of the orientations of the squeezing ellipses with respect to the phasor. Hence when the degree of shot noise suppression within an entire pulse is evaluated one must average over the orientations of the ellipses, namely the angle ϑ in (12.44). The optimum adjustment is achieved when the local-oscillator phase is adjusted to coincide with the minor axis of the maximally squeezed ellipse at the peak of the pulse. This means that one sets $\vartheta = 0$ at the peak of the pulse by proper choice of the phase of the local oscillator. The phase Φ then varies with the Gaussian pulse profile

$$\Phi(x) = \Phi_{\max} \exp(-x^2/2x_o^2) , \tag{12.45}$$

and the phases of μ and ν are varied accordingly. In this manner we may evaluate the net squeezing by averaging (12.44) The result is shown in Fig. 12.6. As one can see, a Gaussian pulse cannot produce shot noise reduction better than 7 dB, owing to this misalignment effect.

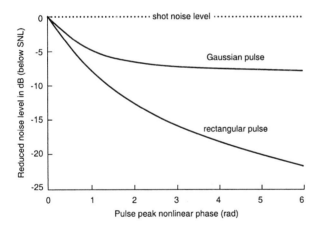

Fig. 12.6. Noise reduction below standard noise limit (SNL) by rectangular pulse and Gaussian pulse

12.6 Squeezing Experiment

In the preceding section we analyzed the generation of squeezed vacuum using a nonlinear Mach–Zehnder interferometer. If a fiber is used for the nonlinear medium of propagation, fiber lengths of several tens of meters are required to produce the required Kerr phase shifts for peak powers of the order of 50 W, as produced by a mode-locked laser operating at 1 GHz repetition rate. If an interferometer were formed from two fibers of such a length, unavoidable environmental changes would produce large fluctuations of the relative phase shift in the two arms of the interferometer, preventing interference at the output mirror. In order to provide stability against such environmental fluctuations, the Mach–Zehnder interferometer was replaced by a Sagnac loop as shown in Fig. 12.7 [162]. The incoming pump pulses are split by the fiber coupler into two equally intense counterpropagating pulses. Within the travel time around the loop of the order of microseconds, environmental fluctuations are negligible, and the two pulses travel through identical optical path lengths. Thus, the Sagnac loop provides an environmentally stabilized realization of the nonlinear Mach–Zehnder interferometer if pulses are used for the excitation. The coupler functions as both the input and the output beam splitter of the Mach–Zehnder interferometer.

The experimental setup is shown in Fig. 12.8 [163]. The Sagnac loop was made of a polarization-maintaining fiber and a 50/50 fiber coupler. The pump, a mode-locked Nd:YAG laser delivering 100 ps pulses at 1.3 μm wavelength and with a repetition rate of 100 MHz, was passed through an isolator to reduce reflections back into the laser. The fiber had zero dispersion at a wavelength of 1.3 μm. A polarizer and a half-wave plate were used to vary the input power level. Before entering the Sagnac loop coupler, the pump was passed through an 85/15 beam splitter that picked off a portion of the reflected pump for use as the local oscillator in the balanced homodyne detector.

The squeezed vacuum emerges from the unexcited port; the pump pulses are recombined and exit in the same fiber through which they entered the interferometer. The local oscillator and squeezed vacuum are mixed in the balanced detector. By varying the phase between the local oscillator and the squeezed vacuum, different noise levels were observed. Figure 12.9 shows the noise in the time domain as the phase between the local oscillator and squeezed vacuum was varied continuously. The noise was filtered with a passband filter at 50 kHz with a 2 kHz bandwidth. One sees clearly time segments of large noise and small noise. The left trace is shot noise, obtained by blocking the entry of the squeezed vacuum into the balanced detector. To make sure that the shot noise level was properly calibrated, illumination by a broadband source was used to produce the same detector current and the noise level was compared with that observed when the squeezed vacuum was blocked. The two readings were in good agreement. With the phase stabilized at the minimum noise level, the degree of reduction of noise below the shot noise level

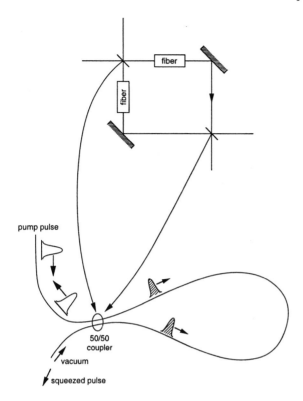

Fig. 12.7. Replacement of Mach–Zehnder interferometer with Sagnac loop

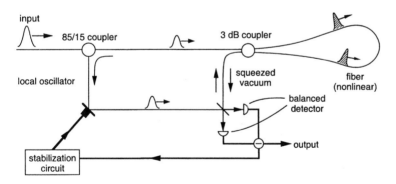

Fig. 12.8. Experimental configuration

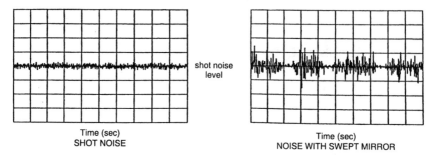

Fig. 12.9. Detector current as a function of phase difference with a "sawtooth" piezo-voltage drive. Center frequency 40 kHz; bandwidth 2 kHz

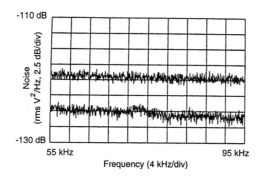

Fig. 12.10. Spectrum of detector noise; the *top trace* is the shot noise

was ascertained with a spectrum analyzer as shown in Fig. 12.10. The noise reduction measured was 5.1 dB.

12.7 Guided-Acoustic-Wave Brillouin Scattering

Guided-acoustic-wave Brillouin scattering (GAWBS) was first discovered by Levenson et al. in their squeezing experiments with c.w. pumps [164, 165]. The cause of this scattering is thermally excited acoustic modes of the fiber near the cutoff of the acoustic modes, when they propagate nearly transversely to the axis of the fiber. These are acoustic resonances of the fiber whose frequencies are determined by the fiber stiffness and geometry. The lowest frequency of the modes is near 10 MHz. The spectrum of these modes extends to about 1 GHz. The acoustic waves couple to the optical waves via the acousto-optic effect, a change of index caused by the strain produced by the acoustic wave. At higher frequencies, the coupling to the optical mode vanishes because their mode profile varies so rapidly over the optical-mode profile that their coupling is negligible, and also because the acoustic propagation losses become so high that their excitation becomes negligible. The axial component of the propagation constant of the acoustic mode is zero at cutoff (purely transverse propagation) and remains small for small deviations from transverse propagation. These modes can phase match an optical wave at frequency ω_o and an acoustic wave at frequency Ω, with the up-shifted and down-shifted optical waves at frequencies $\omega_o \pm \Omega$. Since the acoustic wavelength is about 10^5 times smaller than the optical wavelength, the phase matching occurs only for acoustic waves that are almost entirely transverse.

GAWBS produces sidebands on the optical waves spaced by the acoustic frequencies, ranging from 10 MHz to somewhat below 1 GHz. If squeezing is done with a c.w. pump, and the squeezed radiation is detected in a balanced homodyne detector, the spectrum of the current shows spectral spikes at 10 MHz and higher frequencies that overwhelm the noise reduction due to squeezing. At frequencies below 10 MHz it would be still possible to use the squeezing for noise reduction. It turns out, however, that c.w. excitation can produce stimulated Brillouin scattering (SBS) [166, 167], which is narrow-band and thus preferentially generated by a narrow band pump. The threshold of SBS can be increased if excitation at multiple frequencies is employed.

Pulse excitation has a much higher SBS threshold, since it is broadband. However, the role of GAWBS under pulsed excitation is different from that under c.w. excitation. The pump, as well as the squeezed radiation, acquires GAWBS spectral spikes. In balanced homodyne detection the two excitations are multiplied in the time domain and convolved in the frequency domain. The convolution can place spectral spikes at many combination frequencies. Only under very fortunate conditions does one find spectral windows that are free of the GAWBS spikes. The appearance of GAWBS spikes in the spectrum of the detector current at frequencies of interest for sub-shot noise measurements can be prevented by two methods [168, 169]:

(a) The repetition rate of the pump source is 1 GHz or higher.
(b) The pump pulse is split into two pulses spaced by less than 1 ns apart, and the second pulse is phase reversed when converted into the local-ocillator excitation.

Method (a) is easily understood. If the spectrum of the pump has spectral components spaced 1 GHz or more apart, the sidebands produced by GAWBS, which occupies a spectral range of less than half the spacing, never convolve into the low-frequency window. Moreover, the spectral spikes due to GAWBS at 20 MHz and higher can be observed directly, without distortion by convolution with other spikes. Figure 12.11 shows an example of such a spectrum, achieved with a laser mode-locked at 1 GHz [170].

Method (b) relies on the fact that GAWBS is a process with typical time constants longer than 1 ns. If two pump pulses are used, one delayed with respect to the other by less than 1 ns, as shown in Fig. 12.12, both pulses experience the same change of index. Thus, they carry the same GAWBS

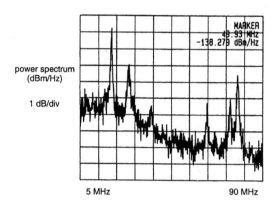

Fig. 12.11. Spectrum of GAWBS

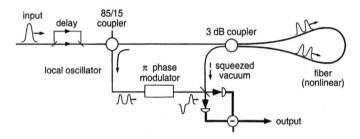

Fig. 12.12. Suppression of GAWBS by phase reversal of local oscillator [168]

signature. Before entering the balanced detector as the local-oscillator excitation, the phase of one of the pulses is reversed. The detector integrates the current over both pulses. The reversal of the phase of the second pulse reverses the phase of the GAWBS excitation of the second pulse. As the currents of the detector are added in the integration, the GAWBS excitation cancels. The quantum fluctuations in the two time slots are uncorrelated and add in the mean square sense. In the next section we describe a phase measurement at a noise level below shot noise that uses this cancellation of GAWBS.

12.8 Phase Measurement Below the Shot Noise Level

The purpose of the generation of squeezed vacuum is its use in measurements below the shot noise level. Quantum theory permits noise-free measurements in principle. An ideal photodetector measures the photon number of wavepackets impinging upon it in a noise-free manner. In principle, an ideal measurement of any observable can be devised that would measure this observable with no uncertainty. The uncertainty in the measurement would be attributable to the preparation of the state and not to the measurement itself.

Lasers produce coherent states, at least in the ideal limit. Such coherent states are minimum-uncertainty states with equal uncertainty in the in-phase and quadrature components. Hence, they are not ideal for the measurement of interferometric phase changes, since they are not in an eigenstate of the observable to be measured. However, the combination of squeezed vacuum and a coherent state fed into the two ports of an interferometer can achieve a measurement of the phase that, in principle, could be made noise-free if the squeezing of the vacuum were perfect. We describe such a measurement in simple terms and then present an actual measurement of phase that employs a modification of the setup of Fig. 12.8.

Consider the Mach–Zehnder interferometer of Fig. 12.13, which has been unbalanced by phase changes $\pm \Delta\theta$ in its two arms. We follow the probe excitation at port (a) and the squeezed-vacuum excitation at port (b) through the interferometer. Since the system is linear, we may analyze one excitation at a time. The probe in a coherent state in Fig. 12.13a has associated noise, which is scaled down by $\sqrt{2}$ along with the amplitude of the probe. The phase imbalance tilts the phasors so that there is an output at the port that would be unexcited in the absence of an imbalance. The noise accompanying the signal is reduced and is negligible if we ignore signal-dependent noise effects. The vacuum fluctuations entering the other input port, as shown in Fig. 12.13b, emerge at the signal output port, and the contribution to the horizontal output beam is negligible. The schematic (Fig. 12.13c) at the bottom of the figure shows the superposition. As we can see, the noise accompanying the signal is due to the zero-point fluctuations entering the vacuum input port.

If one feeds squeezed zero-point fluctuations, represented by an ellipse (of the proper orientation) in Fig. 12.14, into the vacuum input port, an analogous argument shows that the noise accompanying the signal can have a reduced in-phase component. A homodyne detection that is phased along the signal direction sees reduced noise. In constructing the squeezed output noise in the figure, the small imbalance in the interferometer has been neglected (i.e. signal-dependent noise has been neglected).

Figure 12.15 shows an experimental setup used to demonstrate a phase measurement at a noise level below shot noise [168]. The setup consists of a squeezing apparatus, followed by an interferometer whose phase change is to be measured and a homodyne balanced detector. In order to suppress GAWBS, the pump pulse is split into two pulses, one delayed with respect to the other by 500 ps. A phase modulator reverses the phase of the second pulse after passage through the squeezer. The interferometer whose phase imbalance is measured is made of bulk components, with one mirror mounted on a piezoelectric mount. The phase is changed sinusoidally at 50 kHz by a voltage drive of the piezoelectric mount. Figure 12.16 shows the spectrum of

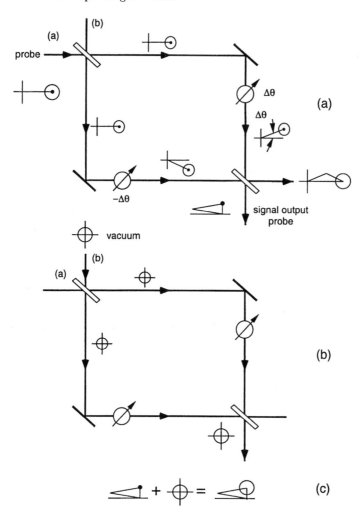

Fig. 12.13. Quantum noise in phase measurement

the homodyne-detector current at and around 50 kHz for two conditions: (a) with the squeezed vacuum blocked from entry into the interferometer and (b) with it unblocked. It is clear that the noise level for case (b) is below that of case (a). Calibration shows a lowering of the noise level by 3 dB. This improvement is less than the 5 dB shot noise reduction level, mainly owing to the additional losses in the interferometer constructed of bulk components. However, this experiment illustrates the possibility of phase measurements performed below the shot noise level.

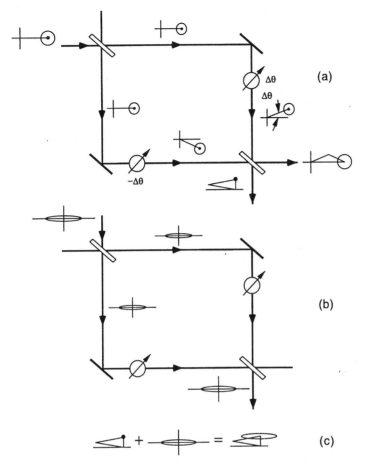

Fig. 12.14. Reduced quantum noise in phase measurement by squeezed-vacuum injection

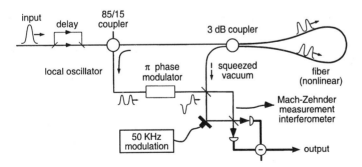

Fig. 12.15. Experimental setup for sub-shot-noise phase measurement with a Mach–Zehnder interferometer whose optical path length is piezoelectrically varied at 50 kHz

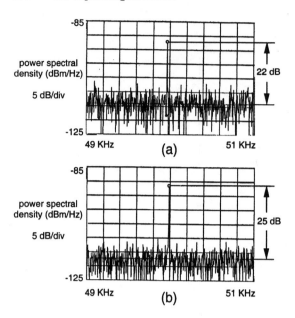

Fig. 12.16. Signal and noise (a) with squeezed radiation blocked, and (b) with squeezed vacuum

12.9 Generation of Schrödinger Cat State via Kerr Effect

Thus far we have studied the very practical aspects of shot noise reduction using a nonlinear Mach–Zehnder interferometer, with the Kerr effect responsible for the nonlinearity. The predictions can be, and have been, tested experimentally. Now we leave this realistic realm for an excursion into a thought experiment that is not realizable in practice, but which is nevertheless intriguing.

We discussed the Schrödinger cat state in conjunction with the definition of the Wigner function in Chap. 7. Now that we have an operator formalism for the Kerr effect, we can show that a Schrödinger cat state could be generated in principle via the Kerr nonlinearity via propagation of a coherent state [67]. In such propagation, the evolution of the state $|\phi\rangle$ is

$$|\psi\rangle = \exp(i\kappa\hat{A}^\dagger\hat{A}^\dagger\hat{A}\hat{A})|\phi\rangle = \exp(-i\kappa\hat{A}^\dagger\hat{A})\exp[i\kappa(\hat{A}^\dagger\hat{A})^2]|\phi\rangle , \qquad (12.46)$$

where κ is an appropriately defined Kerr effect parameter. Suppose we start with a coherent state

$$|\phi\rangle = e^{-|\alpha|^2/2}\sum_n \frac{\alpha^n}{\sqrt{n!}}|n\rangle . \qquad (12.47)$$

The output state is

$$|\psi\rangle = \exp(-i\kappa\hat{A}^\dagger\hat{A})\exp[i\kappa(\hat{A}^\dagger\hat{A})^2]e^{-|\alpha|^2/2}\sum_n \frac{\alpha^n}{\sqrt{n!}}|n\rangle$$

$$= \exp(-i\kappa\hat{A}^\dagger\hat{A})e^{-|\alpha|^2/2}\sum_n \exp(i\kappa n^2)\frac{\alpha^n}{\sqrt{n!}}|n\rangle \ . \tag{12.48}$$

We pick the special case when the coefficient $\kappa = \pi/2$. This means that one single photon produces a phase shift of 90°. We ignore the multiplier $\exp(-i\kappa\hat{A}^\dagger\hat{A})$, since it represents a linear element producing a simple phase shift. Now, note an interesting property of the square of an integer. It is clear that for an even number, its square is a multiple of four. The square of an odd number $n = E+1$, where E is an even number, is $n^2 = (E+1)^2 = E^2+2E+1$. One can see that $n^2 = 1 + \mathrm{mod}(4)$. Hence

$$\frac{\pi}{2}n^2 = \begin{cases} \dfrac{\pi}{2} + \mathrm{mod}(2\pi) & \text{for } n \text{ odd} \\[2mm] \mathrm{mod}(2\pi) & \text{for } n \text{ even} \ . \end{cases}$$

Therefore,

$$e^{i\kappa n^2} = \frac{1}{\sqrt{2}}[e^{i\pi/4} + (-1)^n e^{-i\pi/4}] \ , \tag{12.49}$$

and

$$\exp[i\kappa(\hat{A}^\dagger\hat{A})^2]|\alpha\rangle$$

$$= e^{-|\alpha|^2/2}\frac{1}{\sqrt{2}}\left(e^{i\pi/4}\sum_n \frac{\alpha^n}{\sqrt{n!}}|n\rangle + e^{-i\pi/4}\sum_n(-1)^n\frac{\alpha^n}{\sqrt{n!}}|n\rangle\right) \tag{12.50}$$

$$= \frac{1}{\sqrt{2}}\left(e^{i\pi/4}|\alpha\rangle + e^{-i\pi/4}|-\alpha\rangle\right) \ .$$

This is the Schrödinger cat state of (7.96). The Kerr coefficient would have to be unrealistically high to generate such a state. A pulse one picosecond in duration of one single photon of 1 micron wavelength carries a peak power of roughly 10^{-7} W. A fiber Kerr nonlinearity of 3.2×10^{-16} cm^2/W and a mode profile of 10 μm^2 would call for a *lossless* fiber a million kilometers long to achieve a phase shift of 90°.

Another word of caution is in order. This example of generation of a Schrödinger cat state treats the fiber propagation in terms of a single mode. Even if only one coherent state associated with the propagation constant β is excited at the input, zero-point fluctuations in all the other modes enter the fiber. Four-wave mixing of these zero-point fluctuations with the pump leads, in fact, to infinities, i.e. singularities. In reality, the Kerr medium does not respond instantaneously, i.e. it has a response of finite bandwidth. When this finite response time is taken into account, the singularity is removed. But the response of the fiber is quite different from the idealized model used to show how a Scrödinger cat state could be generated.

12.10 Summary

In this chapter we discussed in detail the x representation of field operators. It is the representation convenient for the analysis of the Kerr effect for pulses with a temporal (or, rather, spatial) profile. In the next chapter it will be used to deal with dispersive propagation as well. Next, we analyzed the generation of squeezed vacuum by a nonlinear fiber Sagnac loop at the zero-dispersion wavelength of the fiber. This was followed by a description of the experiments that verified the predictions of the theory. Guided-acoustic-wave Brillouin scattering was found to be an impediment to unfettered squeezing with optical pulses, an impediment that could be overcome, however, by proper choice of the exciting source or subsequent processing of the pulse(s).

This squeezing with pulses is analogous to the squeezing of continuum radiation in a Kerr medium. Pulse excitation raises the stimulated Brillouin backscattering threshold. The use of a Sagnac loop has the additional advantage of saving the pump power, to be used as local-oscillator power. If the available power is limited, as it most often is, this scheme promises to yield improved interferometric measurement accuracy in systems in which quantum noise is the dominant source of noise.

Squeezing with Gaussian pulses at the zero-dispersion wavelength of the fiber incurs a penalty in the noise reduction ultimately achievable owing to the different orientations of the squeezing ellipse. Even perfect squeezing could not achieve a reduction of the noise below shot noise of better than about 7 dB. The question arises of whether this penalty could be avoided. In the next chapter we shall investigate squeezing with soliton pulses operating at a center wavelength at which the fiber has negative dispersion. A particularly convenient operating wavelength is the 1.54 μm wavelength of erbium-doped fiber lasers, which have been perfected for use in long-distance optical communications. We shall see that squeezing with solitons does not suffer the noise penalty that is encountered with pulses propagating in fibers at zero dispersion.

Problems

12.1* Determine the peak phase shift Φ for a Gaussian pulse of peak intensity 50 W propagating over a fiber of length 50 m with an effective area $A_{\text{eff}} = 80 \, \mu\text{m}^2$; $\lambda = 1.55 \, \mu\text{m}$; $n_2 = 3 \times 10^{-16} \, \text{cm}^2/\text{W}$.

12.2 Evaluate the signal-to-noise ratio of the measurement of phase shown in Fig. 12.13. The probe is in a coherent state $|\alpha\rangle$, the squeezed vacuum entering the other port is characterized by ν.

12.3* Evaluate the shot noise reduction for a Gaussian pulse as a function of the angle Φ_o, as shown in Fig. 12.6.

12.4 Evaluate the shot noise reduction for a hyperbolic secant pulse of peak phase shift Φ_o.

12.5 By the same approach as we used to quantize the Kerr effect in the x representation, quantize the response of a second-order nonlinearity involving signal and idler propagation.

12.6 The state $|\phi\rangle = (1/\sqrt{2})(|1\rangle + |2\rangle)$ is passed through the Kerr medium of Sect. 12.9. Find the output state.

Solutions

12.1 Use the meter as the unit of length. The phase shift for a peak power P is

$$\frac{2\pi}{\lambda} n_2 \frac{P}{A_{\text{eff}}} \ell = \frac{2\pi}{1.55 \times 10^{-6}} \times 3 \times 10^{-20} \times \frac{50 \times 50}{80 \times 10^{-12}}$$

$$= 3.8 \text{ radians}$$

at the peak.

12.3 The squeezing is characterized by

$$|\mu|^2 + |\nu|^2 - 2|\mu\nu| \cos\varphi = 1 + 2\Phi^2 - 2\Phi\sqrt{1 + \Phi^2} \cos\varphi \, ,$$

where $\varphi = \arg(\mu\nu) - \arg(2\alpha_L)$.

The pump phase is chosen so that for the maximum squeezing at the phase angle Φ_o, optimum projection is achieved, i.e.

$$|\mu|^2 + |\nu|^2 - 2|\mu\nu| \cos\varphi$$

$$= 1 + 2\Phi_o^2 - 2\Phi_o\sqrt{1 + \Phi_o^2} = (\sqrt{1 + \Phi_o^2} - \Phi_o)^2 \, .$$

When the squeezing is less, the pump phase is not optimum. We have

$$\cos\varphi = \cos(\tan^{-1}\Phi - \tan^{-1}\Phi_o)$$

$$= \frac{1}{\sqrt{1 + \Phi^2}} \frac{1}{\sqrt{1 + \Phi_o^2}} + \frac{\Phi}{\sqrt{1 + \Phi^2}} \frac{\Phi_o}{\sqrt{1 + \Phi_o^2}} \, .$$

Thus

$$|\mu|^2 + |\nu|^2 - 2|\mu\nu| \cos\varphi = 1 + 2\Phi^2 - \frac{2\Phi}{\sqrt{1 + \Phi_o^2}} (1 + \Phi\Phi_o) \, .$$

The noise is weighted by the local oscillator pulse shape, which is also a Gaussian. The integrated noise suppression is

$$\frac{1}{\sqrt{\pi}} \int d\tau \left\{ 1 + 2\Phi^2(\tau) - \frac{2\Phi(\tau)}{\sqrt{1 + \Phi_o^2}} [1 + \Phi(\tau)\Phi_o] \right\} \exp(-\tau^2) ,$$

with $\Phi(\tau) = \Phi_o \exp(-\tau^2)$. The integral evaluates to

$$1 + \frac{2\Phi_o^2}{\sqrt{3}} \left(1 - \frac{\Phi_o}{\sqrt{1 + \Phi_o^2}} \right) - \frac{2\Phi_o}{\sqrt{2}\sqrt{1 + \Phi_o^2}} .$$

In the limit $\Phi_o \to \infty$, the value of the function becomes

$$1 - \sqrt{2} + \frac{1}{\sqrt{3}} = 0.16 .$$

13. Quantum Theory of Solitons and Squeezing

In Chap. 12 we studied the generation of squeezed vacuum in a Sagnac fiber loop at the zero-dispersion wavelength of the fiber. For practical reasons, pulses were used. In a dispersionless fiber, the analysis proceeds by subdividing the pulse into time intervals containing rectangular segments of intensity, each of which generates squeezed radiation within its time segment. The governing equation was linearized. If the Kerr effect is treated as instantaneous, the full nonlinear analysis runs into singularities [171]. A kind of ultraviolet catastrophe is produced because the zero-point fluctuations at all frequencies mix eventually. The linearized analysis avoids this singularity, and can be proven to be adequate for reasonable distances of propagation and amounts of squeezing [109]. A more careful model of the Kerr nonlinearity that takes the finite response time of the Kerr medium into account also avoids the singularity [171]. A third approach, that of quantizing the time in terms of shortest allowable time intervals [172], avoids the singularity but leads to unphysical periodicities. The quantized soliton equations introduce a bandwidth limitation via dispersion. As a consequence, the quantum analysis of solitons avoids entirely the singularities associated with an instantaneous Kerr response.

Squeezing of solitons is of interest because solitons maintain a uniform phase across their intensity profile. The amplitude–phase fluctuation ellipse is thus a property of the entire soliton, with a fixed phase angle of its minor axis across the entire soliton pulse. In detection, the projection of the squeezed fluctuations does not experience the averaging over different orientations of the squeezing ellipse that occurs in the case of a pulse at zero dispersion. Hence, the shot noise reduction is not limited to 7 or so dB, which was shown to be the limit when squeezing was effected with nonsoliton pulses at the zero-dispersion wavelength of the fiber.

Quantum analyses of solitons have been presented by several authors. A quantum analysis of soliton propagation and soliton detection based on stochastic differential equations has been carried out numerically by Carter et al. [173] and by Drummond et al. [174,175]. Haus et al. [176] started from the classical inverse scattering theory using Kaup's quantization procedure [177]. The transition to classical stochastic differential equations calls for the introduction of noise sources along the fiber. A full quantum treatment has been presented by Lai and Haus [178, 179] using the time-dependent Hartree approximation followed by an exact analysis based on the Bethe ansatz [180]. The soliton states were constructed from a superposition of eigenstates of the Hamiltonian. Since the propagation along the fiber is described by a Hamiltonian, no Langevin noise sources appear in this approach. This seems to contradict the approach of [174], which contains distributed noise sources. This apparent contradiction was the topic of a paper by Fini et al. [181], in which it was shown that the noise sources of [174] do not contribute to the expectation values of the operators. A fully analytic treatment based on linearization of the quantum form of the nonlinear Schrödinger equation offers the simplest approach [109]. It is this approach that forms the basis of the present chapter. The outcome of the analysis has a simple physical interpretation. A soliton behaves like a wave *and* a particle. The particle nature is represented by momentum and position operators obeying the standard commutation relations. The wave nature is represented by photon number and phase operators (or more precisely the in-phase and quadrature field operators). The expectation values of these operators can be measured in homodyne detection with properly shaped local-oscillator waveforms.

We generalize the Hamiltonian derived in Chap. 12 for the case of dispersive propagation in Sects. 13.1 and 13.2. Then we set up the quantized nonlinear Schrödinger equation using the developments of Sects. 13.1 and 13.2. Next we linearize the equation. Once the equation is linearized, no ordering of the operators is required and the solution is that of a classical equation with c-number variables. The classical perturbation analysis treated in Chap. 10 can be applied to the quantum problem. In Sect. 13.5 we consider the theory of measurement of the soliton perturbation parameters, which is then applied to a phase measurement in which the probe consists of a train of soliton pulses. An increased signal-to-noise ratio can be achieved with squeezed solitons, as described in Sect. 13.6.

Thus far, generation of squeezed vacuum using squeezed solitons has not demonstrated large amounts of squeezing [182]. The soliton pulse width is inversely proportional to the square of the peak intensity. With available fiber dispersions, it is necessary to use subpicosecond pulses in order to arrive at acceptable peak pulse intensities. The much broader bandwidths of these pulses introduce new effects that have not been fully characterized. The achievement of larger amounts of squeezing with solitons is still a goal of ongoing research.

13.1 The Hamiltonian and Equations of Motion of a Dispersive Waveguide

In Sect. 12.1 we considered the equation of a dispersion-free waveguide, such as a standard fiber in the wavelength regime of 1.3 μm. When the waveguide is dispersive, the Taylor expansion of the frequency as a function of the propagation constant β must be carried out to higher order:

$$\omega(\beta) = \omega_o(\beta_o) + \sum_{n=1}^{\infty} \frac{1}{n!} \frac{d^n \omega}{d\beta^n} \delta\beta^n \;, \tag{13.1}$$

where $\beta = \beta_o + \delta\beta$. First we write down the standard Hamiltonian and introduce the expansion (13.1). We take note of the fact that the spectrum occupies a finite interval $\beta_o - \Delta < \beta < \beta_o + \Delta$:

$$\begin{aligned}
\hat{H} &= 2\pi\hbar \int_{\beta_o-\Delta}^{\beta_o+\Delta} d\beta\, \omega(\beta) \hat{a}^\dagger(\beta)\hat{a}(\beta) \\
&= 2\pi\hbar \int_{\beta_o-\Delta}^{\beta_o+\Delta} d\beta \left(\omega_o + \sum_{n=1}^{\infty} \frac{1}{n!} \frac{d^n \omega}{d\beta^n} \delta\beta^n \right) \hat{a}^\dagger(\beta)\hat{a}(\beta) \;.
\end{aligned} \tag{13.2}$$

Next, we introduce the Fourier-transformed creation and annihilation operators:

$$\begin{aligned}
\hat{H} = \frac{\hbar}{2\pi} \int_{-\Delta}^{\Delta} d\delta\beta \left(\omega_o + \sum_{n=1}^{\infty} \frac{1}{n!} \frac{d^n \omega}{d\beta^n} \delta\beta^n \right) \\
\times \int dx\, \hat{a}^\dagger(x) \int dx'\, \hat{a}(x') e^{i\delta\beta(x-x')} \;.
\end{aligned} \tag{13.3}$$

An interchange of the orders of integration gives

$$\begin{aligned}
\hat{H} &= \frac{\hbar}{2\pi}\omega_o \int dx\, \hat{a}^\dagger(x) \int dx'\, \hat{a}(x') \int_{-\Delta}^{\Delta} d\delta\beta\, e^{i\delta\beta(x-x')} \\
&\quad + \frac{\hbar}{2\pi} \int dx\, \hat{a}^\dagger(x) \int dx'\, \hat{a}(x') \int_{-\Delta}^{\Delta} d\delta\beta \left(\sum_{n=1}^{\infty} \frac{1}{n!} \frac{d^n \omega}{d\beta^n} \delta\beta^n \right) e^{i\delta\beta(x-x')} \\
&= \hbar\omega_o \int dx\, \hat{a}^\dagger(x) \int dx'\, \hat{a}(x')\delta(x-x') \\
&\quad + \hbar \int dx\, \hat{a}^\dagger(x) \int dx'\, \hat{a}(x') \sum_{n=1}^{\infty} i^n \frac{d^n \omega}{d\beta^n} \frac{\partial^n}{\partial x^n} \delta(x-x') \\
&= \hbar\omega_o \int dx\, \hat{a}^\dagger(x)\hat{a}(x) + \hbar \sum_{n=1}^{\infty} \frac{(-i)^n}{n!} \frac{d^n \omega}{d\beta^n} \int dx\, \hat{a}^\dagger(x) \frac{\partial^n}{\partial x^n} \hat{a}(x) \;.
\end{aligned} \tag{13.4}$$

Here we have replaced the Nyquist function by a delta function, a legitimate step if the excitation extends over a time interval much smaller than $1/\Delta\beta$. The Heisenberg equation of motion for the operator $\hat{a}(x)$ is

$$\frac{\partial}{\partial t}\hat{a}(x) = -i\left(\omega_o + \sum_{n=1}^{\infty} \frac{(-i)^n}{n!} \frac{d^n\omega}{d\beta^n} \frac{\partial^n}{\partial x^n}\right)\hat{a}(x) .$$ (13.5)

Again we may simplify this equation in two steps:

(a) we make the replacement $\hat{a}(x) \rightarrow \hat{a}(x)\exp(-i\omega_o t)$, and
(b) we make a change of variables that transforms the coordinates, $t \rightarrow t - (d\beta/d\omega)x$ and $x \rightarrow x$.

Without changing notation, we obtain

$$\frac{\partial}{\partial t}\hat{a}(x) = \sum_{n=2}^{\infty} \frac{(-i)^{n+1}}{n!} \frac{d^n\omega}{d\beta^n} \frac{\partial^n}{\partial x^n}\hat{a}(x) .$$ (13.6)

If we retain only the first term in the summation, thus including only the simplest form of group velocity dispersion, (13.6) is the Heisenberg equation of motion of the Hamiltonian

$$\hat{H} = -\frac{\hbar}{2}\frac{d^2\omega}{d\beta^2}\hat{a}^\dagger(x)\frac{\partial^2}{\partial x^2}\hat{a}(x) .$$ (13.7)

Integration by parts leads to a more familiar form,

$$\hat{H} = \frac{\hbar}{2}\frac{d^2\omega}{d\beta^2}\frac{\partial}{\partial x}\hat{a}^\dagger(x)\frac{\partial}{\partial x}\hat{a}(x) .$$ (13.8)

This is the second-quantized Hamiltonian of particles with mass. Thus, dispersion imparts mass to the photon. The photon is coupled to the material and the combination of electromagnetic field and material excitation produces this effective mass for what may be called a "dressed" photon. If we specialize to the simple case of GVD represented by the second derivative, the equation of motion (13.6) simplifies to [see Appendix A.19]

$$\frac{\partial}{\partial t}\hat{a}(x) = i\frac{1}{2!}\frac{d^2\omega}{d\beta^2}\frac{\partial^2}{\partial x^2}\hat{a}(x) .$$ (13.9)

This equation for the envelope amplitude operator bears a close resemblance to the classical equation for a wavepacket envelope. There are differences, however. The classical equation involves the first derivative with respect to the spatial coordinate z, not the time t. A replacement of t by z/v_g can fix this discrepancy. In the classical equation the second derivative with respect to x is replaced by a time derivative. This can be fixed as well by the replacement $x \rightarrow v_g t$. With these changes of notation (13.9) reads

$$\frac{\partial}{\partial z}\hat{a}(v_g t) = i\frac{1}{2!}\left(\frac{d\beta}{d\omega}\right)^3\frac{d^2\omega}{d\beta^2}\frac{\partial^2}{\partial t^2}\hat{a}(v_g t) .$$ (13.10)

Correspondence with the classical form is established if one can argue that $(d\beta/d\omega)^3 d^2\omega/d\beta^2 = -d^2\beta/d\omega^2$. But this relation is a simple consequence of differential calculus. Indeed,

$$\frac{d^2\beta}{d\omega^2} = \frac{d}{d\omega}\left(\frac{1}{d\omega/d\beta}\right) = -\frac{1}{(d\omega/d\beta)^2}\frac{d^2\omega}{d\beta^2}\frac{d\beta}{d\omega} = -\left(\frac{d\beta}{d\omega}\right)^3\frac{d^2\omega}{d\beta^2} .$$

Thus we have shown that the Hamiltonian (13.7) leads to an equation of motion for the operator $\hat{a}(x)$ that is in one-to-one correspondence with the equation of motion of the complex field amplitude in a dispersive waveguide.

13.2 The Quantized Nonlinear Schrödinger Equation and Its Linearization

The Hamiltonian of the Kerr effect in the x representation has been derived in Sect. 12.3:

$$\hat{H}_K = -\hbar\frac{K}{2}\int dx\,\hat{a}^\dagger(x)\hat{a}^\dagger(x)\hat{a}(x)\hat{a}(x) . \tag{13.11}$$

When the Kerr Hamiltonian (13.11) is added to the Hamiltonian (13.7), the Heisenberg equation of motion becomes

$$\frac{\partial}{\partial t}\hat{a}(x) = \frac{i}{2}C\frac{\partial^2}{\partial x^2}\hat{a}(x) + iK\hat{a}^\dagger(x)\hat{a}(x)\hat{a}(x) , \tag{13.12}$$

where $C \equiv d^2\omega/d\beta^2$. This is the quantized nonlinear Schrödinger equation. A few words of caution are in order. The mode patterns of modes on a fiber are not independent of frequency or β. The present formalism ignores this dependence. This is an approximation, but a good one, since pulses as short as a picosecond contain thousands of wavelengths at an optical (infrared) wavelength of one micron or so. This means that pulses of one picosecond are very narrow-band and the assumption of β independence of the mode profile is an excellent one over the range of β involved.

The quantized form of the nonlinear Schrödinger equation was solved rigorously using the Bethe ansatz [178, 179]. The analysis is complicated and analytic results can only be obtained when certain limits are taken. An approach that leads to simple analytic expressions and permits physical insight is based on the linearization approximation [109]. We set for the operator $\hat{a}(x)$

$$\hat{a}(x) = a_o(x) + \Delta\hat{a}(x) , \tag{13.13}$$

where the first term is a c number, and the second is an operator that takes over the commutation relation of $\hat{a}(x)$. Thus,

$$[\Delta\hat{a}(x), \Delta\hat{a}^\dagger(x')] = \delta(x - x') \, . \tag{13.14}$$

The replacement (13.13) is rigorous, and by itself does not imply any approximation. Approximations are made when the Schrödinger equation is linearized in terms of $\Delta\hat{a}(x)$. Thus, $a_o(x)$ obeys the equation

$$-i\frac{\partial}{\partial t}a_o = \frac{C}{2}\frac{\partial^2}{\partial x^2}a_o + Ka_o^*a_oa_o \, . \tag{13.15}$$

The solution is

$$a_o(t,x) = A_o \exp\left[i\left(\frac{KA_o^2}{2}t - \frac{C}{2}p_o^2t + p_ox + \theta_o\right)\right]$$
$$\times \mathrm{sech}\left(\frac{x - x_o - Cp_ot}{\xi}\right) , \tag{13.16}$$

with the constraint

$$A_o^2\xi^2 = \frac{C}{K} \tag{13.17}$$

(see Fig. 13.1). The solution has four arbitrary integration constants, $n_o(= 2\xi|A_o|^2), p_o, \theta_o$, and x_o. These have been chosen in anticipation of their interpretation as average photon number, momentum, phase and position. In the classical form of the equation, p_o had the meaning of carrier frequency deviation. (Note the change of sign convention: $p_o > 0$ corresponds to $\Delta\omega < 0$.) In the quantum formalism, it will become the conjugate variable to position, and hence interpreting it as momentum is more appropriate. Yet it is in the classical sense that the solution (13.16) is most easily understood. A frequency deviation causes a change in the propagation constant that accounts for the

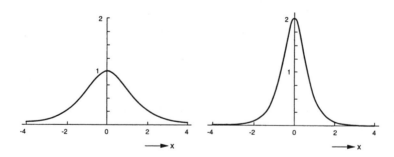

Fig. 13.1. The amplitude of the soliton as a function of x at $t = 0$ for two different values of n_o; $q_o = x_o = p_o = 0$. The narrower pulse has twice the photon number of the wider pulse.

phase accumulation $p_o^2 \tau / 2$ as the pulse propagates and is responsible for a group velocity change $C p_o$, which accounts for the shift of position under propagation.

The envelope $a_o(t, x)$ is so normalized that its magnitude squared gives the photon number:

$$\int dx |a_o(t, x)|^2 = \int |A_o|^2 \operatorname{sech}^2 \left(\frac{x - x_o - C p_o t}{\xi} \right) dx = 2 |A_o|^2 \xi = n_o .$$

(13.18)

In the subsequent analysis, we shall set $p_o = \theta_o = x_o = 0$ and $|A_o| = \sqrt{n_o / 2\xi}$ which simply means that we have chosen a coordinate system whose origin is at the pulse center, we have set the phase equal to zero, and we have picked a momentum (or carrier frequency) that coincides with the nominal carrier frequency ω_o.

When the ansatz (13.13) is introduced into the nonlinear Schrödinger equation, and terms of order higher than first in $\Delta \hat{a}$ and $\Delta \hat{a}^\dagger$ are dropped, we obtain a linear equation of motion for these two operators:

$$-i \frac{\partial}{\partial t} \Delta \hat{a} = \frac{C}{2} \frac{\partial^2}{\partial x^2} \Delta \hat{a} + 2K |a_o|^2 \Delta a + K a_o^2 \Delta \hat{a}^\dagger .$$

(13.19)

The equation couples $\Delta \hat{a}$ and $\Delta \hat{a}^\dagger$ in a way characteristic of a parametric process as described in Chap. 11. It is worth reiterating that linear equations of motion of an operator are in one-to-one correspondence with linear equations of motion of the classical evolution equation. In the integration of such equations one does not encounter products of operators, for the inclusion of which one would have to use the commutation relations. Hence, the integration can proceed "classically", as if the operators were c numbers. The classical transfer functions apply directly to the quantum problem.

We note that $\Delta \hat{a}$ must consist of two parts: a part $\Delta \hat{a}_{\text{sol}}$ that describes the change of the soliton parameters, i.e. a part that is associated with the soliton, and a part $\Delta \hat{a}_{\text{cont}}$ that is not associated with the soliton, the continuum part:

$$\Delta \hat{a} = \Delta \hat{a}_{\text{sol}} + \Delta \hat{a}_{\text{cont}} .$$

(13.20)

The soliton perturbation is with respect the four degrees of freedom of the soliton: the photon number, the phase, the momentum, and the position. These perturbations are now all operators. They are functions of t. As in the classical case, we attempt a solution of (13.19) through separation of variables, using the solutions of the classical form of the nonlinear Schrödinger equation as a guide. We write the perturbation as a superposition of operators with associated functions of x. The operators for photon number and phase, $\Delta \hat{n}$ and $\Delta \hat{\theta}$, have the usual interpretation. The operator of the position, $\Delta \hat{x}$, is associated with the position displaced from x_o, the operator for momentum with the shift from the carrier frequency p_o. A carrier frequency shift Δp corresponds to a change of propagation constant $\Delta \beta$, $\hbar \beta$ being the momentum.

It is important to note that the change of momentum of a wavepacket with an average number of photons n_o is equal to $n_o \Delta p$. Hence, it is natural to write the perturbation in the form

$$\Delta \hat{a}_{\text{sol}} = [\Delta \hat{n}(t) f_n(x) + \Delta \hat{\theta}(t) f_\theta(x) + \Delta \hat{x}(t) f_x(x) + n_o \Delta \hat{p}(t) f_p(x)]$$

$$\times \exp \left(i \frac{K A_o^2}{2} t \right) .$$

$$(13.21)$$

The functions $f_i(x)$ are chosen, as in the classical analysis, as derivatives of the soliton evaluated at $t = 0$. We choose the phase of A_o to be zero, i.e. A_o is real and positive. We have

$$f_n(x) = \frac{1}{2 A_o \xi} \left[1 - \frac{x}{\xi} \tanh(x/\xi) \right] \text{sech}(x/\xi) . \qquad (13.22)$$

In taking the derivative with respect to A_o, account has been taken of the area theorem, which ties changes of amplitude to changes of pulse width:

$$f_\theta(x) = i A_o \text{ sech}(x/\xi) , \qquad (13.23)$$

$$f_x(x) = \frac{A_o}{\xi} \tanh(x/\xi) \text{ sech}(x/\xi) , \qquad (13.24)$$

$$f_p(x) = i \frac{1}{2 A_o \xi} x \text{ sech}(x/\xi) . \qquad (13.25)$$

The four functions are shown in Fig. 13.2. When the ansatz (13.20) is introduced into the linearized nonlinear Schrödinger equation we find that no new functions are generated, just as in the classical case. Equating the coefficients of the functions $f_Q(x), Q = n, \theta, x, p$, we find the following equations of motion for the operator soliton perturbations:

$$\frac{d}{dt} \Delta \hat{n} = 0 , \qquad (13.26)$$

$$\frac{d}{dt} \Delta \hat{\theta} = \frac{1}{2} K \frac{\Delta \hat{n}}{\xi} , \qquad (13.27)$$

$$\frac{d}{dt} \Delta \hat{x} = C \Delta \hat{p} , \qquad (13.28)$$

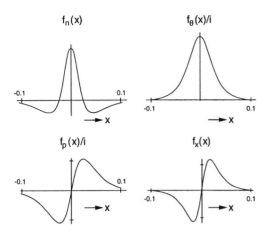

Fig. 13.2. The shape of the four functions $f_Q(x)$, $Q = n, \theta, p, x$

$$\frac{d}{dt}\Delta\hat{p} = 0 \ . \tag{13.29}$$

These equations of motion for the operators are in complete correspondence with the classical analysis. They make good sense. A perturbation of photon number propagates unperturbed, but affects the phase owing to a change of the Kerr phase shift. Similarly, a perturbation of the momentum (carrier frequency) propagates undisturbed, but affects the displacement owing to the change of group velocity with carrier frequency.

We find that the commutator of $\Delta\hat{n}$ and $\Delta\hat{\theta}$ on one hand, and that of $\Delta\hat{x}$ and $\Delta\hat{p}$ on the other hand, are constants of motion. Indeed, if we consider $[\Delta\hat{n}, \Delta\hat{\theta}]$ as an example, we find from (13.26) and (13.27)

$$\frac{d}{dt}[\Delta\hat{n}, \Delta\hat{\theta}] = \left[\frac{d\Delta\hat{n}}{dt}, \Delta\hat{\theta}\right] + \left[\Delta\hat{n}, \frac{d\Delta\hat{\theta}}{dt}\right] \propto [\Delta\hat{n}, \Delta\hat{n}] = 0 \ . \tag{13.30}$$

In the same way, we can show conservation of $[\Delta\hat{x}, \Delta\hat{p}]$. Even though the equations of the operators have been obtained by a linearization approximation, the expectation values of the phase and timing perturbation operators need not remain small. They are driven cumulatively by a photon number perturbation and by a momentum perturbation, respectively. The accumulated changes may become large, the only requirement is that the perturbation per unit length be small to permit the linearization of the equations.

13.3 Soliton Perturbations Projected by the Adjoint

We have developed equations of motion for the perturbations of solitons. In order to determine the initial amplitudes of the perturbations, we must be able to determine the four perturbation operators with the equations of motion (13.26)–(13.29) from a given initial condition, $\Delta \hat{a}(x)$ at $t = 0$. This is accomplished by the adjoints already developed in the context of the classical equations of motion. They are

$$\underline{f}_n(x) = 2A_o \, \text{sech}(x/\xi) \,, \tag{13.31}$$

$$\underline{f}_\theta(x) = \frac{i}{A_o \xi} \left[1 - \frac{x}{\xi} \, \text{sech}(x/\xi) \right] \text{sech}(x/\xi) \,, \tag{13.32}$$

$$\underline{f}_x(x) = \frac{1}{A_o} \frac{x}{\xi} \, \text{sech}(x/\xi) \,, \tag{13.33}$$

$$\underline{f}_p(x) = 2A_o \frac{i}{\xi} \, \tanh(x/\xi) \, \text{sech}(x/\xi) \,. \tag{13.34}$$

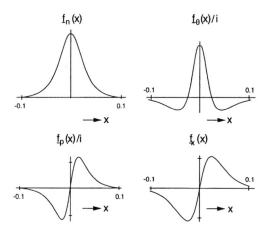

Fig. 13.3. The shape of the four adjoint functions $\underline{f}_Q(x)$, $Q = n, \theta, p, x$

Here the adjoints have been normalized so that their products with the original perturbation functions integrate to unity. The adjoints are shown in Fig. 13.3. They obey the self- and cross-orthonormality condition

$$\text{Re} \left[\int \underline{f}_P^*(x) f_Q(x) dx \right] = \delta_{PQ} \qquad \text{for} \quad P = n, \theta, x, p \,. \tag{13.35}$$

Next, we note that the initial condition $\Delta \hat{a}(x)$ can be separated into Hermitian operators

$$\Delta \hat{a}(x) = \Delta \hat{a}^{(1)}(x) + \mathrm{i}\Delta \hat{a}^{(2)}(x) , \tag{13.36}$$

with the commutation relation

$$\left[\Delta \hat{a}^{(1)}, \Delta \hat{a}^{(2)} \right] = \frac{\mathrm{i}}{2} . \tag{13.37}$$

We find that the operators are related pairwise to the in-phase and quadrature fluctuations:

$$\Delta \hat{n}(0) = \int \underline{f}_n^*(x) \Delta \hat{a}^{(1)}(x) dx , \tag{13.38}$$

$$\Delta \hat{\theta}(0) = \mathrm{i} \int \underline{f}_\theta^*(x) \Delta \hat{a}^{(2)}(x) dx , \tag{13.39}$$

$$\Delta \hat{x}(0) = \int \underline{f}_x^*(x) \Delta \hat{a}^{(1)}(x) dx , \tag{13.40}$$

$$\Delta \hat{p}(0) = \mathrm{i} \int \underline{f}_p^*(x) \Delta \hat{a}^{(2)}(x) dx . \tag{13.41}$$

The commutator of $\Delta \hat{n}(x)$ and $\Delta \hat{\theta}(x)$ is found using (13.37):

$$[\Delta \hat{n}, \Delta \hat{\theta}] = \frac{\mathrm{i}}{2} \int dx \, \underline{f}_n(x) \underline{f}_\theta(x) = \mathrm{i} . \tag{13.42}$$

The commutator of $\Delta \hat{x}(x)$ and $\Delta \hat{p}(x)$ is

$$[\Delta \hat{x}, n_o \Delta \hat{p}] = \frac{\mathrm{i}}{2} \int dx \, \underline{f}_x(x) \underline{f}_p(x) = \mathrm{i} . \tag{13.43}$$

$\Delta \hat{n}$ and $\Delta \hat{\theta}$ commute with both $\Delta \hat{x}$ and $\Delta \hat{p}$. These are the commutation relations of the photon number and phase of a wave, and the position and momentum of a particle. The soliton combines properties of wave and particle and possesses pairs of operators describing both properties. As shown in the preceding section, the commutators are invariants of the equations of motion (13.26)–(13.29).

A soliton in a uniform zero-point fluctuation background does not form a minimum-uncertainty packet. Indeed, when we set

$$\langle |\Delta \hat{a}(x) \Delta \hat{a}^\dagger(x')| \rangle = \delta(x - x') \tag{13.44}$$

and

$$\langle |\Delta \hat{a}^\dagger(x) \Delta \hat{a}(x')| \rangle = 0 \tag{13.45}$$

as appropriate for a vacuum state, then the fluctuations of the in-phase and quadrature components, in terms of which the soliton perturbations are expressed, can be evaluated as follows:

$$\langle \Delta \hat{a}^{(1)}(x) \Delta \hat{a}^{(1)}(x') \rangle$$

$$= \frac{1}{4} [\langle \Delta \hat{a}(x) \Delta \hat{a}(x') \rangle + \langle \Delta \hat{a}^\dagger(x) \Delta \hat{a}^\dagger(x') \rangle$$

$$+ \langle \Delta \hat{a}^\dagger(x) \Delta \hat{a}(x') \rangle + \langle \Delta \hat{a}(x) \Delta \hat{a}^\dagger(x') \rangle]$$

$$(13.46)$$

$$= \frac{1}{4} [\langle \Delta \hat{a}(x) \Delta \hat{a}^\dagger(x') \rangle] = \frac{1}{4} \delta(x - x') ,$$

where we have used (13.44) and (13.45). The expectation values of the products of the creation operators and of the annihilation operators vanish because of the stationarity of the zero-point fluctuations. In a similar way we find

$$\langle \Delta \hat{a}^{(2)}(x) \Delta \hat{a}^{(2)}(x') \rangle = \frac{1}{4} [\langle \Delta \hat{a}(x) \Delta \hat{a}^\dagger(x') \rangle] = \frac{1}{4} \delta(x - x') \tag{13.47}$$

and

$$\frac{1}{2} \langle \Delta \hat{a}^{(1)}(x) \Delta \hat{a}^{(2)}(x') + \Delta \hat{a}^{(2)}(x) \Delta \hat{a}^{(1)}(x') \rangle = 0 . \tag{13.48}$$

With these expressions it is easy to evaluate the fluctuations of photon number and phase:

$$\langle \Delta \hat{n}^2(0) \rangle = \frac{1}{4} \int |\underline{f}_n(x)|^2 dx = n_o . \tag{13.49}$$

These are the fluctuations of a Poisson process. The phase fluctuations are

$$\langle \Delta \hat{\theta}^2(0) \rangle = \frac{1}{4} \int |\underline{f}_\theta(x)|^2 dx = \frac{0.607}{n_o} . \tag{13.50}$$

The product is

$$\langle \Delta \hat{n}^2(0) \rangle \langle \Delta \theta^2(0) \rangle = 0.607 > 0.25 . \tag{13.51}$$

The phase fluctuations are larger than those of a minimum-uncertainty state of photon number and phase, for which the uncertainty product would be equal to 1/4.

As mentioned earlier in Chap. 10 in the development of the adjoint of the linearized NLSE, the linearized equation does not conserve energy. The linearized equation describes a parametric process in the presence of a pump which is capable of generation or annihilation of photons. A consequence of this fact is that the linearized equation is not self-adjoint, requiring the pairing of its solutions with those of the adjoint equation. Conservation of the cross-energy of the solutions with their adjoint then leads to the orthogonality relations.

Equations derived from a Hamiltonian conserve energy and, in the quantized form, the commutator brackets. Equations that do not conserve energy are not derivable from a Hamiltonian. Therefore, one may not assume a priori that the quantized form of such equations conserves commutator brackets. It is easy to show that (13.19) in fact conserves the commutator $[\Delta\hat{a}, \Delta\hat{a}^\dagger]$. Hence, the linearization of the NLSE does not call for the introduction of noise sources. This fact could have been anticipated from the eminently reasonable conservation relations of the commutator brackets $[\Delta\hat{n}, \Delta\hat{\theta}]$ and $[\Delta\hat{p}, \Delta\hat{x}]$. We have mentioned that in [174] Drummond and Carter developed a formalism of soliton squeezing that arrived at classical stochastic equations of motion for numerical solutions of the soliton-squeezing phenomenon. Their formalism contains noise sources. Hence one must ask the question of how our formalism, free of noise sources, can agree with [174]. This question was asked and answered in [181]. It turns out that the noise sources do not contribute to the perturbation operators in the limit when the linearized analysis applies.

13.4 Renormalization of the Soliton Operators

In the preceding chapter, we have studied squeezing of pulses in dispersion-free fibers by splitting the pulse into segments of quasiconstant excitation; the different evolutions of the in-phase and quadrature components of these quasi-c.w. waves resulted in reduction of one component of the noise. If an analogy is to be established with this process, we have to arrive at equivalent in-phase and quadrature components of the soliton.

Thus far, we have used operators representing the perturbations of photon number, phase, position, and momentum. The perturbation operator $\Delta\hat{a}(x)$ has the commutator $[\Delta\hat{a}(x), \Delta\hat{a}^\dagger(x')] = \delta(x - x')$, and thus has dimensions of inverse length to the power of one-half. The photon number perturbation Δn is given by $\Delta n = \Delta(2A_o^2\xi) = 4A_o\,\Delta A_o\,\xi + 2A_o^2\,\Delta\xi = 2A_o\,\Delta A_o\,\xi$, where we have used the area theorem to relate the pulse width change to the pulse amplitude change. Consider a continuous wave of amplitude A_o and its associated photon number $n_o = A_o^2$. The change in photon number is $\Delta n = 2A_o\,\Delta A_o$. When quantized, the perturbation ΔA_o would be replaced by the in-phase operator, $\Delta A_o \to \Delta\hat{A}_1$. This fact and the dimensions of the $\Delta\hat{a}(x)$ operator suggest that the soliton perturbation $\Delta A_o\sqrt{\xi}$ is to be replaced by

$$\Delta A_o\sqrt{\xi} \to \Delta\hat{A}_1 \ . \tag{13.52}$$

Its associated expansion function is changed by the renormalization from the expansion function of the photon number perturbation to the following:

$$f_1(x) = \frac{1}{\sqrt{\xi}}\left[1 - \frac{x}{\xi}\tanh\left(\frac{x}{\xi}\right)\right]\mathrm{sech}\left(\frac{x}{\xi}\right). \tag{13.53}$$

The adjoint function with the property $\int dx\, f_1(x) \underline{f}_1^*(x) = 1$ is

$$\underline{f}_1(x) = \frac{1}{\sqrt{\xi}}\,\mathrm{sech}\left(\frac{x}{\xi}\right). \tag{13.53a}$$

The same approach suggests the definition of the quadrature component as

$$A_o\,\Delta\hat{\theta}\sqrt{\xi} \to \Delta\hat{A}_2. \tag{13.54}$$

We find for the expansion function

$$f_2(x) = \frac{i}{\sqrt{\xi}}\,\mathrm{sech}\left(\frac{x}{\xi}\right) \tag{13.55}$$

and the adjoint function is

$$\underline{f}_2(x) = \frac{i}{\sqrt{\xi}}\left[1 - \frac{x}{\xi}\tanh\left(\frac{x}{\xi}\right)\right]\mathrm{sech}\left(\frac{x}{\xi}\right). \tag{13.55a}$$

A similar renormalization is possible for the perturbation operators of position and momentum. As we shall see, it is convenient to change the commutation relation by a factor of $1/2$. This is accomplished by the identification of the new operators $\Delta\hat{X} = A_o\,\Delta\hat{x}/\sqrt{\xi}$ and $\Delta\hat{P} = n_o\,\Delta\hat{p}\sqrt{\xi}/2A_o$. The commutator is now

$$[\Delta\hat{X}, \Delta\hat{P}] = \frac{i}{2}. \tag{13.56}$$

The respective perturbation functions become

$$f_X(x) = \frac{1}{\sqrt{\xi}}\tanh\left(\frac{x}{\xi}\right)\mathrm{sech}\left(\frac{x}{\xi}\right) \tag{13.57}$$

and

$$f_P(x) = \frac{i}{\sqrt{\xi}}\frac{x}{\xi}\,\mathrm{sech}\left(\frac{x}{\xi}\right). \tag{13.58}$$

The adjoint functions are

$$\underline{f}_X(x) = \frac{1}{\sqrt{\xi}}\frac{x}{\xi}\,\mathrm{sech}\left(\frac{x}{\xi}\right) \tag{13.57a}$$

and

$$\underline{f}_P(x) = \frac{i}{\sqrt{\xi}}\tanh\left(\frac{x}{\xi}\right)\mathrm{sech}\left(\frac{x}{\xi}\right). \tag{13.58a}$$

The expansion (13.21) of the pulse is now in the form

$$\Delta \hat{a}_{\text{sol}} = [\Delta \hat{A}_1(t) f_1(x) + \Delta \hat{A}_2(t) f_2(x) + \Delta \hat{X}(t) f_X(t) + \Delta \hat{P}(t) f_P(x)]$$

$$\times \exp\left(i(K A_o^2/2) t \right) .$$

(13.59)

The commutator of the in-phase and quadrature components is

$$[\Delta \hat{A}_1, \Delta \hat{A}_2] = i \int dx \, \underline{f}_1^*(x) \int dx' \underline{f}_2^*(x') [\Delta \hat{a}^{(1)}(x), \Delta \hat{a}^{(2)}(x')]$$

(13.60)

$$= -\frac{1}{2} \int dx \, \underline{f}_1^*(x) \int dx' \underline{f}_2^*(x') \delta(x - x') = \frac{i}{2} ,$$

and has the expected value.

It is clear from the preceding discussions that the expansion of a pulse excitation into a soliton part and a continuum part is an expansion into orthogonal modes. These modes are phase-dependent; the components in phase with the pulse $a_o(t,x)$ are different from those in quadrature. They form an orthonormal set into which any excitation can be decomposed and whose amplitudes are quantized. Of course, the decomposition makes physical sense only when the expansion represents perturbations of a hyperbolic secant pulse. But the pulse need not be a soliton; for example, it could be a hyperbolic secant pulse produced in the output of a beam splitter with a soliton impinging on one of its input ports.

Next, it is of interest to determine the mean square fluctuations of the soliton perturbation parameters, if the background is zero-point-fluctuations. We find

$$\langle (\Delta \hat{A}_1)^2 \rangle = \int dx \, \underline{f}_1^*(x) \int dx' \underline{f}_1(x') \langle \Delta \hat{a}^{(1)}(x) \Delta \hat{a}^{(1)}(x') \rangle$$

$$= \frac{1}{4} \int dx \, \underline{f}_1^*(x) \int dx' \underline{f}_1(x') \delta(x - x')$$

(13.61)

$$= \frac{1}{4} \int dx \, |\underline{f}_1^{*2}(x)|^2 = \frac{1}{2} .$$

The mean square fluctuations are twice the minimum-uncertainty value for equal in-phase and quadrature fluctuations. The remaining three fluctuations can be computed analogously. It is clear that they involve the values of the integrals

$$\int dx \, |\underline{f}_Q(x)|^2 = 2, \ 1.214, \ \frac{\pi^2}{6}, \ \frac{2}{3}; \quad Q = 1, 2, X, P .$$

(13.62)

The uncertainty products are

$$\langle(\Delta\hat{A}_1)^2\rangle\langle(\Delta\hat{A}_2)^2\rangle = \frac{1}{16} \times 2.43 \tag{13.63}$$

and

$$\langle\Delta\hat{X}^2\rangle\langle\Delta\hat{P}^2\rangle = \frac{1}{16} \times 1.09 . \tag{13.64}$$

The in-phase and quadrature fluctuations are uncorrelated:

$$\langle\Delta\hat{A}_1\Delta\hat{A}_2 + \Delta\hat{A}_2\Delta\hat{A}_1\rangle = 0 . \tag{13.65}$$

The renormalization has changed the uncertainty ellipse. In the photon number–phase description, the photon number fluctuations were at the Poisson value; the phase fluctuations were larger than the minimum uncertainty. In the in-phase and quadrature description, the amplitude fluctuations are excessive, whereas the quadrature fluctuations are close to the minimum value. This shows that the description of squeezing is dependent upon the representation. In fact, the minimum-uncertainty ellipse of the momentum and position of a particle is plotted along axes of different dimensions and thus the shape of the ellipse is not an indication of "squeezing". It is only when the noncommuting variables are of the same dimensions *and* of the same character, such as the in-phase and quadrature components of the electric field, that squeezing can be identified. The stationary character of the standard fluctuations dictates a circular locus of uncertainty. Squeezing produces nonstationary statistics that manifest themselves in an elliptic uncertainty locus.

In order to appreciate better the significance of the in-phase fluctuations, we return to (13.52) and take note of the fact that the photon number fluctuations are given by

$$\Delta n = 2A_0\,\Delta A_0\,\xi \rightarrow 2A_0\sqrt{\xi}\,\Delta\hat{A}_1 .$$

Thus, the mean square photon number fluctuations are

$$\langle\Delta\hat{n}^2\rangle = 2A_0^2\xi 2\langle\Delta\hat{A}_1^2\rangle = \langle n\rangle .$$

They have the Poisson value. Hence, the in-phase fluctuations of a soliton with twice the minimum value are, in fact, the fluctuations associated with a Poisson distribution of photons.

The position and momentum operators do not obey the standard commutator relation, but a new one, in one-to-one correspondence with the commutator of the in-phase and quadrature components. This renormalization also introduces a welcome symmetrization to the equations of motion of the operators. In lieu of (13.26)–(13.29) we now have

$$\frac{d}{dt}\Delta\hat{A}_1 = 0 , \tag{13.66}$$

$$\frac{d}{dt}\Delta \hat{A}_2 = \frac{C}{\xi^2}\Delta \hat{A}_1 \, , \tag{13.67}$$

$$\frac{d}{dt}\Delta \hat{X} = \frac{C}{\xi^2}\Delta \hat{P} \, , \tag{13.68}$$

$$\frac{d}{dt}\Delta \hat{P} = 0 \, , \tag{13.69}$$

where we have used the area theorem $KA_o^2 = C/\xi^2$.

13.5 Measurement of Operators

The definition of operators is clarified when measurements can be described that determine these operators. Figure 13.4 shows a soliton (hyperbolic-secant pulse) source, followed by a beam splitter. Part of the output is modulated in a "soliton modifier" that produces the soliton perturbation; the other part is used as the local oscillator, with a proper pulse shape change and phase adjustment. The two signals are combined in a balanced detector. The pulse shape changer is a filter that produces an output coherent with its input. Such optical pulse-shaping functions have been demonstrated by Liu et al. [183] with a scheme in which a grating spatially disperses the spectrum of an incoming pulse, a spatially distributed absorber and phase shifter modifies the spectrum, and finally the spectrum is spatially superimposed and reassembled by another grating. The pulse shaper produces a local-oscillator pulse $f_L(x)$, which can be treated classically if the usual linearization approximation is used in the analysis of the balanced detector. The soliton modifier produces an excitation $\Delta \hat{a}(x)$ from an incoming pulse $\hat{a}(x)$. We model the detector as an ideal photon flux detector with a response much slower than the inverse

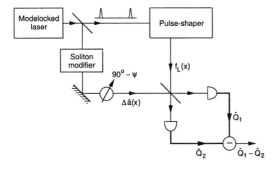

Fig. 13.4. Setup for measurement of operators

pulse widths of the local-oscillator and incident pulses. The difference charge is

$$\hat{Q} \equiv (\hat{Q}_1 - \hat{Q}_2) = iq \int dx \left[f_L(x) \Delta \hat{a}^\dagger(x) - f_L^*(x) \Delta \hat{a}(x) \right] , \qquad (13.70)$$

where q is the electron charge. Note that the operation on the field operators that produces the difference charge is identical to the operation that projects out the four operators of the soliton perturbation. Since the difference charge of homodyne detection contains no fluctuations from the local oscillator, it is a noise-free measurement of the signal incident upon the signal port of the beam splitter. Various choices of the pulse shape and phase of the local oscillator give different responses. Thus, for example, if the local-oscillator pulse is chosen so that $if_L(x) = -f_\theta(x)/2$, the expectation value of the signal is the phase change of the soliton:

$$\frac{\langle \hat{Q} \rangle}{q} = \langle \Delta \hat{\theta} \rangle . \qquad (13.71)$$

In this measurement, only the perturbation of the phase contributes to the signal. Note that in an actual measurement, the local-oscillator pulse would be chosen to be many times larger in order to achieve a gain greater than unity. In a similar way we find that the expectation values of the timing perturbation and the carrier frequency perturbation are picked out by the choices $if_L(x) = f_x(x)/2$ and $if_L(x) = -f_p(x)/2$.

These three choices of the local-oscillator pulse all result in a measurement of an observable perturbation. The main pulse is orthogonal to the projections via the local oscillator. If the local oscillator is chosen so that $if_L(x) = \frac{1}{2} f_n(x)$, the balanced detector measures both the photon number and its perturbation:

$$\frac{\langle Q \rangle}{q} = n_o + \langle \Delta \hat{n} \rangle . \qquad (13.72)$$

The reader may have noticed that some of the perturbations $\Delta \hat{a}$ considered here were very simple soliton perturbations, the generation of which does not require a sophisticated filter. A phase shift can be produced by a phase shifter, a time delay by a delay line, and a photon number change by attenuation followed by propagation through a fiber to reestablish the height–width ratio of the soliton.

13.6 Phase Measurement with Soliton-like Pulses

Pulses, and in particular solitons, can be used as probes in a Mach–Zehnder interferometer for the measurement of the phase imbalance of the interferometer. Here we determine the signal-to-noise ratio of such a measurement in the

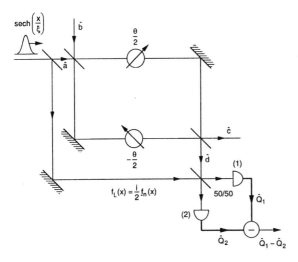

Fig. 13.5. Phase measurement with soliton pulses

case when the probe solitons have fluctuations given by (13.63) and (13.64). A schematic of the experiment is shown in Fig. 13.5. A Mach–Zehnder interferometer is unbalanced by small phase shifts $\theta/2$ in one arm and $-\theta/2$ in the other arm. Into the input port (a) are fed hyperbolic-secant-shaped pulses $a_o(x)$ with zero-point fluctuation background. Vacuum fluctuations enter the vacuum port (b). Owing to the interferometer imbalance the output from port (d) is composed of a signal part and vacuum fluctuations. We neglect the signal-dependent noise contribution. We have

$$\hat{d} = \frac{1}{2}[a_o(x)e^{i\theta/2} - a_o(x)e^{-i\theta/2}] + \Delta\hat{d}(x)$$

$$= ia_o(x)\sin\frac{\theta}{2} + \Delta\hat{d}(x) \qquad (13.73)$$

$$\approx ia_o(x)\frac{\theta}{2} + \Delta\hat{d}(x) .$$

The operator $\Delta\hat{d}(x)$ represents pure vacuum fluctuations. The fluctuations are not changed by the interferometer imbalance, since the contribution of the vacuum fluctuations of port (b), lost owing to the imbalance, is made up by vacuum fluctuations from port (a).

The signal is a hyperbolic secant. The simplest procedure is to project it out with the pulse itself, thus choosing the local-oscillator function $f_L(x) = A_o\operatorname{sech}(x/\xi)$. The detector charge difference is

$$\frac{\hat{Q}}{q} = \frac{\hat{Q}_1 - \hat{Q}_2}{q}$$

$$= iq \int dx \, [f_L(x)\hat{d}^\dagger(x) - f_L^*(x)\hat{d}(x)] \tag{13.74}$$

$$= q \int dx \, A_o \, \mathrm{sech}(x/\xi)[\theta A_o \, \mathrm{sech}(x/\xi) - i\,\Delta\hat{d}(x) + i\,\Delta\hat{d}^\dagger(x)] \, .$$

We find for the expectation value of the measurement

$$\frac{\langle \hat{Q} \rangle}{q} = \int dx \, \theta A_o^2 \, \mathrm{sech}^2(x/\xi) = \theta n_o \, . \tag{13.75}$$

The mean square fluctuations are

$$\frac{\langle \Delta\hat{Q}^2 \rangle}{q^2}$$

$$= \int dx \int dx' A_o^2 \, \mathrm{sech}(x/\xi) \, \mathrm{sech}(x'/\xi)$$

$$\times \langle |[-i\,\Delta\hat{d}(x) + i\,\Delta\hat{d}^\dagger(x)][-i\,\Delta\hat{d}(x') + i\,\Delta\hat{d}^\dagger(x')]| \tag{13.76}$$

$$= \int dx \int dx' A_0^2 \, \mathrm{sech}(x/\xi) \, \mathrm{sech}(x'/\xi)\delta(x - x')$$

$$= \int dx \, A_o^2 \, \mathrm{sech}^2(x/\xi) = n_o \, .$$

Thus, we find for the signal-to-noise ratio

$$\frac{S}{N} = \frac{\langle \hat{Q} \rangle^2}{\langle \Delta\hat{Q}^2 \rangle} = \theta^2 n_o \, . \tag{13.77}$$

Let us express the signal-to-noise ratio in terms of the photon number of the signal. According to (13.73),

$$\text{signal photon number} = \int dx \, |a_o(x)|^2 \left(\frac{\theta}{2}\right)^2 = n_o \frac{\theta^2}{4} \, . \tag{13.78}$$

We find for the signal-to-noise ratio

$$\frac{S}{N} = 4 \times \text{signal photon number} \, . \tag{13.79}$$

The signal-to-noise ratio is twice that obtained for homodyne detection in Sect. 8.5. The factor of two improvement in the present case comes from the fact that the signal is fixed, reflecting the constant phase of the interferometer, not averaged over a cosine variation associated with a time-dependent signal.

13.7 Soliton Squeezing in a Fiber

Propagation of a soliton along a lossless dispersive fiber leads to squeezing of the soliton fluctuations. We note, first of all, that the evolutions of the photon number and phase on one hand, and of the position and momentum on the other hand, proceed independently, as illustrated by (13.66)–(13.69). These evolutions represent the separate natures of the soliton as both a wave and a particle. Squeezing occurs owing to the coupling between operators that conserves the commutation relations. With the renormalization of Sect. 13.4 the squeezing can be expressed as the evolution of an uncertainty ellipse of constant area in the two-dimensional space of the complementary variables.

We repeat the renormalized Heisenberg equations of motion below:

$$\frac{d}{dt}\Delta \hat{A}_1 = 0 \, , \tag{13.80}$$

$$\frac{d}{dt}\Delta \hat{A}_2 = \frac{C}{\xi^2}\Delta \hat{A}_1 \, , \tag{13.81}$$

$$\frac{d}{dt}\Delta \hat{X} = \frac{C}{\xi^2}\Delta \hat{P} \, , \tag{13.82}$$

$$\frac{d}{dt}\Delta \hat{P} = 0 \, . \tag{13.83}$$

The solutions of (13.80) and (13.81) are

$$\Delta \hat{A}_1(t) = \Delta \hat{A}_1(0) \tag{13.84}$$

and

$$\Delta \hat{A}_2(t) = \Delta \hat{A}_2(0) + 2\Phi(t)\Delta \hat{A}_1(0) \, , \tag{13.85}$$

where

$$\Phi(t) \equiv \frac{1}{2}KA_o^2 t = \frac{1}{2}\frac{C}{\xi^2}t$$

is the classical soliton phase shift.

These two equations describe the evolution of the uncertainty ellipse in the plane of the in-phase and quadrature components. The mean square deviations along the in-phase and quadrature directions are

$$\langle [\Delta \hat{A}_1(t)]^2 \rangle = \langle [\Delta \hat{A}_1(0)]^2 \rangle \tag{13.86}$$

and

$$\langle [\Delta \hat{A}_2(t)]^2 \rangle = \langle [\Delta \hat{A}_2(0)]^2 \rangle + 4\Phi^2 \langle [\Delta \hat{A}_1(0)]^2 \rangle , \tag{13.87}$$

respectively, with the cross-correlation

$$\frac{1}{2} \langle \Delta \hat{A}_1(0) \Delta \hat{A}_2(t) + \Delta \hat{A}_2(0) \Delta \hat{A}_1(t) \rangle = 2\Phi \langle \Delta \hat{A}_1^2(0) \rangle . \tag{13.88}$$

Since the input is assumed to be white noise, the probability distribution of the two variables at the input is a two-dimensional Gaussian with mean square deviations along the two orthogonal axes along the in-phase and quadrature directions. The $1/\sqrt{e}$ points of the Gaussian probability distribution lie on an ellipse with its major and minor axes parallel to the in-phase and quadrature component axes. Propagation along the fiber couples the in-phase component to the quadrature component; fluctuations in amplitude are transformed into fluctuations of the quadrature component or phase. The new ellipse of $1/\sqrt{e}$ points has new major and minor axes that are rotated relative to the original axes. The area of the ellipse is preserved in the process. The mathematical proof can be developed using the formalism presented in Appendix A.18. Here we present a simple geometric argument.

Consider the original Gaussian probability distribution, with its major and minor axes as shown in Fig. 13.6. Take a set of sample points that lie at ΔA_1, within the differential range $d\Delta A_1$. Because these points have the same amplitude, they experience the same quadrature phase shift. The slice

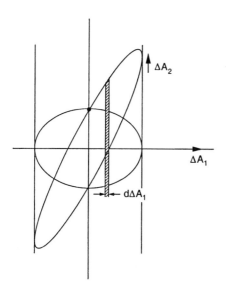

Fig. 13.6. The squeezing ellipse as constructed from changes of $\Delta \hat{A}_2(t)$ proportional to $\Delta \hat{A}_1(0)$

of the ellipse within that range is translated as shown in Fig. 13.6. In this manner, the entire ellipse can be split into slices, and each slice shifted by the appropriate amount. Since the shift is proportional to ΔA_1, one sees that a new elliptic region is formed, with the same area as the original one. The greater the shift, the narrower the ellipse and the smaller the minor axis, and thus the greater the squeezing.

The process of squeezing can be treated in a more formal way by establishing the correspondence of the solutions (13.84) and (13.85) with the Bogolyubov transformation that described squeezing in Chap. 11. We have

$$\Delta\hat{A}(t) \equiv \Delta\hat{A}_1(t) + i\,\Delta\hat{A}_2(t) = \mu(t)\Delta\hat{A}(0) + \nu(t)\Delta\hat{A}^\dagger(0) , \qquad (13.89)$$

with

$$\mu = 1 + i\Phi(t) \quad \text{and} \quad \nu = i\Phi(t) . \qquad (13.90)$$

The perturbation (13.89) accompanies the soliton pulse $a_o(t,x)$. We are now ready to analyze the generation of squeezed soliton vacuum by the setup illustrated in Fig. 12.8, repeated in a slightly modified version in Fig. 13.7. The hyperbolic secant pulse incident upon one of the input ports of the Sagnac interferometer is split into two pulses of the appropriate area, so that the two pulses propagate as solitons in the loop. In the process, they each squeeze the accompanying vacuum fluctuations in the manner indicated by (13.89). The squeezed vacua in the two arms are incoherent with each other. At the output of the Sagnac loop, they superimpose incoherently to emerge from the loop in the two ports of the beam splitter. The classical part of the pulse emerging from one of the output ports is reshaped, and is reused as the local oscillator. We assume that the reshaping produces the local-oscillator waveform

$$if_L(x) = \frac{1}{2}[\cos\psi\underline{f}_1(x) + \sin\psi\underline{f}_2(x)]\exp i\Phi(t) , \qquad (13.91)$$

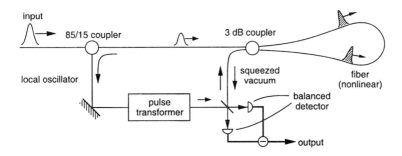

Fig. 13.7. The squeezing apparatus

which can be put into the form ideal for the purpose of projecting out a linear combination of $\Delta\hat{A}_1(t)$ and $\Delta\hat{A}_2(t)$. The noise of the local oscillator is suppressed by the balanced detector. The squeezed-vacuum fluctuations emerging from the other port are projected out in the balanced detector, resulting in the net charge operator

$$
\begin{aligned}
\frac{\Delta\hat{Q}}{q} &= -\mathrm{i}\int dx[f_L^*(x)\Delta\hat{a}_{\text{sol}} - f_L(x)\Delta\hat{a}_{\text{sol}}^\dagger] \\[2mm]
&= \cos\psi\,\Delta\hat{A}_1(t) + \sin\psi\,\Delta\hat{A}_2(t) \\[2mm]
&= \frac{1}{2}[e^{-\mathrm{i}\psi}\Delta\hat{A}(t) + e^{\mathrm{i}\psi}\Delta\hat{A}^\dagger(t)] \\[2mm]
&= \frac{1}{2}\{e^{-\mathrm{i}\psi}[\mu\,\Delta\hat{A}(0) + \nu\,\Delta\hat{A}^\dagger(0)] + e^{\mathrm{i}\psi}[\mu^*\,\Delta\hat{A}^\dagger(0) + \nu^*\,\Delta\hat{A}(0)]\} \,.
\end{aligned}
$$
$$(13.92)$$

Equation (13.92) expresses the normalized difference charge in two ways: (a) as the projection of a vector with components $\Delta\hat{A}_1(t), \Delta\hat{A}_2(t)$ onto an axis inclined at an angle ψ with respect to the (1) axis, and (b) as the sum of the phase-shifted squeezed input excitations $\mu\,\Delta\hat{A}(0) + \nu\,\Delta\hat{A}^\dagger(0)$. The two representations are equivalent, but in particular applications one may be more convenient than the other. We shall determine the degree of squeezing and antisqueezing from representation (a). The mean square fluctuations of the charge are

$$
\frac{\langle\Delta\hat{Q}^2(t)\rangle}{q^2} = \cos^2\psi\langle\Delta\hat{A}_1^2(t)\rangle + \sin^2\psi\langle\Delta\hat{A}_2^2(t)\rangle
$$
$$(13.93)$$
$$
+ \sin(2\psi)\frac{1}{2}\langle\Delta\hat{A}_1(t)\Delta\hat{A}_2(t) + \Delta\hat{A}_2(t)\Delta\hat{A}_1(t)\rangle \,.
$$

If the projection $\sqrt{\langle|\Delta\hat{Q}^2(t)|\rangle}$ is plotted in the (1)–(2) plane as a function of the orientation angle ψ, an ellipse is traced out, the locus of the root mean square deviation of the Gaussian distribution of $\langle|\Delta\hat{Q}^2(t)|\rangle$. According to (13.85), the component in direction (1) remains unchanged, whereas the component in direction (2) shifts proportionally to $\sqrt{\langle|\Delta\hat{A}_1^2(0)|\rangle}$ as shown schematically in Fig. 13.6. The mean square fluctuations of the normalized difference charge along the two axes are given by (13.86) and (13.87), and the cross-correlation by (13.88). The probability distribution of the normalized difference charge in the (1)–(2) plane, with coordinates ξ_1 and ξ_2, is given by

$$
p(\xi_1, \xi_2, t) \propto \exp-\frac{1}{2}\left(\frac{\xi_1^2}{\sigma_{11}(t)} + \frac{\xi_2^2}{\sigma_{22}(t)} + \frac{2\xi_1\xi_2}{\sigma_{12}(t)}\right) \,,
$$
$$(13.94)$$

where

$$
\begin{bmatrix} \sigma_{11}(t) & \sigma_{12}(t) \\ \sigma_{21}(t) & \sigma_{22}(t) \end{bmatrix} = \langle \Delta \hat{A}_1^2(0) \rangle \begin{bmatrix} 1 & 2\Phi(t) \\ 2\Phi(t) & \eta + 4\Phi^2(t) \end{bmatrix} , \tag{13.95}
$$

and $\eta = \langle |\Delta \hat{A}_2^2(0)| \rangle / \langle |\Delta \hat{A}_1^2(0)| \rangle = 0.607$. The Fourier transform of the probability distribution expressed in k space, the characteristic function, is of the form

$$
C(k_1, k_2, t) \propto \exp{-\frac{1}{2}[\sigma_{11}(t)k_1^2 + \sigma_{22}(t)k_2^2 + 2\sigma_{12}(t)k_1 k_2]} . \tag{13.96}
$$

The quadratic form in the exponent of the characteristic function can be diagonalized by a reorientation of the axes. A coordinate transformation into new orthogonal coordinates k_1' and k_2' finds the mutually orthogonal directions along which the fluctuations are uncorrelated. These are the major and minor axes of an ellipse. The transformation is a unitary transformation of the matrix which leaves the eigenvalues of the matrix (13.95) invariant. The eigenvalues are

$$
\lambda_{\pm} = \langle \Delta \hat{A}_1^2(0) \rangle \left(\frac{1 + \eta + 4\Phi^2}{2} \pm \sqrt{\left(\frac{1 + \eta + 4\Phi^2}{2} \right)^2 - \eta} \right) . \tag{13.97}
$$

These eigenvalues are the squares of the major and minor axes of the uncertainty ellipse. The product of the eigenvalues is

$$
\lambda_+ \lambda_- = \eta \langle \Delta \hat{A}_1^2(0) \rangle^2 \tag{13.98}
$$

and is constant, independent of the degree of squeezing. The squeezing and antisqueezing are illustrated in Fig. 13.8. With zero phase shift, the fluctuations in the (1) direction are shot noise fluctuations. These are equal to twice the zero-point fluctuations of $1/4$. In the orthogonal direction, the fluctuations are less, but they are still larger than $1/4$. As the nonlinear phase shift increases, the branch that represents shot noise at $\Phi = 0$ shows monotonically increasing fluctuations, whereas the orthogonal direction decreases and reaches zero asymptotically. Figure 13.9 shows the fluctuations as a function of the phase angle ψ for different degrees of squeezing and antisqueezing. This figure shows that the phase angle regime within which a large degree of squeezing is observed becomes narrower and narrower as the degree of squeezing is increased. The greater the degree of squeezing, the harder it is to find the squeezing angle and stabilize the system at that angle.

13.8 Summary

The Heisenberg representation of pulse propagation through a dispersive nonlinear Kerr medium (a fiber) leads to operator equations that resemble their

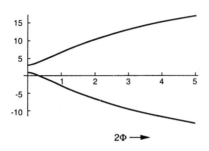

Fig. 13.8. Squeezing and antisqueezing (the minor and major axes of the squeezing ellipse) in dB, normalized to the zero-point fluctuations of $1/4$, as functions of 2Φ

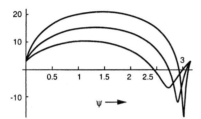

Fig. 13.9. Fluctuations as a function of phase angle ψ for different degrees of squeezing and antisqueezing; $2\Phi = 2, 4, 8$

classical counterparts. When these equations are linearized, an assumption justified in those cases in which the fluctuations of the amplitude are small compared with its expectation value, the distinction between the classical and quantum problem all but disappears. The solution of the linearized equations does not face the problem of ordering of the operators. Therefore, the solution of the classical form of the linearized equations is also the solution of the quantum problem. Differences between the classical and quantum problems appear only when expectation values of squares and products are taken. This fact alone, however, is easily taken into account.

We have approached the quantization of the soliton problem via linearization. We found that a soliton of the NLSE can be described as a wave–particle complex sharing quantum properties of both. Four operators describe the soliton: position and momentum, giving it particle properties; and photon number and phase (or in-phase and quadrature amplitudes), giving it wave properties. The operators representing these obey the usual commutation relations. We found that all four operators were independently measurable using a homodyne detector excited by an appropriately chosen local-oscillator pulse. The measurement suppressed fluctuations associated with the continuum.

The propagation of these operators along the fiber is described by total differential equations. These equations are free of noise sources, because they conserve the commutator brackets. It is easy to generalize these equations to the case of loss in the fiber compensated by gain. In this case noise sources are introduced. The long-distance propagation of solitons in fibers whose loss is compensated by distributed gain can be analyzed in this manner. The analysis of Gordon and Haus [6] that led to the so-called "Gordon–Haus" limit of soliton propagation is consistent with the quantization detailed in this chapter.

Squeezing using rectangular pulses at the zero-dispersion wavelength can use minimum-uncertainty states as input states. We have found that solitons in a zero-point fluctuation background are not in minimum-uncertainty states. This leads to a penalty in the amount of noise reduction that can be achieved, but one that is not overly serious.

In the measurement of the squeezing we assumed a pulse shape for the local oscillator that was an ideal projector of the in-phase and quadrature components, a combination of f_1 and f_2. In practice it is more convenient to use the sech-shaped pulse of the pump. This choice of local oscillator not only deviates from the ideal pulse shape for projection of the soliton fluctuations, but also couples to the continuum. An analysis of this case has been carried out [184] which shows that the shot noise reduction is not affected seriously, and is less the larger the squeezing angle Φ is. The reason for this is that the ideal local-oscillator pulse shape itself approaches a simple sech for large squeezing angles.

The generation of squeezed states using pulse excitation is just beginning at the time this chapter is being written. It is hoped that the analysis presented here will serve to stimulate further developments in this promising field.

Problems

13.1 Derive the Heisenberg equation of motion for the Hamiltonian

$$\mathrm{i}\frac{\hbar}{2}\frac{d\omega}{d\beta}\int dx\left\{\hat{a}^\dagger(x)\left[\frac{\partial}{\partial x}\hat{a}(x)\right]-\left[\frac{\partial}{\partial x}\hat{a}^\dagger(x)\right]\hat{a}(x)\right\}.$$

13.2 The Heisenberg equation of motion (13.9) describes the effect of second-order dispersion, or so-called group velocity dispersion (GVD). Derive the equation of motion for combined second- and third-order dispersions.

13.3* Show that the simple Gaussian operator $\hat{a}(x)=\hat{A}_o(1/\sqrt{t+\mathrm{i}b})\exp\left[-\mathrm{i}kx^2/2(t+\mathrm{i}b)\right]$, with proper constraints on the parameter k, is a solution of (13.9).

13.4* Show that the linearized nonlinear Schrödinger equation (13.9) conserves the commutator bracket.

13.5 A soliton with the uncertainties (13.49) and (13.50) is incident upon a 50/50 beam splitter. The input from the other port is vacuum. Determine the uncertainties in the output ports.

13.6 A soliton is incident upon a balanced detector with the local-oscillator pulse shape i$f_L(x,t) = (1/2)[\cos\psi\, \underline{f}_X(x,t) + \sin\psi\, \underline{f}_P(x,t)]$. Determine the charge and the mean square fluctuations of the charge. Determine the major and minor axes of the uncertainty ellipse of the charge.

13.7 Show that simple vacuum fluctuations accompanying a soliton lead to excessive fluctuations of three out of four soliton parameters.

Solutions

13.3 The operator equation is linear. Therefore standard classical calculus can be applied to the operator equation:

$$\frac{\partial}{\partial t}\hat{a} = \left[-\frac{1}{2(t + ib)} + \frac{ikx^2}{2(t + ib)^2}\right]\hat{a},$$

$$\frac{\partial}{\partial x}\hat{a} = -\frac{ikx}{t + ib}\hat{a},$$

$$\frac{\partial^2}{\partial x^2}\hat{a} = \left[\left(\frac{ikx}{t + ib}\right)^2 - \frac{ik}{t + ib}\right]\hat{a}.$$

Equation (13.9) is balanced when $1/k = -d^2\omega/d\beta^2$.

13.4 The linearized NLSE is

$$\frac{\partial}{\partial t}\Delta\hat{a} = \frac{i}{2}\frac{\partial^2}{\partial x^2}\Delta\hat{a} + 2i|a_o|^2\Delta\hat{a} + ia_o^2\,\Delta\hat{a}^\dagger.$$

The time rate of change of the commutator is

$$\frac{\partial}{\partial t}[\Delta\hat{a}, \Delta\hat{a}^\dagger] = \frac{i}{2}\left[\frac{\partial^2}{\partial x^2}\Delta\hat{a}, \Delta\hat{a}^\dagger\right] - \frac{i}{2}\left[\Delta\hat{a}, \frac{\partial^2}{\partial x^2}\Delta\hat{a}^\dagger\right].$$

All other terms cancel. The conservation is to be interpreted in the integral sense. (Compare Appendix A.19.) Integration by parts leads to

$$\frac{i}{2}\left[\frac{\partial^2}{\partial x^2}\Delta\hat{a}, \Delta\hat{a}^\dagger\right] \rightarrow -\frac{i}{2}\left[\frac{\partial}{\partial x}\Delta\hat{a}, \frac{\partial}{\partial x}\Delta\hat{a}^\dagger\right],$$

and similarly for the other term. In this form they cancel.

14. Quantum Nondemolition Measurements and the "Collapse" of the Wave Function

This book has dealt extensively with the interaction of optical apparatus with electromagnetic waves. With this background it is possible to analyze quantum measurements in the domain of optics. However, the point of view represented here is not confined to optics; it is an attempt to clarify some fundamental issues of the theory of quantum measurement. The interpretation of quantum measurements has long been the subject of controversy. The von Neumann postulate [8] "that the act of measurement projects the state of the observable into an eigenstate of the measurement apparatus" is a good working hypothesis for the interpretation of a quantum measurement. However, it has been criticized by Bell [9, 185] as being an add-on to quantum theory, which describes the evolution of the states of physical observables by the Schrödinger equation. The "suddenness" of the collapse of the wave function into an eigenstate also contradicts physical intuition.

We take the following stand on the meaning of quantum mechanics and on the act of measurement.

(a) Quantum theory is fundamentally a statistical theory, in analogy with statistical mechanics, except that the probabilistic nature of the initial conditions of an evolving quantum system is a fundamental property, rather than an attribute traceable to incomplete knowledge of a system with many degrees of freedom.

(b) Quantum theory predicts the evolution of an ensemble of systems and does not predict the outcome of a single measurement (except in those special cases in which the system is in an eigenstate of the measurement apparatus).

(c) The von Neumann postulate stating that a measurement collapses the observable into an eigenstate of the measurement apparatus cannot be taken literally, since quantum theory cannot predict the outcome of a single measurement, but can only give the probabilities of measurements on an ensemble of systems.

(d) It is misleading to state that the act of measurement perturbs the quantity (observable) measured. Indeed, this statement implies that there exists a well-defined observable in the absence of the measurement. Bohr [186] always maintained that an observable can only be defined when the measurement apparatus is properly accounted for, a point of view to which we fully subscribe.

The point of view we take is that a quantum observable can be described only when the measurement apparatus itself is quantized and treated quantum mechanically. The measurement apparatus is a system with many degrees of freedom, which cause decoherence of quantum interferences, thus enabling an outcome that can be described classically. This approach will be illustrated further in the optical domain, for which quantization of the measurement apparatus is relatively simple.

An ideal photodetector measures the number of photons incident upon the detector without noise or uncertainty. In the process of measurement, the photons are converted into photoelectrons and the wavepacket carrying the photons is annihilated. A quantum nondemolition (QND) measurement [143, 187, 188], in which an observable is measured without destroying it, was first proposed for the purpose of improving the sensitivity in the detection of gravitational waves [143, 189]. The concept of QND measurements is of great help in the study of the theory of quantum measurements, because it allows repeated measurements on the same observable and thus permits the determination of the effects of the measurements on the wave function, or density matrix, of an observable.

In Sect. 14.1 we describe the QND measurement of photons in general, and show the properties of the interaction Hamiltonian of the system containing the observable and the measurement apparatus. Section 14.2 analyzes the QND measurement of photon number in a "signal" (the observable) using a nonlinear Kerr medium. We determine the range of uncertainty of the "signal" photon number, measured by a probe, and show that the greater the accuracy of the measurement, the larger the perturbation of the phase of the "signal". The product of photon number uncertainty and phase uncertainty obeys Heisenberg's uncertainty principle with an equality sign. Section 14.3 goes through the "which path" analysis of a linear interferometer with a QND measurement apparatus inserted in one of the two paths. It is found that increased knowledge of the number of photons in one of the two paths of the interferometer decreases the fringe contrast. Section 14.4 studies the evolution of the wave function of the observable and measurement apparatus of a QND measurement. It is shown that a QND apparatus that provides precise knowledge of the photon number passing through it renders diagonal the density matrix of the signal, traced over the apparatus states. In Sect. 14.5 we analyze two QND measurements in cascade. We show that the conditional probability of measuring m photons in the second measurement with n photons measured in the first approaches a Kronecker delta. This means that, for all practical purposes (to paraphrase Bell [9]), the first measurement has projected the state of the observable into an eigenstate of the measurement apparatus. This we consider to be a derivation, from quantum theory, of von Neumann's projection postulate. It must be pointed out, however, that the analysis itself does not lead directly to the collapse of the wave function. All one can conclude is that an intelligent observer of the first measurement can predict the outcome of future measurements by starting his or her calculation with the assumption that the observable is in a photon eigenstate $|n\rangle$, if n photons were measured by the first apparatus. Finally, we look at the so-called Schrödinger cat thought experiment, which has led to much controversy. We show that the cat is alive or dead with probability $1/2$ each, if the apparatus follows properly the specification in the case described.

14.1 General Properties of a QND Measurement

In a general quantum measurement, the observable $\hat{O}_s$, the value of the "signal", is inferred from a change of a probe observable $\hat{O}_p$. The probe is coupled to the signal by an interaction Hamiltonian $\hat{H}_I$. The total Hamiltonian is expressed as [190]

$$\hat{H} = \hat{H}_s + \hat{H}_p + \hat{H}_I , \tag{14.1}$$

where $\hat{H}_s$ is the Hamiltonian of the signal system and $\hat{H}_p$ is that of the probe apparatus. Heisenberg's equations of motion for $\hat{O}_s$ and $\hat{O}_p$ are

$$\frac{d}{dt}\hat{O}_s = \frac{i}{\hbar}\{[\hat{H}_s, \hat{O}_s] + [\hat{H}_I, \hat{O}_s]\} \tag{14.2}$$

and

$$\frac{d}{dt}\hat{O}_p = \frac{i}{\hbar}\{[\hat{H}_p, \hat{O}_p] + [\hat{H}_I, \hat{O}_p]\} . \tag{14.3}$$

The first commutators in (14.2) and (14.3) describe the free motion of the operators, whereas the second commutators express the coupling between the observable (signal) and probe that is designed to measure the observable. In order that the probe can perform a measurement, the commutator $[\hat{H}_I, \hat{O}_p]$ must be nonzero, and $\hat{H}_I$ must be a function of $\hat{O}_s$. This does not necessarily imply that the measurement must affect the observable. Indeed, the commutator $[\hat{H}_I, \hat{O}_s]$ can be zero. In fact, the commutator must be zero if the measurement is to be of the "nondemolition" kind. But since $\hat{H}_I$ is a function of $\hat{O}_s$, it cannot commute with the observables conjugate to $\hat{O}_s$. Hence a measurement *must* perturb the variable conjugate to $\hat{O}_s$.

14.2 A QND Measurement of Photon Number

A QND measurement of photon number can be accomplished by measuring the phase shift induced in a probe beam by a signal beam, both of which pass through a nonlinear Kerr medium (see Fig. 14.1) [191, 192]. The two beams are of different carrier frequencies and are separated via dichroic mirrors that are reflecting for the signal beam and transmitting for the probe beam. The probe beam is derived from a source via a beam splitter with reflectivity σ and transmissivity $\sqrt{1-\sigma^2}$. The intent is to make $\sigma \ll 1$, so that the beam E is powerful and acts as the local oscillator for detection of the probe beam F. In the balanced detector, which forms the output of the interferometer, the beam splitter is a 50/50 one. The probe source is assumed to be in a coherent state. According to the analysis of Sect. 7.5, beam F is in the coherent state $|i\sigma\alpha\rangle$ and beam E in the coherent state $|\sqrt{1-\sigma^2}\alpha\rangle$.

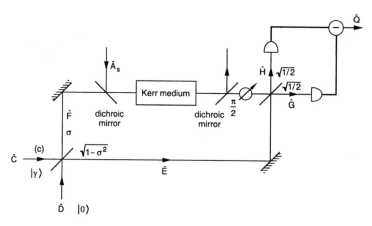

Fig. 14.1. A QND measurement of the photon number of a signal $\hat{A}_s$

The general operation of the measurement apparatus can be grasped without a detailed analysis. The signal changes the phase of the probe. The energy in the signal (pulse) does not change, since the Kerr medium is lossless and thus conserves energy. The signal produces a probe beam imbalance which is measured by the balanced detector. From the imbalance, the amount of energy in the signal (i.e. the signal photon number) can be inferred. The measurement perturbs the conjugate observable of the signal photon number, namely the phase, via the unavoidable fluctuations in the amplitude of the probe, which cause index fluctuations in the Kerr medium. Thus, one would expect that knowledge of the signal photon number results in increased fluctuations of the phase of the signal.

We have discussed extensively the Kerr nonlinearity, both in its classical description in Chap. 10 and in its quantum reformulation in Chap. 12. The Kerr effect produces self-phase modulation of each mode and cross-phase modulation between the two modes. In an ideal quantum nondemolition measurement, the role of the apparatus is to couple the "signal" observable to a "probe" observable without affecting the "signal" observable itself. In order to accomplish such a measurement, the Kerr effect has to be one with a resonant response in which the index is modulated only when the beat frequency between the two modes lies within the bandwidth of the resonance of the medium. Thus, if mode A has the resonance frequency ω_a, and mode F the resonance frequency ω_f, then the index varies at the frequency $\omega_a - \omega_f$, which is assumed to be the resonance frequency of the medium. We describe the mode amplitudes by annihilation operators assigned to single modes over lengths L or time intervals $T = L/v_g$ (see Chap. 6). The Hamiltonian for a Kerr medium that responds resonantly to optical beats of frequency $\omega_a - \omega_f$ is

$$\hat{H} = \hbar K \hat{A}_s^\dagger \hat{A}_s \hat{F}^\dagger \hat{F} \ . \tag{14.4}$$

Of course, in order to avoid losses, the excitation frequency must lie in the wings of the medium resonance, the transitions in the Kerr medium must be virtual.

Equation (14.4) does not address the issue of the response time of the Kerr medium. On the face of it, the response appears instantaneous. In Chap. 11, in connection with the generation of Schrödinger cat states via propagation in a fiber, we have pointed out the conceptual difficulties associated with a Kerr medium with an instantaneous response; the zero-point fluctuations of all frequencies are coupled nonlinearly, provoking a kind of ultraviolet catastrophe. The investigations in this chapter, starting with the Hamiltonian (14.4), do not encounter the same difficulty. The response is resonant, and thus is spectrally limited. The signal mode $\hat{A}_s$ and the probe mode $\hat{F}$ may be considered to be modes of a resonator. The interaction is a resonant interaction via these modes and involves only the spectra of these individual modes. The only constraint is that the rate of change of these modes is slow compared with the response rate of the Kerr medium. The equation of motion for the operator $\hat{A}_s$ is

$$\frac{d}{dt} \hat{A}_s = iK \hat{F}^\dagger \hat{F} \hat{A}_s \ . \tag{14.5}$$

Similarly, the equation of motion for mode F is

$$\frac{d}{dt} \hat{F} = iK \hat{A}_s^\dagger \hat{A}_s \hat{F} \ . \tag{14.6}$$

Note that both $\hat{A}_s^\dagger \hat{A}_s$ and $\hat{F}^\dagger \hat{F}$ are constants of motion. Thus, they may be evaluated at the input to the Kerr medium, at the time $t = 0$. Integration of the two equations of motion over the interaction time T_K in the Kerr medium gives

$$\hat{A}_s(T_K) = e^{iK \hat{F}^\dagger \hat{F} T_K} A_s(0) \ , \tag{14.7}$$

$$\hat{F}(T_K) = e^{\mathrm{i}K\hat{A}_s^\dagger\hat{A}_s T_K}\hat{F}(0) \ . \tag{14.8}$$

The two operators have been phase-shifted by the operator phases $\hat{\Phi}_A = KT_K\hat{F}^\dagger\hat{F}$ and $\hat{\Phi}_F = KT_K\hat{A}_s^\dagger\hat{A}_s$, respectively. The probe interferometer in Fig. 14.1 contains a 90° phase shifter. The scattering matrices of the two beam splitters have been chosen as

$$\begin{bmatrix} \sqrt{1-\sigma^2} & \mathrm{i}\sigma \\ \mathrm{i}\sigma & \sqrt{1-\sigma^2} \end{bmatrix} \quad \text{and} \quad \frac{1}{\sqrt{2}}\begin{bmatrix} 1 & \mathrm{i} \\ \mathrm{i} & 1 \end{bmatrix} \ .$$

The beams G and H at the output of the second beam splitter, with a 50/50 splitting ratio, are

$$\hat{G} = \frac{\mathrm{i}}{\sqrt{2}}(e^{\mathrm{i}\hat{\Phi}_F}\hat{F} + \hat{E}) \tag{14.9}$$

and

$$\hat{H} = \frac{1}{\sqrt{2}}(-e^{\mathrm{i}\hat{\Phi}_F}\hat{F} + \hat{E}) \ . \tag{14.10}$$

We ignore the identical phase shifts in the two arms of the interferometer since they cancel upon detection. Next we look at the detector charge of the experimental arrangement of Fig. 14.1. The phase shift is measured by detecting the phase shifted beam in a balanced detector, with the beam E acting as the local oscillator. The charge operator of the balanced detector is then

$$\hat{Q} = q(\hat{H}^\dagger\hat{H} - \hat{G}^\dagger\hat{G}) = -q(\hat{F}^\dagger\hat{E}e^{-\mathrm{i}\hat{\Phi}_F} + e^{\mathrm{i}\hat{\Phi}_F}\hat{E}^\dagger\hat{F}) \ . \tag{14.11}$$

Since the beams E and F are in the coherent states $|\sqrt{1-\sigma^2}\gamma\rangle$ and $|\mathrm{i}\sigma\gamma\rangle$, respectively, the expectation value of the charge operator obtained by tracing over the measurement apparatus is

$$\langle\hat{Q}\rangle_M = 2q\sigma\sqrt{1-\sigma^2}|\gamma|^2\langle\sin(\hat{\Phi}_F)\rangle \ . \tag{14.12}$$

If the signal is in a photon eigenstate, $\hat{\Phi}_F$ is fluctuation-free. If the signal is in any other state, there are fluctuations of $\hat{\Phi}_F$ that cause fluctuations of the detector charge (see analysis in next section). Here we shall ignore these fluctuations, assuming that either the signal is in a photon state or that the signal beam and its coupling to the probe beam are very weak compared with the local-oscillator beam.

The mean square fluctuations of the charge are

$$\langle\hat{Q}^2\rangle - \langle\hat{Q}\rangle^2 = q^2|\gamma|^2 \ . \tag{14.13}$$

This result is obtained by casting the square of the operator $\hat{Q}$ into normal order. When this is done, there result terms of fourth order in $\hat{E}$ and $\hat{F}$ and their Hermitian conjugates, and owing to the use of the commutation relations there appear terms of second order, $\hat{E}^\dagger\hat{E}$ and $\hat{F}^\dagger\hat{F}$. Using the properties of creation and annihilation operators, all terms of fourth order cancel against $-\langle\hat{Q}\rangle^2$. The mean square fluctuations are due to the second-order terms. If the fluctuations of the phase $\hat{\Phi}_F$ are ignored, the difference current has no induced fluctuations due to $\hat{\Phi}_F$ and one finds pure shot noise associated with the sum of the two detector currents, $q^2|\gamma|^2$.

Equation (14.12) relates the trace of the charge operator over the Hilbert space of the measurement apparatus to a function of the photon number operator of the signal pulse. If the measurement system has large amplification, i.e. in the case when $\sigma\sqrt{1-\sigma^2}|\gamma|^2 \gg 1$, the charge can be observed on a macroscopic scale. Furthermore, if the interaction between the signal and the measurement apparatus is weak, i.e. $KT_K \ll 1$, we obtain a linear scale for the photon number operator. Therefore, if the signal has photon states only up to a number N such that $KT_K N \ll 1$, then we can linearize (14.12) and introduce the abbreviation $KT_K \equiv \kappa$:

$$\langle\hat{Q}\rangle_M = 2q\sigma\sqrt{1-\sigma^2}|\gamma|^2|\hat{\Phi}_F = 2q\sigma\sqrt{1-\sigma^2}|\gamma|^2|\kappa\hat{A}_s^\dagger\hat{A}_s . \tag{14.14}$$

The deviation Δn_A of the signal photon number from a nominal number n_A can be measured only if the fluctuations of the charge are smaller than, or at most equal to, the change in the charge caused by Δn_A. From (14.14) we obtain

$$4q^2\sigma^2(1-\sigma^2)|\gamma|^4\langle|\Delta\Phi_F|^2\rangle = 4q^2\sigma^2(1-\sigma^2)|\gamma|^4\kappa^2\langle|\Delta n_A|^2\rangle \geq q^2|\gamma|^2 . \tag{14.15}$$

Thus, the smallest change of photon number that can be measured is given by

$$\langle|\Delta n_A|^2\rangle \geq \frac{1}{4\sigma^2(1-\sigma^2)|\gamma|^2\kappa^2} . \tag{14.16}$$

The resolution of the measurement of photon number is the finer the greater the photon number $|\gamma|^2$ of the probe beam. An increase in the probe beam photon number is accompanied by a cost. The larger the photon number in the probe beam, the greater the fluctuations of the probe photon number n_F, and hence the greater the perturbation of the phase of the signal beam.

The photon number fluctuations in the probe induce phase fluctuations in the signal with a mean square value

$$\langle|\Delta\Phi_A|^2\rangle = \kappa^2(\langle n_F^2\rangle - \langle n_F\rangle^2) = \kappa^2\langle n_F\rangle = \kappa^2|\sigma\gamma|^2 . \tag{14.17}$$

The product of the mean square spread in the measured photon number and the mean square phase fluctuations induced in the signal is obtained from (14.15) and (14.15):

$$\langle|\Delta n_A|^2\rangle\langle|\Delta\Phi_A|^2\rangle \geq \frac{1}{4(1-\sigma^2)} \; . \tag{14.18}$$

We find that the mean square deviation of the measured photon number times the mean square phase fluctuations induced by the measurement obeys the Heisenberg uncertainty relation with an equality sign when $\sigma \ll 1$ (see Appendix A.8). Note that the measurement need not yield the photon number with precision. The mean square spread of the measured photon number decreases with increasing probe beam intensity. An accurate measurement of photon number is only possible for sufficiently high probe beam intensities that the root mean square noise fluctuations are much smaller than the difference between the signal registered for n_A and $n_A \pm 1$ photons. Let us look at this issue more carefully. We prepare the signal so that it possesses a definite photon number n_A. Then, the charge registered by the apparatus is $q\eta n_A$, where η is the sensitivity per photon

$$\eta = 2\sigma\sqrt{1-\sigma^2}|\gamma^2|\kappa \; . \tag{14.19}$$

For a given signal photon state the detector charge number deviates from the mean value (14.14) from measurement to measurement with an r.m.s. deviation $\delta = |\gamma|^2$. The probability is approximately Gaussian-distributed when the probe beam is intense. If the observed charge lies in the interval

$$\langle\hat{Q}\rangle_M \in q\left[\eta\left(n-\frac{1}{2}\right), \quad \eta\left(n+\frac{1}{2}\right)\right]$$

and we decide that the measured photon number is n, this decision has the error probability

$$P_{\text{error}} \approx 2\int_{\eta/2}^{\infty} \frac{1}{\sqrt{2\pi\delta}} e^{-x^2/2\delta} dx$$

$$= \text{erfc}\left(\frac{\eta}{2\sqrt{2}|\gamma|}\right) \approx 2\sqrt{\frac{2}{\pi}}\frac{|\gamma|}{\eta}e^{-\eta^2/8|\gamma|^2} \tag{14.20}$$

for

$$\frac{\eta}{|\gamma|} = 2\sigma\sqrt{1-\sigma^2}\kappa|\gamma| \gg 1 \; .$$

The conditional probability that the charge is contained in the above interval when a state of photon number m has been sent is

$$P(n|m) = \int_{-\eta/2}^{\eta/2} \frac{1}{\sqrt{2\pi\delta}} e^{-[x-\eta(m-n)]^2/2\delta^2} dx$$

$$\approx \frac{\eta}{\sqrt{2\pi\delta}} e^{-[\eta(m-n)]^2/2\delta^2} \tag{14.21}$$

$$= \frac{\eta}{2\sqrt{2\pi}|\gamma|} e^{-[\eta(m-n)]^2/2|\gamma|^2} \quad \text{for} \quad \frac{\eta}{|\gamma|} \gg 1 \; .$$

In order to be able to resolve the photon number of the signal reliably, the sensitivity per photon must be much larger than the variance of the noise, i.e. $\eta \gg |\gamma|$. For a value of $\eta/|\gamma| = 10$ we get already an error probability as low as 10^{-6}.

14.3 "Which Path" Experiment

In a Mach–Zehnder interferometer, particles behave like waves and interfere if no knowledge has been acquired about the path taken in the interferometer. If one arm of the interferometer is blocked by a detector, the fringes disappear. A QND measurement in one arm of the interferometer provides knowledge about the path taken. The virtue of the QND measurement described in the preceding section is that its sensitivity can be varied from zero (zero-intensity probe) to perfect sensitivity (the probe is intense enough to distinguish single photons). Hence, one may study the contrast of the fringes as a function of the amount of knowledge about the path taken by the photon(s). In this way, the QND measurement permits a variety of choices with regard to the amount of knowledge gained as to which path the photons have taken. This should be contrasted with the either/or situation in which either a measurement is made with a photodetector of the photons in one arm or no knowledge is acquired at all. Figure 14.2 shows the setup. The Mach–Zehnder interferometer contains a phase shifter in one arm and a QND measurement apparatus in the other arm. It is excited by a coherent state in input (a) and vacuum in input (b). The difference current of the two detectors illuminated by the outputs E and F is measured. The difference current is recorded as a function of the phase shift θ. Of interest is the change of the fringe contrast of the difference current with increasing intensity of the probe G. The excitations in the two arms, C and D, are

$$\hat{C} = \frac{1}{\sqrt{2}}(\hat{A} + i\hat{B}) , \tag{14.22}$$

$$\hat{D} = \frac{1}{\sqrt{2}}(i\hat{A} + \hat{B}) . \tag{14.23}$$

The excitations C and D are phase-shifted and then combined at the output beam splitter into the outputs E and F:

$$\hat{E} = \frac{1}{\sqrt{2}}(e^{i\theta}\hat{D} + ie^{i\hat{\Phi}_G}\hat{C}) , \tag{14.24}$$

$$\hat{F} = \frac{1}{\sqrt{2}}(ie^{i\theta}\hat{D} + e^{i\hat{\Phi}_G}\hat{C}) , \tag{14.25}$$

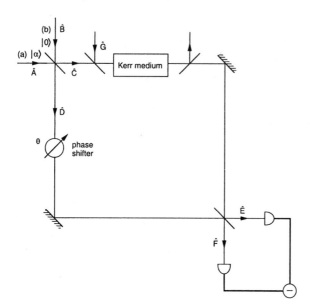

Fig. 14.2. Interferometer with photon measurement in one arm

where the induced phase shift $\hat{\Phi}_G$ is given by $\hat{\Phi}_G = \kappa \hat{n}_G$. The difference charge is

$$\hat{Q} = q(\hat{E}^\dagger \hat{E} - \hat{F}^\dagger \hat{F}) = qi \left(e^{-i(\theta - \hat{\Phi}_G)} \hat{D}^\dagger \hat{C} - e^{i(\theta - \hat{\Phi}_G)} \hat{C}^\dagger \hat{D} \right) . \tag{14.26}$$

In order to evaluate the expectation value of the charge, we need the expectation value of the function $\exp(i\hat{\Phi}_G)$. The probe G is assumed to be in a coherent state with a Poissonian probability distribution of photons:

$$\langle e^{i\hat{\Phi}_G} \rangle = \sum_n e^{-\langle n_G \rangle} \frac{\langle n_G \rangle^n}{n!} e^{i\kappa n}$$

$$= e^{-\langle n_G \rangle} \sum_n \frac{1}{n!} \left(\langle n_G \rangle e^{i\kappa} \right)^n \tag{14.27}$$

$$= e^{-\langle n_G \rangle \{1 - [\cos(\kappa) + i \sin(\kappa)]\}} .$$

When this expression is introduced into the expectation value of the charge, we find:

$$\langle \hat{Q} \rangle = q e^{-\langle n_G \rangle [1 - \cos(\kappa)]} |\alpha|^2 \sin[\theta + \langle n_G \rangle \sin(\kappa)] . \tag{14.28}$$

Figure 14.3a shows the fringe contrast as a function of the Kerr nonlinearity $\kappa = KT_K$ for a fixed average probe photon number $\langle n_G \rangle = 1$. From the

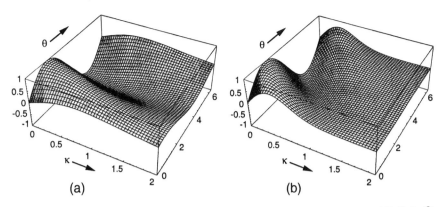

Fig. 14.3. Fringe contrast as a function of κ and θ. The ordinate is $\langle Q \rangle / \langle q | \alpha |^2 \rangle$. (a) $\langle n_G \rangle = 1$; (b) $\langle n_G \rangle = 3$

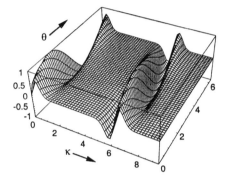

Fig. 14.4. Fringe contrast for $\langle n_G \rangle = 3$ and a larger range of κ

analysis of the preceding section we know that single photons can be distinguished when $2\kappa\sqrt{\langle n_G \rangle} \gg 1$. When this becomes possible, the photons behave as particles and thus the fringe pattern disappears when $\kappa \gg 0.5$, as can be seen from the figure.

An interesting property of (14.28) does not show up in these two graphs. Figure 14.4 is plotted for a larger range of κ values. This graph shows a reappearance of the fringes. The reason for this is not hard to find. When $\kappa = 2\pi$, every signal photon shifts the interferometer phase by a multiple of 2π, so that no information as to the number of the signal photons can be obtained. Under this condition, the beat pattern reappears. It is of interest to determine the fluctuations of the measurement variable Q. We have

$$\langle \hat{Q}^2 \rangle - \langle \hat{Q} \rangle^2$$

$$= q^2 \langle \hat{C}^\dagger \hat{C} \hat{D} \hat{D}^\dagger + \hat{D}^\dagger \hat{D} \hat{C} \hat{C}^\dagger - e^{2i(\theta - \hat{\Phi}_G)} \hat{C}^{\dagger 2} \hat{D}^2 - e^{-2i(\theta - \hat{\Phi}_G)} \hat{C}^2 \hat{D}^{\dagger 2} \rangle$$

$$+ q^2 \langle e^{i(\theta - \hat{\Phi}_G)} \hat{C}^\dagger \hat{D} - e^{-i(\theta - \hat{\Phi}_G)} \hat{C} \hat{D}^\dagger \rangle^2$$

$$= q^2 \left[|\alpha|^2 + \frac{1}{4} |\alpha|^4 e^{2i\theta} (\langle e^{-2i\hat{\Phi}_G} \rangle - \langle e^{-i\hat{\Phi}_G} \rangle^2) \right.$$

$$\left. + \frac{1}{4} |\alpha|^4 e^{-2i\theta} \left(\langle e^{2i\hat{\Phi}_G} \rangle - \langle e^{i\hat{\Phi}_G} \rangle^2 \right) \right] ,$$

$$(14.29)$$

where we have set $\arg(\alpha) = 0$. The noise is shown in Fig. 14.5. The noise is shot noise due to the signal beam, except near the phase shifts of $\theta = 0$ and 2π. At these angles, the probe contributes significantly to the noise by interfering with the signal.

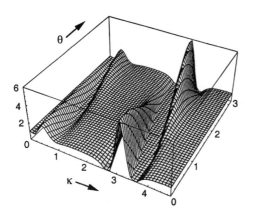

Fig. 14.5. The mean square fluctuations of the charge as a function of κ and θ; $\langle n_G \rangle = 5$. The fluctuations are normalized to $q^2 |\alpha|^2$

14.4 The "Collapse" of the Density Matrix

In Sect. 14.2 we used the Heisenberg representation and derived the charge operator for a QND measurement of photon number. From the statistical distribution of the charge we were able to infer the choice of parameters necessary to measure the photon number of a signal beam. In this section we study the evolution of the wave function [191,192]. The system wave function starts out in a product state of the signal, of the probe in a coherent state at port (c), and of a vacuum state entering through port (d):

$$|\psi\rangle_I = |\psi\rangle_s \otimes |\gamma\rangle_c \otimes |0\rangle_d \ . \tag{14.30}$$

In the Heisenberg representation, the operators evolve as one proceeds through the system. The analogy with the classical propagation of wavepackets through the system is unmistakable. At any cross section, the different beams are identified by the corresponding operators. In the specific case of Fig. 14.1, we assigned different letters to the operators at different cross sections. In the Schrödinger representation, the operators remain fixed, whereas the wave functions evolve with time. Passage through any element of the system changes the wave function. In order to clarify the notation used in this section, let us look in more detail at the passage of the wave function through the first beam splitter, the passage from reference cross section I to reference cross section II (see Fig 14.6). The product state (14.30) remains a product state, since $|\psi_s\rangle$ is unaffected, and a beam splitter preserves product states of coherent states. After the beam splitter, the wave function is

$$|\psi\rangle_{II} = |\psi\rangle_s \otimes |\sqrt{1-\sigma^2}\gamma\rangle_c \otimes |i\sigma\gamma\rangle_d \ . \tag{14.31}$$

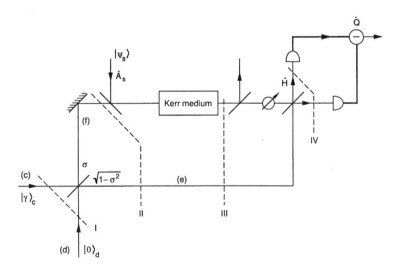

Fig. 14.6. Schematic of QND measurement for the Schrödinger representation

The coherent state $|\sqrt{1-\sigma^2}\gamma\rangle$ is assigned to arm (e) of the interferometer, according to the definition of the beam splitter ratio. Note that we retain the subscripts "c" and "d" to indicate the port at which the wave function was defined originally. Since the operators remain unchanged in the Schrödinger representation, the subscripts indicate which operators operate on which wave function. Thus, if we asked for the photon number in arm (e) of Fig. 14.6, we would form the following expectation value:

$$_{\mathrm{II}}\langle\psi|\hat{C}^\dagger\hat{C}|\psi\rangle_{\mathrm{II}} = {}_c\langle\sqrt{1-\sigma^2}\gamma|\hat{C}^\dagger\hat{C}|\sqrt{1-\sigma^2}\gamma\rangle_c = (1-\sigma^2)|\gamma|^2 . \quad (14.32)$$

This example illustrates the notation used as we proceed through the nonlinear interferometer. After passage of the signal and probe beam (d) through the Kerr medium the state is

$$|\psi\rangle_{\mathrm{III}} = e^{i\kappa\hat{A}_s^\dagger\hat{A}_s\hat{D}^\dagger\hat{D}}|\psi\rangle_s \otimes |\sqrt{1-\sigma^2}\gamma\rangle_c \otimes |i\sigma\gamma\rangle_d , \quad (14.33)$$

where the operator $\exp(i\kappa\hat{A}_s^\dagger\hat{A}_s\hat{D}^\dagger\hat{D})$ represents the interaction in the Kerr medium. We now write the signal state in the photon number representation:

$$|\psi\rangle_s = \sum_n c_n|n\rangle_s . \quad (14.34)$$

Action upon $|\psi_s\rangle$ by the operator $\exp(i\kappa\hat{A}_s^\dagger\hat{A}_s\hat{D}^\dagger\hat{D})$ produces

$$e^{i\kappa\hat{A}_s^\dagger\hat{A}_s\hat{D}^\dagger\hat{D}}|\psi\rangle_s = \sum_n c_n|n\rangle_s e^{i\kappa n_s\hat{D}^\dagger\hat{D}} . \quad (14.35)$$

Next, consider the operation of $e^{i\kappa n_s\hat{D}^\dagger\hat{D}}$ on the coherent state $|i\sigma\gamma\rangle_d \equiv |\delta\rangle_d$. We express $|\delta\rangle_d$ in the photon number representation and operate on it with the operator $\exp(i\kappa n_s\hat{D}^\dagger\hat{D})$:

$$e^{i\kappa n_s\hat{D}^\dagger\hat{D}}\sum_{n_d}e^{-|\delta|^2/2}\frac{\delta^{n_d}}{\sqrt{n_d!}}|n\rangle_d = \sum_{n_d}e^{i\kappa n_s n_d}e^{-|\delta|^2/2}\frac{\delta^{n_d}}{\sqrt{n_d!}}|n\rangle_d$$

$$= \sum_{n_d}e^{-|\delta|^2/2}\frac{(\delta e^{i\kappa n_s})^{n_d}}{\sqrt{n_d!}}|n_d\rangle = |\delta e^{i\kappa n_s}\rangle_d . \quad (14.36)$$

The coherent wave function has been multiplied by $\exp(i\kappa n_s)$. When this result is introduced into (14.33) we find

$$|\psi\rangle_{\mathrm{III}} = \sum_n c_n|n\rangle_s \otimes |\sqrt{1-\sigma^2}\gamma\rangle_c \otimes |i\sigma e^{i\kappa n_s}\gamma\rangle_d . \quad (14.37)$$

Every component of the signal wave function written in terms of photon number states has one associated coherent state that has been phase-shifted by the phase κn_s. Finally, we phase-shift beam (d) by 90° and propagate the wave functions through the output beam splitter with a 50/50 splitting ratio to obtain

$$|\psi\rangle_{\mathrm{IV}} = \sum_n c_n |n\rangle_s \otimes \left| \frac{1}{\sqrt{2}} (\sqrt{1 - \sigma^2} - i\sigma e^{i\kappa n_s})\gamma \right\rangle_c$$

$$\otimes \left| \frac{1}{\sqrt{2}} (i\sqrt{1 - \sigma^2} - \sigma e^{i\kappa n_s})\gamma \right\rangle_d .$$

(14.38)

With the signal in a photon state, $c_m = 1$ and $c_n = 0$ for $m \neq n$, and the system is in the simple product state

$$|\psi\rangle_{\mathrm{IV}} = c_m |m\rangle_s \otimes \left| \frac{1}{\sqrt{2}} (\sqrt{1 - \sigma^2} - i\sigma e^{i\kappa m})\gamma \right\rangle_c$$

$$\otimes \left| \frac{1}{\sqrt{2}} (i\sqrt{1 - \sigma^2} - \sigma e^{i\kappa m})\gamma \right\rangle_d .$$

(14.39)

In the general case (14.38), however, the wave function is in an entangled state; the wavefunction is a sum over product states. Note that each of the signal photon number states is associated with a pair of characteristic coherent states involving phase shifts proportional to the photon number, the factors $e^{i\kappa n_s}$. If the probe intensity $|\gamma|^2$ is large enough, these coherent states do not overlap in the γ plane. A measurement of the probe by the balanced detector can resolve individual photons. The overlap of the probe wave functions is expressed conveniently in terms of the density matrix. The density matrix is (Appendix A.10)

$$\rho = |\psi\rangle_{\mathrm{IV}\ \mathrm{IV}}\langle\psi| = \sum_{m,n} c_m c_n^* |m\rangle_s\ {}_s\langle n|$$

$$\otimes \left| \frac{1}{\sqrt{2}} (\sqrt{1 - \sigma^2} - i\sigma e^{i\kappa m})\gamma \right\rangle_c\ {}_c\left\langle \frac{1}{\sqrt{2}} (\sqrt{1 - \sigma^2} - i\sigma e^{i\kappa n})\gamma \right|$$

$$\otimes \left| \frac{1}{\sqrt{2}} (i\sqrt{1 - \sigma^2} - \sigma e^{i\kappa m})\gamma \right\rangle_d\ {}_d\left\langle \frac{1}{\sqrt{2}} (i\sqrt{1 - \sigma^2} - \sigma e^{i\kappa n})\gamma \right| .$$

(14.40)

It is of interest to ask for the reduced density matrix obtained from ρ by tracing it over the probe beams that are to be detected by the two detectors. We have

$$\rho_R = \mathrm{Tr}_{c,d}(\rho) = \sum_{m,n} c_m c_n^* |m\rangle_s\ {}_s\langle n| R_{m,n,c} R_{m,n,d} ,$$

(14.41)

where $R_{m,n,c}$ and $R_{m,n,d}$ are the reduction factors

$$R_{m,n,c} \equiv {}_c\left\langle \frac{1}{\sqrt{2}} (\sqrt{1 - \sigma^2} - i\sigma e^{i\kappa n})\gamma \left| \frac{1}{\sqrt{2}} (\sqrt{1 - \sigma^2} - i\sigma e^{i\kappa m})\gamma \right\rangle_c \right. ,$$

$$R_{m,n,d} \equiv {}_d\left\langle \frac{1}{\sqrt{2}}(i\sqrt{1-\sigma^2} - \sigma e^{i\kappa n})\gamma \middle| \frac{1}{\sqrt{2}}(i\sqrt{1-\sigma^2} - \sigma e^{i\kappa m})\gamma \right\rangle_d .$$

Next, we show that the reduction factors approach Kronecker deltas when the probe intensity is made large. The reduction factors involve the scalar product of two coherent states. Now, it is easy to show that

$$\langle \alpha | \beta \rangle = \exp -\frac{1}{2}(|\alpha|^2 + |\beta|^2 - 2\alpha^*\beta) . \tag{14.42}$$

Indeed,

$$\langle \alpha | \beta \rangle = \sum_n e^{-|\alpha|^2/2} \frac{\alpha^{*n}}{\sqrt{n!}} \langle n| \sum_m e^{-|\beta|^2/2} \frac{\beta^m}{\sqrt{m!}} |m\rangle$$

$$= e^{-(|\alpha|^2+|\beta|^2)/2} \sum_n \frac{(\alpha^*\beta)}{n!} \tag{14.43}$$

$$= e^{-(|\alpha|^2+|\beta|^2-2\alpha^*\beta)/2} .$$

Thus we obtain for the reduction factors

$$R_{m,n,c} = R_{m,n,d} = \exp -\frac{\sigma^2}{2}|\gamma|^2[1 - \cos\kappa(m-n)]e^{i\phi_{mn}} \tag{14.44}$$

with

$$\phi_{mn} = \frac{|\gamma|^2}{2}\left[\sigma\sqrt{1-\sigma^2}(\cos\kappa n - \cos\kappa m) + \sigma^2 \sin\kappa(m-n)\right] . \tag{14.45}$$

For $|\kappa(m-n)| \ll 1$ we may expand $R_{m,n,c}$ and $R_{m,n,d}$:

$$|R_{m,n,c}| = |R_{m,n,d}| \propto \exp -\frac{|\sigma\gamma|^2}{4}\kappa^2(m-n)^2 , \tag{14.46}$$

which can be made to approach a Kronecker delta when $|\kappa\sigma\gamma/2| \gg 1$ (compare with (14.20)). In this limit, the reduced density matrix becomes diagonal:

$$\rho_R = \sum_n |c_n|^2 |n\rangle_s {}_s\langle n| . \tag{14.47}$$

A trace over the probe takes an average over all probe measurements. The state of the signal as inferred from the probe measurements is represented by a diagonal density matrix that can be interpreted as a superposition of signal number states occurring with probabilities $p(n) = |c_n|^2$.

We have seen in Sect. 14.2 that single photons can be distinguished when the sensitivity per photon, η, is much greater than the amplitude of the probe:

$$\sigma\sqrt{1-\sigma^2}\kappa|\gamma|^2 \gg |\gamma| .$$

It follows that an accurate measurement of the photon number collapses the reduced density matrix into diagonal form. Further, we observe that the collapse of the density matrix is not sudden; it requires the time necessary for the QND measurement to take place. Finally, it should be noted that the collapse is not truly total. Indeed, if we did not carry out the measurement and did not trace over the measurement apparatus, we could propagate the density matrix through another nonlinear Mach–Zehnder interferometer with a Hamiltonian that was the inverse of the Hamiltonian of the first Mach–Zehnder interferometer, as shown in Fig. 14.7. The second Mach–Zehnder interferometer has a Kerr medium of negative Kerr coefficient, equal in magnitude to that of the first. After the passage through both interferometers, the original input state is recovered. This shows that the full information of the input state is still contained in the density matrix of the signal and probe, even though the reduced density matrix appears diagonal.

This finding deserves more attention. In the setup of Fig. 14.7 the probe is not detected. The balanced detector has its own Hamiltonian evolution, in which the photons generate carriers. Could one recover the original state by transforming the density matrix through a Hamiltonian inverse to that of the balanced detector? A Hamiltonian inverse to that of the detector would take the carriers generated by the photons and use them to emit coherent

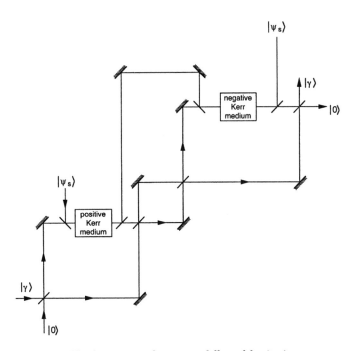

Fig. 14.7. Nonlinear interferometer followed by its inverse

light. Such a system boggles the imagination. The reason that this appears so difficult to realize is because detection is an irreversible process. But the objection could still be made that quantum mechanics is reversible at the truly basic level. In principle one ought to be able to construct a Hamiltonian inverse to that of the detector. Here we touch on an issue that is analogous to the discovery of Poincaré cycles in statistical mechanics. Boltzmann's H theorem, which served as the underpinning of the entropy increase in statistical mechanics, was put into question by the existence of Poincaré cycles, i.e. by the proof that a system can and will return to its initial state arbitrarily closely if left to evolve long enough. It turned out that systems of any complexity take astronomic times for the completion of a cycle. So, the Poincaré cycles seem to be an artifact that has no bearing on the evolution of a system over reasonable lengths of time. The situation in quantum mechanics seems to be analogous. An inverse Hamiltonian may in fact exist for any physical system. But if the system is of any complexity, the evolution via the inverse Hamiltonian will take times that are too long to affect predictions for the foreseeable future.

In the next section we shall consider two QND measurements of photons in cascade and shall derive the conditional probability of measuring m photons in the second apparatus when n photons have been measured by the first. The conditional probability will enable us to make further inferences on the effect of an individual measurement event.

14.5 Two Quantum Nondemolition Measurements in Cascade

In his mathematical formulation of quantum mechanics, von Neumann introduced into quantum mechanics a discontinuous evolution of the Schrödinger wave function [8] with his projection postulate. This states that the measurement process projects the measured state of the observable into an eigenstate of the measurement equipment. However, this postulate raises some fundamental questions. How does this sudden projection take place? Further, the postulate assigns meaning to the outcome of a single measurement, whereas quantum theory, in the statistical interpretation of Max Born, predicts probabilities of outcomes, and not the outcome of a single measurement.

In Sect. 14.4 we showed that a measurement puts the density matrix traced over the measurement apparatus into diagonal form, thus permitting a classical probabilistic interpretation of the outcome of the measurement. The derivation also made it clear that the process is a continuous one evolving during the process of measurement, not a sudden "collapse" into diagonal form. No observer need be present; the diagonalization is caused by the measurement apparatus. In this section we broach the question as to whether meaning can be attached to a single measurement event in the spirit of von Neumann's postulate, in contrast to Born's probabilistic interpretation of quantum mechanics as a predictor of the probability of the outcomes of measurements on an ensemble of identically prepared systems. We answer this question by studying two QND measurements in cascade. As before, we quantize the measurement apparatus. We evaluate the conditional probability of measuring n photons in the second measurement if m photons have been measured in the first. We shall be able to show that this conditional probability approaches a Kronecker delta δ_{nm}. With this result we shall accomplish a kind of "proof" of the von Neumann postulate. We shall not, in fact, show that the measurement projects the state of the observable into an eigenstate. Quantum theory, within its statistical interpretation, is not equipped to arrive at such a conclusion. However, by showing that the conditional probability is a Kronecker delta, we have proven that a measurement of n in the first apparatus is followed with certainty by a measurement of n in the second apparatus. For all practical purposes (to paraphrase Bell), *an intelligent observer of the first measurement can set up his or her calculations predicting the outcome of a further experiment by assuming that the state of the observable is in the eigenstate n after the first measurement apparatus.*

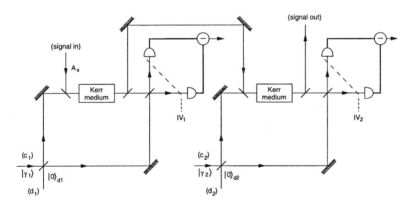

Fig. 14.8. Two QND measurements in cascade

By a method analogous to the analysis of the preceding section, we may derive the state of the entire system by following the input product wave function through the system (see Fig. 14.8). We assume that the two QND measurements are identical and that the probe beams are in coherent states $|\gamma_1\rangle$ and $|\gamma_2\rangle$, respectively. The initial wave function is the product state

$$|\psi\rangle_I = |\psi\rangle_s |\gamma_1\rangle_{c1} \otimes |0\rangle_{d1} \otimes |\gamma_2\rangle_{c2} \otimes |0\rangle_{d2} . \tag{14.48}$$

After passage through both systems, the wave function is

$$|\psi\rangle_{\mathrm{IV}_2} = \sum_{m,n} c_n |n\rangle_s \otimes \left| \frac{1}{\sqrt{2}} \left(\sqrt{1-\sigma^2} - i\sigma e^{i\kappa n} \right) \gamma_1 \right\rangle_{c1}$$

$$\otimes \left| \frac{1}{\sqrt{2}} \left(i\sqrt{1-\sigma^2} - \sigma e^{i\kappa n} \right) \gamma_1 \right\rangle_{d1}$$

$$\otimes \left| \frac{1}{\sqrt{2}} \left(\sqrt{1-\sigma^2} - i\sigma e^{i\kappa n} \right) \gamma_2 \right\rangle_{c2}$$

$$\otimes \left| \frac{1}{\sqrt{2}} \left(i\sqrt{1-\sigma^2} - \sigma e^{i\kappa n} \right) \gamma_2 \right\rangle_{d2} . \tag{14.49}$$

From the above one may form a density matrix and proceed through the same steps as before, by tracing it over the equipment. It is clear that after the double measurement the reduced density matrix is still of the form (14.47) if the probe beams satisfy the conditions for (practically) error-free measurements of the individual photon states:

$$\rho_R = \sum_n |c_n|^2 |n\rangle_s \, _s\langle n| . \tag{14.50}$$

The probability distribution of the second measurement is the same as that of the first. This finding determines the conditional probability of measuring n photons in the second measurement if m photons have been measured in the first. Indeed, the probability distribution of the second measurement is

$$p_2(n) = \sum_m p(n|m) p_1(m) . \tag{14.51}$$

If and only if the conditional probability is a Kronecker delta,

$$p(n|m) = \delta_{mn} , \tag{14.52}$$

can we have

$$p_2(n) = p_1(n) , \tag{14.53}$$

as is the case here.

When the gain of the QND measurement apparatus is not high enough, i.e.

$$\sigma \kappa |\gamma| \leq 1 ,$$

then the QND measurement does not have perfect resolution; no classical interpretation of the outcome of the interaction of the signal with the measurement apparatus is possible.

14.6 The Schrödinger Cat Thought Experiment

Schrödinger devised a thought experiment in order to demonstrate the strangeness of quantum states when set in a classical environment. We describe here a version of the experiment presented by John Gribbin in his book *Schrödinger Kittens and the Search for Reality* [193]. The entire operation takes place in an enclosure that is inaccessible to outside inspection. An electron is put into a closed box inside the enclosure. A partition is introduced into the box, dividing it into two compartments of equal volume. The electron is now in a coherent superposition of states "left" and "right". Next, the left side of the box is opened to let the electron, if present, escape into the larger enclosure containing the box. This enclosure contains an electron detector which, when triggered by the electron, will flood the enclosure with a poisonous gas. A cat trapped in the enclosure is killed by exposure to the gas. The paradox is that the electron is in a quantum mechanical superposition state, or so it is asserted. It follows from this assertion that the cat is in a superposition state of being either dead or alive. Its state is not determined until the box is opened by an outside observer.

This scenario has been the source of much controversy. Here we shall argue that the system is not in a quantum superposition state when the poison gas is released. Briefly stated, we shall show that the detection process of the particle (the electron in the above scenario) destroys the quantum nature of the state. We shall also argue that no additional observation is necessary to determine the probability of the cat's demise. The enclosure can remain closed. The analysis will be carried out in a scenario in which the particle in question is a photon, not an electron, because we can base the analysis on the simple formalisms developed in the preceding sections.

It should be stated at the outset that we do not imply that Schrödinger cat states are inconsistent with quantum mechanics. Much recent work has been done on the generation of more and more sophisticated quantum mechanical superposition states, which have been called by this name. The thrust of our argument is that the apparatus necessary to kill a macroscopic cat destroys the superposition, i.e. puts the density matrix of the system into diagonal form.

First, let us define the scenario that we shall analyze. We start with the famous enclosure and do not permit inspection of its contents after the experiment has been initiated. A single photon is passed through a beam splitter (see Fig. 14.9). At the output reference plane of the beam splitter the quantum state is in a superposition of the states $|1\rangle_a|0\rangle_b$ and $|0\rangle_a|1\rangle_b$. A quantum nondemolition apparatus is attached to port (c) to determine whether the photon appears in that port (Fig. 14.10). When the passage of the photon is detected, the deadly contraption is activated. We shall show that the detection of the passage of the photon removes the quantum coherence.

The passage of a photon through a beam splitter that couples the two entering radiations has been analyzed in Sect. 7.3. With proper choice of $|M|T$ and ϕ,

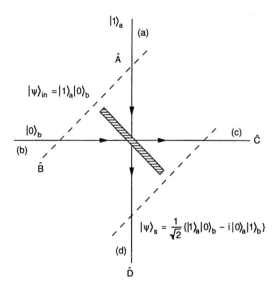

Fig. 14.9. Entangled state generated by passage of photon state through beam splitter

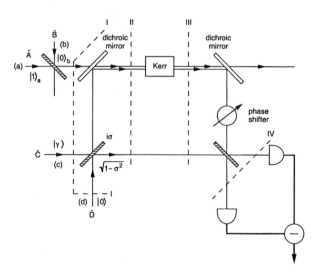

Fig. 14.10. The detection apparatus for the Schrödinger cat experiment considered here

we can put the signal into the state

$$|\psi\rangle_s = \frac{1}{\sqrt{2}}(|1\rangle_a|0\rangle_b - i|0\rangle_a|1\rangle_b) \ . \tag{14.54}$$

The evolution of the entangled state through the measurement apparatus need not be reanalyzed in detail. Indeed, Sect. 14.4 takes the state $c_m|m\rangle$ through the nonlinear interferometer, ending up at plane IV with the wave function (14.39). In the present case we take the states $|1\rangle_a|0\rangle_b$ and $|0\rangle_a|1\rangle_b$ through the system. The Schrödinger evolution is linear; a sum of two states evolves as the sum of the individual evolutions. We may use the previous results if we interpret properly the nature of these two states. First of all, we note that the state $|1\rangle_a|0\rangle_b$ implies that the photon has gone through the beam splitter and stayed in the output port that impinges on the QND apparatus. This state will unbalance the interferometer. The state $|0\rangle_a|1\rangle_b$ corresponds to a photon in the other output port of the beam splitter, which does not feed into the interferometer. Thus, using (14.39), we derive the state at plane IV:

$$
\begin{aligned}
|\psi\rangle_{\text{IV}} = {} & 1/\sqrt{2}\,|1\rangle_a \otimes |0\rangle_b \otimes |1/\sqrt{2}(\sqrt{1-\sigma^2}\gamma - i\sigma e^{i\kappa}\gamma)\rangle_c \\
& \otimes |1/\sqrt{2}\,(i\sqrt{1-\sigma^2}\gamma - \sigma e^{i\kappa}\gamma)\rangle_d \\
& - 1/\sqrt{2}\,i|0\rangle_a \otimes |1\rangle_b \times |1/\sqrt{2}(\sqrt{1-\sigma^2}\gamma - i\sigma\gamma)\rangle_c \\
& \otimes |1/\sqrt{2}\,(i\sqrt{1-\sigma^2}\gamma - \sigma\gamma)\rangle_d \ .
\end{aligned}
\tag{14.55}
$$

The wave function is in an entangled state. We study the density matrix

$$
\begin{aligned}
\rho = {} & |\psi\rangle_{\text{IV}}\ _{\text{IV}}\langle\psi| \\
= {} & 1/2\,|1\rangle_a\ _a\langle 1| \otimes |0\rangle_b\ _b\langle 0| \otimes R_{aa,cc} \otimes R_{aa,dd} \\
& - 1/2\,i|1\rangle_a\ _a\langle 0| \otimes |0\rangle_b\ _b\langle 1| \otimes R_{ab,cc} \otimes R_{ab,dd} \\
& + 1/2\,i|0\rangle_a\ _a\langle 1| \otimes |1\rangle_b\ _b\langle 0| \otimes R_{ba,cc} \otimes R_{ba,dd} \\
& + 1/2\,|0\rangle_a\ _a\langle 0| \otimes |1\rangle_b\ _b\langle 1| \otimes R_{bb,cc} \otimes R_{bb,dd} \ ,
\end{aligned}
\tag{14.56}
$$

with

$$
\begin{aligned}
R_{aa,cc} = {} & \frac{1}{2}\left| \frac{1}{\sqrt{2}}(\sqrt{1-\sigma^2} - i\sigma e^{i\kappa})\gamma \right\rangle_c\ _c\left\langle \frac{1}{\sqrt{2}}(\sqrt{1-\sigma^2} - i\sigma e^{i\kappa})\gamma \right| \\
& = R_{ab,cc} = R_{ba,cc} = R_{bb,cc}
\end{aligned}
\tag{14.57}
$$

and

$$R_{aa,dd} = \frac{1}{2}\left|\frac{1}{\sqrt{2}}(i\sqrt{1-\sigma^2} - \sigma e^{i\kappa})\gamma\right\rangle_d {}_d\left\langle\frac{1}{\sqrt{2}}(i\sqrt{1-\sigma^2} - \sigma e^{i\kappa})\gamma\right|$$

$$= R_{ab,dd} = R_{ba,dd} = R_{bb,dd} \ .$$

$$(14.58)$$

Detection calls for tracing over the measurement apparatus. The partial trace over the density matrix (14.56) is carried out as before, using (14.42):

$$\mathrm{Tr}(R_{ab,cc}) = \mathrm{Tr}(R_{ba,cc})^* = \mathrm{Tr}(R_{ab,dd}) = \mathrm{Tr}(R_{ba,dd})^*$$

$$= \exp\left\{-\frac{\sigma^2}{2}|\gamma|^2[1-\cos(\kappa)]\right\}\exp(i\phi) \ ,$$

$$(14.59)$$

with $\phi = |\gamma^2/2|[\sigma\sqrt{1-\sigma^2}(\cos\kappa - 1) - \sigma^2\sin\kappa]$.

The requirement that the off-diagonal elements of the density matrix vanish is:

$$\sigma^2|\gamma|^2(1-\cos\kappa)/2 = \sigma^2|\gamma|^2\sin^2(\kappa/2) \gg 1 \ .$$

$$(14.60)$$

When this inequality is obeyed, the reduced matrix becomes diagonal

$$\mathrm{Tr}(\rho) = \frac{1}{2}(|1\rangle_a {}_a\langle 1| \otimes |0\rangle_b {}_b\langle 0| + |0\rangle_a {}_a\langle 0| \otimes |1\rangle_b {}_b\langle 1|) \ .$$

$$(14.61)$$

The requirement is that the apparatus uses a sufficient number of photons that a significant phase shift is produced by one single signal photon.

The apparatus is noisy. If a decision has to be made whether a photon has passed or not, the signal must be much larger than the noise. Now we show that a good signal-to-noise ratio is only achieved when the inequality (14.60) is satisfied, i.e. when the reduced density matrix is diagonal.

The expectation value of the charge is

$$_{\mathrm{IV}}\langle\psi|\hat{Q}|\psi\rangle_{\mathrm{IV}} = q_{\mathrm{IV}}\langle\psi|\hat{C}^\dagger\hat{C} - \hat{D}^\dagger\hat{D}|\psi\rangle_{\mathrm{IV}} \ .$$

$$(14.62)$$

The charge will fluctuate if the signal fluctuates, but it already has fluctuations solely because the apparatus is excited with a coherent-state probe. In the absence of a signal the wave function is

$$|\psi_o\rangle_{\mathrm{IV}} = \left[|0\rangle_a \otimes |0\rangle_b \otimes \left|\frac{1}{\sqrt{2}}\left(\sqrt{1-\sigma^2} - i\sigma\right)\gamma\right\rangle_c\right.$$

$$\left.\otimes \left|\frac{1}{\sqrt{2}}(i\sqrt{1-\sigma^2} - \sigma)\gamma\right\rangle_d\right] \ .$$

$$(14.63)$$

We find for the charge

$$_{\text{IV}}\langle\psi_o|\hat{Q}|\psi_o\rangle_{\text{IV}} = 0 \, , \tag{14.64}$$

and for the mean square fluctuations

$$_{\text{IV}}\langle\psi_o|\hat{Q}^2|\psi_o\rangle_{\text{IV}} = q^2|\gamma|^2 \, . \tag{14.65}$$

In the presence of a signal, represented by the wave function (14.55), the charge is

$$_{\text{IV}}\langle\psi|\hat{Q}|\psi\rangle_{\text{IV}} = q\sigma\sqrt{1-\sigma^2}\sin\kappa|\gamma|^2 \, . \tag{14.66}$$

The signal-to-noise ratio is

$$\text{SNR} = \frac{_{\text{IV}}\langle\psi|\hat{Q}|\Psi\rangle_{\text{IV}}^2}{_{\text{IV}}\langle\psi_o|\hat{Q}|\psi_o\rangle_{\text{IV}}^2} \approx \sigma^2(1-\sigma^2)|\gamma|^2\sin^2\kappa \, . \tag{14.67}$$

If the signal-to-noise ratio is to be made much greater than one, we require

$$\sigma^2(1-\sigma^2)|\gamma|^2\sin^2\kappa \gg 1 \, . \tag{14.68}$$

This is essentially the same condition as required for the collapse of the density matrix.

Tracing over a subsystem performs an average over all the states of the subsystem. Tracing over the measurement apparatus thus expresses an average taken over all measurements. The fact that the reduced density matrix decoheres (becomes diagonal) for system parameters that yield a large signal-to-noise ratio, and thus provide accuracy of the measurement, shows that the cat cannot be in a superposition state of dead or alive. When the system is able to decide that a photon has passed through the measurement apparatus, the reduced density matrix is rendered diagonal.

14.7 Summary

We have taken the point of view that quantum theory is a statistical theory that predicts only the outcome of an ensemble of measurements. The outcome of a single measurement is described only probabilistically. In this sense quantum theory resembles statistical mechanics, in which detailed knowledge of the initial state of the system is unavailable because of the complexity of a system with many degrees of freedom. In the case of quantum theory, the detailed knowledge of the initial state is unavailable in principle, because of Heisenberg's uncertainty principle. Born espoused this interpretation of quantum theory. In his book *Natural Philosophy of Cause and Chance* [54], he approached his probabilistic interpretation of quantum theory with the following words:

"Now the curious situation arises after this code of rules (of science), which ensures the possibility of scientific laws, in particular of the cause effect relations, contains besides many other prescriptions those related to observational errors, a branch of theory of probability. This shows that the conception of chance enters into the first steps of scientific activity, by virtue of the fact that no observation is absolutely correct. I think chance is a more fundamental conception than causality; for whether in a concrete case a cause–effect relation holds or not can only be judged by applying the laws of chance to observations."

With this probabilistic interpretation of quantum mechanics, we studied the properties of quantum nondemolition measurements, quantizing the measurement apparatus as well. We showed that a QND measurement of photon number can be carried out while imparting no more than the minimum uncertainty to the measured signal as required by Heisenberg's uncertainty principle. With a QND measurement apparatus in one arm of an interferometer, we could show the gradual disappearance of the fringes in proportion to the degree of knowledge that could be gained about the photon number by the measurement.

We have shown that the density matrix of the system composed of the observable (the signal photon number) and the QND measurement apparatus, traced over the measurement apparatus, becomes diagonal when the number of signal photons passing through the measurement apparatus can be discerned. Next, we found from the study of two QND measurements in cascade that the conditional probability of measuring m photons in the second measurement when n photons have been measured in the first apparatus approaches a Kronecker delta if both measurements are performed with sufficient accuracy. This is consistent with the von Neumann postulate stating that a measurement casts the state of the observable into an eigenstate of the measurement apparatus. It is not a proof of the postulate, but only suggests that an intelligent observer could predict the outcome of the second measurement by assuming that the observable is in the photon eigenstate $|n\rangle$ if n photons were observed by the first apparatus.

Finally, we addressed the Schrödinger cat paradox. A photon was put into an entangled state which was passed through a measurement apparatus. If the apparatus, which triggered a contraption that killed the cat, was to register the passage of a photon, then we found that the measurement destroyed the entangled state. The photon was registered with a probability of $1/2$; the probability of the cat being dead was $1/2$, the probability of it being alive was $1/2$.

Problems

14.1 Redo the calculations of Sect. 14.3 for the "which path" experiment with one QND measurement in each arm. The two probe beams are of equal intensities.

14.2* A Schrödinger cat state $N(|\alpha\rangle + |-\alpha\rangle)$ is incident upon an ideal photodetector. What are the photon statistics? Assume α to be real and positive.

14.3* The state $|\alpha\rangle|\beta\rangle$ is incident upon a beam splitter characterized by $MT = \phi$. Each output port is fed into an ideal detector. Find the probability generating function of the photon count. Find the joint probabilities.

14.4 The state $|1\rangle|1\rangle$ is incident upon a 50/50 beam splitter. Each output port has a detector. Find the joint probability generating function. Determine the joint probability distribution.

14.5 The state $|2\rangle|0\rangle$ is incident upon a beam splitter characterized by $MT = \phi$. Each output port is fed into an ideal detector. Find the probability generating function of the photon count. Find the joint probabilities.

14.6* Find the output state when a Schrödinger cat state $N(|\alpha\rangle + |-\alpha\rangle)$, with α *real and positive*, is incident upon a beam splitter, with vacuum incident upon the other port.

14.7 The output state of the preceding problem is detected by two ideal detectors. Find the probabilities of the photon count in the two detectors.

14.8 Derive the probability generating function for the process of the preceding problem.

14.9 Find the falling-factorial-moment generating function for the preceding two problems.

14.10 What is the probability $p(m)$ of detecting m photons with the detector in output (1) of Prob. 14.7. Determine its falling-factorial-moment distribution.

14.11 Derive the falling factorial moment of the preceding problem by the method of Sect. 9.2.

Solutions

14.2 We start with the falling-factorial-moment-generating function

$$F(\xi) = \sum_R \frac{\langle\psi|\hat{A}^{\dagger r}\hat{A}^r|\psi\rangle}{r!}\xi^r \,,$$

where $|\psi\rangle = N(|\alpha\rangle + |-\alpha\rangle)$. Note that α is real and positive. Since $\langle\psi|\psi\rangle = 1$, we find for $1/N^2$

$$\frac{1}{N^2} = (\langle\alpha| + \langle-\alpha|)(|\alpha\rangle + |-\alpha\rangle)$$

$$= 1 + 1 + 2\langle\alpha| - \alpha\rangle = 2(1 + e^{-2|\alpha|^2}) \,.$$

The rth element of $F(\xi)$ contains

$$\langle\psi|\hat{A}^{\dagger r}\hat{A}^r|\psi\rangle = [\alpha^r\langle\alpha| + (-\alpha)^r\langle-\alpha|][\alpha^r|\alpha\rangle + (-\alpha)^r|-\alpha\rangle]$$

$$= 2|\alpha|^{2r} + (-1)^r 2|\alpha|^{2r}e^{-2|\alpha|^2} .$$

We obtain

$$F(\xi) = \frac{1}{1 + e^{-2|\alpha|^2}}\left(e^{\xi|\alpha|^2} + e^{-\xi|\alpha|^2}e^{-2|\alpha|^2}\right) .$$

The probability generating function is

$$P(\xi) = F(\xi - 1) = \frac{\cosh(\xi|\alpha|^2)}{\cosh|\alpha|^2} .$$

We can check that $P(1) = 1$, as it must be. The probabilities are proportional to

$$\frac{|\alpha|^{2n}}{n!}$$

for n even, and are zero for n odd. Only even photon numbers are detected.

14.3 The output state is

$$|\psi(T)\rangle = |\gamma\rangle|\delta\rangle ,$$

with

$$\gamma = \alpha\cos\phi - \mathrm{i}\,\beta\sin\phi , \quad \delta = -\mathrm{i}\,\alpha\sin\phi + \beta\cos\phi .$$

The falling-factorial-moment generating function is

$$F(\xi,\eta) = \sum_{p,q}\frac{\xi^p\eta^q}{p!q!}\langle\hat{A}^{\dagger p}\hat{A}^p\hat{B}^{\dagger q}\hat{B}^q\rangle .$$

We find for the expectation value

$$\langle\hat{A}^{\dagger p}\hat{A}^p\hat{B}^{\dagger q}\hat{B}^q\rangle = |\gamma|^{2p}|\delta|^{2q} .$$

The falling-factorial-generating function becomes

$$F(\xi,\eta) = \sum_{p,q}\frac{(|\gamma|^2\xi)^p(|\delta|^2\eta)^q}{p!q!} = \exp(|\gamma|^2\xi)\exp(|\delta|^2\eta) .$$

The probability generating function is:

$$P(\xi,\eta) = F(\xi - 1,\eta - 1) = \exp(-|\gamma|^2)\exp(|\gamma|^2\xi)\exp(-|\delta|^2)\exp(|\delta|^2\eta) .$$

We find for the probabilities

$$P(m,n) = p(m)p(n) = \frac{|\gamma|^{2m}}{m!}e^{-|\gamma|^2}\frac{|\delta|^{2n}}{n!}e^{-|\delta|^2} .$$

The counts in the two detectors are independent and Poisson-distributed. This is consistent with the interpretation in which two Poisson-distributed signals are acted upon by a binomial process. The binomial process preserves Poisson distributions! The two distributions are statistically independent; the detections are uncorrelated.

14.6 The cat state is a superposition of two coherent states. We can take advantage of the simple transformation law for a coherent state passing through a beam splitter. If the beam splitter matrix is

$$\begin{bmatrix} r & t \\ t & r \end{bmatrix} \quad \text{with} \quad t = -i\sqrt{1 - r^2}\,,$$

the output state is

$$|\psi\rangle = N(|r\alpha\rangle|t\alpha\rangle + |-r\alpha\rangle|-t\alpha\rangle)\,.$$

Epilogue

The topics discussed in this book reflect one of my main research interests over 48 years at MIT. The progress from the study of electromagnetic noise in electronic amplifiers to the investigation of noise in optical amplifiers was motivated by the fact that electronic amplifiers do not possess a fundamental limit on their "noise measure", whereas optical amplifiers do. The emphasis on practical engineering amplifiers and systems that operate with a large number of photons (50 photons per bit or more) permits the use of the linearization approximation, in which the noise is additive to the signal. Within this approximation, the phenomena can be explained by a semiclassical theory that is in close analogy with classical physics. One may attach "physical reality" to an observable, the signal, in the absence of the measurement. Fluctuations in the measured observable are attributed to the additive noise of the measurement apparatus and the noise (uncertainty) "accompanying" the signal. In order to illustrate situations in which this simple picture fails, we looked at a few examples involving photon states of a few photons. In these cases the Wigner function ceases to be positive definite, no classical joint-probability description is feasible in such cases. The noise is not additive. "Physical reality" can be defined only after a full specification of the measurement apparatus.

Fluctuations set limits to the accuracy of measurements and the distances of reliable communications. Quantum mechanics is intimately connected with uncertainty, which manifests itself through fundamental, unavoidable noise. A single quantum observable can be measured, in principle, without uncertainty. If the observable is prepared in an eigenstate of the measurement apparatus, every ideal measurement yields the same value, the eigenvalue of the state. Linear amplifiers permit the simultaneous measurement of the in-phase and quadrature components of the electric field, two noncommuting observables. Such amplifiers must add unavoidable noise to the measurement. Fiber amplifiers operate very near the fundamental limit set by quantum mechanics on the uncertainty of a simultaneous measurement of two noncommuting variables. Even though fiber communications today utilize only intensity modulation, and hence phase-sensitive parametric amplification could be employed without adding noise in the amplification process, long-distance fiber communications will not be able to utilize this form of amplification. The technical difficulties of locking the amplifier phase to the signal phase, which is randomized by environmentally imposed fluctuations, are too severe.

The study of optical amplifiers led us to a discussion of the noise figure. The definition of noise figure currently in use is based on signal-to-noise ratios defined after optical detection. This definition does not take advantage of the fact that optical amplifiers are linear amplifiers of the electromagnetic field. The definition employed for linear electronic amplifiers can be adapted for optical amplifiers. Time will tell whether engineering practice will find this alternative definition better suited for measurement and prediction of the noise performance of optical-amplifier systems. The author prefers it, since it meshes well with the concept of noise measure that proved so useful in the discussion of electronic-amplifier performance.

We concluded with a detailed description of an optical quantum measurement. The groundwork was laid with an analysis of squeezing. The Hamiltonian of a signal and a probe interacting in a Kerr medium was developed and justified as a self-consistent model of a measurement apparatus. No linearization approximation was used. Several key interpretations of a quantum measurement were proffered:

(a) The measurement *occupies a finite time interval* and leads to an entanglement of the wave functions of the observable and of the apparatus.
(b) Tracing over the measurement apparatus at the end of this time interval leads to a *diagonal reduced density matrix*.
(c) The conditional probability of observing the same outcome in two quantum nondemolition experiments in cascade is unity.
(d) This fact entitles an intelligent observer to predict outcomes of future measurements by an analysis that starts with the observable in an eigenstate of the measurement apparatus, *as if* the measurement had projected the state of the observable into an eigenstate of the measurement apparatus.
(e) Finally, it is the author's view, shared by many physicists, that quantum theory is fundamentally a probabilistic theory that is *complete* in the sense defined by John Bell.

Some readers may see, find, or know of better ways to address these important issues. I have tried to do my best and I feel privileged that I have been given the opportunity to assemble these ideas,

"verso la fine del cammin di nostra vita".

Appendices

A.1 Phase Velocity and Group Velocity of a Gaussian Beam

A diffracting Gaussian beam experiences a phase advance $\phi = \tan^{-1}(z/b)$ in addition to the plane-wave phase delay kz. The net phase delay is thus $kz - \phi = kz - \tan^{-1}(z/b)$. The phase advance has a limit of a value of $\pi/2$ for large values of z/b. Hence, at large values of z, the propagation constant approaches k, and the phase velocity and group velocity approach c. On the other hand, for small values of z the net phase delay is a linear function of z, giving the effective propagation constant

$$\beta_{\text{eff}} = k - \frac{1}{b} = k - \frac{\lambda}{\pi w_o^2} = \frac{\omega}{c} - \frac{c}{\omega} \frac{2}{w_o^2} \, . \tag{A.1.1}$$

Thus, the phase velocity is

$$v_p \equiv \frac{\omega}{\beta_{\text{eff}}} = \frac{c}{1 - 2c^2/w_o^2 \omega^2} \, . \tag{A.1.2}$$

The phase velocity is greater than the speed of light. On the other hand, the inverse group velocity is

$$\frac{1}{v_g} \equiv \frac{d\beta_{\text{eff}}}{d\omega} = \frac{d}{d\omega} \left(\frac{\omega}{c} - \frac{2}{w_o^2} \frac{c^2}{\omega} \right) = \frac{1}{c} + \frac{2}{w_o^2} \frac{c^2}{\omega^2} \, , \tag{A.1.3}$$

and the group velocity is

$$v_g = \frac{c}{1 + 2c^2/w_o^2 \omega^2} \, . \tag{A.1.4}$$

To the order $\lambda^2/\pi^2 w_o^2$, the product of the group velocity and phase velocity is equal to the speed of light squared:

$$v_g v_p = c^2 \, . \tag{A.1.5}$$

This is the order to which the paraxial wave approximation is valid. The reader will note that (A.1.5) holds rigorously for the dispersion of metallic waveguides discussed in Chap. 2.

A.2 The Hermite Gaussians and Their Defining Equation

A.2.1 The Defining Equation of Hermite Gaussians

The Hermite Gaussian functions are best understood with the aid of the differential equation for which they form a complete set of solutions, the Schrödinger equation of the one-dimensional quantum mechanical harmonic oscillator [194–196], which is, in normalized form,

$$\frac{d^2\phi}{d\xi^2} + (\lambda - \xi^2)\phi = 0 \ . \tag{A.2.1}$$

An orthogonal set of functions $\psi_m(\xi)$ of a single independent variable ξ is generated by this differential equation in the sense that

$$\int_{-\infty}^{\infty} \phi_m(\xi)\phi_n^*(\xi)d\xi = 0 \quad \text{for } m \neq n \ . \tag{A.2.2}$$

To prove this, and gain further understanding of the solutions of (A.2.1), we study the geometric interpretation of (A.2.1) by means of Fig. A.2.1. Because the coefficients of (A.2.1) are symmetric with respect to ξ, the solutions must be either symmetric or antisymmetric. A symmetric solution that starts out from the center, $\xi = 0$, with zero slope has a prescribed slope and curvature for ever after, going to the right, as determined by the second-order differential equation. The solution will be concave toward the ξ axis in region I and convex in region II. It will shoot off toward $\pm\infty$ as $\xi \to \infty$, unless λ is carefully chosen – at the so-called eigenvalue of λ corresponding to a bounded solution. The lowest-order solution has the lowest curvature, the lowest eigenvalue, and only one extremum. It is the Gaussian $e^{-\xi^2/2}$, with

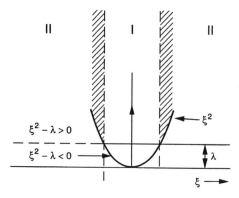

Fig. A.2.1. Regions of positive and negative net energy

$\lambda = 1$. The next solution is antisymmetric with two extrema, the following one is symmetric with three extrema, and so forth.

Next, we investigate how the higher-order solutions are related to the lower-order ones. For this purpose it is convenient to introduce the "creation" and "annihilation" operators, or "raising" and "lowering" operators, $d/d\xi \mp \xi$, where the $-$ sign goes with raising and the $+$ sign with lowering. Consider a function $\phi(\xi)$ which is assumed to obey (A.2.1) and which vanishes for $\xi \to \pm\infty$. Operate on (A.2.1) with $d/d\xi \mp \xi$ and rearrange the terms so that $(d/d\xi \mp \xi)$ is brought to the right of $d^2/d\xi^2$ and ξ^2. For this purpose we note that

$$\left(\frac{d}{d\xi} \mp \xi\right)\frac{d^2\phi}{d\xi^2} = \frac{d^2}{d\xi^2}\left[\left(\frac{d}{d\xi} \mp \xi\right)\phi\right] \pm 2\frac{d\phi}{d\xi} , \tag{A.2.3}$$

$$\left(\frac{d}{d\xi} \mp \xi\right)\xi^2\phi = \xi^2\left(\frac{d}{d\xi} \mp \xi\right)\phi + 2\xi\phi . \tag{A.2.4}$$

Using (A.2.3) and (A.2.4) in (A.2.1) operated on by $d/d\xi \mp \xi$, we obtain

$$\frac{d^2}{d\xi^2}\left[\left(\frac{d}{d\xi} \mp \xi\right)\phi\right] + [(\lambda \pm 2) - \xi^2]\left[\left(\frac{d}{d\xi} \mp \xi\right)\phi\right] = 0 . \tag{A.2.5}$$

We have recovered the original equation, where the new solution $(d/d\xi \mp \xi)\phi$ has the eigenvalue $\lambda \pm 2$. Consider the lowest-order solution $\exp(-\xi^2/2)$, with $\lambda = 1$. This has the lowest possible negative curvature in the range where $\lambda - \xi^2$ is positive, and hence the lowest possible value of λ. The next solution obtained by operating with the raising operator, $(d/d\xi - \xi)\exp(-\xi^2/2) = -2\xi\exp(-\xi^2/2)$, has two extrema. Each successive application produces one more extremum. Hence we collect *all* possible solutions by successive application of the raising operator. The mth eigenvalue λ_m is given by $\lambda_m = 2(m + 1/2)$.

Conversely, operation by the lowering operator produces a lower-order solution from a higher-order one by "climbing down" the eigenvalue "ladder" in increments of 2, producing a solution on one "lower rung" of the "ladder". The solutions of (A.2.1) for the different discrete eigenvalues are the *Hermite Gaussians*.

A.2.2 Orthogonality Property of Hermite Gaussian Modes

The Hermite Gaussians $\phi_m(\xi)$ are orthogonal in the sense that [194, 195]

$$\int_{-\infty}^{\infty} d\xi\, \phi_m(\xi)\phi_n(\xi) = 0 , \tag{A.2.6}$$

if $m \neq n$. To show this we use their defining equation (A.2.1):

$$\frac{d^2\phi_m}{d\xi^2} + \lambda_m\phi_m - \xi^2\phi_m = 0 \ , \tag{A.2.7}$$

where

$$\phi_m \equiv H_m(\xi)e^{-\xi^2/2} \ . \tag{A.2.8}$$

Multiplying (A.2.7) by ϕ_n and subtracting (A.2.7) applied to ϕ_n multiplied by ϕ_m, we find

$$(\lambda_m - \lambda_n)\int_{-\infty}^{\infty}\phi_m\phi_n \, d\xi = \int_{-\infty}^{\infty}\frac{d}{d\xi}\left(\phi_m\frac{\partial\phi_n}{\partial\xi} - \phi_n\frac{\partial\phi_m}{\partial\xi}\right)d\xi = 0 \ , \tag{A.2.9}$$

because ϕ_m and ϕ_n vanish at $\xi = \pm\infty$. Thus the orthogonality condition

$$\int_{-\infty}^{\infty}\phi_m\phi_n \, d\xi = 0 \tag{A.2.10}$$

is obeyed when

$$\lambda_m \neq \lambda_n \ .$$

Further, note that (A.2.8) introduced into (A.2.7) leads to the differential equation obeyed by the Hermite polynomials:

$$\frac{d^2H_m}{d\xi^2} - 2\xi\frac{dH_m}{d\xi} + 2mH_m = 0 \ . \tag{A.2.11}$$

A.2.3 The Generating Function and Convolutions of Hermite Gaussians

The generating function of the Hermite Gaussians $\phi_n(\xi)$ is (as we prove below)

$$F(x,\xi) \equiv \exp\left(-s^2 + 2s\xi - \frac{\xi^2}{2}\right) = \sum_{n=0}^{\infty}\frac{s^n}{n!}H_n(\xi)e^{-\xi^2/2} \tag{A.2.12}$$

$$= \sum_{n=0}^{\infty}\frac{s^n}{n!}\phi_n(\xi) \ .$$

The Hermite Gaussians are the "coefficients" of the Taylor expansion in s of $F(x,\xi)$. Comparison of the two sides of (A.2.12) for $s = 0$ gives $\phi_o(\xi) = \exp(-\xi^2/2)$. We shall now show, through application of the lowering operator $\partial/\partial\xi + \xi$ to both sides of (A.2.12), that all terms in the series are solutions of (A.2.1). We obtain

$$\left(\frac{\partial}{\partial\xi} + \xi\right)F(s,\xi) = 2sF(s,\xi) = 2\sum_{n=0}^{\infty}\frac{s^{n+1}}{n!}\phi_n(\xi)$$

$$= \sum_{n=0}^{\infty}\frac{s^n}{n!}\left(\frac{\partial}{\partial\xi} + \xi\right)\phi_n(\xi) ,$$

(A.2.13)

where we have differentiated the exponential function $F(s,\xi)$ directly, observing that the operator $\partial/\partial\xi + \xi$ operating on $F(s,\xi)$ is equivalent to multiplication by $2s$. Then we replace $F(s,\xi)$ by its defining expansion, and finally equate the result to the operation of $\partial/\partial\xi + \xi$ on the defining expansion. By comparing equal powers of s, we obtain

$$\left(\frac{d}{d\xi} + \xi\right)\phi_{n+1}(\xi) = 2(n+1)\phi_n(\xi) .$$

(A.2.14)

The lowering operator transforms the $(n+1)$th function $\phi_{n+1}(\xi)$ into the nth function $\phi_n(\xi)$. Because the function $\phi_o(\xi)$ is a simple Gaussian, the function $\phi_1(\xi)$ must be the first higher-order solution of the differential equation (A.2.1). The remaining eigenfunctions along the "ladder" may be identified by induction.

We may use the generating function to evaluate $\phi_n(\xi) \equiv H_n(\xi)e^{-\xi^2/2}$. Expanding $F(s,\xi)$ in powers of its exponent in s, and equating terms in (A.2.12), we have

$$H_0(\xi) = 1 ,$$

(A.2.15)

$$H_1(\xi) = 2\xi ,$$

(A.2.16)

$$H_2(\xi) = 4\xi^2 - 2 .$$

(A.2.17)

The three lowest-order Hermite Gaussians are shown in Fig. A.2.2.

The generating function can be used to relate $dH_n/d\xi$ to H_{n-1}. This is accomplished by taking a derivative with respect to ξ of (A.2.12) and rewriting the result, $(2s - \xi)F(s,\xi)$, in terms of the defining sums. Equating terms of the same powers of s, we obtain

$$\frac{d}{d\xi}H_n = 2nH_{n-1} .$$

(A.2.18)

If we differentiate the above and use the differential equation obeyed by H_n (A.2.11) and (A.2.18) to eliminate the derivatives, we obtain the recursion formula

$$H_{n+1} - 2\xi H_n + 2nH_{n-1} = 0 .$$

(A.2.19)

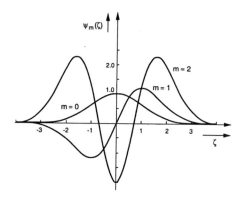

Fig. A.2.2. The three lowest order Hermite Gaussians

Another very important use of the generating function is evaluation of convolutions and Fourier transforms of eigenfunctions $\phi_n(\xi)$. Consider first the Fourier transform of the Taylor expansion of the generating function:

$$\frac{1}{2\pi} \int_{-\infty}^{\infty} \sum_{n=0}^{\infty} \frac{s^n}{n!} H_n(\xi) e^{-s^2/2} e^{ik\xi} d\xi$$

$$= \frac{1}{2\pi} \int_{-\infty}^{\infty} \exp\left(-s^2 + 2s\xi - \frac{\xi^2}{2} + ik\xi\right) d\xi$$

$$= \frac{1}{2\pi} \int_{-\infty}^{\infty} \exp\left(-\frac{\xi^2}{2} + (2s + ik)\xi - \frac{1}{2}(2s + ik)^2\right) d\xi$$

$$\times \exp\left(s^2 + 2isk - \frac{1}{2}k^2\right) .$$

The integral evaluates to $\sqrt{2\pi}$ and we recognize the factor in the last expression to be the generating function $F(is, k)$. Thus

$$\text{F.T.}\left[\sum_{n=0}^{\infty} \frac{s^n}{n!} H_n(\xi) e^{-\xi^2/2}\right] = \frac{1}{\sqrt{2\pi}} \sum_{n=0}^{\infty} \frac{(is)^n}{n!} H_n(k) e^{-k^2/2} . \qquad (A.2.20)$$

The Fourier transform of $H_n(\xi) e^{-\xi^2/2}$ is $(1/\sqrt{2\pi}) i^n$ times the same function of k. Next, consider the convolution of $\phi_n(\xi)$ with the Gaussian $\exp(-a\xi^2/2)$:

$$\sum_{n=0}^{\infty} \frac{s^n}{n!} \int_{-\infty}^{\infty} \phi_n(\xi_o) e^{-(a/2)(\xi-\xi_o)^2} d\xi_o$$

$$= \int_{-\infty}^{\infty} d\xi_o \exp\left[-s^2 + 2s\xi_o - \frac{\xi_o^2}{2} - \frac{a}{2}(\xi-\xi_o)^2 \right]$$

$$= \int_{-\infty}^{\infty} d\xi_o \exp\left[-\frac{a+1}{2}\xi_o^2 + 2\left(s + \frac{a}{2}\xi\right)\xi_o - \frac{2}{a+1}\left(s + \frac{a}{2}\xi\right)^2 \right]$$

$$\times \exp\left[-\left(s\sqrt{\frac{a-1}{a+1}}\right)^2 + 2\left(s\sqrt{\frac{a-1}{a+1}}\right)\frac{a\xi}{\sqrt{a^2-1}} - \frac{a}{2(a+1)}\xi^2 \right].$$

$$(A.2.21)$$

In the first step we convolve the entire series of functions $\phi_n(\xi_o)$, equate them to the convolution of the generating function, and then evaluate the convolution of the latter by completion of the square. The integral evaluates to $\sqrt{2\pi/(a+1)}$. The remaining exponential factor is the generating function of a Hermite Gaussian. Equating the first expression in (A.2.21) to the last one expanded as a series of functions ϕ_n, we obtain

$$\sum_{n=0}^{\infty} \frac{s^n}{n!} \int_{-\infty}^{\infty} \phi_n(\xi_o) e^{-(a/2)(\xi-\xi_o)^2} d\xi_o$$

$$= \sqrt{\frac{2\pi}{a+1}} \sum_{n=0}^{\infty} \frac{1}{n!}\left(s\sqrt{\frac{a-1}{a+1}}\right)^n \phi_n\left(\frac{a\xi}{\sqrt{a^2-1}}\right) \exp\left[\frac{a\xi^2}{2(a^2-1)}\right].$$

$$(A.2.22)$$

Term-by-term identification gives

$$\int_{-\infty}^{\infty} \phi_n(\xi_o) e^{-(a/2)(\xi-\xi_o)^2} d\xi_o$$

$$(A.2.23)$$

$$= \sqrt{\frac{2\pi}{a+1}}\left(\frac{a-1}{a+1}\right)^{n/2} \phi_n\left(\frac{a\xi}{\sqrt{a^2-1}}\right) \exp\left[\frac{a\xi^2}{2(a^2-1)}\right].$$

Note that the square root in the argument of ϕ_n has to be interpreted so as to yield solutions decaying with increasing $|\xi|$. This same interpretation has to be given to

$$\sqrt{\frac{a-1}{a+1}} = \frac{a-1}{\sqrt{a^2-1}}.$$

Finally, consider the product of two generating functions for the purpose of evaluating the normalization integral:

$$\int_{-\infty}^{\infty} e^{s^2+2s\xi-\xi^2/2} e^{-p^2+2p\xi-\xi^2/2} d\xi$$

$$= \sum_{m=0}^{\infty} \sum_{n=0}^{\infty} \frac{s^m p^n}{m!n!} \int_{-\infty}^{\infty} H_m(\xi) H_n(\xi) e^{-\xi^2} d\xi .$$

(A.2.24)

The left-hand side is easily evaluated to give

$$\sqrt{\pi} e^{2sp} = \sqrt{\pi} \sum_n \frac{(2sp)^n}{n!} .$$

(A.2.25)

If equal powers of s and p are equated after substitution of the value of the integral (A.2.25) into (A.2.24), we obtain

$$\int_{-\infty}^{\infty} H_n^2(\xi) e^{-\xi^2} d\xi = \sqrt{\pi} 2^n n! ,$$

(A.2.26)

and the orthogonality condition for $m \neq n$,

$$\int_{-\infty}^{\infty} H_m(\xi) H_n(\xi) e^{-\xi^2} d\xi = 0 .$$

(A.2.27)

A.3 Recursion Relations of Bessel Functions

Given a function $Z(x)$ that is a linear superposition of a Bessel function and a Neumann function of order p, then the following recursion formula holds:

$$\frac{dZ_p(x)}{dx} = -\frac{p}{x} Z_p(x) + Z_{p-1}(x) .$$

(A.3.1)

If we apply this formula to the Bessel function of zeroth order, we obtain

$$\frac{dJ_0}{dx} = J_0'(x) = -J_1(x) .$$

(A.3.2)

A modified Bessel function of zeroth order is an ordinary Bessel (or Neumann) function of imaginary argument. Thus, we have

$$\frac{dK_0}{dx} = i\frac{dZ_0(ix)}{d(ix)} = (-iZ_1(ix)) = K_1(x) .$$

(A.3.3)

The functions of first order give respectively

$$\frac{dJ_1}{dx} = -\frac{1}{x} J_1(x) + J_0(x) ,$$

(A.3.4)

$$\frac{dZ_1(ix)}{d(ix)} = -\frac{1}{ix} Z_1(ix) + Z_0(x) ,$$

(A.3.5)

and therefore

$$\frac{dK_1}{dx} = -\frac{1}{x} K_1(x) + K_0(x) .$$

(A.3.6)

A.4 Brief Review of Statistical Function Theory

A brief account is given here of the concepts of spectral density and auto-correlation function for stationary statistical time functions. Stationary statistical time functions that obey the ergodic theorem are particularly simple to analyze. The ergodic theorem states that averages over time of these statistical functions are equal to averages over an ensemble of such functions. A consequence of the ergodic theorem is that one may collect an ensemble of such functions from one single source by collecting samples over a sufficiently long (ideally infinitely long) time interval. In order to analyze them it is convenient to treat them as if they were periodic with a period T. For such an analysis a Fourier series is convenient. Thus, consider the statistical time function $f(t)$. It has associated with it the Fourier transform

$$f_n = \frac{1}{T} \int_{-T/2}^{T/2} dt\, f(t) \exp(\mathrm{i}\omega_n t) \,, \tag{A.4.1}$$

where $\omega_n = (2\pi/T)n$ and n is an integer. The inverse Fourier transform is

$$f(t) = \sum_n f_n \exp(-\mathrm{i}\omega_n t) \,. \tag{A.4.2}$$

The limit $T \to \infty$ is implied throughout. We allow for complex functions, such as the amplitude envelopes of signals with a given carrier frequency. Consider the ensemble average of the "power" associated with $f(t)$, indicated by angle brackets:

$$\begin{aligned} \text{``power''} &= \frac{1}{T} \int_{-T/2}^{T/2} dt\, \langle |f(t)|^2 \rangle \\ &= \frac{1}{T} \sum_{m,n} \int_{-T/2}^{T/2} dt\, \langle f_n f_m^* \rangle \exp[\mathrm{i}(\omega_m - \omega_n)t] \,. \end{aligned} \tag{A.4.3}$$

For a stationary function, the power cannot depend on time. We thus have for a stationary function

$$\langle f_n f_m^* \rangle = \langle |f_m|^2 \rangle \delta_{mn} \,, \tag{A.4.4}$$

where δ_{mn} is the Kronecker delta. Thus, continuing with (A.4.3), we have

$$\text{``power''} = \frac{1}{T} \int_{-T/2}^{T/2} dt\, \langle |f(t)|^2 \rangle$$

$$= \frac{1}{T} \sum_{m,n} \int_{-T/2}^{T/2} dt \, \langle f_n f_m^* \rangle \exp[\mathrm{i}(\omega_m - \omega_n)t]$$

$$= \sum_m \frac{1}{T} \int_{-T/2}^{T/2} dt \, \langle |f_m^2| \rangle \tag{A.4.5}$$

$$= \sum_m \langle |f_m|^2 \rangle = \sum_m \frac{2\pi}{T} \frac{T}{2\pi} \langle |f_m|^2 \rangle \; .$$

The last expression can be transformed into an integral in the limit of infinite T, because then the Fourier components become infinitely closely spaced, $2\pi/T = \Delta\omega \to d\omega$. We find

$$\text{"power"} = \sum_m \frac{2\pi}{T} \frac{T}{2\pi} \langle |f_m|^2 \rangle = \int_{-\infty}^{\infty} d\omega \, \Phi(\omega) \; , \tag{A.4.6}$$

where $\Phi(\omega)$ is the spectral density, defined by

$$\Phi(\omega) = \lim_{T \to \infty} \left[\frac{T}{2\pi} \langle |f_m|^2 \rangle \right] . \tag{A.4.7}$$

The integral of the spectral density gives the power of the statistical process. The autocorrelation function is defined by

$$\langle f^*(t) f(t-\tau) \rangle = R_f(\tau) \; . \tag{A.4.8}$$

For a stationary process the autocorrelation function is a function only of the time shift τ. It is related to the spectral density:

$$R_f(\tau) = \langle f^*(t) f(t-\tau) \rangle = \sum_{m,n} \langle f_n^* f_m \rangle \exp(\mathrm{i}\omega_n t) \exp[-\mathrm{i}\omega_m(t-\tau)]$$

$$= \sum_m \langle |f_m|^2 \rangle \exp(\mathrm{i}\omega_m \tau) = \sum_m \frac{2\pi}{T} \frac{T}{2\pi} \langle |f_m|^2 \rangle \exp(\mathrm{i}\omega_m \tau)$$

$$= \int d\omega \, \Phi(\omega) \exp(\mathrm{i}\omega\tau) \; .$$

$$\tag{A.4.9}$$

The autocorrelation function is the Fourier transform of the power spectral density.

A.5 The Different Normalizations of Field Amplitudes and of Annihilation Operators

In this book we use several different normalizations for the complex mode amplitudes in the classical domain and for creation and annihilation operators in the quantum domain. This appendix summarizes the different normalizations and reviews the motivations for their choice.

A.5.1 Normalization of Classical Field Amplitudes

Statistical mechanics assigns an energy of $(1/2)k\theta$ to the excitation energy of every degree of freedom at equilibrium at temperature θ. An electromagnetic, acoustic, or mechanical vibrational mode has two degrees of freedom, and hence possesses an energy with expectation value $k\theta$. Hence, in the analysis of systems at thermodynamic equilibrium it makes sense to define mode amplitudes A_m whose square is equal to the *energy* in a transmission medium of length L, so that their excitation energy at temperature θ is equal to $k\theta$. Since equilibrium conditions are stationary, and the energy cannot vary with time, the expectation values of different mode amplitudes are uncorrelated:

$$\langle A_m A_n^* \rangle = k\theta \delta_{nm} \ . \tag{A.5.1}$$

The length L, taken as very large (ideally infinitely long), should not enter into the evaluation of relevant physical quantities. Hence, it is desirable to normalize the amplitudes so that the length does not appear explicitly in the answers. The length L defines the spacing of the propagation constants of the modes, $\Delta\beta = 2\pi/L$. We choose renormalized variables $a(\beta)$ such that their integrals evaluate to the energy per unit length:

$$\sum_{n,m} \left\langle \frac{A_n^* A_m}{L} \right\rangle = \int d\beta \int d\beta' \langle a^*(\beta) a(\beta') \rangle \ . \tag{A.5.2}$$

The expectation value is

$$\langle a(\beta) a^*(\beta') \rangle = \frac{k\theta}{2\pi} \delta(\beta - \beta') \ . \tag{A.5.3}$$

In problems involving the excitation of linear multiports, equal frequencies couple to each other, not equal propagation constants. Thus, it is more appropriate to use amplitudes $a(\omega)$ assigned to frequency intervals $\Delta\omega$, rather than $a(\beta)$ assigned to propagation-constant intervals. Further, the quantity of interest is *power flow* in a frequency interval $\Delta\omega$:

$$\langle a(\omega) a^*(\omega') \rangle \Delta\omega \Delta\omega' = \frac{k\theta}{2\pi} \delta(\omega - \omega') \Delta\omega \Delta\omega' \to \frac{k\theta}{2\pi} \Delta\omega \ . \tag{A.5.4}$$

The amplitude $a(\omega)$ is related to $a(\beta)$ by

$$a(\omega) = \frac{1}{\sqrt{v_g}} a(\beta) \; . \tag{A.5.5}$$

Finally, we introduced another normalization of the complex mode amplitude, namely a without parentheses, which simplified the analysis of noise in multiports. The fact that excitations at different frequencies were uncorrelated was subsumed, and the amplitudes were written so that $|a|^2$ was equal to the power within the frequency increment $\Delta\omega$:

$$\langle aa^* \rangle = \frac{k\theta}{2\pi} \Delta\omega \; . \tag{A.5.6}$$

The relationships among the different amplitudes may be gleaned from these equilibrium relations. We have

$$2\pi \langle a(\beta) a^*(\beta) \rangle \Delta\beta = 2\pi \langle a(\omega) a^*(\omega) \rangle \Delta\omega = 2\pi \langle aa^* \rangle / \Delta\omega = |A_n|^2 \; , \quad (A.5.7)$$

and thus

$$\frac{2\pi}{\sqrt{L}} a(\beta) = \frac{2\pi}{\sqrt{L/v_g}} a(\omega) = \sqrt{L/v_g} \, a = A_n \; . \tag{A.5.8}$$

A.5.2 Normalization of Quantum Operators

In the quantum analysis, the definition of the operators follows closely the renormalizations of the classical amplitudes. In the quantum analysis we do not deal with energies, but rather with photon numbers, which are related to the energies through division by $\hbar\omega_o$. Further, the commutator brackets play a similar role to the equipartition theorem in classical statistical mechanics at thermal equilibrium. In this spirit, the above relations can be rewritten using commutator brackets instead:

$$[\hat{A}_m, \hat{A}_n^\dagger] = \delta_{nm} \; , \tag{A.5.9}$$

and, for the commutator defined per unit length,

$$[\hat{a}(\beta), \hat{a}^\dagger(\beta')] = \frac{1}{2\pi} \delta(\beta - \beta') \; . \tag{A.5.10}$$

The power flow becomes a commutator flow. Here we have taken two approaches. We noted that a linear multiport excited by incoming waves couples waves of equal frequencies, not of equal propagation constants. The quantum operators are related in a similar way to (A.5.8):

$$\frac{2\pi}{\sqrt{L}} \hat{a}(\beta) = \frac{2\pi}{\sqrt{L/v_g}} \hat{a}(\omega) = \sqrt{L/v_g} \, \hat{a} = \hat{A}_n \; . \tag{A.5.11}$$

The different renormalized operators are related to each other in the same way as their classical excitation amplitudes. With regard to (A.5.11), note should be taken of the fact that the characterization is still in terms of prop-agation constants; the only change is that the interval designation $\Delta\beta$ has been replaced by $\Delta\omega/v_g$. This is a departure from the classical analogy. Classically, the frequency spectrum is the transform of the time dependence of an excitation. In quantum theory, on the other hand, the evolution of operators (in the Heisenberg representation) or states (in the Schrödinger representation) occurs in time; a wavepacket with a certain time dependence in the classical sense is described as a superposition of mode excitations in the β representation. This representation can be Fourier transformed into the x representation, thus preserving the special role of the time variable in the Heisenberg equation. We have, instead of (A.5.10),

$$[\hat{a}(x), \hat{a}^\dagger(x')] = \delta(x - x') .\tag{A.5.12}$$

The creation operator $\hat{a}^\dagger(x)$ generates a photon in the spatial interval Δx.

A.6 Two Alternative Expressions for the Nyquist Source

We found that a termination impedance Z at thermal equilibrium has an associated noise source

$$\langle |E_s|^2 \rangle = 4\operatorname{Re}(Z)k\theta B .\tag{A.6.1}$$

In the wave formalism, the wave source delivers a power $(1 - |\Gamma|^2)k\theta B$. Clearly, the two results must be consistent. In this appendix we derive (A.6.1) using the expressions for the power of the wave source. The wave source is a composite of a voltage source E and a current source J that is fully correlated with E and equal to $Y_o E$. The source s is related to E and J by

$$s = \frac{1}{2}(\sqrt{Y_o}E + \sqrt{Z_o}J) = \sqrt{Y_o}E .\tag{A.6.2}$$

The mean square value of E is thus

$$\langle |E|^2 \rangle = Z_o \left(1 - |\Gamma|^2\right) k\theta B = Z_o \left(1 - \left|\frac{Z - Z_o}{Z + Z_o}\right|^2\right) k\theta B .\tag{A.6.3}$$

Next, consider the equivalent voltage source in series with Z. The wave generator consists of a current source in parallel and a voltage source in series, as shown in Fig. A.6.1. These can be converted into a single voltage source in series using the Thévenin equivalent, as shown in Fig. A.6.2. Hence, the equivalent noise source in series with the impedance has the mean square value

$$E_s = E + ZJ = E\left(1 + \frac{Z}{Z_o}\right) .\tag{A.6.4}$$

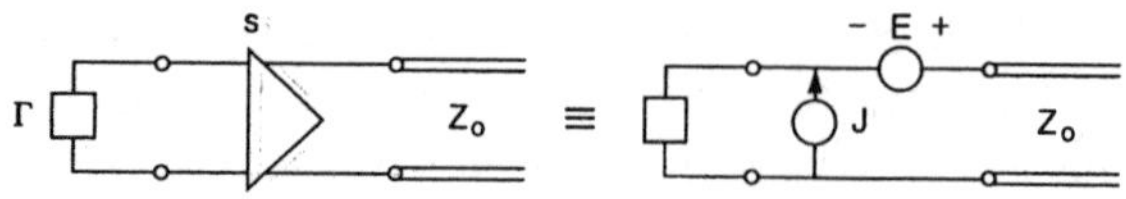

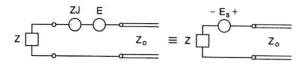

Fig. A.6.1. Representation of wave generator

Fig. A.6.2. Conversion of termination

Its mean square value is

$$|E_s|^2 = |E|^2 \left|1 + \frac{Z}{Z_o}\right|^2 = Z_o \left(1 - \left|\frac{Z - Z_o}{Z + Z_o}\right|^2\right) k\theta B \left|1 + \frac{Z}{Z_o}\right|^2 \tag{A.6.5}$$

$$= 4\,\mathrm{Re}(Z)k\theta B ,$$

which agrees with the alternative derivation.

A.7 Wave Functions and Operators in the n Representation

It is convenient to express the wave functions and the operators defined in the text in the number state representation. Thus, a general state of an electromagnetic field is given by

$$|\psi\rangle = \sum_n c_n |n\rangle , \tag{A.7.1}$$

where the $|n\rangle$ are photon number states in Dirac notation. The c_n are complex coefficients. In particular, we have introduced the coherent state for which the c_n are given by

$$c_n = e^{-|\alpha|^2/2} \frac{\alpha^n}{n!} . \tag{A.7.2}$$

The identity operator is clearly

$$I = \sum_n |n\rangle\langle n| . \tag{A.7.3}$$

Indeed, operation of the identity operator on a state leaves the state invariant:

$$I|\psi\rangle = \sum_n |n\rangle\langle n| \sum_m c_m |m\rangle = \sum_n c_n |n\rangle = |\psi\rangle . \qquad (A.7.4)$$

The number state operator is given by

$$\hat{n} = \sum_n n|n\rangle\langle n| . \qquad (A.7.5)$$

The expectation value of the number operator is the average photon number:

$$\langle\psi|\hat{n}|\psi\rangle = \sum_n c_n^* \langle n| \sum_m m|m\rangle\langle m| \sum_p c_p|p\rangle = \sum_n n|c_n|^2 = \langle n\rangle . \qquad (A.7.6)$$

We see that the coefficients $|c_n^2|$ function as probabilities. Next, we consider the annihilation operator. This is

$$\hat{A} = \sum_n \sqrt{n+1}|n\rangle\langle n+1| . \qquad (A.7.7)$$

This operator, operating on a state $|\alpha\rangle$, gives

$$\hat{A}|\alpha\rangle = \sum_n \sqrt{n+1}|n\rangle\langle n+1| \sum_m e^{-\alpha^2/2}\frac{\alpha^m}{\sqrt{m!}}|m\rangle$$

$$= \alpha \sum_n e^{-\alpha^2/2}\frac{\alpha^n}{\sqrt{n!}}|n\rangle = \alpha|\alpha\rangle . \qquad (A.7.8)$$

Hence, the coherent state is indeed an eigenstate of the annihilation operator.

Carruthers and Nieto [197] have introduced operators that can be viewed as cosine and sine operators. Let us start with their definition and then show the plausibility of this identification. The operators are

$$\hat{C} = \frac{1}{2}\left(\frac{1}{\sqrt{\hat{n}+1}}\hat{A} + \hat{A}^\dagger\frac{1}{\sqrt{\hat{n}+1}}\right) , \qquad (A.7.9)$$

$$\hat{S} = \frac{1}{2i}\left(\frac{1}{\sqrt{\hat{n}+1}}\hat{A} - \hat{A}^\dagger\frac{1}{\sqrt{\hat{n}+1}}\right) . \qquad (A.7.10)$$

These are Hermitian operators. The inverse square root of the operator $n+1$ is interpreted in terms of a Taylor expansion in powers of $n+1$. Note that the classical interpretation of $\sqrt{\hat{n}+1} \approx \sqrt{\hat{n}}$ is the amplitude of an α state. Hence, if we interpreted the annihilation and creation amplitudes as complex phasors, we would interpret (A.7.9) and (A.7.10) as the cosine and sine. The operators (A.7.9) and (A.7.10) have simple appearances in the number representation. The operator $(1/\sqrt{\hat{n}+1})\hat{A}$ is simply

$$\frac{1}{\sqrt{\hat{n}+1}}\hat{A} = \sum_n |n\rangle\langle n+1| . \qquad (A.7.11)$$

The cosine and sine operators are thus

$$\hat{C} = \frac{1}{2}\left(\sum_n |n\rangle\langle n+1| + \sum_n |n+1\rangle\langle n| \right) , \tag{A.7.12}$$

$$\hat{S} = \frac{1}{2i}\left(\sum_n |n\rangle\langle n+1| - \sum_n |n+1\rangle\langle n| \right) . \tag{A.7.13}$$

The commutators of these operators with the number state operator are of interest. For this purpose, let us look at the commutator of $\sum_n |n\rangle\langle n+1|$ with $\hat{n}$. We have

$$[\hat{n}, \sum_n |n\rangle\langle n+1|]$$

$$= \sum_n n|n\rangle\langle n| \sum_m |m\rangle\langle m+1| - \sum_m |m\rangle\langle m+1| \sum_n n|n\rangle\langle n|$$

$$= \sum_n n|n\rangle\langle n+1| - \sum_m (m+1)|m\rangle\langle m+1| \tag{A.7.14}$$

$$= -\sum_n |n\rangle\langle n+1| .$$

In a similar way, we find

$$[\hat{n}, \sum_n |n+1\rangle\langle n|]$$

$$= \sum_n |n\rangle\langle n| \sum_m |m+1\rangle\langle m| - \sum_m |m+1\rangle\langle m| \sum_n n|n\rangle\langle n| \tag{A.7.15}$$

$$= \sum_n |n\rangle\langle n+1| .$$

In this way we find

$$[\hat{n}, \hat{C}] = -i\hat{S} \tag{A.7.16}$$

and

$$[\hat{n}, \hat{S}] = i\hat{C} . \tag{A.7.17}$$

The cosine and sine operators are referred to the real and imaginary axes in the complex phasor plane. It is often convenient to pick operators referenced to a particular α state so that their expectation values yield the in-phase and quadrature components of the α state. Suppose that the α state has phase ϕ, i.e. $\arg(\alpha) = \phi$. We may define

$$\hat{C}(\phi) = \hat{C}\cos\phi + \hat{S}\sin\phi \qquad (A.7.18a)$$

and

$$\hat{S}(\phi) = \hat{S}\cos\phi - \hat{C}\sin\phi . \qquad (A.7.18b)$$

These new operators have the same commutation relations as the original ones. They are referred to the phase ϕ. Let us consider the expectation values of the operators for an α state with $\arg[\alpha] = \phi$. We have

$$\langle\alpha|\hat{C}(\phi)|\alpha\rangle = \langle\alpha|\hat{C}\cos\phi + \hat{S}\sin\phi|\alpha\rangle$$

$$= \sum_n e^{-|\alpha|^2/2}\frac{\alpha^{*n}}{\sqrt{n!}}|n\rangle \begin{bmatrix} \frac{1}{2}\left(\sum_m |m\rangle\langle m+1| + \sum_m |m+1\rangle\langle m|\right)\cos\phi \\ \\ +\frac{1}{2i}\left(\sum_m |m\rangle\langle m+1| - \sum_m |m+1\rangle\langle m|\right)\sin\phi \end{bmatrix}$$

$$\times \sum_p e^{-|\alpha|^2/2}\frac{\alpha^p}{\sqrt{p!}}|p\rangle$$

$$= \frac{(\alpha+\alpha^*)\cos\phi - i(\alpha-\alpha^*)\sin\phi}{2}\sum_n \frac{e^{-|\alpha|^2}}{\sqrt{n+1}}\frac{|\alpha|^2}{n!}$$

$$= |\alpha|\sum_n \frac{e^{-\langle n\rangle}}{\sqrt{n+1}}\frac{\langle n\rangle^n}{n!} = |\alpha|\left\langle\frac{1}{\sqrt{n+1}}\right\rangle .$$

$$(A.7.19)$$

The expectation value of the operator is equal to the product of the square root of the average photon number and the average of $1/\sqrt{(n+1)}$. For a large photon number, the product approaches unity. The relative mean square fluctuations approach zero, as we now proceed to show. The analysis is simplified if we set the argument of α equal to zero, treating α as a real number and setting $\phi = 0$. Then

$$\langle\alpha|\hat{C}^2|\alpha\rangle = \frac{1}{4}\sum_n e^{-|\alpha|^2/2}\frac{\alpha^n}{\sqrt{n!}}\langle n|$$

$$\times \left(\sum_m |m\rangle\langle m+1| + \sum_m |m+1\rangle\langle m|\right)$$

$$\times \left(\sum_p |p\rangle\langle p+1| + \sum_p |p+1\rangle\langle p| \right) \sum_q e^{-|\alpha|^2/2} \frac{\alpha^q}{\sqrt{q!}} |q\rangle$$

$$= \frac{1}{2} e^{-\alpha^2} \sum_m \frac{\alpha^{2m}}{m!} \left(\frac{\alpha^2}{\sqrt{(m+1)(m+2)}} \right)$$

$$= \frac{1}{2} \left(1 + \langle n\rangle \left\langle \frac{1}{\sqrt{(n+1)(n+2)}} \right\rangle \right).$$

$$(A.7.20)$$

The mean square fluctuations are

$$\langle\alpha|\hat{C}^2|\alpha\rangle - \langle\alpha|\hat{C}|\alpha\rangle^2 = \frac{1}{2} \left(1 + \langle n\rangle \left\langle \frac{1}{\sqrt{(n+1)(n+2)}} \right\rangle \right)$$

$$(A.7.21)$$

$$- \langle n\rangle \left\langle \frac{1}{\sqrt{n+1}} \right\rangle^2.$$

If we expand the fractions and square roots, we find that

$$\langle\alpha|\hat{C}^2|\alpha\rangle - \langle\alpha|\hat{C}|\alpha\rangle^2 \approx \frac{1}{4} \frac{1}{\langle n\rangle}. \tag{A.7.22}$$

The fluctuations are small compared with unity for large photon numbers. When an operator acquires mean square deviations much smaller than its expectation value, it can be replaced approximately by a c number. Thus, we may introduce an approximate operator applicable to coherent states and states that have a large photon number and a relatively small spread of phase. If we consider an α state with a small phase ϕ, then the commutation relation

$$[\hat{n}, \hat{S}(\phi)] = i\hat{C}(\phi) \tag{A.7.23}$$

can be approximated by

$$[\hat{n}, \hat{S}(\phi)] = i, \tag{A.7.24}$$

with $C(\phi)$ replaced by unity. The approximate operator $\hat{S}(\phi)$, obeying the commutation relation (A.7.24) and usually denoted by the operator symbol $\hat{\theta}$, is Heitler's phase operator. It is applicable to states of large photon number for which a probabilistic phase distribution can be defined unequivocally.

A.8 Heisenberg's Uncertainty Principle

The Heisenberg uncertainty principle applies to the mean square fluctuations of two noncommuting observables. If two observables are represented by the operators $\hat{A}$ and $\hat{B}$ and the commutator of the two operators is a c number iC, where C is real,

$$[\hat{A}, \hat{B}] = iC , \tag{A.8.1}$$

then the mean square fluctuations $\langle \Delta\hat{A}^2 \rangle$ and $\langle \Delta\hat{B}^2 \rangle$ obey the inequality

$$\langle \Delta\hat{A}^2 \rangle \langle \Delta\hat{B}^2 \rangle \geq C^2 . \tag{A.8.2}$$

The proof of this relation proceeds as follows. We first introduce the Schwarz inequality. Consider two states $|u\rangle$ and $|v\rangle$. Any state has a real, nonnegative norm,

$$\langle u|u \rangle > 0 . \tag{A.8.3}$$

The scalar product of $|u\rangle$ with $|v\rangle$ is the complex conjugate of the scalar product in reverse order:

$$\langle u|v \rangle = \langle v|u \rangle^* . \tag{A.8.4}$$

From these properties follows the inequality

$$|\langle u|v \rangle|^2 \leq \langle u|u \rangle \langle v|v \rangle . \tag{A.8.5}$$

Now, define the deviation operators $\hat{a}$ and $\hat{b}$ by $\hat{a} = \hat{A} - \langle \hat{A} \rangle$ and $\hat{b} = \hat{B} - \langle \hat{B} \rangle$. Then the new operators obey the commutations relation

$$[\hat{a}, \hat{b}] = iC . \tag{A.8.6}$$

The mean square fluctuations are $\langle \Delta\hat{A}^2 \rangle = \langle \hat{a}^2 \rangle$ and $\langle \Delta\hat{B}^2 \rangle = \langle \hat{b}^2 \rangle$. If the state of the system is $|\Psi\rangle$, then $\hat{a}|\Psi\rangle$ is a new state $|u\rangle$ and $\hat{b}|\Psi\rangle$ is a new state $|v\rangle$. Using the Schwarz inequality, we obtain

$$\langle \Psi|\hat{a}^2|\Psi \rangle \langle \Psi|\hat{b}^2|\Psi \rangle = \langle u|u \rangle \langle v|v \rangle \geq |\langle u|v \rangle|^2 = |\langle \Psi|\hat{a}\hat{b}|\Psi \rangle|^2 . \tag{A.8.7}$$

Here we have used the fact that $\hat{a}$ and $\hat{b}$ are Hermitian operators. Now, separate the operator product $\hat{a}\hat{b}$ into Hermitian and anti-Hermitian parts:

$$\hat{a}\hat{b} = \frac{1}{2}(\hat{a}\hat{b} + \hat{b}\hat{a}) + \frac{1}{2}(\hat{a}\hat{b} - \hat{b}\hat{a}) = \frac{1}{2}(\hat{a}\hat{b} + \hat{b}\hat{a}) + \frac{iC}{2} . \tag{A.8.8}$$

Decompose $\langle \Psi|\hat{a}\hat{b}|\Psi \rangle = \langle u|v \rangle$ into real and imaginary parts:

$$\langle \Psi|\hat{a}\hat{b}|\Psi \rangle = \frac{1}{2}\langle \Psi|\hat{a}\hat{b} + \hat{b}\hat{a}|\Psi \rangle + \frac{iC}{2} . \tag{A.8.9}$$

Rewriting the Schwarz inequality, we obtain:

$$\langle|\hat{a}^2|\rangle\langle|\hat{b}^2|\rangle \geq \left\langle\left|\frac{\hat{a}\hat{b}+\hat{b}\hat{a}}{2}\right|\right\rangle^2 + \frac{|C|^2}{4} , \tag{A.8.10}$$

and, of course, even more strongly,

$$\langle|\hat{a}^2|\rangle\langle|\hat{b}^2|\rangle = \langle|\Delta A^2|\rangle\langle|\Delta\hat{B}^2|\rangle \geq \frac{|C|^2}{4} = \frac{1}{4}|[\hat{A},\hat{B}]|^2 . \tag{A.8.11}$$

This is Heisenberg's uncertainty principle. If a state $|\Psi\rangle$ is prepared, then measurements of $\hat{A}$ on an ensemble of identically prepared states and measurements of $\hat{B}$ on another ensemble of similarly prepared states yield a scatter of data that obeys the inequality (A.8.11). If it is found that the inequality is obeyed with the equality sign, the corresponding states are called minimum-uncertainty states.

A.9 The Quantized Open-Resonator Equations

In Sect. 6.5 we showed how the decay rate of an open resonator can be evaluated from the coupling of the resonator to the modes of a waveguide, ideally infinitely long. The coupling of the waveguide modes back into the resonator accounted for the Langevin noise sources that maintain the commutator of the resonator mode amplitude.

The formalism can be carried further to derive the full quantum equations of the open resonator We note that (6.90), repeated below,

$$\frac{d\hat{U}}{dt} = -i\left(\omega_o - \frac{i}{\tau_e}\right)\hat{U} - i\sum_j K_j V_j , \tag{A.9.1}$$

contains the information on the excitation of the resonator by the waveguide modes incident upon the resonator in the sum $\sum_j K_j\hat{V}_j$. This excitation is caused by the traveling-wave component $\hat{a}$ of the standing-wave modes propagating in the direction of the resonator. Power conservation arguments, as presented in Sect. 2.12, identify this component with $\sqrt{2/\tau_e}\hat{a}(t)$:

$$-i\sum_j K_j\hat{V}_j \rightarrow \sqrt{\frac{2}{\tau_e}}\int d\omega\, e^{-i\omega t}\hat{a}(\omega) , \tag{A.9.2}$$

where we have adhered to the assumption that all coupling coefficients are imaginary and equal.

The identification (A.9.2) can be checked by evaluating the commutators. We have for the commutator of the left hand side:

$$\left[\sum_j K_j \hat{V}_j, \sum_k K_k^* \hat{V}_k^\dagger \right] = |K|^2 \left[\sum_j \hat{V}_j, \sum_k \hat{V}_k^* \right] \tag{A.9.3}$$

$$= |K|^2 \sum_k = \frac{v_g}{\tau_e} \frac{1}{L} \sum_k .$$

The commutator of the right-hand side is

$$\frac{2}{\tau_e} \left[\int d\omega \, e^{-i\omega t} \hat{a}(\omega), \int d\omega' \, e^{-i\omega' t} \hat{a}^\dagger(\omega') \right]$$

$$= \frac{2}{\tau_e} \frac{1}{2\pi} \left[\int d\omega \int d\omega' \, \delta(\omega - \omega') \right] \tag{A.9.4}$$

$$= \frac{1}{\pi \tau_e} \int_{\text{band}} d\omega = \frac{1}{\pi \tau_e} \Delta\omega \sum_k = \frac{v_g}{\tau_e} \frac{1}{L} \sum_k ,$$

where $\Delta\omega = v_g \pi / L$ is the frequency separation of the modes. The commutators indeed agree. Hence, we find that (A.9.1) can be cast into the form

$$\frac{d\hat{U}}{dt} = -\left(i\omega_o + \frac{1}{\tau_e} \right) \hat{U} + \sqrt{\frac{2}{\tau_e}} \hat{a}(t) , \tag{A.9.5}$$

the form which is derived in (6.12) as a result of the quantization of the classical open-resonator equation. Next consider the excitation of the waveguide by the resonator. From (6.84) we have

$$\frac{d\hat{V}_j}{dt} = -i\omega_j V_j - iK_j^* \hat{U} . \tag{A.9.6}$$

From this equation it follows that the resonator excites the superposition of modes

$$\sum_j \hat{V}_j^{(U)} = -\sum_j \frac{K_j^* \hat{U}}{\omega_j - \omega} \approx -K^* \hat{U} \frac{1}{\Delta\omega_j} \int \frac{d\omega_j}{\omega_j - \omega} . \tag{A.9.7}$$

This superposition of modes forms a traveling wave propagating away from the resonator, a wave $\hat{b}^{(U)}$ that, in accordance with (A.9.2), is related to $\sum_j \hat{V}_j^{(U)}$ by

$$-i \sum_j K_j^* \hat{V}_j^{(U)} \rightarrow \sqrt{\frac{2}{\tau_e}} \int d\omega \, e^{-i\omega t} \hat{b}^{(U)}(\omega) . \tag{A.9.8}$$

From (A.9.8) and (A.9.7) we find

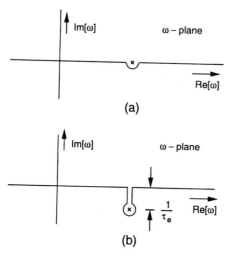

Fig. A.9.1. Contour in complex ω plane: (a) for real excitation frequency; (b) for complex frequency

$$\sqrt{\frac{2}{\tau_e}} \int d\omega\, e^{-i\omega t} \hat{b}^{(U)}(\omega) = -i\sum_j K_j^* V_j^{(U)} = -iK^{*2} \sum_j \frac{\hat{U}}{\omega_j - \omega}$$

(A.9.9)

$$\simeq i\frac{1}{\pi\tau_e}\hat{U} \int \frac{d\omega_j}{\omega_j - \omega}.$$

Again, we face an integral expressing a summation over all the modes. It looks like the integral that led to (6.88). However, there is a subtle difference. Equation (6.88) is a determinantal equation derived for a superposition of waveguide modes, all excited at a real frequency. The contour of the integral passes around the pole in a semicircle, as shown in Fig. A.9.1. The side on which the pole is passed is determined by the fact that a Laplace transform starts in the upper half of the complex ω plane and the pole reaches the real axis from above. The integral (A.9.9) is written for the complex frequency at which the resonator mode decays. The pole moves through the real axis into the lower half-plane and is fully encircled by the contour. This gives an additional factor of 2, so that we obtain

$$\int \frac{d\omega_j}{\omega_j - \omega} = 2\pi i\,.$$

(A.9.10)

Combining (A.9.9) and (A.9.10), we find for the wave emitted by the resonator

$$\int d\omega\, e^{-i\omega t} \hat{b}^{(U)}(\omega) = \hat{b}^{(U)}(t) = -\sqrt{\frac{2}{\tau_e}}\hat{U}\,.$$

(A.9.11)

In the absence of the resonator, the boundary condition on the standing-wave waveguide modes of a magnetic short at the reference plane imposes the constraint $\hat{b} = \hat{a}$. The presence of the resonator changes this relation to

$$\hat{b} = \hat{a} - \sqrt{\frac{2}{\tau_e}}\hat{U} \ . \tag{A.9.12}$$

Equations (A.9.5) and (A.9.11) are the quantum equivalents of the classical equations of the open resonator that were derived from time reversal and energy conservation. The sign change is the result of redefinition of the reference plane in the waveguide. A quarter-wave shift of the reference plane and a redefinition of the phase of $\hat{U}$ establishes full correspondence.

A.10 Density Matrix and Characteristic Functions

Any state of the electromagnetic field can be described by a superposition of photon number states

$$|\Psi\rangle = \sum_n c_n |n\rangle \ . \tag{A.10.1}$$

Equation (A.10.1) is a so-called pure state, if the complex numbers c_n are all specifiable. If the process under consideration is a member of a statistical ensemble, then the system cannot be in a single pure state. To express such a statistical superposition, the density matrix is used. It is defined as a sum of the operators $|n\rangle\langle m|$

$$\hat{\rho} = \sum_{m,n} \overline{c_m^* c_n} |n\rangle\langle m| \ , \tag{A.10.2}$$

where $\overline{c_m^* c_n}$ are the statistically averaged products of the coefficients. When the coefficients are independent

$$\overline{c_m^* c_n} = |c_n|^2 \delta_{mn} \ , \tag{A.10.3}$$

and the density matrix simplifies to

$$\hat{\rho} = \sum_n |c_n|^2 |n\rangle\langle n| \ . \tag{A.10.4}$$

The trace of the density operator is unity

$$\mathrm{Tr}(\rho) = \sum_n |c_n|^2 = 1 \ . \tag{A.10.5}$$

The coefficients $|c_n|^2$ may be interpreted as probabilities. But even in the general case of a nondiagonal density matrix, the expectation value of an operator is the trace of the product of the density matrix with the operator:

$$\langle \hat{O} \rangle = \sum_m \langle m | c_m^* \hat{O} \sum_m c_n | n \rangle = \mathrm{Tr}\left(\hat{\rho}\hat{O} \right) . \tag{A.10.6}$$

Thermal equilibrium is represented by a density matrix that is diagonal in the photon number representation because thermal equilibrium is fully described by the probability distribution of the system's energy (e.g. photon number).

Even if the process is not a statistical one, the density matrix formulation may be preferred to the description of a quantum process in terms of state amplitudes c_n, since it removes an arbitrary phase of the state, which has no physical meaning (say the phase of c_o, all other phases being referred to that of c_o).

Equation (A.10.6) illustrates the fact that quantum theory contains, in general, two "averaging" operations: one is with respect to the quantum states; the other is with respect to the statistical ensemble representing the process.

The definition of the density matrix is analogous to the way correlation matrices are defined classically. Thus, if the E_i denotes amplitudes of noise voltages as in Chap. 5, the correlation matrix is defined as $\overline{E_i E_j^*}$. The overbar indicates a statistical average. Taking a single number state as an example, one may define its density matrix as $|n\rangle |c_n|^2 \langle n|$. If the state is made up of a statistical superposition of number states, its density matrix is $\rho = |n\rangle \overline{c_n c_m^*} \langle m|$, where we indicate the statistical average by an overbar to distinguish it from the quantum evaluation of an expectation value. The matrix $\overline{c_n c_m^*}$ is in complete analogy with the correlation matrix.

A.10.1 Example 1. Density Matrix of Bose–Einstein State

A Bose–Einstein state has the density matrix $\rho = |n\rangle \overline{c_n c_m^*} \langle m|$ with

$$\overline{c_n c_m^*} = \frac{1}{1 + \langle n \rangle} \left(\frac{\langle n \rangle}{1 + \langle n \rangle} \right)^n \delta_{nm} .$$

The correlation matrix is diagonal, since the excitations of different photon number states are statistically independent.

A.10.2 Example 2. Density Matrix of Coherent State

A coherent state is a deterministic superposition of number states and hence a statistical average is not taken. The "correlation matrix" is not diagonal and is of the form (note the absence of statistical averaging)

$$c_n c_m^* = \exp(-|\alpha|^2) \frac{\alpha^n \alpha^{*m}}{\sqrt{n!m!}} .$$

Figure A.10.1 shows the amplitude of this matrix versus n and m for $|\alpha|^2 = \langle n \rangle = 100$ and for α real and positive.

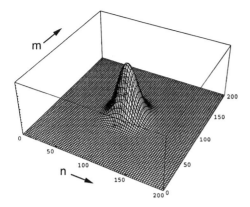

Fig. A.10.1. Amplitude of density matrix of coherent state

A.11 Photon States and Beam Splitters

The quantum properties of a beam splitter emerge clearly when both its inputs are in photon states. Here we go through the analysis of such an excitation. The Hamiltonian of a beam splitter can be written in the form (compare (7.12))

$$\hat{H} = \hbar(M\hat{A}^\dagger\hat{B} + M^*\hat{B}^\dagger\hat{A}) + \frac{1}{2}\hbar\omega \ . \tag{A.11.1}$$

The equation of motion of the wave function $|\psi(t)\rangle$ is

$$\frac{d}{dt}|\psi\rangle = -\frac{i}{\hbar}\hat{H}|\psi\rangle \ . \tag{A.11.2}$$

The solution of this equation, when integrated over the time T during which the wavepacket interacts with the beam splitter, is

$$|\psi(T)\rangle = \exp[-i(MT\hat{A}^\dagger\hat{B} + M^*T\hat{B}^\dagger\hat{A})]|\psi(0)\rangle \ . \tag{A.11.3}$$

Now suppose that the input is in the state $|\psi(0)\rangle = |1\rangle|1\rangle$. The output can be evaluated by expanding the exponential in (A.11.3) and evaluating the operation of the operator $(MT\hat{A}^\dagger\hat{B} + M^*T\hat{B}^\dagger\hat{A})^n$ on the wave function. Clearly,

$$(MT\hat{A}^\dagger\hat{B} + M^*T\hat{B}^\dagger\hat{A})|1\rangle|1\rangle = \sqrt{2}(MT|2\rangle|0\rangle + M^*T|0\rangle|2\rangle)$$

$$= 2|M|T\frac{1}{\sqrt{2}}(e^{i\phi}|2\rangle|0\rangle + e^{-i\phi}|0\rangle|2\rangle) \ ,$$

$$\tag{A.11.4}$$

where $\phi = \arg(M)$. The next operation gives

$$(MT\hat{A}^{\dagger}\hat{B} + M^{*}T\hat{B}^{\dagger}\hat{A})\sqrt{2}(MT|2\rangle|0\rangle + M^{*}T|0\rangle|2\rangle) = 4|MT|^{2}|1\rangle|1\rangle \ . \tag{A.11.5}$$

An operation by an odd power $2m + 1$ of the operator produces the wave function $(1/\sqrt{2})(e^{i\phi}|2\rangle|0\rangle + e^{-i\phi}|0\rangle|2\rangle)$ with a multiplier $(2|MT|)^{2m+1}$, and an operation by an even power $2m$ produces $(2|MT|)^{2m}|1\rangle|1\rangle$. The result is thus

$$\exp[-\mathrm{i}(MT\hat{A}^{\dagger}\hat{B} + M^{*}T\hat{B}^{\dagger}\hat{A})]|1\rangle|1\rangle$$

$$= \cos(2|M|T)|1\rangle|1\rangle + \sin(2|M|T)\frac{1}{\sqrt{2}}(e^{i\phi}|2\rangle|0\rangle + e^{-i\phi}|0\rangle|2\rangle) \ . \tag{A.11.6}$$

The beam splitter is a 50/50 beam splitter when $2|M|T = \pi/2$. Then the passage through the beam splitter produces the superposition state $1/\sqrt{2}(e^{i\phi}|2\rangle|0\rangle + e^{-i\phi}|0\rangle|2\rangle)$. This means that both photons emerge in either one output port or the other output port, with a probability of one-half. The photons, so to speak, "stick together". In the analysis of a beam splitter illuminated by one photon in one of the inputs we found that the beam splitter sends the photon into either one of the two output ports with a binomial probability distribution. What is new here is the simultaneous arrival of two photons at both inputs. Whereas one may say that each photon exits with a probability of one-half, the two photons always exit in pairs.

A.12 The Baker–Hausdorff Theorem

A.12.1 Theorem 1

Denote by $\hat{A}$ *and* $\hat{B}$ *two noncommuting operators that satisfy the condition*

$$\left[\hat{A}, [\hat{A}, \hat{B}]\right] = \left[\hat{B}, [\hat{A}, \hat{B}]\right] = 0 \ . \tag{A.12.1}$$

If ξ *is a c number, the following relationship holds:*

$$e^{\xi\hat{A}}\hat{B}e^{-\xi\hat{A}} = \hat{B} + \xi[\hat{A}, \hat{B}] \ . \tag{A.12.2}$$

The proof of the theorem is as follows. Define the function $h(\xi)$ as

$$h(\xi) \equiv e^{\xi\hat{A}}\hat{B}e^{-\xi\hat{A}} \ . \tag{A.12.3}$$

Now, differentiate $h(\xi)$ with respect to ξ:

$$\frac{dh}{d\xi} = e^{\xi\hat{A}}\hat{A}\hat{B}e^{-\xi\hat{A}} - e^{\xi\hat{A}}\hat{B}\hat{A}e^{-\xi\hat{A}} = [\hat{A}, \hat{B}] \ . \tag{A.12.4}$$

Since $h(0) = \hat{B}$, we have, after integrating (A.12.4),

$$h(\xi) = \hat{B} + \xi[\hat{A}, \hat{B}] \ . \tag{A.12.5}$$

This completes the proof.

A.12.2 Theorem 2

If the operators $\hat{A}$ and $\hat{B}$ satisfy (A.12.1) and ξ is a c number, the following relationship holds:

$$\exp\left(\xi(\hat{A}+\hat{B})\right)\exp\left((\xi^2/2)[\hat{A},\hat{B}]\right) = e^{\xi\hat{A}}e^{\xi\hat{B}} \ . \tag{A.12.6}$$

For the proof, define the function

$$f(\xi) = e^{\xi\hat{A}}e^{\xi\hat{B}} \ . \tag{A.12.7}$$

If we differentiate with respect to ξ, we obtain the result

$$\frac{d}{d\xi}f = \hat{A}e^{\xi\hat{A}}e^{\xi\hat{B}} + e^{\xi\hat{A}}e^{\xi\hat{B}}\hat{B} = (\hat{A} + e^{\xi\hat{A}}\hat{B}e^{-\xi\hat{A}})f(\xi) \ , \tag{A.12.8}$$

where we have used the fact that $\exp(-\xi\hat{A})\exp(\xi\hat{A}) = I$. Because of (A.12.2) we have

$$e^{\xi\hat{A}}\hat{B}e^{-\xi\hat{A}} = \hat{B} + \xi[\hat{A},\hat{B}] \ . \tag{A.12.9}$$

Thus (A.12.8) can be rewritten as

$$\frac{d}{d\xi}f = \{(\hat{A}+\hat{B}) + \xi[\hat{A},\hat{B}]\}f(\xi) \ . \tag{A.12.10}$$

Since, according to (A.12.1), $\hat{A}$ and $\hat{B}$ commute with $[\hat{A},\hat{B}]$, the variable $\hat{A}+\hat{B}$ can be treated as a c number and the integration can proceed in the standard way. Since $f(0) = 1$, we obtain after integration

$$f(\xi) = e^{\xi(\hat{A}+\hat{B})}e^{(\xi^2/2)[\hat{A},\hat{B}]} = e^{\xi\hat{A}}e^{\xi\hat{B}} \ . \tag{A.12.11}$$

This is the desired result. These proofs follow derivations by Glauber as presented by Louisell [65].

Thus far the Baker–Hausdorff theorems have been stated for "scalar" operators, as contrasted with the column matrix operators $\boldsymbol{a}$ and $\boldsymbol{a}^\dagger$ used in the text. The Baker–Hausdorff Theorem 2 is used for the derivation of the characteristic function of an observable. The generalization to the column matrix case adapts the theorem for use in the derivation of the characteristic function of a set of observables collected in a column matrix. We proceed as follows:

A.12.3 Matrix Form of Theorem 1

The operators $\hat{A}_i$ and $\hat{B}_j$ are assumed to have the commutator

$$[\hat{A}_i, \hat{B}_j] = \hat{D}_{ij} \ , \tag{A.12.12}$$

where both $\hat{A}_i$ and $\hat{B}_j$ commute with $\hat{D}_{ij}$. *The matrix form of theorem 1 states that if the ξ_i are a column matrix of c numbers the following relationship holds:*

$$\exp\left(\sum_i \xi_i \hat{A}_i\right)\hat{B}_j \exp\left(-\sum_k \xi_k A_k\right) = \hat{B}_j + \sum_i \xi_i[\hat{A}_i, \hat{B}_j]. \qquad \text{(A.12.13)}$$

The proof of the theorem starts with the definition of the functions

$$h_j(\xi_i) \equiv \exp\left(\sum_i \xi_i \hat{A}_i\right)\hat{B}_j \exp\left(-\sum_k \xi_k A_k\right). \qquad \text{(A.12.14)}$$

Differentiate $h_j(\xi_i)$ with respect to ξ_m. The result is

$$\begin{aligned}
\frac{\partial}{\partial \xi_m}h_j(\xi_i) &= \exp\left(\sum_i \xi_i \hat{A}_i\right)\hat{A}_m \hat{B}_j \exp\left(-\sum_k \xi_k A_k\right) \\
&\quad - \exp\left(\sum_i \xi_i \hat{A}_i\right)\hat{B}_j \hat{A}_m \exp\left(-\sum_k \xi_k A_k\right) \\
&= \exp\left(\sum_i \xi_i \hat{A}_i\right)[\hat{A}_m, \hat{B}_j]\exp\left(-\sum_k \xi_k A_k\right) \\
&= [\hat{A}_m, \hat{B}_j].
\end{aligned} \qquad \text{(A.12.15)}$$

But the integral of (A.12.15), with the constraint that $h_j(\xi_i) = \hat{B}_j$ for all $\xi_i = 0$, is (A.12.13), which completes the proof.

A.12.4 Matrix Form of Theorem 2

If the operators $\hat{A}_i$ and $\hat{B}_i$ satisfy (A.12.12) and the ξ_i are all c numbers, the following relationship holds:

$$\begin{aligned}
&\exp\left(\sum_i \xi_i(\hat{A}_i + \hat{B}_i)\right)\exp\left(\frac{1}{2}\sum_{j,k} \xi_j \xi_k[\hat{A}_j, \hat{B}_k]\right) \\
&= \exp\left(\sum_i \xi_i \hat{A}_i\right)\exp\left(\sum_j \xi_j \hat{B}_j\right).
\end{aligned} \qquad \text{(A.12.16)}$$

For the proof, define the function

$$f(\xi_i) = \exp\left(\sum_i \xi_i \hat{A}_i\right)\exp\left(\sum_j \xi_j \hat{B}_j\right). \qquad \text{(A.12.17)}$$

Differentiation with respect to ξ_k gives

$$\frac{\partial}{\partial \xi_k} f(\xi_i) = \hat{A}_k \exp\left(\sum_i \xi_i \hat{A}_i\right) \exp\left(\sum_j \xi_j \hat{B}_j\right)$$

$$+ \exp\left(\sum_i \xi_i \hat{A}_i\right) \exp\left(\sum_j \xi_j \hat{B}_j\right) \hat{B}_k$$

$$= \left[\hat{A}_k + \exp\left(\sum_j \xi_j \hat{A}_j\right) \hat{B}_k \exp\left(-\sum_m \xi_m \hat{A}_m\right)\right] \qquad \text{(A.12.18)}$$

$$\times \exp\left(\sum_i \xi_i \hat{A}_i\right) \exp\left(\sum_j \xi_j \hat{B}_j\right)$$

$$= \left\{\hat{A}_k + \hat{B}_k + \sum_m \xi_m [\hat{A}_m, \hat{B}_i]\right\} f(\xi_i)$$

$$= \left(\hat{A}_k + \hat{B}_k + \sum_m \xi_m \hat{D}_{mi}\right) f(\xi_i) ,$$

where we have used the fact that $\exp(-\sum_j \xi_j \hat{A}_j) \exp(\sum_i \xi_i \hat{A}_i) = I$, the identity, and we have employed (A.12.13). Since the $\hat{A}_i$ and $\hat{B}_i$ commute with their commutator, the differential equation (A.12.18) can be treated as a c-number differential equation. The value of $f(\xi_i)$ is 1 when all $\xi_i = 0$, and thus the integral is

$$f(\xi_i) = \exp\left(\sum_k \xi_k (\hat{A}_k + \hat{B}_k)\right) \exp\left(\frac{1}{2} \sum_{j,k} \xi_j \xi_k [\hat{A}_j, \hat{B}_k]\right) . \qquad \text{(A.12.19)}$$

This completes the proof of theorem 2 in matrix form.

A.13 The Wigner Function of Position and Momentum

In the text we deal with characteristic functions of observables and their Fourier transforms. The Fourier transforms of characteristic functions become probability distributions in the classical regime. In the quantum domain, the Fourier transforms of characteristic functions of noncommuting observables may acquire negative values, and thus cannot be interpreted as probability distributions. They are related to a function introduced by Wigner in 1932 [198]. Here we start with the characteristic function of position and momentum and show how the Wigner function is obtained from it.

Consider the characteristic function of momentum and position for the state $|\psi\rangle$:

$$C(\xi_1, \xi_2) = \langle | \exp(i\xi_1\hat{q} + i\xi_2\hat{p}) | \rangle$$

$$= \int dq \, \langle \psi | \exp(i\xi_1\hat{q} + i\xi_2\hat{p}) | \psi \rangle \ . \tag{A.13.1}$$

The Baker–Hausdorff theorem allows us to write this in the form

$$C(\xi_1, \xi_2) = \exp\left\{ -\frac{1}{2}[i\xi_1\hat{q}, i\xi_2\hat{p}] \right\} \int dq \, \langle \psi | \exp(i\xi_1\hat{q}) \exp(i\xi_2\hat{p}) | \psi \rangle \ . \tag{A.13.2}$$

The exponential $\exp(i\xi_1\hat{q})$ can be treated as a c number in the space of $\hat{q}$. The operator $\exp(i\xi_2\hat{p}) = \exp(\xi_2\hbar\partial/\partial q)$ operating on the wave function following it produces a displacement:

$$C(\xi_1, \xi_2) = \exp\left(\frac{i\hbar\xi_1\xi_2}{2} \right) \int dq \, \psi^*(q)\psi(q + \hbar\xi_2) \exp(i\xi_1 q) \ . \tag{A.13.3}$$

Fourier transformation with respect to ξ_1 gives

$$\frac{1}{2\pi} \int d\xi_1 \int dq \exp(-i\xi_1 x) C(\xi_1, \xi_2)$$

$$= \frac{1}{2\pi} \int d\xi_1 \int dq \exp\left[i\xi_1\left(q - x + \frac{\hbar\xi_2}{2} \right) \right] \psi^*(q)\psi(q + \hbar\xi_2)$$

$$= \int dq \, \delta\left(q - x + \frac{\hbar\xi_2}{2} \right) \psi^*(q)\psi(q + \hbar\xi_2)$$

$$= \psi^*\left(x - \frac{\hbar\xi_2}{2} \right) \psi\left(x + \frac{\hbar\xi_2}{2} \right) \ . \tag{A.13.4}$$

The second integration yields the Wigner function:

$$W(x, s) = \left(\frac{1}{2\pi} \right)^2 \int d\xi_1 \int d\xi_2 \exp(-i\xi_1 x - i\xi_2 s) C(\xi_1, \xi_2)$$

$$= \frac{1}{2\pi} \int d\xi \exp(-i\xi s)\psi^*\left(x - \frac{\hbar\xi}{2} \right) \psi\left(x + \frac{\hbar\xi}{2} \right) \ . \tag{A.13.5}$$

As an example, take the wave function

$$\psi(x) = \mathcal{N} \left[\exp\left(-\frac{(x - a)^2}{\sigma^2} \right) \pm \exp\left(-\frac{(x + a)^2}{\sigma^2} \right) \right] \ , \tag{A.13.6}$$

where $\mathcal{N}$ is a normalization constant. This wave function leads to the Wigner distribution

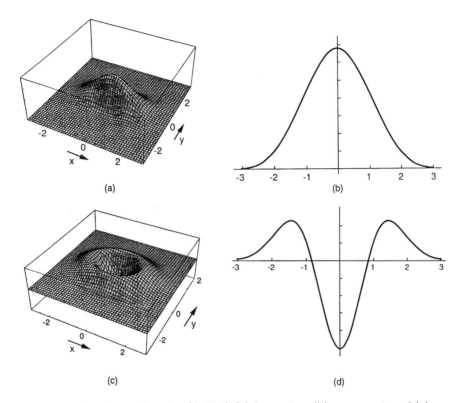

Fig. A.13.1. The Wigner function (A.13.7) (**a**) for + sign; (**b**) cross section of (**a**) at $y = 0$; (**c**) for − sign; (**d**) cross section of (**c**) at $y = 0$

$$W(x, s) \propto \exp\left(-\frac{y^2}{\sigma^2}\right)\left[\exp\left(-\frac{(x-a)^2}{\sigma^2}\right) + \exp\left(-\frac{(x+a)^2}{\sigma^2}\right)\right.$$

$$\left. \pm 2\cos\left(\frac{2ya}{\sigma^2}\right)\exp\left(-\frac{x^2}{\sigma^2}\right)\right],$$

$$(A.13.7)$$

with $y \equiv s/\hbar$. This function is plotted in Fig. A.13.1 for $a = \sigma$, for both signs. This Wigner function exhibits negative values.

A.14 The Spectrum of Non-Return-to-Zero Messages

Non-return-to-zero (NRZ) digital transmission is now widely used. It is the encoding employed in the new "repeaterless" transoceanic links. Ideally, the signal consists of rectangular pulses of width τ_o and height A, representing "ones", and of "zero" intervals of the same width. If two pulses follow each other they form a rectangle of twice the width. Hence the name "non-return-to-zero". Here we wish to evaluate the spectrum of the function

$$f(t) = \sum_r p(r)h(t - t_r) \,, \tag{A.14.1}$$

where $p(r)$ is the probability weighting function of the occurrence of a pulse at time t_r; $p(r)$ assumes the values 0 and 1 with probability $1/2$ each, the usual choice. The time instants t_r can be taken as the positions of the leading edge of the rectangle. The autocorrelation function of $\langle f(t)f(t-\tau)\rangle$ is obtained very similarly to the method employed in Sect. 4.2. The product of the two sums is written as a double sum. Then the terms involving the same event are grouped in one sum, and the terms representing products due to two different events are grouped in another sum.

$$\langle f(t), f(t - \tau)\rangle$$

$$= \left\langle \sum_{r=r'} h(t - t_r)h(t - \tau - t_{r'}) \right\rangle + \left\langle \sum_{r'\neq r} h(t - t_{r'}) \sum_{r\neq r'} h(t - t_r) \right\rangle$$

$$= \sum_{r=r'} p(r)h(t - t_r)h(t - \tau - t_{r'})$$

$$+ \sum_{r'\neq r} p(r')h(t - t_{r'}) \sum_{r\neq r'} p(r)h(t - t_r) \,,$$

$$\tag{A.14.2}$$

where the last expression takes advantage of the independence of events, making an average of a product into the product of the averages.

For the evaluation of the averages, it is helpful to visualize the functions. Consider the first term and pick one of the terms in the sum assigned to the time instant t_r. The individual terms are still functions of t and τ. Figure A.14.1 illustrates the function in the $t - \tau$ plane. As the pulse is moved over by τ the overlap becomes smaller and the rectangle representing the product becomes narrower, going to zero width when $\tau = \tau_o$. The statistical average can be supplemented by a time average. We have

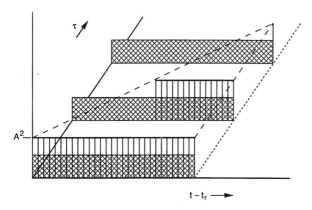

Fig. A.14.1. The function $h(t - t_r)h(t - t_r - \tau)$

$$\left\langle \sum_{r=r'} h(t - t_r)h(t - \tau - t_{r'}) \right\rangle = \frac{1}{T} \int_{-T/2}^{T/2} dt\, p(r)h(t - t_r)h(t - \tau - t_{r'})$$

$$= \begin{cases} \dfrac{1}{2} A^2 \dfrac{\tau_o - |\tau|}{\tau_o} & \text{for } |\tau| \leq \tau_o \\[2ex] 0 & \text{for } |\tau| > \tau_o \,, \end{cases}$$

$$(A.14.3)$$

where A is the amplitude of the rectangle. The averages in the second term can also be evaluated by time integrals of the individual factors:

$$\frac{1}{T} \int_{-T/2}^{T/2} dt \sum_r p(r)h(t - t_r) = A\frac{\tau_o}{T}\frac{T}{\tau_o} = \frac{1}{2}A \,. \qquad (A.14.4)$$

This process has not excluded terms with $r = r'$. The error decreases to zero as the time interval T is taken infinitely long.

We thus obtain for the correlation function

$$\langle f(t)f(t - \tau) \rangle = A^2 \left(\frac{1}{2}\frac{\tau_o - |\tau|}{\tau_o} + \frac{1}{4} \right) \qquad (A.14.5)$$

for $|\tau| < \tau_o$ and $\langle f(t)f(t - \tau) \rangle = \frac{1}{4}A^2$ for $|\tau| > \tau_o$. Fourier transformation gives us the spectrum

$$\frac{1}{2\pi} \int d\tau \, \langle f(t)f(t-\tau) \rangle \exp(-i\omega\tau)$$

$$= A^2 \frac{1}{2\pi} \int d\tau \, \exp(-i\omega\tau) \left(\frac{1}{2} \frac{\tau_o - |\tau|}{\tau_o} + \frac{1}{4} \right) \tag{A.14.6}$$

$$= \frac{1}{2\pi} \tau_o A^2 \frac{1 - \cos(\omega\tau_o)}{\omega^2 \tau_o^2} + \frac{1}{4} A^2 \delta(\omega) \, .$$

The spectrum is illustrated in Fig. A.14.2.

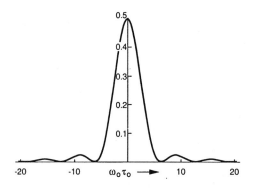

Fig. A.14.2. A plot of the first term in (A.14.6), $[1 - \cos(\omega\tau)]/(\omega\tau_o)^2$

A.15 Various Transforms of Hyperbolic Secants

The hyperbolic secant plays an important role in the perturbation analysis of soliton propagation. For this reason it is useful to have a compendium of mathematical relations for the hyperbolic secant and its powers.

We start with the Fourier transform of sech t,

$$\text{F.T. (sech } t) = \int_{-\infty}^{\infty} dt \, e^{i\Omega t} \text{sech} \, t \, . \tag{A.15.1}$$

This Fourier transform can be evaluated with the help of the residue theorem. The hyperbolic secant turns into a secant for imaginary values of the argument. The secant has an infinite set of poles. In particular, on the positive imaginary axis in the complex Ω plane, the poles are at $t = i[(2n+1)/2]\pi$. The cosh function is expanded as follows:

$$\cosh\left(i\frac{2n+1}{2}\pi + \Delta t\right) = \cos\left(\frac{2n+1}{2}\pi + \frac{\Delta t}{i}\right) = (-1)^{n+1}\frac{\Delta t}{i} \, . \tag{A.15.2}$$

This expansion gives poles in the integral of (A.15.1) with the residues equal to $i(-1)^{n+1} \exp\{-[(2n+1)/2]\pi\Omega\}$. The infinite integral can be closed in the upper half-plane, picking up the integrals around each of the poles. Thus

$$
\text{F.T. } (\operatorname{sech} t) = \int_{-\infty}^{\infty} dt\, e^{i\Omega t} \operatorname{sech} t
$$

$$
= \sum_{0}^{\infty} (-1)^{n+1} 2\pi i \left[i \exp\left(-\frac{2n+1}{2}\pi\Omega \right) \right]
$$

(A.15.3)

$$
= \exp\left(-\frac{\pi}{2}\Omega \right) \sum_{0}^{\infty} (-1)^n 2\pi \exp(-n\pi\Omega)
$$

$$
= 2\pi \frac{\exp[-(\pi/2)\Omega]}{1 + \exp[-(\Omega)]} = \pi \operatorname{sech}\left(\frac{\pi}{2}\Omega \right).
$$

The Fourier transform of the sech function is a sech function. The Fourier transform of the sech-squared function can be evaluated in the same way. The square now introduces second-order poles. The function multiplying $1/\Delta t^2$ must be expanded to first order in Δt. Thus the behavior of the kernel near the nth pole is

$$
-\frac{1}{\Delta t^2} \exp\left(-\frac{2n+1}{2}\pi\Omega \right)(1 + i\Omega\,\Delta t).
$$

(A.15.4)

The integration around each pole gives

$$
-2\pi i(i\Omega) \exp\left(-\frac{2n+1}{2}\pi\Omega \right) = 2\pi\Omega \exp\left(-\frac{2n+1}{2}\pi\Omega \right).
$$

(A.15.5)

The summation over all contour integrals gives

$$
\text{F.T. } (\operatorname{sech}^2 t) = \exp\left(-\frac{\pi\Omega}{2} \right) \sum_{0}^{\infty} 2\pi\Omega \exp(-n\pi\Omega)
$$

(A.15.6)

$$
= 2\pi\Omega \frac{\exp(-\pi\Omega/2)}{1 - \exp(-\pi\Omega/2)} = \frac{\pi\Omega}{\sinh(\pi\Omega/2)}.
$$

The Fourier transforms evaluated at $\Omega = 0$ give the integrals of the functions over all t:

$$
\int_{-\infty}^{\infty} dt\, \operatorname{sech} t = \pi
$$

(A.15.7)

and

Fourier Transforms of Interest

Function:	Fourier Transform Defined by $\int dt\, f(t)\, e^{i\Omega t}$
$\mathrm{sech}\,(t)$	$\pi\,\mathrm{sech}\!\left(\dfrac{\pi}{2}\Omega\right)$
$\tanh(t)\,\mathrm{sech}\,(t)$	$i\Omega\pi\,\mathrm{sech}\!\left(\dfrac{\pi}{2}\Omega\right)$
$\tanh^2(t)\,\mathrm{sech}\,(t)$	$\dfrac{1}{2}(1-\Omega^2)\pi\,\mathrm{sech}\!\left(\dfrac{\pi}{2}\Omega\right)$
$\tanh^3(t)\,\mathrm{sech}\,(t)$	$i\dfrac{\Omega}{6}(5-\Omega^2)\pi\,\mathrm{sech}\!\left(\dfrac{\pi}{2}\Omega\right)$
$\mathrm{sech}^3(t)$	$\dfrac{1}{2}(1+\Omega^2)\pi\,\mathrm{sech}\!\left(\dfrac{\pi}{2}\Omega\right)$
$\tanh(t)\,\mathrm{sech}^3(t)$	$\dfrac{i\Omega}{6}(1+\Omega^2)\pi\,\mathrm{sech}\!\left(\dfrac{\pi}{2}\Omega\right)$
$t\,\mathrm{sech}\,(t)$	$i\,\dfrac{\pi^2}{2}\tanh\!\left(\dfrac{\pi}{2}\Omega\right)\mathrm{sech}\!\left(\dfrac{\pi}{2}\Omega\right)$
$t\tanh(t)\,\mathrm{sech}\,(t)$	$\pi\,\mathrm{sech}\!\left(\dfrac{\pi}{2}\Omega\right)-\Omega\,\dfrac{\pi^2}{2}\tanh\!\left(\dfrac{\pi}{2}\Omega\right)\mathrm{sech}\!\left(\dfrac{\pi}{2}\Omega\right)$
$t\tanh^2(t)\,\mathrm{sech}\,(t)$	$i\Omega\pi\,\mathrm{sech}\!\left(\dfrac{\pi}{2}\Omega\right)+i\,\dfrac{\pi^2}{4}(1-\Omega^2)\tanh\!\left(\dfrac{\pi}{2}\Omega\right)\mathrm{sech}\!\left(\dfrac{\pi}{2}\Omega\right)$
$t\tanh^3(t)\,\mathrm{sech}\,(t)$	$\dfrac{1}{6}(5-3\Omega^2)\pi\,\mathrm{sech}\!\left(\dfrac{\pi}{2}\Omega\right)-\dfrac{\pi^2}{2}\dfrac{\Omega}{6}(5-\Omega^2)\tanh\!\left(\dfrac{\pi}{2}\Omega\right)\mathrm{sech}\!\left(\dfrac{\pi}{2}\Omega\right)$
$t\tanh(t)\,\mathrm{sech}^3(t)$	$\dfrac{1}{6}(1+3\Omega^2)\pi\,\mathrm{sech}\!\left(\dfrac{\pi\Omega}{2}\right)-\dfrac{\pi^2}{2}\dfrac{\Omega}{6}(1+\Omega^2)\tanh\!\left(\dfrac{\pi}{2}\Omega\right)\mathrm{sech}\!\left(\dfrac{\pi}{2}\Omega\right)$
$t\,\mathrm{sech}^3(t)$	$-i\pi\Omega\,\mathrm{sech}\!\left(\dfrac{\pi}{2}\Omega\right)+i\,\dfrac{\pi^2}{4}(1+\Omega^2)\tanh\!\left(\dfrac{\pi}{2}\Omega\right)\mathrm{sech}\!\left(\dfrac{\pi}{2}\Omega\right)$
$\tanh^2(t)\,\mathrm{sech}^3(t)$	$\dfrac{1}{24}(1+\Omega^2)(3-\Omega^2)\pi\,\mathrm{sech}\!\left(\dfrac{\pi}{2}\Omega\right)$
$\mathrm{sech}^5(t)$	$\dfrac{1}{24}(\Omega^2+1)(\Omega^2+9)\pi\,\mathrm{sech}\!\left(\dfrac{\pi}{2}\Omega\right)$

Table A.15.1. Fourier transforms of interest

$$\int_{-\infty}^{\infty} dt \ \text{sech}^2 t = 2 \ . \tag{A.15.8}$$

This latter integral is, of course, simply evaluated by noting that

$$\frac{d}{dt} \tanh t = \text{sech}^2 t \ . \tag{A.15.9}$$

Another useful integral is obtained from the Fourier transform of $\text{sech}^2 t$:

$$\frac{d^2}{d\Omega^2} \left(\frac{\pi\Omega}{\sinh(\pi\Omega/2)} \right)_{\Omega=0} = -\int_{-\infty}^{\infty} t^2 dt \ \text{sech}^2 t = -\frac{\pi^2}{6} \ . \tag{A.15.10}$$

Table A.15.1 lists some other useful Fourier transforms.

A.16 The Noise Sources Derived from a Lossless Multiport with Suppressed Terminals

A lossless quantum mechanical $2N$-port is derivable from a Hamiltonian. Commutator brackets are preserved and hence no noise sources are needed. When information on N of the ports is suppressed, and their excitation is via zero-point fluctuations, the network does not preserve power, and N independent noise sources are introduced, whose states are in the ground state. It is of interest to follow through the analysis of such "suppression". The $2N$-port is described by the scattering process

$$\begin{bmatrix} \hat{b}_s \\ \hat{b}_n \end{bmatrix} = \begin{bmatrix} S_{ss} & S_{sn} \\ S_{ns} & S_{nn} \end{bmatrix} \begin{bmatrix} \hat{a}_s \\ \hat{a}_n \end{bmatrix} \ . \tag{A.16.1}$$

We use the subscript "s" to denote the signal part of the network and the subscript "n" for the part of the network to be suppressed and thus responsible for the noise sources. After suppression of the "n" network, we obtain for the "s" network

$$\hat{b}_s = \hat{S}_{ss}\hat{a}_s + \hat{s}_s, \quad \text{where} \quad \hat{s}_s = S_{sn}\hat{a}_n \ . \tag{A.16.2}$$

We have found noise sources. The commutators of the noise sources are

$$[\hat{s}_s, \hat{s}_s^\dagger] = [S_{sn}\hat{a}_n, \hat{a}_n^\dagger S_{sn}^\dagger] = \frac{\Delta\omega}{2\pi} S_{sn} S_{sn}^\dagger \ . \tag{A.16.3}$$

Conservation of power requires the scattering matrix to be unitary. Thus, we find

$$S_{ss}S_{ss}^\dagger + S_{sn}S_{sn}^\dagger = 1 \ . \tag{A.16.4}$$

We then find from (A.16.3), (A.16.4), and (A.16.5)

$$[\hat{s}_s, \hat{s}_s^\dagger] = \frac{\Delta\omega}{2\pi} S_{sn} S_{sn}^\dagger = \frac{\Delta\omega}{2\pi}(1 - S_{ss}S_{ss}^\dagger) \ . \tag{A.16.5}$$

This is the proper commutator relation for the noise sources of the N-port. Hence, we have shown that the noise sources which provide commutator conservation for a non-Hamiltonian system originate, ultimately, from an incomplete description of the network under consideration.

A.17 The Noise Sources of an Active System Derived from Suppression of Ports

A parametric amplifier is an example of an active quantum mechanical four-port, resulting from the coupling of a signal wave to an idler wave as described in Sect. 10.2. Instead of one signal wave and one idler wave, one may couple N signal waves to N idler waves and obtain a $2N$-port. N ports are excited by signal waves, and N ports are excited by idler waves. If all the idler waves are in the ground state, the noise of the amplifier is at its minimum. In matrix notation,

$$\hat{b} = S\hat{a} \ , \tag{A.17.1}$$

where the S matrix is of rank $2N$,

$$S = \begin{matrix} S_{ss} & S_{si} \\ S_{is} & S_{ii} \end{matrix} \ , \tag{A.17.2}$$

and the submatrices are of rank N. The input and output excitation column matrices consist of annihilation operators for the signal channels, and creation operators for the idler channels:

$$\hat{b} = \begin{bmatrix} \hat{b}_s \\ \hat{b}_i^\dagger \end{bmatrix} \quad \text{and} \quad \hat{a} = \begin{bmatrix} \hat{a}_s \\ \hat{a}_i^\dagger \end{bmatrix} \tag{A.17.3}$$

Suppose we suppress the "unexcited" idler ports and write an equation for the signal ports alone. This results in an N-port with gain, and with noise sources:

$$\hat{b}_s = S_{ss}\hat{a}_s + S_{si}\hat{a}_i^\dagger = S_{ss}\hat{a}_s + \hat{s}_s, \quad \text{with} \quad \hat{s}_s = S_{si}\hat{a}_i^\dagger \ , \tag{A.17.4}$$

where $\hat{s}_s$ must now be interpreted as a noise source column matrix. This equation is indistinguishable from that of a simple multiport amplifier. Its noise sources must obey the commutator relation

$$[\hat{s}_s, \hat{s}_s^\dagger] = \frac{\Delta\omega}{2\pi}(1 - S_{ss}S_{ss}^\dagger) \tag{A.17.5}$$

in order to conserve commutator brackets. It is easily shown that this commutator relation indeed holds. The parametric amplifier generates signal photons and idler photons in pairs. This "conservation" principle is encapsulated in the matrix equation

$$S^\dagger P S = S P S^\dagger = P , \tag{A.17.6}$$

where

$$P = \begin{bmatrix} 1 & 0 \\ 0 & -1 \end{bmatrix} .$$

We thus have

$$S_{ss} S_{ss}^\dagger - S_{si} S_{si}^\dagger = 1 . \tag{A.17.7}$$

From (A.17.4) we find that

$$[\hat{s}_s, \hat{s}_s^\dagger] = [S_{si} \hat{a}_i^\dagger, S_{si}^* \hat{a}_i] = -\frac{\Delta\omega}{2\pi} S_{si} S_{si}^* . \tag{A.17.8}$$

Combining (A.17.8) and (A.17.7), we prove the commutator relation (A.17.5). Thus, we have shown in one special case how the noise sources of an amplifier can be derived from a Hamiltonian description of a parametric amplifier system in which information on the idler channels is suppressed. The derivation also shows that the noise sources associated with annihilation operator excitations of an amplifier are formed from creation operators.

A.18 The Translation Operator and the Transformation of Coherent States from the β Representation to the x Representation

The vacuum state in the β representation is a product state $\prod_j |0\rangle_j$, where the subscript j indicates the vacuum states pertaining to different propagation constants β_j. A superposition of coherent states is given by the following (the Einstein summation convention is used):

$$|\Psi\rangle = \exp(\alpha_k \hat{A}_k^\dagger + \alpha_k^* \hat{A}_k) \prod_j |0\rangle_j . \tag{A.18.1}$$

Now consider the transition from the β representation to the x representation. First of all we note that the ground state is invariant under the transformation. Next we note that the Fourier transform $\exp(i\beta_j x_k)\hat{A}_j$ yields the annihilation operator $\hat{a}(x_k) = \exp(i\beta_j x_k)\hat{A}_j$ in the x representation or, conversely, that the annihilation operator in the β representation can be written

$$\hat{A}_j = \exp(-i\beta_j x_k)\hat{a}(x_k) . \tag{A.18.2}$$

When we introduce (A.18.2) into (A.18.1) we find

$$|\Psi\rangle = \exp[\exp(-\mathrm{i}\beta_j x_k)\alpha_j^* \hat{a}(x_k) + \exp(\mathrm{i}\beta_j x_k)\alpha_j \hat{a}^\dagger(x_k)] \prod_j |0\rangle_j \ . \quad \text{(A.18.3)}$$

We have obtained a new translation operator which shifts every vacuum state to an excitation $|\exp(-\mathrm{i}\beta_j x_k)\alpha_j\rangle$. The excitation of the coherent state has transferred its spatial dependence to the coherent states in the x representation.

   Acknowledgment. The above derivation is the result of very helpful discussions with Dr. F. X. Kärtner.

A.19 The Heisenberg Equation in the Presence of Dispersion

In evaluating the Heisenberg equation of motion for the annihilation operator, one must evaluate the commutator $[(\partial\hat{a}^\dagger/\partial x)(\partial\hat{a}/\partial x), \hat{a}]$. This is done most simply by interpreting the commutator as follows (this approach was pointed out to the author by Dr. F.X. Kärtner):

$$\left[\frac{\partial\hat{a}^\dagger}{\partial x}\frac{\partial\hat{a}}{\partial x}, \hat{a}\right] = \lim_{y\to x, z\to x}\left[\frac{\partial\hat{a}^\dagger(x)}{\partial x}\frac{\partial\hat{a}(y)}{\partial y}, \hat{a}(z)\right]$$

$$= \lim_{y\to x, z\to x}\frac{\partial}{\partial x}\frac{\partial}{\partial y}[\hat{a}^\dagger(x)\hat{a}(y), \hat{a}(z)]$$

$$= -\lim_{y\to x, z\to x}\left[\frac{\partial}{\partial x}\delta(x-z)\frac{\partial\hat{a}(y)}{\partial y}\right] \quad \text{(A.19.1)}$$

$$= \lim_{y\to x, z\to x}\left[\frac{\partial^2}{\partial x\partial y}\hat{a}(y)\delta(x-z)\right]$$

$$= \frac{\partial^2}{\partial x^2}\hat{a}(x)\delta(0) \ .$$

A.20 Gaussian Distributions and Their $e^{-1/2}$ Loci

Zero point fluctuations lead to amplitude distributions that are Gaussian. A general, two-dimensional Gaussian distribution of two random variables x and y can be written compactly as

$$p(x, y) = N\exp\left(-\frac{1}{2}\boldsymbol{x}^\dagger \boldsymbol{A}\boldsymbol{x}\right) \ , \quad \text{(A.20.1)}$$

where

$$x = \begin{bmatrix} x \\ y \end{bmatrix} ,$$

A is a positive definite symmetric matrix of second rank, and N is a normalizing factor. The matrix A contains the information on the mean square spread of the variables x and y. When A is diagonal, the $1/e$ locus of the probability distribution is an ellipse with its major and minor axes along the x and y axes. Consider the characteristic function, defined as the Fourier transform of the probability distribution:

$$C(\xi, \eta) = \int dx \int dy\, p(x, y) e^{i\xi x + i\eta y} . \tag{A.20.2}$$

An expansion of the characteristic function in powers of ξ and η gives the moments of the probability distribution, as follows:

$$C(\xi, \eta) = \sum_{m,n} \int dx \int dy\, p(x, y) x^m y^n \frac{(i\xi)^m (i\eta)^n}{m! n!}$$

$$= \sum_{m,n} M_{mn} \frac{(i\xi)^m (i\eta)^n}{m! n!} , \tag{A.20.3}$$

where $M_{mm} = \int dx \int dy\, p(x, y) x^m y^n$ is the moment of mnth order.

A Fourier transform of a Gaussian is also a Gaussian. Thus, for the probability distribution (A.20.1), the characteristic function is of the form

$$C(\xi, \eta) = M \exp\left(-\frac{1}{2}\xi^\dagger B \xi\right) \quad \text{with} \quad \xi = \begin{bmatrix} \xi \\ \eta \end{bmatrix} , \tag{A.20.4}$$

where M is another normalization factor and B is the inverse of A. To prove this last assertion, let us introduce the unitary transformation U that casts the matrix A into diagonal form:

$$U A U^\dagger = D . \tag{A.20.5}$$

Such a transformation is a rotation, with

$$U = \begin{bmatrix} \cos \beta & \sin \beta \\ -\sin \beta & \cos \beta \end{bmatrix} .$$

It is clear that the matrix D contains the squares of the inverse lengths of the major and minor axes of the $1/e$ locus of the probability distribution, which we denote by σ_x and σ_y:

$$D = \begin{bmatrix} 1/\sigma_x^2 & 0 \\ 0 & 1/\sigma_y^2 \end{bmatrix} . \tag{A.20.6}$$

The Fourier transformation is carried out particularly simply in the diagonal form. We use the fact that $U^\dagger U = UU^\dagger = 1$. Then we may write

$$
\begin{aligned}
C(\xi, \eta) &= \int dx \int dy\, p(x,y) e^{i\xi x + i\eta y} \\
&= N \int dx \int dy \exp\left(-\frac{1}{2} x^\dagger U^\dagger U A U^\dagger U x\right) \exp(ix^\dagger U^\dagger U \xi) \\
&= N \int dx \int dy \exp\left(-\frac{1}{2} x^\dagger U^\dagger D U x\right) \exp(ix^\dagger U^\dagger U \xi) \\
&= N \int dx' \int dy' \exp\left(-\frac{1}{2} x'^\dagger D x'\right) \exp(ix'^\dagger \xi') \,,
\end{aligned}
$$

(A.20.7)

where $x' = Ux$ and $\xi' = U\xi$ are the components in the new, rotated coordinate system. With a diagonal matrix, the Fourier transformation is easily seen to give

$$
\begin{aligned}
C(\xi, \eta) &= N \int dx' \int dy' \exp\left(-\frac{1}{2} x'^\dagger D x'\right) \exp(ix'^\dagger \xi') \\
&\propto \exp\left(-\frac{1}{2}\xi'^\dagger D^{-1}\xi'/2\right) = \exp\left(-\frac{1}{2}\xi^\dagger U^\dagger D^{-1} U \xi\right) \qquad \text{(A.20.8)} \\
&= \exp\left(-\frac{1}{2}\xi^\dagger A^{-1}\xi\right) = \exp\left(-\frac{1}{2}\xi^\dagger B\xi\right) \,.
\end{aligned}
$$

Thus we have shown that the matrix B in the characteristic function is the inverse of the matrix A in the probability distribution. The quadratic term in the expansion of the characteristic function obtained from (A.20.8) yields the second-order moments, according to (A.20.3):

$$
B_{xx}\xi^2 + 2B_{xy}\xi\eta + B_{yy}\eta^2 = \sigma_{xx}^2 \xi^2 + 2\sigma_{xy}^2 \xi\eta + \sigma_{yy}^2 \eta^2 \,.
$$

(A.20.9)

Thus, we find that the matrix B contains the mean square deviations as its matrix elements.

The rotation introduced earlier is useful in determining parameters of the $e^{-1/2}$ ellipse. Indeed, rotation of the coordinates into the major and minor axes of the ellipse gives for its area

$$
\text{area} = \sigma_x \sigma_y \,,
$$

(A.20.10)

which is equal to the product of the two eigenvalues of the matrix B. Since eigenvalues are invariant under rotation, the area of the ellipse can be computed for any arbitrary orientation:

$$
\text{area} = \sqrt{\text{product of eigenvalues}} \,.
$$

(A.20.11)

Now, the eigenvalues of the matrix $\boldsymbol{B}$ are

$$\lambda_\pm = \frac{\sigma_{xx}^2 + \sigma_{yy}^2}{2} \pm \sqrt{\left(\frac{\sigma_{xx}^2 - \sigma_{yy}^2}{2}\right)^2 + \sigma_{xy}^4} \,. \tag{A.20.12}$$

The product of the eigenvalues is

$$\text{area} = \sqrt{\lambda_+ \lambda_-} = \sqrt{\sigma_{xx}^2 \sigma_{yy}^2 - \sigma_{xy}^4} \,. \tag{A.20.13}$$

We may now confirm that the area of the fluctuation ellipse remains invariant in the process of squeezing. We have from (13.86), (13.87), and (13.88)

$$\sigma_{xx}^2 = \langle \Delta \hat{A}_1^2(0) \rangle \,,$$

$$\sigma_{yy}^2 = \langle \Delta \hat{A}_2^2(0) \rangle + 4\Phi^2 \langle \Delta \hat{A}_1^2(0) \rangle \,, \tag{A.20.14}$$

$$\sigma_{xy}^2 = 2\Phi \langle \Delta \hat{A}_1^2(0) \rangle \,.$$

We find from (A.20.13) and (A.20.14)

$$\text{area} = \sqrt{\langle \Delta \hat{A}_1^2(0) \rangle \langle \Delta \hat{A}_2^2(0) \rangle} \,. \tag{A.20.15}$$

References

1. G. Chedd, *SOUND*, from *Communications to Noise Pollution*, Doubleday, Garden City, NY, 1970.
2. W. Schottky, Ann. Physik **57**, 541 (1918).
3. S. Weinberg, *The First Three Minutes, A Modern View of the Origin of the Universe*, Bantam, New York, 1977.
4. A. Hasegawa and F. Tappert, Appl. Phys. Lett. **23**, 142 (1973).
5. A. Hasegawa, Appl. Opt. **23**, 3302 (1984).
6. J. P. Gordon and H. A. Haus, Opt. Lett. **11**, 665 (1986).
7. L. F. Mollenauer, M. J. Neubelt, S. G. Evangelides, J. P. Gordon, J. R. Simpson, and L. G. Cohen, Opt. Lett. **15**, 1203 (1990).
8. J. von Neumann, *Mathematische Grundlagen der Quantenmechanik*, Springer, Berlin, 1932.
9. J. S. Bell, *Speakable and Unspeakable in Quantum Mechanics*, Cambridge University Press, New York, 1987, p. 29.
10. N. Bohr, *Quantum Theory of Measurement*, edited by J. A. Wheeler and W. H. Zurek, Princeton University Press, Princeton, 1983, p. 5.
11. W. H. Zurek, Phys. Rev. D **24**, 1516 (1981); Phys. Today **44**, 36 (1991).
12. R. Omnes, *Interpretation of Quantum Mechanics*, Princeton University Press, Princeton, 1994.
13. H. D. Zeh, Phys. Lett. A **172**, 189 (1993).
14. M. Namiki and S. Pascazio, Phys. Rep. **232**, 301 (1993).
15. E. Arthurs and J. L. Kelly, Jr., Bell Systems Tech. J. **44**, 725 (1965).
16. H. A. Haus and J. A. Mullen, Phys. Rev. **128**, 2407 (1962).
17. Subcommittee 7.9 on Noise, H. A. Haus, Chairman, Proc. IRE **48**, 60–74 (1960).
18. A. L. Schawlow and C. H. Townes, Phys. Rev. **112**, 1940 (1958).
19. P. Penfield and H. A. Haus, *Electrodynamics of Moving Media*, MIT Press, Cambridge, MA, 1967.
20. W. K. H. Panofsky and M. Phillips, *Classical Electricity and Magnetism*, 2nd ed., Addison Wesley, Reading, MA, 1962.
21. R. M. Fano, L. J. Chu, and R. B. Adler, *Electromagnetic Fields, Energy, and Forces*, Wiley, New York, London, 1960.
22. B. D. H. Tellegen, Am. J. Phys. **30**, 650 (1962).
23. H. A. Haus and P. Penfield, Jr., Phys. Lett. A **26**, 412 (1968).
24. W. Shockley and R. P. James, Science **156**, 542 (1967); W. Shockley and R. P. James, Phys. Rev. Lett. **18**, 876 (1967).
25. S. Coleman and J. H. Van Vleck, Phys. Rev. **171**, 1370 (1968).
26. L. D. Landau and E. M. Lifshitz, *Electrodynamics of Continuous Media*, Pergamon, Oxford, 1960.
27. J. C. Slater, *Microwave Electronics*, Van Nostrand, New York, Toronto, London, 1950.

28. R. B. Adler, L. J. Chu, and R. M. Fano, *Electromagnetic Energy Transmission and Radiation*, Wiley, New York, London, 1960.
29. R. E. Collins, *Foundations of Microwave Engineering*, McGraw-Hill, New York, 1966.
30. S. Ramo, J. R. Whinnery, and T. van Duzer, *Fields and Waves in Communication Electronics*, Wiley, New York, London, Sydney, 1965.
31. H. A. Haus, *Waves and Fields in Optoelectronics*, Prentice Hall, Englewood Cliffs, NJ, 1984.
32. H. B. Killen, *Fiber Optic Communications*, Prentice Hall, Englewood Cliffs, NJ, 1991.
33. H. A. Haus, J. Appl. Phys. **46**, 3049 (1975).
34. J. Wilson and J. F. B. Hawkes, *Optoelectronics: An Introduction*, 2nd ed., Prentice Hall, Englewood Cliffs, NJ, 1983.
35. G. P. Agrawal, *Nonlinear Fiber Optics*, Academic Press, New York, 1989.
36. A. Yariv, *Optical Electronics*, Holt, Rinehart, and Winston, New York, 1985.
37. S. E. Miller and A. G. Chynoweth, *Optical Fiber Telecommunications*, Academic Press, New York, 1979.
38. D. Gloge, Appl. Opt. **10**, 2252 (1971).
39. G. P. Agrawal, *Fiber-Optic Communcation Systems*, Wiley Series in Microwave and Optics Engineering, Wiley, New York, 1997.
40. K. S. Kim, R. H. Stolen, W. A. Reed, and K. W. Quoi, Opt. Lett. **19**, 257 (1994).
41. Y. Namihira, A. Miyata, and N. Tanahashi, Electron. Lett. **30**, 1171 (1994).
42. T. Kato, Y. Suetsugu, M. Takagi, E. Sasaoka, and M. Nishimura, Opt. Lett. **20**, 988 (1995).
43. T. Kato, Y. Suetsugu, and M. Nishimura, Opt. Lett. **20**, 2279 (1995).
44. M. Artiglia, E. Ciaramella, and B. Sordo, Electron. Lett. **31**, 1012 (1995).
45. L. Prigent and J.-P. Hamaide, IEEE Photon. Technol. Lett. **5**, 1092 (1993).
46. A. Boskovic, S. Y. Chernikov, J. R. Taylor, L. Gruner-Nielson, and O. A. Levring, Opt. Lett. **21**, 1966 (1996).
47. L. D. Smullin and H. A. Haus, eds., *Noise in Electron Devices*, Wiley, New York, 1959.
48. H. Nyquist, Phys. Rev. **32**, 110 (1978).
49. E. B. A. Saleh and M. C. Teich, *Fundamentals of Photonics*, Wiley, New York, 1991.
50. W. Feller, *An Introduction to Probability Theory and Its Applications*, Vol. I, Wiley, New York, London, 1950.
51. W. B. Davenport and W. L. Root, *An Introduction to the Theory of Random Signals and Noise*, Lincoln Laboratory Publications, McGraw-Hill, New York, 1958.
52. K. Huang, *Statistical Mechanics*, Wiley, New York, London, 1963.
53. R. Q. Twiss, J. Appl. Phys. **26**, 599 (1955).
54. M. Born, *Natural Philosophy of Cause and Chance*, Clarendon Press, Oxford, 1949.
55. H. T. Friis, Proc. IRE **32**, 419 (1944).
56. H. A. Haus and R. B. Adler, "Invariants of linear networks", 1956 IRE Convention Record, Part 2, 53 (1956).
57. H. A. Haus and R. B. Adler, L'Onde Electrique **38**, 380 (1958).
58. H. A. Haus and R. B. Adler, Proc. IRE **46**, 1517 (1958).
59. R. B. Adler and H. A. Haus, IRE Trans. Circuit Theory **CT-5**, 156 (1958).
60. H. A. Haus and R. B. Adler, IRE Trans. Circuit Theory **CT-5**, 161 (1958).
61. H. A. Haus and R. B. Adler, *Circuit Theory of Linear Noisy Networks*, Technology Press Research Monograph, Wiley, New York, 1959.

62. H. Rothe and W. Dahlke, Proc. IRE **44**, 811 (1956).
63. H. Statz, H. A. Haus, and R. A. Pucel, IEEE Trans. Electron. Devices **ED-21**, 549 (1974).
64. A. Messiah, *Quantum Mechanics*, Vols. I and II, Wiley, New York, 1963.
65. W. Louisell, *Radiation and Noise in Quantum Electronics*, Physical and Quantum Electronics Series, McGraw-Hill, New York, London, San Francisco, Toronto, 1964.
66. R. J. Glauber, Phys. Rev. **131**, 2766 (1963).
67. D. F. Walls and G. J. Milburn, *Quantum Optics*, Springer Study Edition, Springer, Berlin, Heidelberg, 1994.
68. M. W. P. Strandberg, Phys. Rev. **106**, 617 (1957); Phys. Rev. **107**, 1483 (1957).
69. D. Leibfried, T. Pfau, and C. Monroe, Phys. Today **51**, 22 (1998).
70. E. Schrödinger, Naturwissenschaften **23**, pp. 807–812, 823–828, 844–849 (1935). Translation in J. A. Wheeler and W. H. Zurek, *Quantum Theory and Measurement*, Princeton Series in Physics, Princeton University Press, Princeton, 1983.
71. B. M. Oliver, Proc. IRE (Correspondence) **49**, 1960 (1961).
72. H. A. Haus and C. H. Townes, Proc. IRE (Correspondence) **50**, 1544 (1962).
73. A. Einstein, B. Podolsky, and N. Rosen, Phys. Rev. **47**, 777 (1935).
74. H. A. Haus and J. A. Mullen, "Photon noise", in *Proceedings of the International Conference on Microwave Tubes*, The Hague, 1962.
75. T. Li and M. C. Teich, Electron. Lett. **27**, 598 (1991).
76. B. E. A. Saleh, *Photoelectron Statistics*, Springer, New York, 1978.
77. W. S. Wong, P. B. Hansen, T. N. Nielsen, M. Margalit, S. Namiki, E. P. Ippen, and H. A. Haus, J. Lightwave Technol. **16**, 1768 (1998).
78. A. H. Gnauch and C. R. Giles, Photonics Technol. Lett. **4**, 80 (1992).
79. J.-M. P. Delavaux, R. J. Nuyts, O. Mizuhara, J. A. Nagel, and D. J. di Giovanni, Photonics Technol. Lett. **6**, 376 (1994).
80. J. C. Livas, "High sensitivity optically preamplified 10 Gb/s receivers", OFC'96 postdeadline paper.
81. E. Desurvire, *Erbium Doped Fiber Amplifiers: Principles and Applications*, Wiley, New York, 1994.
82. J. P. Gordon, Proc. IRE **50**, 1898 (1962).
83. C. E. Shannon, Bell Systems Tech. J. **27**, 379 (1948).
84. L. Brillouin, *Science and Information Theory*, Academic Press, New York, 1956.
85. P. A. Humblet and M. Azizoglu, J. Lightwave Technol. **9**, 1576 (1991).
86. J. Korn, "Propagation of a 10 Gbps RZ bit stream in a circulating loop using TrueWave and DCF with 100 km EDFA spacing", in *LEOS'98*, Orlando, FL, paper TUA4, p. 10 (1998).
87. M. Nakazawa, Y. Kimura, K. Suzuki, and H. Kubota, J. Appl. Phys. **66**, 2803 (1989).
88. V. E. Zakharov and A. B. Shabat, Zh. Eksp. Teor. Fiz. **61**, 118 (1971) [Sov. Phys. JETP **34**, 62 (1972)].
89. D. J. Kaup, SIAM J. Appl. Math. **31**, 121 (1976).
90. L. F. Mollenauer, J. P. Gordon, and S. Evangelides, Opt. Lett. **17**, 1575 (1992).
91. J. P. Gordon, Opt. Lett. **8**, 596 (1983).
92. J. Satsuma and N. Yajima, Prog. Theor. Phys. (Suppl.) **55**, 284 (1974).
93. K. Tai, A. Tomita, J. L. Jewell, and A. Hasegawa, Appl. Phys. Lett. **49**, 236 (1986).
94. E. M. Dianov, P. V. Mamyshev, A. M. Prokhorov, and S. V. Chernikov, Opt. Lett. **14**, 1008 (1989).

552   References

95. S. V. Chernikov, D. J. Richardson, R. I. Laming, E. M. Dianov, and D. N. Payne, Electron. Lett. **28**, 1210 (1992).
96. S. V. Chernikov, J. R. Taylor, and R. Kashyap, Electron. Lett. **29**, 1788 (1993).
97. S. V. Chernikov, J. R. Taylor, and R. Kashyap, Electron. Lett. **30**, 433 (1994).
98. E. A. Swanson, S. R. Chinn, K. Hall, K. A. Rauschenbach, R. S. Bondurant, and J. W. Miller, IEEE Photonics Technol. Lett. **6**, 1194 (1994).
99. E. A. Swanson and S. R. Chinn, IEEE Photonics Technol. Lett. **7**, 114 (1995).
100. D. J. Richardson, R. P. Chamberlin, L. Dong, D. N. Payne, A. D. Ellis, T. Widdowson, and D. M. Spirit, Electron. Lett. **31**, 395 (1995).
101. F. M. Mitschke and L. F. Mollenauer, Opt. Lett. **12**, 355 (1987).
102. V. I. Karpman, Sov. Phys. JETP **46**, 281 (1971).
103. D. J. Kaup and A. C. Newell, Proc. R. Soc. Lond. A **361**, 413 (1978).
104. V. I. Kaupman and V. V. Solovév, Physica **3D**, 487 (1981).
105. D. J. Kaup, Phys. Rev. A **42**, 5689 (1990).
106. Y. S. Kivshar and B. A. Malomed, Rev. Mod. Phys. **61**, 763 (1989).
107. Y. S. Kivshar and B. A. Malomed, Rev. Mod. Phys. **63**, 211 (1991).
108. D. J. Kaup, Phys. Rev. A **44**, 4582 (1991).
109. H. A. Haus and Y. Lai, J. Opt. Soc. Am. B. **7**, 386 (1990).
110. K. Tamura, E. P. Ippen, H. A. Haus, and L. E. Nelson, Opt. Lett. **18**, 1080 (1993).
111. H. A. Haus and A. Mecozzi, IEEE J. Quantum Electron. **29**, 983 (1993).
112. A. Mecozzi, J. D. Moores, H. A. Haus, and Y. Lai, Opt. Lett. **16**, 1841 (1991).
113. Y. Kodama and A. Hasegawa, Opt. Lett. **17**, 31 (1992).
114. L. F. Mollenauer, M. J. Neubelt, and G. T. Harvey, Electron. Lett. **29**, 910 (1993).
115. E. Desurvire, C. R. Giles, J. R. Simpson, and J. L. Zyskind, Opt. Lett. **14**, 1266 (1989).
116. K. Nakagawa, K. Aida, and E. Yoneda, J. Lightwave Technol. **9**, 198 (1991).
117. W. J. Miniscalco, J. Lightwave Technol. **9**, 234 (1991).
118. L. F. Mollenauer, S. G. Evangelides, and H. A. Haus, J. Lightwave Technol. **9**, 194 (1991).
119. A. Hasegawa and Y. Kodama, Opt. Lett. **16**, 1385 (1991).
120. L. F. Mollenauer, B. M. Nyman, M. J. Neubelt, G. Raybon, and S. G. Evangelides, Electron. Lett. **27**, 178 (1991).
121. E. M. Dianov, A. V. Luchnikov, A. N. Pilepetskii, and A. N. Starodumov, Opt. Lett. **15**, 314 (1990).
122. E. M. Dianov, A. V. Luchnikov, A. N. Pilepetskii, and A. N. Staromudov, Sov. Lightwave Commun. **1**, 37 (1991).
123. C. R. Menyuk, IEEE J. Quant. Electron. **QE-23**, 174 (1987).
124. P. K. A. Wai, C. R. Menyuk, and H. H. Chen, Opt. Lett. **16**, 1231 (1991).
125. S. V. Manakov, Zh. Eksp. Teor. Fiz. **65**, 1394 [Sov. Phys. JETP **38**, 248 (1974)].
126. M. G. Taylor, IEEE Photonics Technol. Lett. **5**, 1244 (1993).
127. V. J. Mazurczyk and J. L. Zyskind, IEEE Photonics Technol. Lett. **6**, 616 (1994).
128. M. G. Taylor, IEEE Photonics Technol. Lett. **6**, 860 (1994).
129. D. R. Nicholson and M. V. Goldman, Phys. Fluids **19**, 1621 (1976).
130. H. A. Haus, W. S. Wong, and F. I. Khatri, J. Opt. Soc. Am. B **14**, 304 (1997).
131. J. P. Gordon, J. Opt. Soc. Am. B **9**, 91 (1992).
132. L. F. Mollenauer, "Massive WDM in ultra long-distance soliton transmission", in *OFC'98, Optical Fiber Communication Conference*, San Jose, CA, 1998, Technical Digest Series, Vol. 2 (Optical Society of America, Washington, DC, 1998, paper Th15).

133. P. V. Mamyshev and L. F. Mollenauer, "NRZ-to-soliton data conversion by a filtered transmission line", in *OFC'95, Optical Fiber Communication Conference*, San Diego, CA, 1995, Technical Digest Series, Vol. 8 (Optical Society of America, Washington, DC, 1995, paper FB2).

134. M. Matsumoto, H. Ikeda, and A. Hasegawa, Opt. Lett. **19**, 183 (1994).

135. N. J. Smith and N. J. Doran, Electron. Lett. **30**, 1084 (1994).

136. E. Yamada and M. Nakazawa, IEEE J. Quant. Electron. **30**, 1842 (1994).

137. K. Rottwitt, W. Margulis, and J. R. Taylor, Electron. Lett. **31**, 395 (1995).

138. R. A. Jensen, C. R. Davidson, D. L. Wilson, and J. K. Lyons, "Novel technique for monitoring long-haul undersea optical-amplifier systems", in *OFC'94, Optical Fiber Communication: Summaries of Papers Presented at the Conference on Optical Fiber Communications*, San Jose, CA, Technical Digest Series, Vol. 4 (Optical Society of America, Washington, DC, paper ThR3, 1994).

139. S. G. Evangelides, H. A. Haus, F. I. Khatri, P. V. Mamyshev, and B. M. Nyman, *Soliton Line-Monitoring System*, patent pending, submitted November 30, 1995.

140. H. P. Yuen, Phys. Lett. A **56**, 105 (1976).

141. V. B. Braginsky, C. M. Caves, and K. S. Thorne, Phys. Rev. D **15**, 2047 (1977).

142. V. B. Braginsky and B. Y. Khalili, Sov. Phys. JETP **57**, 1124 (1983).

143. V. B. Braginsky, Y. I. Vorontsov, and K. S. Thorne, Science **209**, 547 (1980).

144. V. B. Braginsky and S. P. Vyatchamin, Dokl. Akad. Nauk SSSR **257**, 570 [Sov. Phys. Dokl. **26**, 686 (1981)].

145. V. B. Braginsky and S. P. Vyatchamin, Dokl. Akad. Nauk SSSR **264**, 1136 [Sov. Phys. Dokl. **27**, 478 (1982)].

146. D. F. Walls, Nature (London) **301**, 14 (1983).

147. J. M. Manley and H. E. Rowe, Proc. IRE **44**, 804 (1956).

148. M. T. Weiss, Proc. IRE **45**, 1012 (1957).

149. H. A. Haus and D. L. Bobroff, J. Appl. Phys. **28**, 694 (1957).

150. A. Guth, *The Inflationary Universe*, Addison Wesley, Reading, MA, 1997.

151. H. A. Haus and Y. Yamamoto, Phys. Rev. D **29**, 1261 (1984).

152. Y. Yamamoto and H. A. Haus, Rev. Mod. Phys. **58**, 1001 (1986).

153. D. Middleton, *An Introduction to Statistical Communication Theory*, McGraw Hill, New York,1960.

154. T. S. Jaseja, A. Javan, and C. H. Townes, Phys. Rev. Lett. **10**, 165 (1963).

155. C. Freed and H. A. Haus, Appl. Phys. Lett. **6**, 85 (1965).

156. E. D. Hinkley and C. Freed, Phys. Rev. Lett. **23**, 277 (1969).

157. C. Freed, J. W. Bielinski, and W. Lo, Appl. Phys. Lett. **43**, 629 (1983).

158. Y. Yamamoto and S. Machida, Phys. Rev. A **34**, 4025 (1986).

159. R. Szipocs, K. Ferencz, C. Spielmann, and F. Krausz, Opt. Lett. **19**, 201 (1994).

160. I. D. Jung, F. X. Kärtner, N. Matuschek, D. H. Suter, F. Morier-Genoud, G. Zhang, U. Keller, V. Scheuer, M. Tilsch, and T. Tschudi, Opt. Lett. **22**, 1009 (1997).

161. U. Morgner, F. X. Kärtner, S. H. Cho, Y. Chen, H. A. Haus, J. G. Fujimoto, E. P. Ippen, V. Scheuer, A. Angelow, and T. Tschudi, Opt. Lett. **24**, 411 (1999).

162. M. Shirasaki and H. A. Haus, J. Opt. Soc. Am. B **7**, 30 (1990).

163. K. Bergman and H. A. Haus, Opt. Lett. **16**, 663 (1991).

164. R. M. Shelby, M. D. Levenson, and P. W. Bayer, Phys. Rev. B **31**, 5244 (1985).

165. S. H. Perlmutter, M. D. Levenson, R. M. Shelby, and M. B. Weissman, Phys. Rev. B **42**, 5294 (1990).

166. E. P. Ippen and R. Stolen, Appl. Phys. Lett. **21**, 539 (1972).

167. A. Yariv, *Quantum Electronics*, 2nd ed., Wiley, New York, 1975.
168. K. Bergman, C. R. Doerr, H. A. Haus, and M. Shirasaki, Opt. Lett. **18**, 643 (1993).
169. K. Bergman, H. A. Haus, E. P. Ippen, and M. Shirasaki, Opt. Lett. **19**, 290 (1994).
170. K. Bergman, H. A. Haus, and M. Shirasaki, Appl. Phys. B **55**, 242 (1992).
171. R. K. John, J. H. Shapiro, and P. Kumar, "Classical and quantum noise transformations produced by self-phase modulation", in *International Quantum Electronics Conference*, Vol. 21 of 1987 OSA Technical Digest Series (Optical Society of America, Washington, DC, 1987), p. 204.
172. L. Joneckis and J. Shapiro, J. Opt. Soc. Am. B **10**, 1102 (1993).
173. S. J. Carter, P. D. Drummond, M. D. Reid, and R. M. Shelby, Phys. Rev. Lett. **58**, 1841 (1987).
174. P. D. Drummond and S. J. Carter, J. Opt. Soc. Am. B **4**, 1565 (1987).
175. P. D. Drummond, S. J. Carter, and R. M. Shelby, Opt. Lett. **14**, 373 (1989).
176. H. A. Haus, K. Watanabe, and Y. Yamamoto, J. Opt. Soc. Am. B **6**, 1138 (1989).
177. K. J. Kaup, J. Math. Phys. **16**, 2036 (1975).
178. Y. Lai and H. A. Haus, Phys. Rev. A **40**, 844 (1989).
179. Y. Lai and H. A. Haus, Phys. Rev. A **40**, 854 (1989).
180. H. A. Bethe, Z. Phys. **71**, 205 (1931).
181. J. M. Fini, P. L. Hagelstein, and H. A. Haus, Phys. Rev. A **57**, 4842 (1998).
182. C. R. Doerr, I. Lyubomirski, G. Lenz, J. Paye, H. A. Haus, and M. Shirasaki, "Optical squeezing with a short fiber", in *Quantum Electronics and Laser Science Conference*, Optical Society of America, 1993, pp. 112–113.
183. Y. Liu, S.-G. Park, and A. M. Weiner, IEEE J. Sel. Topics Quantum Electron. **2**, 709 (1996).
184. H. A. Haus and C. Yu, "Soliton squeezing and the continuum", J. Opt. Soc. Am. B, accepted for publication.
185. J. S. Bell, Phys. World **3**, 33 (1990).
186. N. Bohr, in *Quantum Theory and Measurement*, J. A. Wheeler and W. H. Zurek, eds., Princeton Series in Physics, Princeton, NJ, 1983, p. 3.
187. V. B. Braginsky and Y. I. Vorontsov, Usp. Fiz. Nauk. **114**, 41 (1974) [Sov. Phys. Usp. **17**, 644 (1975)].
188. W. G. Unruh, Phys. Rev. D **19**, 2888 (1979).
189. C. M. Caves, K. S. Thorne, R. W. P. Drever, V. D. Sandberg, and M. Zimmerman, Rev. Mod. Phys. **52**, 341 (1980).
190. N. Imoto, H. A. Haus, and Y. Yamamoto, Phys. Rev. A **32**, 2287 (1985).
191. F. X. Kärtner and H. A. Haus, Phys. Rev. A **47**, 4585 (1993).
192. H. A. Haus and F. X. Kärtner, Phys. Rev. A **53**, 3785 (1996).
193. J. Gribbin, *Schrödinger's Kittens and the Search for Reality*, Back Bay Books, Boston, New York, 1995.
194. L. I. Schiff, *Quantum Mechanics*, McGraw-Hill, New York, 1968.
195. C. Cohen-Tannoudji, B. Diu, and F. Laloë, *Quantum Mechanics*, Wiley, New York, 1977.
196. H. Bateman, *Tables of Integral Transforms*, Vol. 2, McGraw-Hill, New York, 1954, p. 290.
197. P. Carruthers and M. M. Nieto, Phys. Rev. **14**, 387 (1965).
198. E. P. Wigner, Phys. Rev. **40**, 749 (1932).

Index

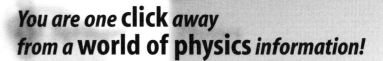

Production: Druckhaus Beltz, Hemsbach